Advances in Science, Technology & Innovation

IEREK Interdisciplinary Series for Sustainable Development

Advances in Science, Technology & Innovation (ASTI) is a series of peer-reviewed books based on important emerging research that redefines the current disciplinary boundaries in science, technology and innovation (STI) in order to develop integrated concepts for sustainable development. It not only discusses the progress made towards securing more resources, allocating smarter solutions, and rebalancing the relationship between nature and people, but also provides in-depth insights from comprehensive research that addresses the **17 sustainable development goals (SDGs)** as set out by the UN for 2030.

The series draws on the best research papers from various IEREK and other international conferences to promote the creation and development of viable solutions for a **sustainable future and a positive societal** transformation with the help of integrated and innovative science-based approaches. Including interdisciplinary contributions, it presents innovative approaches and highlights how they can best support both economic and sustainable development, through better use of data, more effective institutions, and global, local and individual action, for the welfare of all societies.

The series particularly features conceptual and empirical contributions from various interrelated fields of science, technology and innovation, with an emphasis on digital transformation, that focus on providing practical solutions to **ensure food, water and energy security to achieve the SDGs.** It also presents new case studies offering concrete examples of how to resolve sustainable urbanization and environmental issues in different regions of the world.

The series is intended for professionals in research and teaching, consultancies and industry, and government and international organizations. Published in collaboration with IEREK, the Springer ASTI series will acquaint readers with essential new studies in STI for sustainable development.

ASTI series has now been accepted for Scopus (September 2020). All content published in this series will start appearing on the Scopus site in early 2021.

Haşim Altan · Samad Sepasgozar ·
Abdul Lateef Olanrewaju ·
Francisco José García Peñalvo ·
Alessandro Gaetano Severino ·
Tiko Iyamu · Ju Hyun Lee
Editors

Advances in Architecture, Engineering and Technology

Smart Techniques in Urban Planning & Technology

Second Edition

Springer

Editors
Haşim Altan
Department of Architecture, Faculty of Design
Arkin University of Creative Arts and Design (ARUCAD)
Kyrenia, Cyprus

Abdul Lateef Olanrewaju
Department of Construction Management
Faculty of Engineering and Green Technology
Universiti Tunku Abdul Rahman
Kampar, Perak, Malaysia

Alessandro Gaetano Severino
Department of Civil Engineering and Architecture
University of Catania
Catania, Italy

Ju Hyun Lee
School of Research, Built Environment
Faculty of Arts, Design and Architecture
University of New South Wales
Kensington, NSW, Australia

Samad Sepasgozar
School of Research, Built Environment
Faculty of Arts, Design and Architecture
University of New South Wales
Kensington, NSW, Australia

Francisco José García Peñalvo
Department of Computer Science and Automation
University of Salamanca
Salamanca, Spain

Tiko Iyamu
Faculty of Informatics and Design
Cape Peninsula University of Technology
Cape Town, South Africa

ISSN 2522-8714 ISSN 2522-8722 (electronic)
Advances in Science, Technology & Innovation
IEREK Interdisciplinary Series for Sustainable Development
ISBN 978-3-031-11231-7 ISBN 978-3-031-11232-4 (eBook)
https://doi.org/10.1007/978-3-031-11232-4

This Springer imprint is published by the registered company Springer Nature Switzerland AG
The registered company address is: Gewerbestrasse 11, 6330 Cham, Switzerland

Scientific Committee

Preface

In this book, advances in architecture, engineering, and technology have been covered through studies concerning both architecture and urbanization as well as the trends in information and communications technology and innovative digital technologies with systems integration into the architecture, engineering, and construction (AEC) sector. The book provides various examples linking studies from building level to urban level involving the social, cultural, environmental, and economic conditions. It aims to improve our understanding of the advances and the challenges in different contexts and contribute to enhancing sustainability in the built environment. Chapters in this book illuminate the experiences of different localities facing current challenges due to their specific characteristics in geographic, historical legacies, environmental, economic, social, and cultural conditions. Sharing local interventions and responses to global issues such as climate change, urbanization, and pollution will support the development of a connected global–regional–local understanding of sustainable design of the built environment. This book is divided into three parts.

The first part addresses questions of "Architectural and Urban Studies", which contains seven chapters. Chapters "Integrated Design Concept for Identifiable Commercial Heritage Tourism in *Tunjungan* District of Surabaya City"–"Assessing the Environmental Aspects of Road Network Resiliency" present different approaches to architectural and urban design through a series of local case studies.

Chapter "Integrated Design Concept for Identifiable Commercial Heritage Tourism in *Tunjungan* District of Surabaya City" explores the concept of integrated design for identifiable commercial heritage tourism in Tunjungan District of Surabaya City in Indonesia, where the form of adaptation and resilience is embodied to conserve the architectural order of its historic environment and persevere in fighting spirit values in the midst of the digital era. Chapter "Environmental Product Declarations for Building Materials: Advantages, Limits, Developments" covers environmental product declarations for building materials and discusses their advantages, limitations, and developments. The authors' contribution is to the international debate on the environmental matter, which deals with a tool that can guide professionals and users in choosing low environmental impact building materials with similar functional requirements, considering environmental labels and product declarations.

Chapter "Public-Private Partnerships in Romanian New Urban Developments—A Potential Valuable Instrument in the Privatization Era" describes the public–private partnerships in Romanian new urban developments with a potential valuable instrument in the privatization era. The authors analyse the legal and institutional framework of two European Union member states with demonstrated experience in implementing public–private partnerships (France and Poland) to determine the development premises of this collaboration model in the new urban areas of Romania. Chapter "Interaction of Prehistoric Heritage with Today's Landscape: The Case Study of Arslantepe Mound in Malatya, TURKEY" centres around the interaction of prehistoric heritage with today's landscape demonstrated through the case study of Arslantepe Mound in Malatya, Turkey. The authors investigate the interaction between the Arslantepe Mound from the prehistoric age with today's modern landscape including the natural and cultural landscape source values. Chapter "Hygromorphs: Exploring Morphologies of Architectural Adaptive Systems Using Hygroscopic Properties of Wood" investigates hygromorphs that explore morphologies of architectural adaptive systems using hygroscopic properties of wood. The authors present an experimental methodology for identifying the different parameters affecting the actuation of natural materials like wood based on the induced flexible morphologies that regulate different deformation parameters and motion behaviour utilized to respond to varying environmental conditions. Chapter "Analyzing Public Transport System in Twin Cities of Rawalpindi, Islamabad" analyses the public transport system in twin cities of Rawalpindi and Islamabad, Pakistan, considering the development, construction, and operations of Mass Transit System (MTS) in old and large cities without MTS and to examine also whether this development is feasible and sustainable with reference and comparison to the development of MTS in cities of Lahore, Peshawar, and Karachi. Chapter "Assessing the Environmental Aspects of Road Network Resiliency" assesses the environmental aspects of road network resiliency defining optimal planning steps of urban green infrastructure and the most effective sequence of the planning process in relation to climate change adaptation in existing cities. The authors offer a seven-step planning process based on a cross-analysis of different urban green infrastructure project types.

The second part discusses the "Trends in Information and Communications Technology (ICT) and Innovative Digital Technologies (IDT)" in architecture, engineering, and construction (AEC) and consists of seven chapters. Chapters "Building Information Modeling on Construction Safety: A Literature Review"–"The Application of Digital Technologies into Utilizing Urban Voids" present several approaches to applications of ICT and IDT demonstrated through a series of local case studies in different contexts.

Chapter "Building Information Modeling on Construction Safety: A Literature Review" addresses Building Information Modelling (BIM) in connection to construction safety such as the automated detection of possible safety hazards and the prevention of potential risks that may be likely to happen. This technology has been implemented in the AEC industry for decades, and in this chapter, the authors are introducing construction safety management demonstrated through understanding the technology, application, and challenges for better utilization of BIM as a safety tool. Chapter "A Digital Twin-Based System for Smart Management of Office Spaces" describes a digital twin-based system for smart management of office spaces for the attention of the architecture, engineering, construction, and operation (AECO) industry that has been shifting from a great interest in both the design and construction phases to the facility management (FM) and operational phases over the last decade. In this chapter, the authors discuss disruptive technologies such as ICT, Internet of things (IoT), and BIM, as promising applications to achieve a connected and effective management of buildings. Chapter "Application of Virtual Reality to Enhance the Interpretation Dong Ho Folk Paintings in Museums of Fine Arts in Ho Chi Minh City" presents the application of Virtual Reality (VR) to enhance the interpretation of Dong Ho folk paintings in Museums of Fine Arts in Ho Chi Minh City, Vietnam. The authors discuss the power of VR, not only the transfer of the architectural aspect, but also to revive the cultural values hidden within art folk painting to the community as the best way to preserve the culture for the next generations without sacrificing the development potential of the country.

Chapter "Design a Computer Program for Sustainable Materials to Apply in Egyptian Housing" focuses on the design of a computer program for sustainable materials to apply in Egyptian housing. The authors investigate modern methods to reach construction sustainability and to devise ways to measure the applications and determine the degree of sustainability while applying technology and organizing management, sustainability, and environmental conservation for sustainable materials for appropriate solutions in buildings in Egypt. Chapter "Digital Methods and Tools for Mapping Public–Private Partnership Opportunities and Evaluating Proposals in the Regeneration Plans of Building Assets" looks at the digital methods and tools for mapping public–private partnership (PPP) opportunities and evaluates proposals in the regeneration plans of Building Assets. The authors present the traditional method of making information available in the context of the Preliminary Market Exhibition with the definition of the data provided to potential stakeholders, following mapping and analysis of the elements. Chapter "Evaluating the Daylighting and Energy Performance of Container Housing Systems in Coastal Regions: An Iterative Modeling Approach" presents an evaluation of daylighting and energy performance of container housing systems in coastal areas of Chittagong, Bangladesh, using an iterative modelling approach. The authors propose a container housing system as a sustainable solution for the resilience of the low-income population of this region with the goal to lessen environmental effects of structures not only from the ecological point of view, but also from economic, sociocultural, and health-safety perspectives. Chapter "The Application of Digital Technologies into Utilizing Urban Voids" discusses the role of digital technologies in utilizing abandoned urban voids. The authors present the potential for applications of the landscape urbanism principles demonstrated through a case study, analysing urban void exploitation projects and their practical approaches.

The third part, "Systems Integration in Architecture, Engineering and Construction", consists of nine chapters with focus on urban sustainability challenges in different contexts. Chapters "Energy Efficiency Analysis in the Residential Sector of the Urban Area of Canton Saraguro, Ecuador"–"Ceiling Sound Reflectors to Optimize Acoustic Performance Using Parametric Tools" present different system integrations in the AEC sector demonstrated through a series of local case studies in different contexts.

Chapter "Energy Efficiency Analysis in the Residential Sector of the Urban Area of Canton Saraguro, Ecuador" presents an energy efficiency analysis in the residential sector of the urban area of Canton Saraguro, Ecuador. The authors work on the descriptive-inferential statistical analysis of the energy resources of a certain locality, such as the urban area of the Saraguro canton of the Loja Province, investigating the consumption of the four types of energy (electric, LPG, water, and firewood) as the main energy resources used in this parish, showing the reality of consumption in the residential sector, thus identifying the greatest energy resource used in homes. In Chapter "Forecast of Short-Term Energy Production in Ecuador 50 MW Huascachaca Mines Wind Farm", the authors examine the forecast of short-term energy production of 50 MW Huascachaca mines wind farm in Ecuador using computational tools such as the Windographer for the modelling of the wind flow, energy production, and also the IBM SPSS statistics software for the short-term prediction through the classic auto-regressive statistical models like Box and Jenkins. Chapter "Sustainable Architecture from Proper Recycling: Renewable Energy Integration and Housing Automation" explores sustainable architecture with proper recycling, renewable energy integration, and home automation. The increase in energy consumption from non-renewable resources causes large amounts of polluting gasses emitted into the environment, and the authors present a proposal to change certain paradigms with the introduction of alternative energies, with the implementation of self-sustaining homes that can supply their consumption by significantly reducing the costs of buying electricity. Chapter "Research in Policy and Management of Urban Underground Space Utilization" presents research on policy and management of urban underground space utilization looking at Canada, Japan, the USA, and Finland. Combined with specific cases, the authors discuss the characteristics of the four regions considering three aspects, namely planning, regulations and policies, and management system.

Chapter "A Chronological Exploration of Initiatives in an Automobile Cluster: A Case of Pithampur, Madhya Pradesh, India" describes a chronological exploration of initiatives in an automobile cluster through a case of Pithampur, Madhya Pradesh, in India, taking into account the different initiatives employed at national and state levels, which have contributed directly and indirectly in the development of the cluster, extracting evidences from archival records and expert opinion survey. Chapter "Implementation and Evaluation of a MLaaS for Document Classification with Continuous Deep Learning Models" focuses on the implementation and evaluation of machine learning as a service (MLaaS) for document classification with continuous deep learning models. The authors analysed the quality effect of a continuously learning neural network algorithm for document classification by taking user feedback into account where active learning increases the precision and thus makes the process of document classification more efficient. Chapter "Reflections on New Urbanism at the COVID-19 Pandemic Background—Urban Health Research" explores the reflections on new urbanism at the COVID-19 pandemic. The authors describe the influence of new urbanism on urban health from different spatial scales and reflect on whether these design methods still apply to the post-pandemic era through a series of proposed recommendations for urban systems at different levels for further improving the health of cities and residents. Chapter "Additive Formwork: Examining Design, Fabrication Space, and Resolution for Bespoke Concrete Elements" describes Additive Formwork by examining design, fabrication space, and resolution for bespoke concrete elements. Combining the latest advances in computational design and Fused Deposition Modelling (FDM) 3D printing technology, the authors discuss an exploration and assessment of the higher design freedom given by the use of Additive Formwork for architecture through an experimental set-up where the design and fabrication of a series of morphologically diverse concrete panels have been undertaken. Chapter "Ceiling Sound Reflectors to Optimize Acoustic Performance Using Parametric Tools " presents ceiling sound reflectors to optimize acoustic performance using parametric tools to integrate acoustic simulations and parametric geometry modelling into the design process of architectural acoustic design of an existing conference room as a case study located in the Universidad de las Américas Puebla, in Cholula, Mexico.

The topics addressed in this book are unified and contribute to the detailed analysis of architectural and urban studies. Taken together, the chapters in this book suggest systems integration in architecture, engineering, and construction. Different studies compiled in this book provide references for the trends in information and communications technology and the applications of innovative digital technologies in the AEC sector in different contexts around the world. Moreover, this book could become a milestone for changing the built environment for a more sustainable future.

Kyrenia, Cyprus

Haşim Altan

Acknowledgments

We would like to thank the authors of the research papers that were selected for addition in this book. We would also like to thank the reviewers who contributed with their knowledge and constructive feedback in hopes of ensuring the manuscript is of the best quality possible. A special thank goes to the editors of this book for their foresight in organizing this volume and diligence in doing a professional job in editing it. Finally, we would like to express our appreciation to the IEREK team for supporting the publication of the best research papers submitted to the conference.

Contents

Architectural and Urban Studies

Integrated Design Concept for Identifiable Commercial Heritage Tourism in *Tunjungan* District of Surabaya City

Yohanes Karyadi Kusliansjah and Anneke Clauvinia Patriajaya

Abstract

Tunjungan, a commercial district in downtown Surabaya, shows the potential for becoming a commercial heritage tourist site. Besides its identity as an entrance gate during the *Mataram* Empire era, *Tunjungan* has a Colonial artefacts heritage of unique commercial architecture typology, a pedestrian shopping street, and it is well known as a historic site of a heroic deed, namely the Hotel Yamato Incident. However, over the past years, the physical-spatial layout of *Tunjungan* district has been transforming and degrading, triggered by economic development. Key factors that support its integrated development as an identifiable commercial heritage tourism area, such as the attractiveness of historical buildings (currently lost, after their replacement by multi-storey buildings), clarity and ease of use of public–private transportation and pedestrian-friendly accessibility systems are no longer found, and amenities in information technology systems are unavailable. This paper aims to provide an integrated design concept solution for the revitalisation of the *Tunjungan* district of Surabaya. Through a qualitative–descriptive–comparative method, this study identifies problems, values, and proposes the possible designs of the *Tunjungan* district of Surabaya as an integrated identifiable commercial heritage tourism area. The results envisage *Tunjungan* as an entrance gate of the commercial heritage tourism area with improved public–private transportation accessibility systems by clarifying lanes, freeing lanes from parking, accommodating special modes of transportation that connect site attractions, and providing transportation stops with digital areas for Internet access and charging spots. The pedestrian accessibility system maintains its continuity in movement patterns, and the elements of streetscape are provided. The benefits of this academic paper can be found in its contribution to the government and developers to revitalise the *Tunjungan* district of Surabaya as an identifiable commercial heritage tourism area by utilising technological advances. The referenced integrated design concept is a form of adaption and resilience of Surabaya in maintaining the architectural order of its historic environment and continuing the spirit of fighting values in the midst of the digital era.

Keywords

Integrated design concept • Identifiable commercial heritage tourism • Pedestrian shopping street • Digital era • *Tunjungan* district of Surabaya

1 Introduction

Tunjungan, an old district of downtown Surabaya, is one of the areas affected by the conurbation phenomenon[1] of Greater Surabaya. This phenomenon has triggered the unplanned rapid growth and development of Central Business Districts (CBDs) which are scattered throughout several new axes of Surabaya City and the convergence of public transportation systems with the result that the CBDs of *Tunjungan* are experiencing agglomeration and its spatial pattern is transformed from linear to rhizome[2] and is not evolved in an integrated manner. The identity problem affecting *Tunjungan* arises when the attractiveness as a commercial, heritage, and heroic site that it once enjoyed is lost (Patriajaya & Kusliansjah, 2019) along with the

[1] Patrick Geddes coined the term "conurbation" in *Cities In Evolution* (1915).

[2] "Rhizome" is a philosophical concept introduced by Gilles Deleuze and Félix Guattari in their published journal *Ideology and Conciousness* (1981). They compared a city to a network.

Y. K. Kusliansjah (✉) · A. C. Patriajaya
Department of Architecture, Faculty of Engineering, Parahyangan Catholic University, Bandung, Indonesia
e-mail: karyadi@unpar.ac.id

H. Altan et al. (eds.), *Advances in Architecture, Engineering and Technology*,
Advances in Science, Technology & Innovation, https://doi.org/10.1007/978-3-031-11232-4_1

development of Surabaya. As such, it follows the general architectural trends in commercial areas (contemporary architecture styles in universal CBDs), substituted with the multi-level typology of mixed-use commercial clusters, such as *Tunjungan* Plaza, Surabaya Plaza, Grand City, and others. This research paper explains the exploration process and the conceptual steps in an effort to integrate the uniqueness of *Tunjungan*'s identity in its development process in line with the Urban Land-use Plan of Greater Surabaya. Its objective is to provide an integrated design concept solution for the revitalisation of the *Tunjungan* district of Surabaya as an identifiable commercial heritage tourism site. The benefits of this study are the referenced integrated design concept that can be used to embody the form of adaption and resilience of Surabaya in order to conserve the architectural order of its historic environment and persevere in the fighting spirit values in the midst of the digital era.

2 Theoretical and Methodological Approach

The term "urban" is analogised to a living organism (Corbusier, 1967). As a living organism, an urban area experiences birth, development, disintegration, or even death (Bally & Marshall, 2009). Urban development involves a complete change in urban society socially, economically, culturally, and physically in a planned or unplanned manner (Kostof & Tobias, 1991).

Physical-spatial changes in urban space are considered as a transformation that can occur because of the power that controls its order configuration (Habraken, 1983). The transformation in urban space is dominated by the merger or division of the order configuration of building lots that can change or even eliminate their past building typology. Based on the Form and Control theory (Habraken, 1998), transformation can be identified from physical order (urban form), territorial order (place hierarchy), and cultural order (understanding that the behaviour and habit in urban space form a heritage that embraces physical artefacts and intangible characteristics of society that are passed down from previous generations, valuable, preserved in the present, and entrusted to future generations (UNESCO, 1972).

This is based on the values and memories of the uniqueness of the context that are certain and significant (by experiencing activities, artefacts, and the site itself) and genuinely stand for the reminiscences and society of the past and present through heritage tourism (National Trust for Historic Preservation, 2015). Three key factors that determine the success of a tourism site (Samsuridjal & Kaelany, 1997) are as follows: attractions, accessibility, and amenity.

This research focuses on the problems and opportunities in a commercial urban space that has the potential for development into a heritage tourism site, based on local values. Employing qualitative methods, this study explores the urban form through history that has undergone a transformation due to its development in a case study and explains the form and control of commercial, heritage, and tourism-related elements to arrive at a conclusion that provides concepts, solutions, and design alternatives to give feedback on goals and benefits of the research conducted (Fig. 1.1).

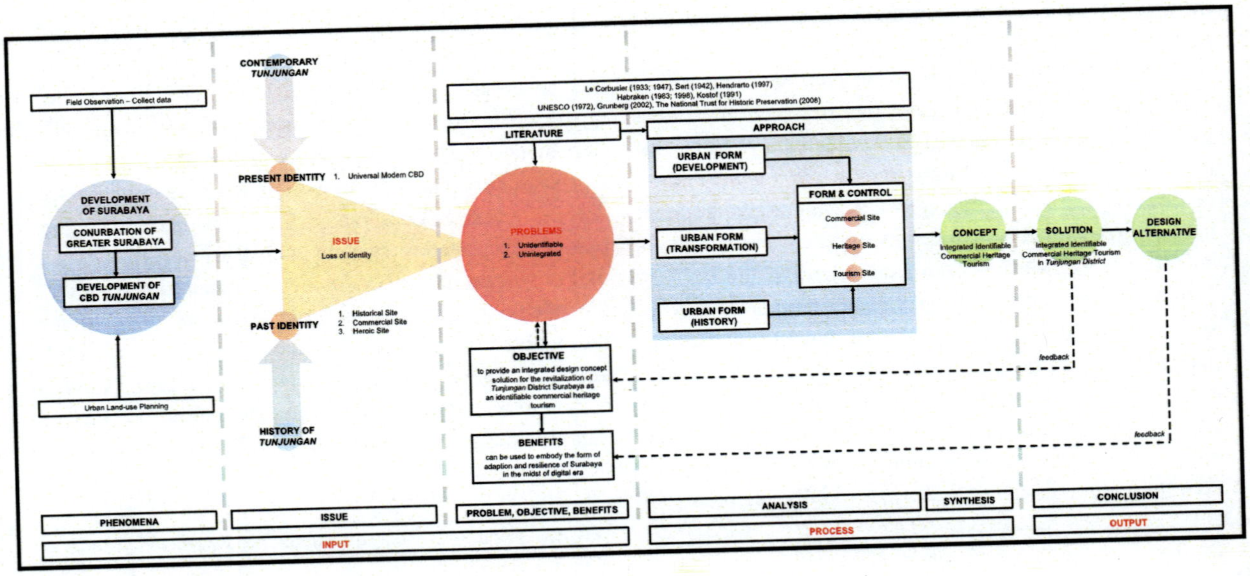

Fig. 1.1 Research framework. *Source* Authors (2019)

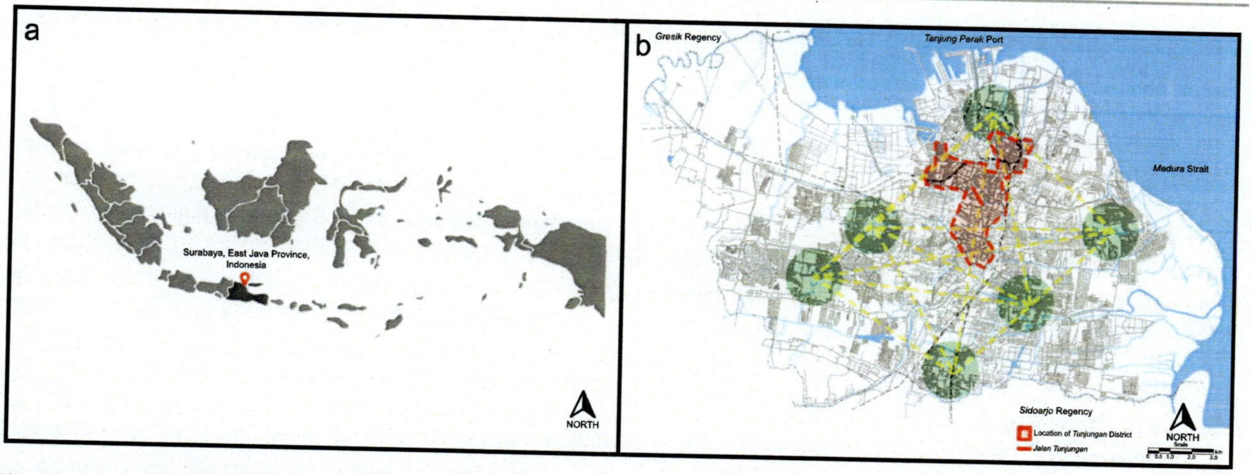

Fig. 1.2 **a** Location of Surabaya city. *Illustration* Authors (2019). **b** Location of *Tunjungan* district in Surabaya city. *Source* Bappeda, Surabaya (2017) *Illustration* Authors (2019)

3 Result and Discussion

Surabaya, the capital city of East Java province in Indonesia, is one of the oldest port cities (established in 1513) in Southern Asia which is still growing and developing rapidly in the present (Borschberg, 2001; Ricklefs, 1993). The *Tunjungan* district is located in the centre of the north–south axis of Surabaya towards the *Tanjung Perak* port which serves as a CBD (Fig. 1.2).

3.1 History of *Tunjungan*

The history of *Tunjungan* can be traced from three milestones, namely the Pre-Colonial era (from the late fifteenth to the early nineteenth century), the Colonial era (1808–1945), and the Post-Colonial era (1945–present) (Santoso, 2006).

According to various historical sources of Surabaya, during the Pre-Colonial era (around the sixteenth century), *Tunjungan* was known as the southern entrance of the Duchy of Surabaya (*Kadipaten Surabaya*), decorated with a white lotus flower, in the time of the *Mataram* Kingdom era. As a local toponym, the white lotus flower is called *tunjung* which later became the etymological forerunner to the naming of *Tunjungan* (Silas, 1994).

Under Colonial rule, Surabaya was established as a *Gemeente* (Dutch for Township) from early 1900 to late 1940. *Tunjungan* was a city centre and built with all the amenities and ease of use in terms of accessibility so that it attracted the bourgeoisie as a Colonial élite pedestrian shopping street area (Fig. 1.3) (Oswan & Arifin, 2013; Poerbantanoe, 1999).

The *Tunjungan* district is known as the site of two historical incidents for the people of Surabaya, namely.

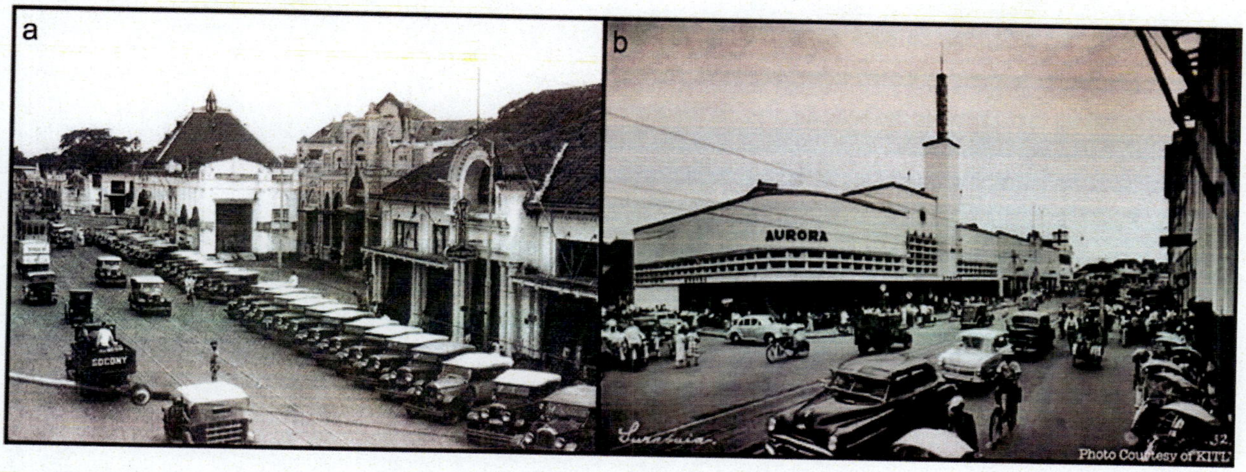

Fig. 1.3 **a** Old *Tunjungan* in 1905. *Source* KITLV and **b** old Tunjungan in 1955. *Source* KITLV

- 1942, the Allied bombing of the *Siola* building during the Japanese occupation.
- 1945, the Hotel *Yamato* Incident where Indonesian youth revolutionaries tore down the Colonial flag at the *Yamato* Hotel (now *Hotel Majapahit*) which led to the Battle of Surabaya.

These three milestones of the *Tunjungan* described above have become an indelible memory, attracting domestic and foreign tourists alike to visit this area.

3.2 Contemporary Architecture of CBD *Tunjungan*

The CBD of *Tunjungan* is located in the core and inner ring of the *Tunjungan* district with areas of 216 ha (covering the *Genteng*, *Embong Kaliasin*, and *Kedungdoro* sub-districts), whose main axis is *Tunjungan* Street to the south towards the *Gubernur Suryo* street and to the north towards *Ketabang Kali* street and is edged with the *Kalimas* River. The CBD *Tunjungan* area has been growing and developing with characteristics of agglomeration and conurbation (change in land-use and regulation of FAR, Building Coverage, and Building Height) (Fig. 1.4). The development of *Tunjungan*'s CBD can be seen from the transformation of the type of building lots, building forms, access and changes in the concept of technology and materials, and urban architecture style, following the recent trends in contemporary architecture. As a result, the old identity of the *Tunjungan* district as a heroic site and a pedestrian shopping street area, with the characteristics of pedestrian streets and storefront-friendly spaces, has been lost.

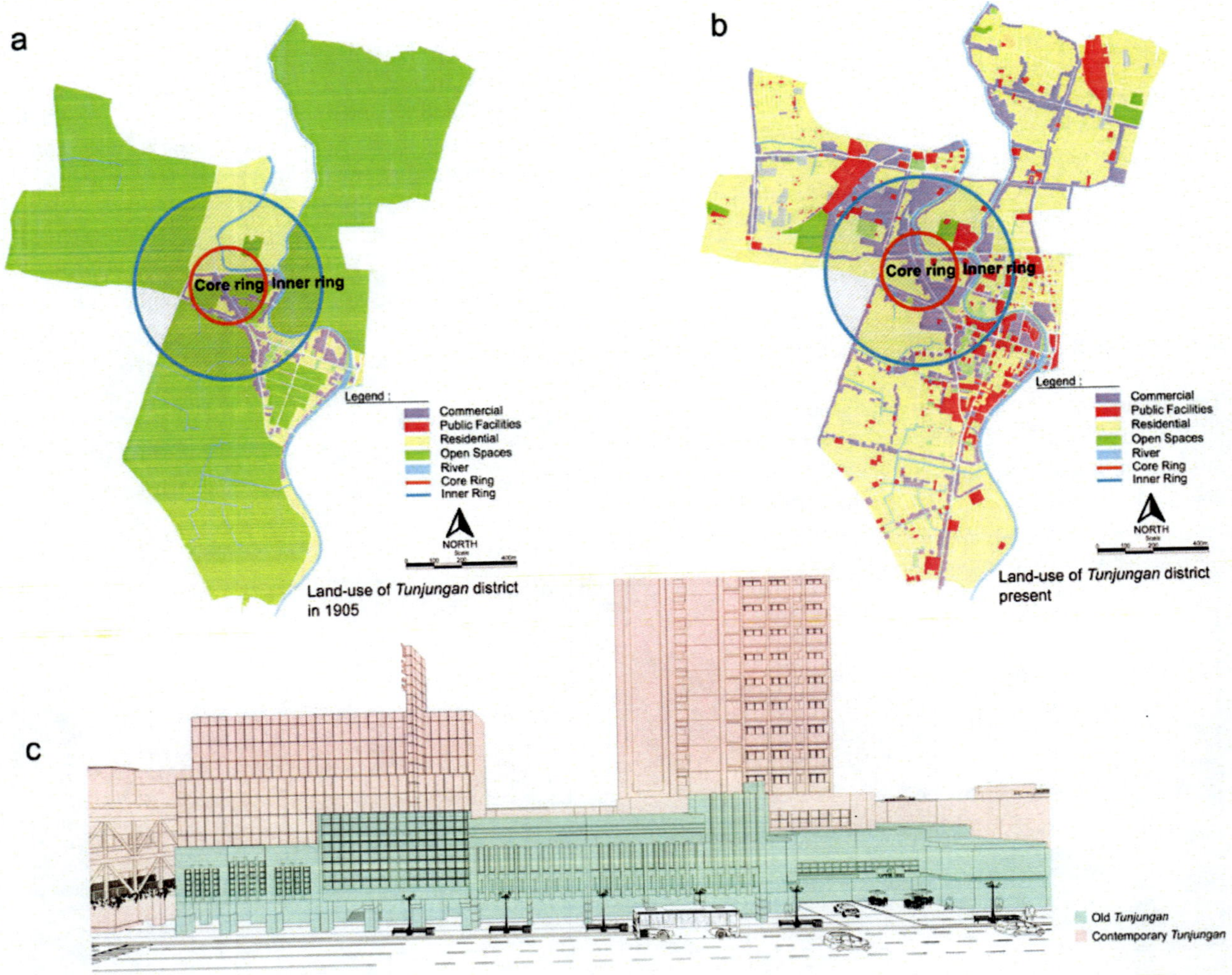

Fig. 1.4 **a** Land-use of *Tunjungan* in 1905. *Source* KITLV *Illustration* Authors (2019), **b** land-use of *Tunjungan* present. *Source* Bappeda (2017), and (c) Change in FAR, Building Coverage, and Building Height in *Tunjungan* district can be seen from the transformation of the buildings. *Source* Research document (2019)

3.3 Integrated Design Concept

The proposed integrated design concept is an academic-based research notion for the future development of the *Tunjungan* district. The identification of historical and present values is a crucial prospective in finding solutions to the problems that this area is facing.

3.3.1 Potentials of the Area to Solve the Problems

The potentials of the *Tunjungan* district, provided they are explored and managed in an integrated manner, can be the key solutions to the problems and make this area into an identifiable commercial heritage tourism site (Table 1.1).

3.3.2 Design Criteria

The spirit of an integrated identifiable commercial heritage tourism concept is to create adaptive and resilient built environments for the *Tunjungan* district by understanding the form and setting through its historical potentials and cultural values of the place, so that it becomes productive, effective, and efficient in a sustainable manner and makes this place an educational laboratory for visitors and future generations. Thus, the design criteria for the commercial heritage tourism design of the *Tunjungan* district are as follows:

Attractiveness

1. Major commercial buildings are designated to be the main attractions.
2. Objects and potentials of the site that have historical values are introduced in an integrated manner with the main attractions through digital information advances.

Accessibility

1. Provide various public transportation systems and build adequate public transportation stops.
2. Provide parking spaces in the specific areas to maintain the continuity of the pedestrian-way and enhance its convenience for the users.
3. Facilitate the pedestrian-way with a pedestrian bridge and/or an underground crossing.

Amenity

1. Clarify the lanes division to create a sense of comfort and security for the users.
2. Equip the pedestrian-way with adequate elements of streetscape and digital spaces.
3. Provide spaces for seasonal festivals or events.

3.3.3 Model

The delineation of the proposed identifiable commercial heritage tourism area in the *Tunjungan* district is determined from four commercial function nodes which form the main attractions of this area (Fig. 1.5a). From the determined boundaries, the supporting function nodes are positioned as secondary attractions (Fig. 1.5b). Thus, the combination of these two attractions becomes a tourist activity generator (Fig. 1.5c).

Table 1.1 Potentials and problems of *Tunjungan* district

Variables	Potentials	Problems
Attractiveness	– A centre of commercial buildings (the dominant typology is one-stop shopping malls), located in the axis of Surabaya – Historical buildings and artefacts and heroic site are situated in *Tunjungan* Street (outdoors)	– Lack of information about the distribution of commercial buildings – The one-stop shopping malls system forms individual characteristics (built-in indoor facilities) and accomplishes partial promotion by themselves, so that the historical value of this area (outdoor experiences) is not widely introduced
Accessibility	– Strategically located in the city centre, connecting north–south and east–west of Surabaya – The primary roads are served by various modes of public transportations (buses, minivans or *angkot*, taxis, and online transportations) – The use of bicycle and traditional transportation, such as a cycle rickshaw (pedicab) or *becak*, on secondary roads	– The transportation system has not been managed in an integrated manner to make it easier for users – This area is not equipped with adequate public transportation stops (the vehicles can stop anywhere), the pedestrian facilities are lacking in the elements of the streetscape (landscape, street furniture, etc.), there is a lack of parking areas, and pedestrian crossing patterns are only on the main road layer (the pedestrian bridge is not functioning)
Amenity	– Commercial and service areas (one-stop shopping (pedestrian shopping street–multi-malls)), hotels, banks, offices, and the like – As a place for holding seasonal festivals or events	– There is no clear lane division for private and public vehicles and pedestrians – The unavailability of environmental digital spaces that facilitate communication and visitor information

Source Author's research document (2019)

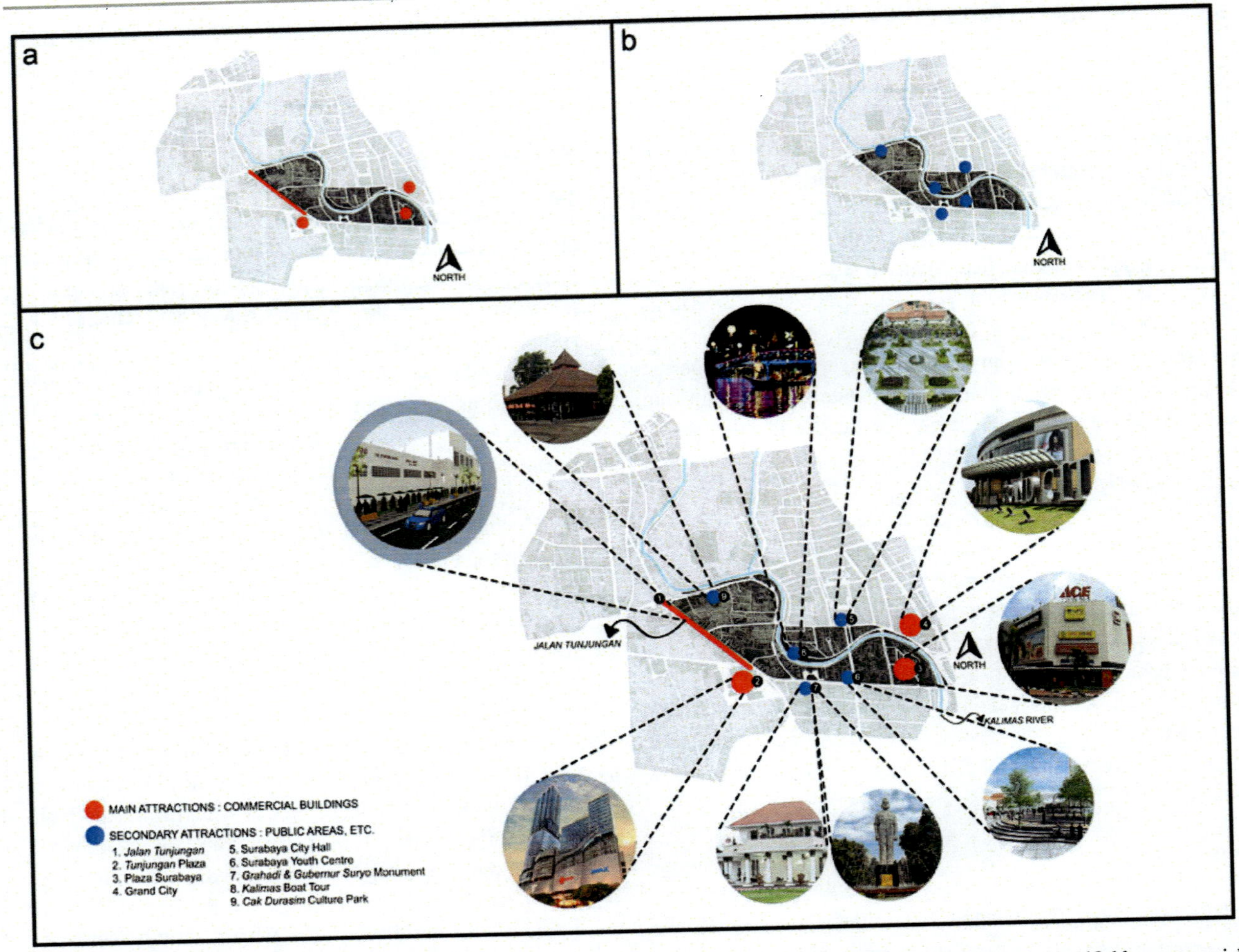

Fig. 1.5 **a** Proposed main attractions map, **b** proposed secondary attractions map, and **c** attractions in *Tunjungan* as identifiable commercial tourism. *Source* Authors (2019)

Tunjungan Street, located on the north-west side, is positioned as the main entrance to the corridor of the proposed identifiable commercial heritage tourism area in the *Tunjungan* district, which is formed from a row of Dutch East Indies heritage buildings in the Indies Empire style, the New Indies style, and the *Nieuwe Bouwen* style, some of which possess high historical value. The entrance gate is marked by two main heritage buildings, located at both ends of *Tunjungan*, namely the *Siola* Building (a) and *Hotel Majapahit* (2) (Fig. 1.6).

The on-street parking system is replaced by the on-site parking system, centred on eight parking spaces in the proposed identifiable commercial heritage tourism area in the *Tunjungan* district (Fig. 1.7). The proposed parking system consists of the existing multi-storey car parks and underground car park, owned by the government and the private sector.

The public transportation system from outside the proposed identifiable commercial heritage tourism area in the *Tunjungan* district uses the existing transportation systems and routes, according to Surabaya City government regulations. Intended specifically for the mobility of the proposed area, three new transportation modes are suggested (Table 1.2).

The pedestrian system is designed in line with the linearity of the roads in the *Tunjungan* district and is integrated with public transportation stops and equipped with a digital communication technology system to enhance the amenities in response to the demands of the digital era (Fig. 1.8 and Table 1.3).

The proposed model design includes the arrangement of vehicles and pedestrian-ways to increase the convenience and attractiveness of the *Tunjungan* district (Fig. 1.9). In principle, these two lanes are separated to serve the needs of

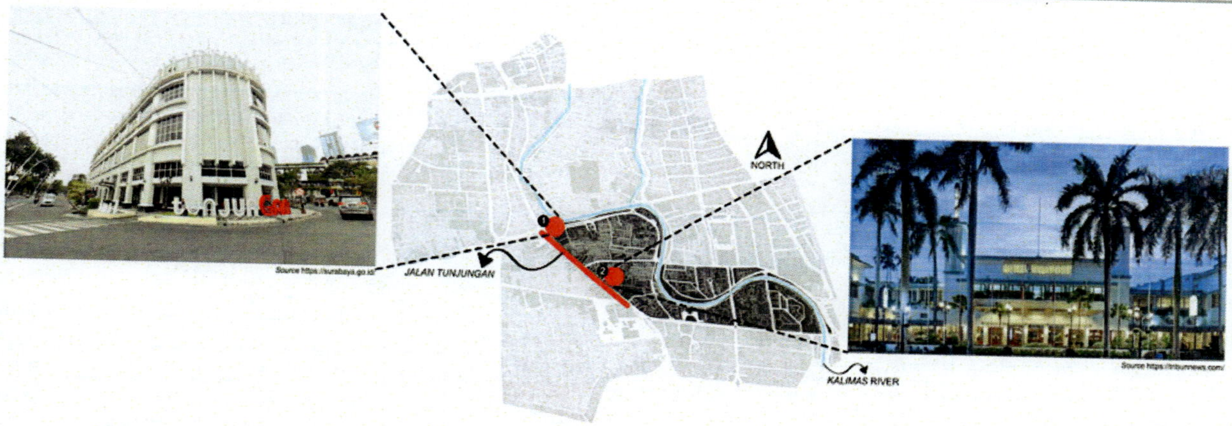

Fig. 1.6 Entrance gate to commercial heritage tourism in *Tunjungan* district is marked by (1) *Siola* building and (2) *Hotel Majapahit*. *Source* Authors (2019)

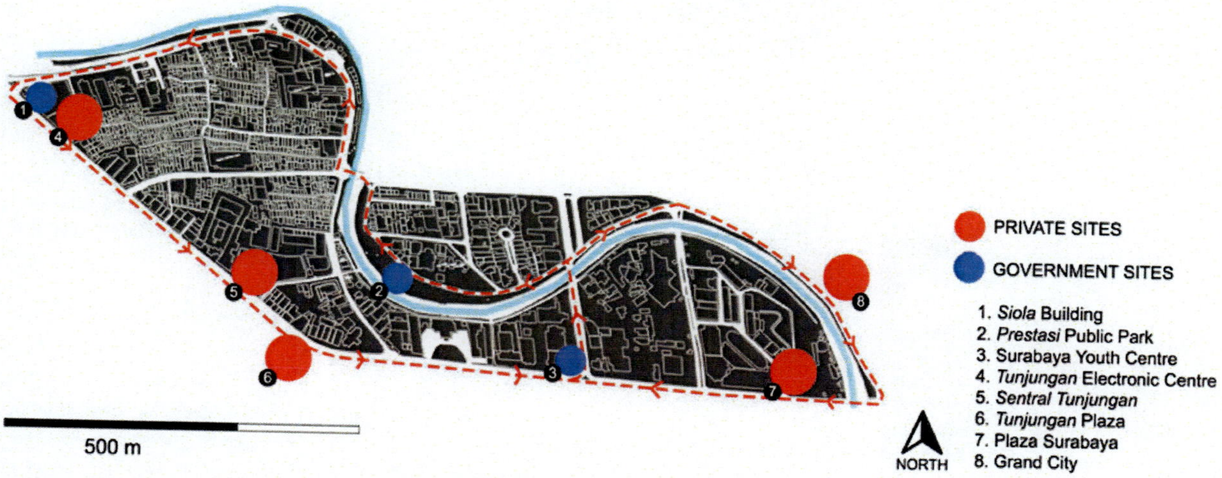

Fig. 1.7 Distribution map of proposed parking spaces for identifiable commercial heritage tourism in *Tunjungan* district. *Source* Authors (2019)

all road users. This area is also equipped with special lanes for disabled users, which are provided with materials and colours according to international standards (Fig. 1.10). Classic street furniture has been chosen to restore the old atmosphere. The street landscape element of choice is a *Tabebuia* tree with white flowers to strengthen the identity of the *Tunjungan* district as a white flower garden, and it has been equipped with street lights that support the decoration theme of seasonal festivals.

4 Conclusions

The development of the *Tunjungan* district as a CBD requires an integrated design concept in order to turn this area into an identifiable commercial heritage tourism area in the midst of the digital era. An integrated design concept is a solution for the unintegrated transformation in the physical-spatial layout of the *Tunjungan* district. Thus, the

Table 1.2 Proposed transportation modes for identifiable commercial heritage tourism in the *Tunjungan* district

Transportation modes	Route	Benefits
Shuttle car	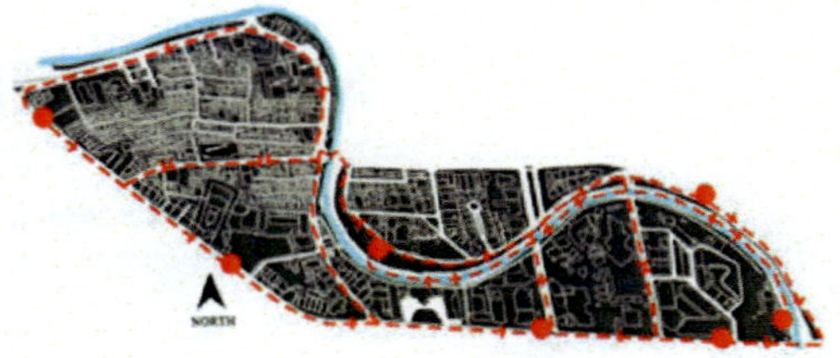 Main route: Jl. Tunjungan–Jl. Gubernur Suryo–Jl. Yos Sudarso–Jl. Ketabang Kali/Jl. Gubeng Pojok–Jl. Pemuda–Jl. Genteng Kali [N.B. The abbreviation *Jl.* stands for *Jalan* = Street)	The local minivan *or angkot* is chosen as the type of shuttle car transportation mode to increase the value of the economic aspects of the local community
Bike or scooter	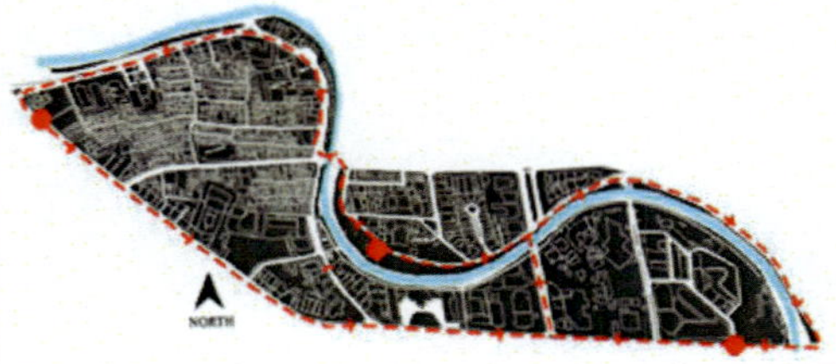 Main route: Jl. Tunjungan–Jl. Gubernur Suryo–Jl. Yos Sudarso–Jl. Ketabang Kali/Jl. Gubeng Pojok–Jl. Pemuda–Jl. Genteng Kali	Folding bikes or scooters are chosen to increase the value of the environmental aspects of the Tunjungan district (low carbon)
Local boat	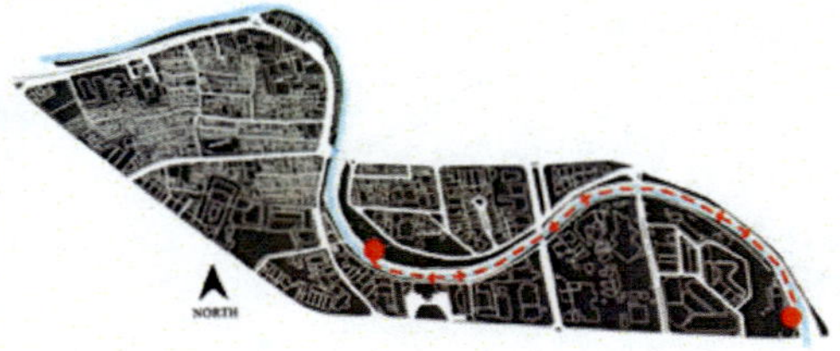 Main route: Kalimas River (Jl. Pemuda–Jl. Ketabang Kali)	Local boats are being reinstated as a mode of transportation to increase the value of the economic and environmental aspects of the local community

Source Authors' Research Document (2019)

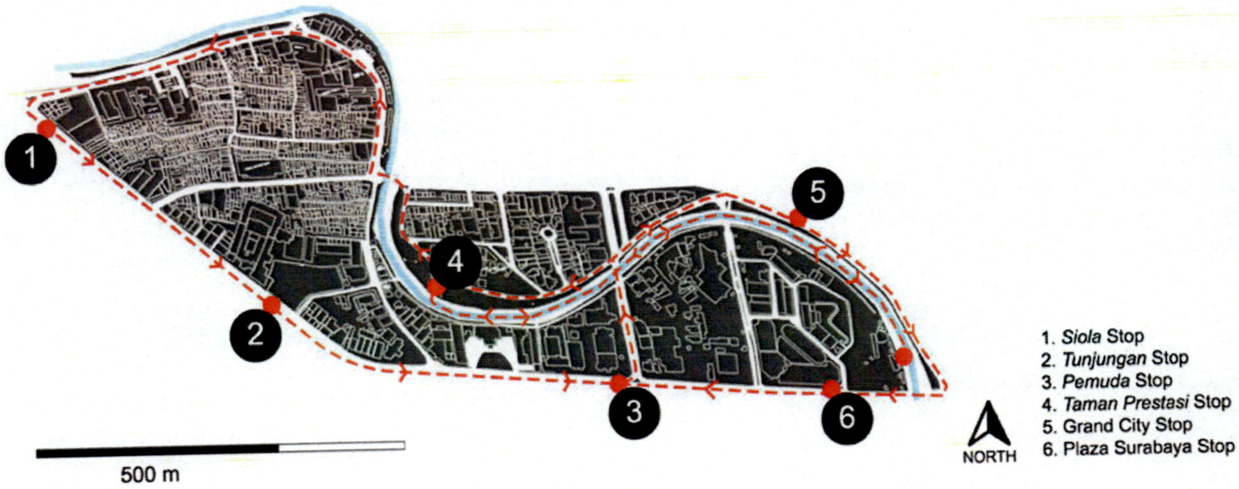

Fig. 1.8 Distribution map of proposed public transportation stops in *Tunjungan* district. *Source* Authors (2019)

Table 1.3 Proposed types of transportation stops for identifiable commercial heritage tourism in the *Tunjungan* district

Variables	Main transit stop	Small transit stop
Map	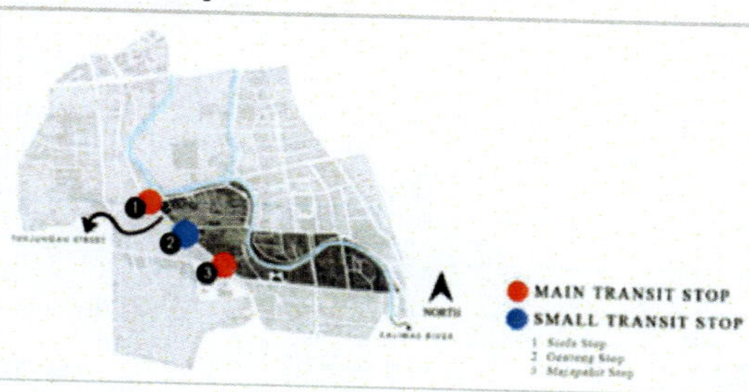	
Key plan	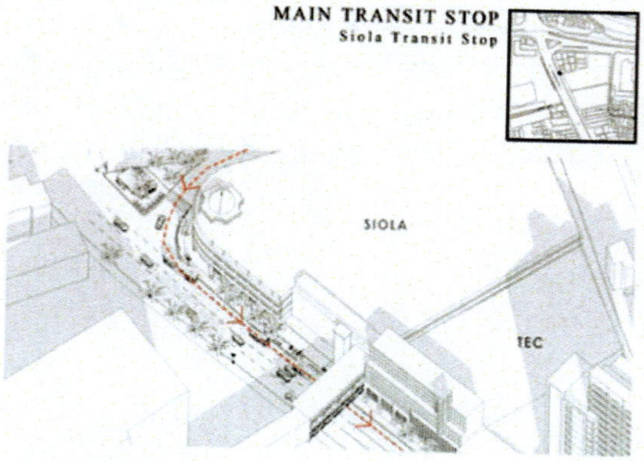	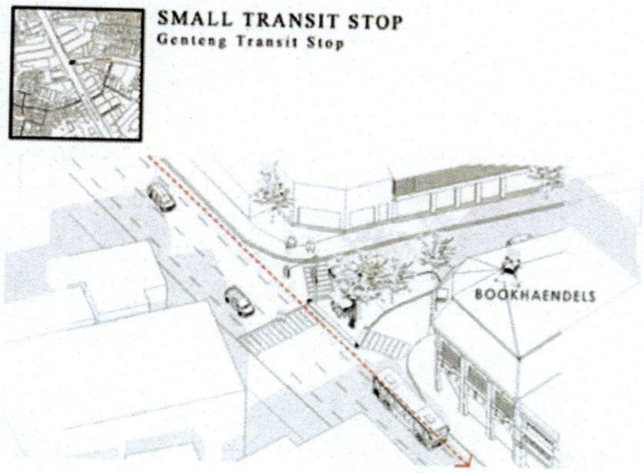
Minimum dimension	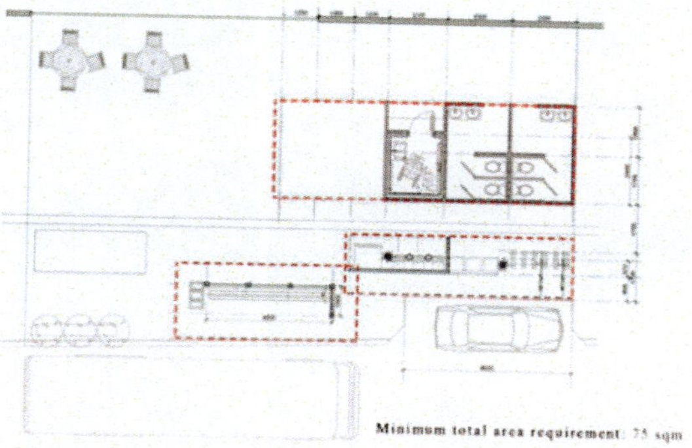	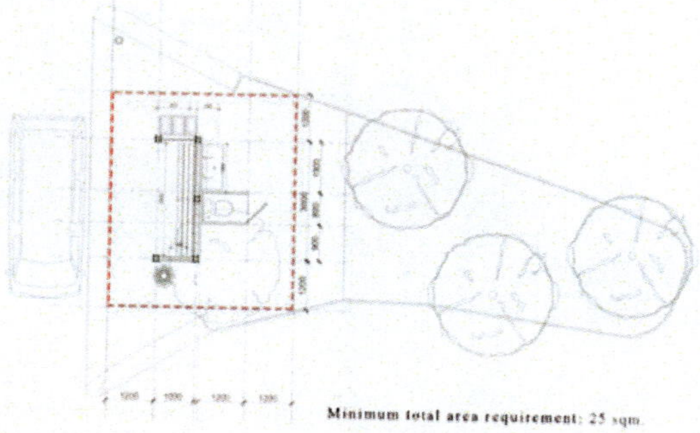

(continued)

Table 1.3 (continued)

<table>
<tr><th>Variables</th><th>Main transit stop</th><th>Small transit stop</th></tr>
<tr><td>Facilities</td><td>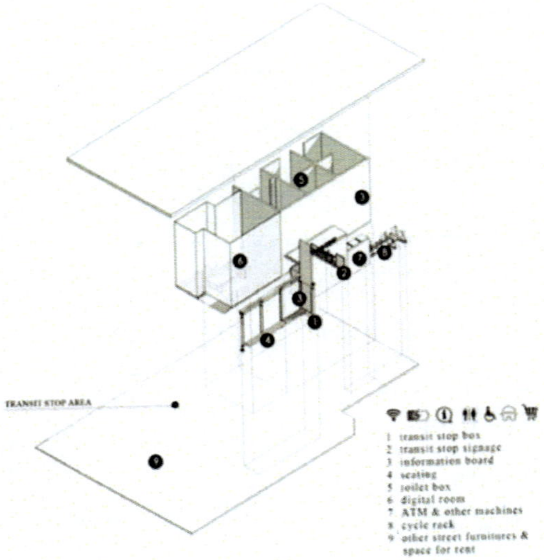
</td><td>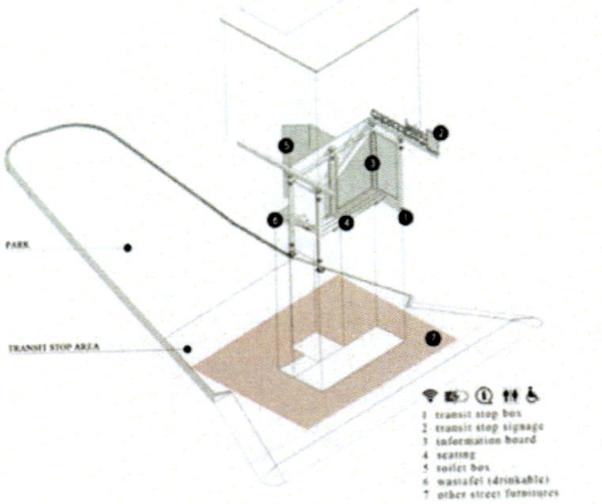
</td></tr>
<tr><td>Perspective</td><td>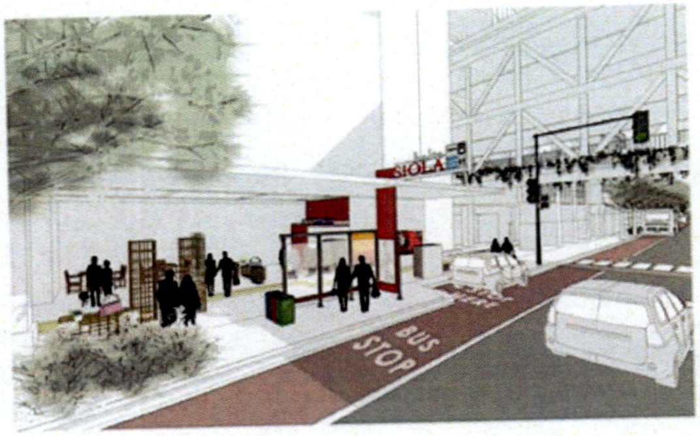
</td><td>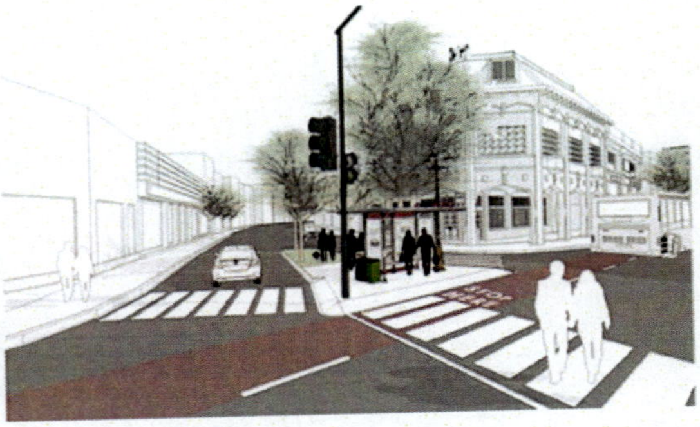
</td></tr>
</table>

Source Authors' Research Document (2019)

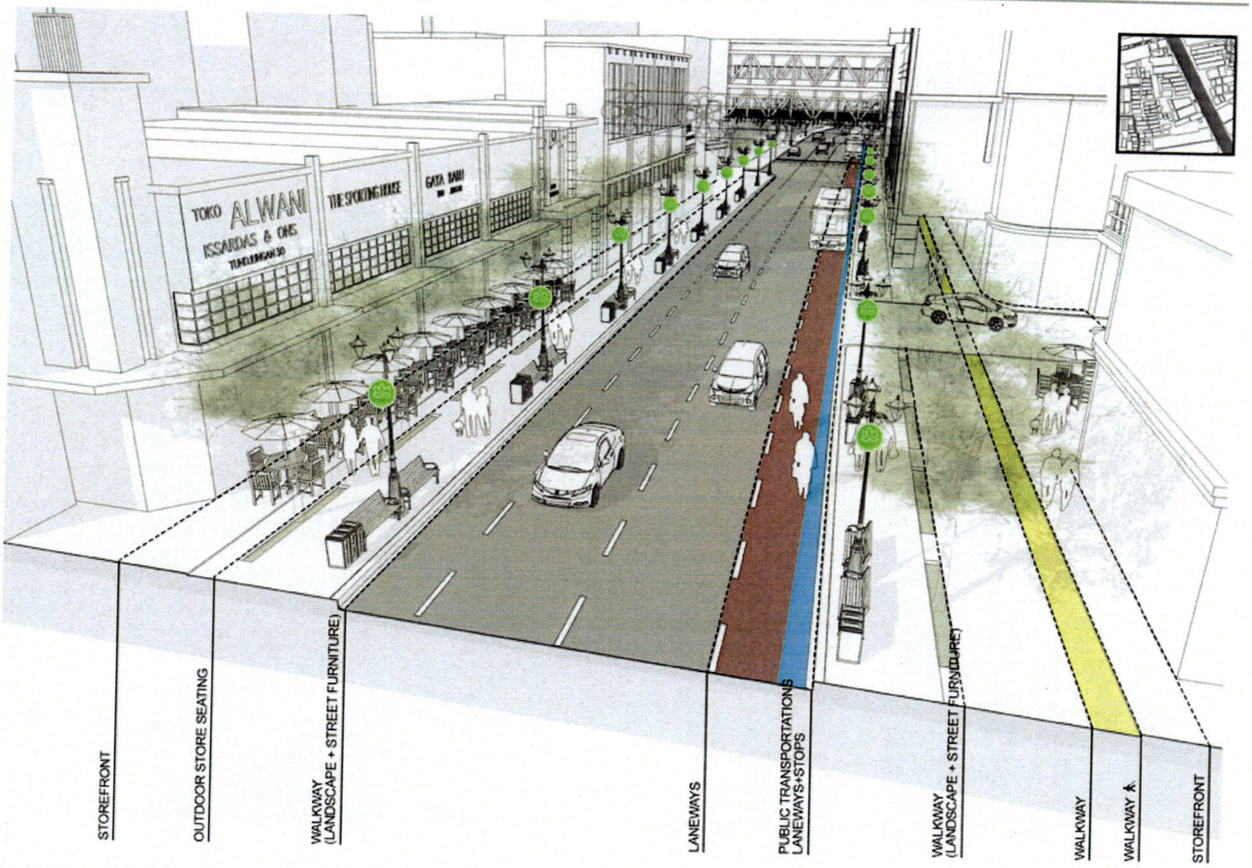

Fig. 1.9 Proposed design of clear lane division in *Tunjungan* district. *Source* Authors (2019)

necessary criteria and alternative design for *Tunjungan* district are as follows:

- Four major commercial street and buildings in this area (*Tunjungan* Street, *Tunjungan* Plaza, Plaza Surabaya, and Grand City) are identified as the main attractions, and the supporting functions (*Kalimas*, Surabaya Youth Centre, and others) are regarded as the secondary attractions.
- The gate to the commercial heritage tourism area in the *Tunjungan* district is *Tunjungan* Street which has the highest historical value derived from a set row of heritage buildings.
- Public transportation stops, equipped with adequate elements of streetscape and digital spaces, aim to serve various public transportation systems and users.
- Parking spaces are provided in specific areas to maintain the continuity of pedestrian-ways and to create sense of comfort for the pedestrian users, and a pedestrian bridge and/or underground crossing is provided.
- The lanes are divided clearly and make use of materials, according to international requirements, to serve the needs of all road users and to provide spaces for seasonal festivals.

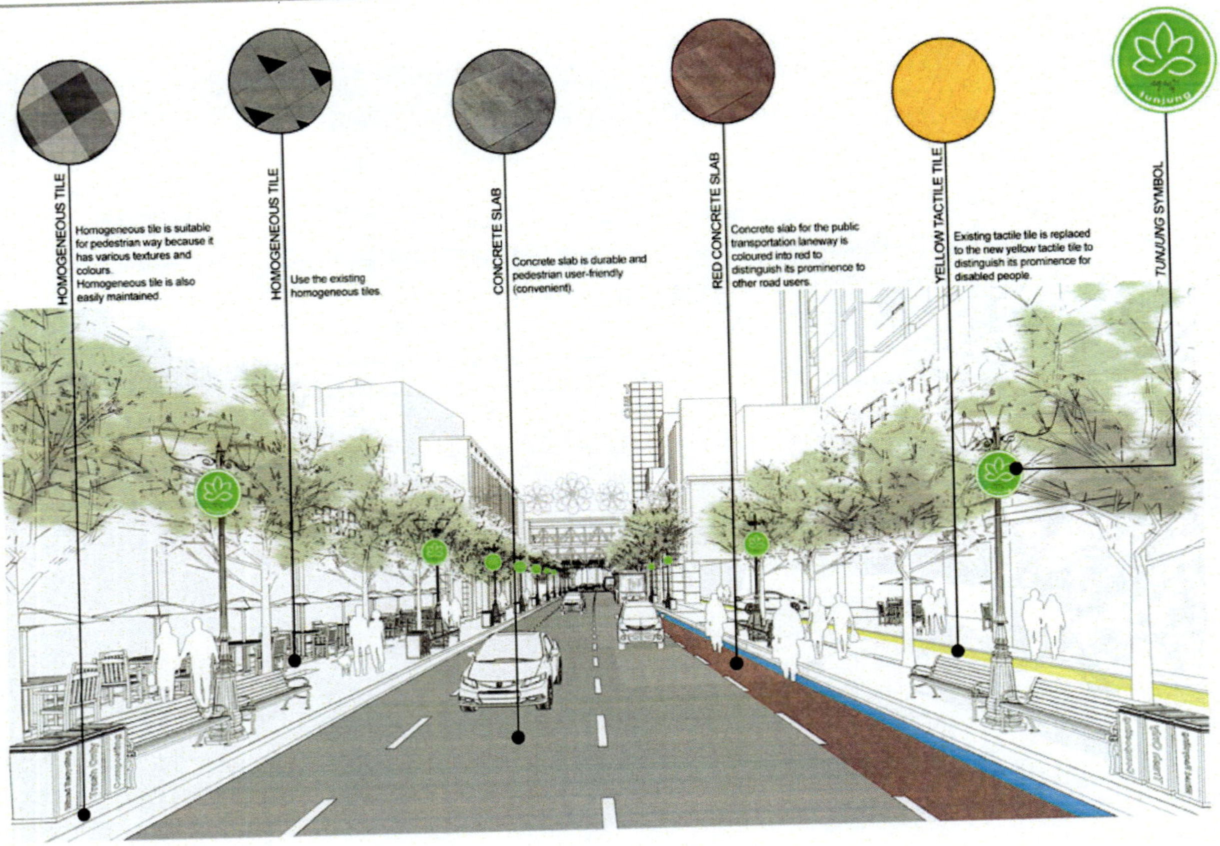

Fig. 1.10 Proposed materials in *Tunjungan* district. *Source* Authors (2019)

Acknowledgements We would like to thank the Architecture Postgraduate Programme, Architecture Department, Engineering Faculty of Parahyangan Catholic University. The content of this paper is a part of the thesis by Anneke C. Patriajaya *"An Integrated Design Concept Implementation for an Identifiable Urban Tourism Area in the Tunjungan District of Surabaya"* for the Master's Degree in Architecture at Parahyangan Catholic University, 2019. Lastly, we would like to thank IEREK for choosing this academic research paper to be published at the international level.

References

Bally, M., & Marshall, S. (2009). The evolution of cities: Geddes, Abercrombie and the new physicalism. *Town Planning Review*. https://doi.org/10.3828/tpr.2009.12

Borschberg, P. (2001). South East Asia colonial history. In P. H. Kratoska (Ed.) Vol. 1: Imperialism before 1800. Routledge.

Corbusier, L. (1967). *The Radiant City*. The Orion Press.

Habraken, N. J. (1983). *Transformation of the Site* (Private ed). A Water Press.

Habraken, N. J. (1998). The structure of the ordinary, form and control in the built environment (J. Teicher (Ed.); Paperback). MIT Press.

Kostof, S., & Tobias, R. (1991). The city shaped: Urban patterns and meanings through history. Thames and Hudson.

Oswan, D. B., & Arifin, L. S. (2013). Fungsi Pedestrian Jalan Tunjungan Dari Sirkulasi Ke Rekreasi: Studi Kasus Berdasarkan Kesejarahan. *DIMENSI (journal of Architecture and Built Environment), 40*(2), 99–111. https://doi.org/10.9744/dimensi.40.2.99-112

Patriajaya, A. C., & Kusliansjah, Y. K. (2019). Hilangnya karakter pedestrian shopping street jalan Tunjungan akibat transformasi Surabaya sebagai Kota Metropolitan. *ARTEKS : Jurnal Teknik Arsitektur*. https://doi.org/10.30822/arteks.v4i1.81

Poerbantanoe, B. (1999). the lost-city dan lost-space karena perkembangan pengembangan tata-ruang kota Kasus : Koridor Komersial Jalan Tunjungan, Kotamadya Surabaya. *Dimensi Teknik Arsitektur, 27*(2), DESEMBER 1999, 31–39.

Ricklefs, M. C. (1993). A history of modern Indonesia since c. 1300. In *A History of Modern Indonesia since c. 1300*. https://doi.org/10.1007/978-1-349-22700-6

Samsuridjal, D., & Kaelany, H. D. (1997). *Peluang di bidang pariwisata*. Mutiara Sumber Widya.

Santoso, J. (2006). *Kota tanpa Warga*. KPG dan Centropolis.

Silas, J. (1994). *Surabaya: Lintas dan Langkah*. Pemerintah Kotamadya Daerah Tingkat II Surabaya.

Environmental Product Declarations for Building Materials: Advantages, Limits, Developments

Francesca Scalisi

Abstract

This paper explains one of the environmental certifications tools that the construction sector can use to communicate the environmental performance of materials: the environmental product declaration (EPD) or type III environmental label. The need to be more transparent on environmental information is mandatory and can no longer be avoided. The paper describes the main environmental product declarations (EPDs), highlighting that over the last years the number of environmental product declarations (EPDs) published by the main European EPD programmes has increased. One of the main characteristics of the environmental product declaration (EPD) is the possibility to compare the environmental aspects of the considered products. The paper compares two case studies on the same material building type, published by two different EPD programmes, highlighting the difficulties of this comparison.

Keywords

Environmental product declaration (EPD) • Product construction rules (PCR) • LCA • ISO 14025 • Material construction

1 Introduction

In general, process, production/creation and product innovations are fostered by the (European, national and local) legislation—nowadays mostly on the environmental subject—and by market mechanisms that stimulate a productive competition for the professionals on the continuous enhancement of the product compared to a constantly changing demand made by planners, builders, buyers and final users. The difficulty of the subject corresponds to a quite complex framework of environmental directives—somehow still fragmented—requiring quite a massive effort to be implemented and most of all to verify the environmental effectiveness of the strategies enforced over the last few years. A first attempt at a systemic approach can be found in the Europe 2020 Strategy and in particular in a resource efficient Europe objective (European Commission, 2011a; b), which provides for the delineation of a set of economic-financial tools to assess the real costs and benefits of the use of resources and to encourage the use, in the long run, of solutions designed to an efficient use of natural resources.

It is established that the environmental subject is heavily affected by the building sector, which can play a decisive role in containing CO_2 emissions in the atmosphere given that globally in 2017, the building industry consumed 36% of energy and was responsible for almost 40% of carbon dioxide emissions (UN Environment, 2018). These percentages are referred to the total energy used by the building that is the sum of operational energy and embodied energy that contribute to the calculation of the total energy (Barucco et al., 2016; Gonzalez & Navarro, 2006; Treloar et al., 2001). Specifically, building emissions can be divided into three fields: direct emissions, coming from the burning of fossil fuels in buildings; indirect emissions, coming from the production of electrical and thermal energy; the embodied carbon or CO_2 emissions, coming from the production of materials. While direct and indirect emissions tend to decrease, the emissions coming from the production stage of materials are becoming increasingly important, especially those related to steel and cement which, from 1.8 gigatons of carbon dioxide ($GtCO_2$) released in 2017, will increase up to around 40% by 2060, according to the recent projections provided by the International Energy Agency (IEA, 2019).

F. Scalisi (✉)
DEMETRA Ce.Ri.Med. (Euro- Mediterranean Documentation and Research Center), Palermo, Italy
e-mail: demetracerimed.scalisi@gmail.com

H. Altan et al. (eds.), *Advances in Architecture, Engineering and Technology*,
Advances in Science, Technology & Innovation, https://doi.org/10.1007/978-3-031-11232-4_2

Therefore, the environmental issue, if referred to the building sector, traditionally well-structured and complex, highlights that it is increasingly necessary to create a system of the regulatory framework and of support decisional tools and to evaluate the environmental efficiency of materials and building components, innovative technologies and building techniques that can make an important contribution to the sustainability and eco-efficiency of the building industry (Pacheco-Torgal & Jalali, 2012). In recent years, the need for a simplified approach, quick to use by professionals and certain to be verified by inspectors, has favoured the spread of tools with checklists (protocols or rating systems such as LEED, BREEM, ITACA, etc.) or minimum environmental criteria for Green Public Procurement, based on a very detailed list of environmental criteria-requirements to be met in order to be awarded or to access tenders, in the case of the general public administration (Ganassali et al., 2016). Overcoming this typically qualitative attitude—on the basis of reasoned and grouped into class parameters— and the need for a verification of environmental effectiveness in strictly quantitative terms, in recent years, have pushed the standardization structures at international (ISO), European (CEN) and national (UNI) level to the implementation of tools, such as life cycle assessment (LCA), capable of quantitatively measuring resource flows and environmental impacts throughout all stages of the life cycle and the increase or reduction of environmental impacts related to process and product innovations.

The goal of reaching high performances of the materials concerning for to environmental indicators, despite the complexity of the required framework to which the project must respond, gives to the project a double challenge. The first one is the relationship between the project and the matter: the research started over the last few years on the bio-based materials (Sposito & Scalisi, 2019; Maskell et al., 2015; Onchiri et al., 2014) are emblematic of the possibility of designing the characteristics of the materials not only from a technical performance —as for the advent of composite materials—and aesthetic point of view, but also from the environmental performance point of view. The second challenge concerns the opportunity to optimize the production processes of materials to reduce the most expensive stages from the point of view of resource consumption and generated impacts (Campioli et al., 2018), enhancing 'low energy consumption' solutions. In this context, a useful tool for assessing the environmental impacts of materials is the environmental product declaration (EPD) which aims to contribute to reducing the impact on climate of the building sector by encouraging planners and designers to use the LCA while planning and designing buildings (Bovea et al., 2014; Pacheco-Torgal, 2014; Del Borghi, 2013). This paper illustrates the main features of the EPD and its dissemination, focusing in particular on the analysis of two environmental product declarations relating to autoclaved aerated concrete, produced by two different EPD programmes, highlighting their strengths and limitations.

2 The Environmental Product Declaration (EPD) and the Product Category Rules (PCR)

The environmental product declaration (EPD) is one of the most recommended methods to report the environmental impacts of building materials. EPD is a certified environmental product declaration that provides environmental data on the life cycle of products in accordance with the international standard ISO 14025 (2006), is "[...] the input for a holistic building assessment considering the functional and technical performances in a building context. For a producer, this also means that his contribution to higher sustainability (environmentally, socially and economically) of course should be done by big and small improvements on every step in the building chain" (Gagari et al., 2013, p. 107). It is a voluntary project decided by the companies that through this tool can communicate the environmental data of their products. These data are processed by one or more organizations, based on life cycle assessment (LCA) in accordance with the ISO 14040 (2006) series of standards and are independently verified.

Therefore, the LCA method can provide important information on the stages of the whole life cycle to reduce environmental loads and impacts, also through the use of open access software (OpenLCA, 2019; SimaPro, 2019). This is why many green building rating systems have added LCA indicators in the criteria relating to materials and have fostered the use of products with EPD certification, allowing to identify the best material during the design stage based on verified environmental information. It should be noted that only one protocol, the German DGNB (2019), was born with the LCA evaluation of the building amongst the first criteria of the protocol, virtuously activating the whole supply chain and leading to a quick increase of EPD certified products, favouring the elimination of irrelevant and misleading data and enhancing primary data strictly linked to the specific product used in the building, stimulating the production sector to direct competition and environmental innovation of products. The general objective of EPD is to encourage the demand and supply of products entailing less stress on the environment while allowing the comparison between products that have the same function. The ISO 14025 (2006) standard establishes nine guidelines of the EPD in the following points: relationship with the ISO 14020 Standard (2000), voluntary nature, life cycle basis, modularity, the involvement of the interested parties, comparability, verification, flexibility and transparency.

The environmental product declaration is also known as a Type III environmental label, according to the classification of the international standards organization (ISO) which divides environmental labels into three types: the type I label describes the impact of products or services on the environment, its acquisition is voluntary-based and is regulated by ISO 14024 (1999); the type II label consists of self-declared environmental declarations of companies and organizations without third-party verification and is regulated by ISO 14021 (1999) and type III label or environmental product declaration label is regulated by ISO 14025 (2006). The developing process of an EPD consists of four stages: (1) the selection of type II environmental declaration programmes; (2) the research of product category rule (PCR) for the product that must be declared (if there is not a product category rule for the product category, it must be created); (3) the creation of a EPD draft based on the implementation of the LCA method, abide by the rules of the EPD programme and the outlining of a specific PCR for that product category and (4) the verification process that has to prove, before the EPD publication that the data collection and the enforcement of the LCA method are made according to the PCR and meet all ISO requirements.

The EN 15804 (2012) standard specifies the basic rules for EPDs relating to the category of building materials, as a guarantee that all declarations are uniformly represented and verified. Specifically, the product category rule (PCR) must list the stages of the life cycle to be included, the parameters to comply with and how the parameters must be collected and reported. The life cycle stages are established by the EN 15804 (2012): product stage, construction stage, use stage, end of life stage and an optional module reuse-recovery (D). In Fig. 1, the mandatory and optional stages are listed, according to the system limit considered: the 'cradle to gate' analysis evaluates only the Product stage (A1–A3), which is therefore mandatory; in the 'cradle to gate with options' analysis, the Product stage (A1–A3) is mandatory while all other stages are optional; in the 'cradle to grave' analysis all stages are mandatory except D) which is optional.

3 Contents of the Environmental Product Declaration and EPD Programmes

The environmental product declaration (EPD) has to facilitate the comparison between products' environmental characteristics that meet equivalent functional requirements. The information included basically concerns: identification and description of the organization that makes the declaration; the product description; the name of the programme and the address of the programme manager and, if necessary, the logo and website; the identification of the PCR; the publication date and the validity period; life cycle analysis data (LCA); additional environmental information; the presence of materials and substances that must be declared, for example, substances that can negatively affect public health

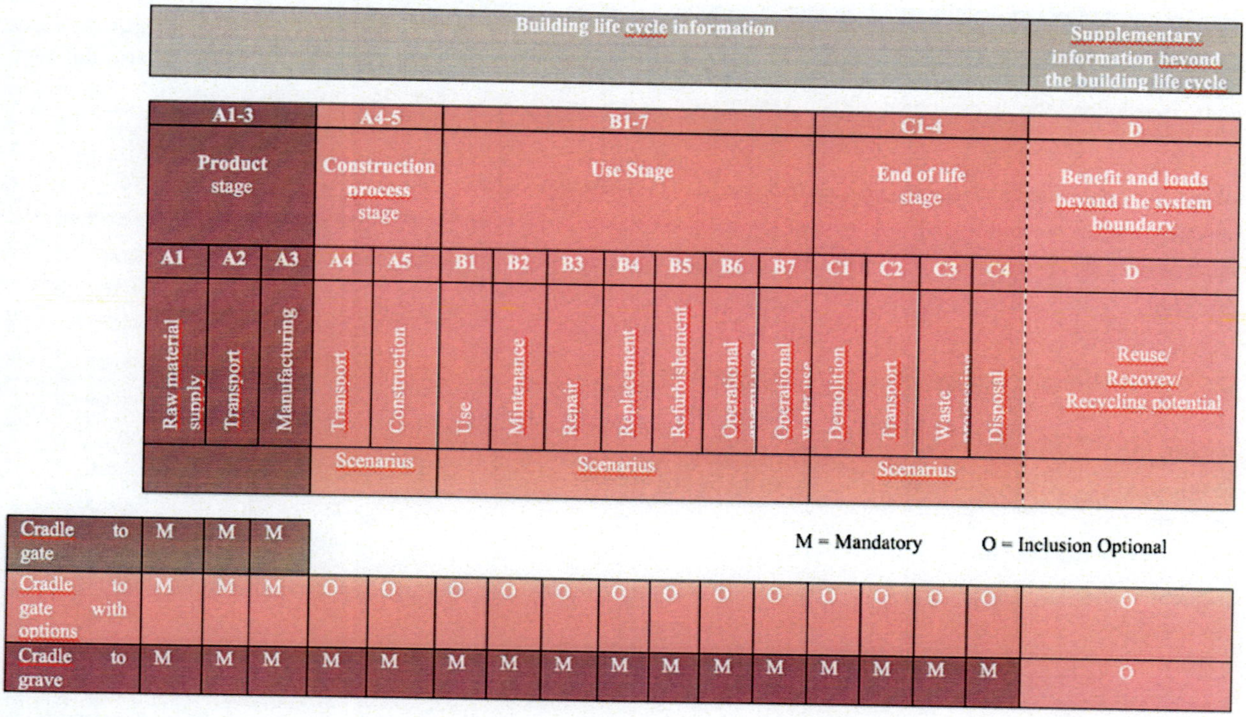

Fig. 1 Information modules for construction products, adapted from EN 15804:2012

and the environment at all stages of its life cycle; information on the stages that have not been taken into consideration, if the declaration is not based on a 'cradle to grave' LCA.

The LCA results are divided into three categories on environmental impact, use of resources, output flows and waste categories. The environmental impact includes the parameters relating to the global warming potential, stratospheric ozone layer depletion, soil and water acidification potential, eutrophication potential, tropospheric ozone and photochemical oxidants formation potential and abiotic degradation of non-fossil resources potential, abiotic degradation of fossil resources potential.

The use of resources includes the parameters on the use of: renewable primary energies as an energy source; renewable primary energy resources as raw materials; renewable primary energy resources; non-renewable primary energies as energy sources; non-renewable primary energy resources such as raw materials; non-renewable primary energy resources; secondary materials; secondary renewable fuels; non-renewable secondary fuels and net consumption of water resources. In the output flows and waste categories, the following parameters are included: hazardous waste disposed of, non-hazardous waste disposed of, radioactive waste disposed of, components destined for re-use, materials destined for recycling, materials destined for energy recovery, exported electricity and exported thermal energy. Finally, additional information on environmental issues can be provided such as the potential impact on biodiversity or the assessment of risks to public health and the environment.

Amongst the many independent organizations or highly organized structures that develop the EPD, it is certainly worth mentioning the International EPD® System which, established in 1999, was the first EPD programme developed on a global scale and still is one of the most widespread in Europe (Hunsager et al., 2014). Table 1 lists the 14 most important EPD programmes in Europe—sorted by the number of EPDs processed—almost all created for the building materials certification, except for the EPD-Norge and the International EPD System. The data in the Table, if compared with the numbers shown in previous studies (Bovea et al., 2014), show a significant increase—from 2014 to date—of the EPDs that have increased from 249 to 1087 in the International EPD® System and from 280 to 1613 in the IBU-EPD, as a reaction from the world of producers to a new sensitivity towards eco-oriented materials and building components by the users.

4 Methodology

This paper has examined two of the main EPD programmes, the International EPD® System and the IBU-EPD. The choice has quantitative reasons justified by the fact that the first one is the most widespread in Europe while the latter has a considerable number of EPDs—the second after the French FDES INIES—but above all, it reports the greatest number of PCR elaborated by an EPD programme in Europe. The first difference between the two programmes can be found in the field of the processed products: while the IBU-EPD deals exclusively with building materials, the International EPD® System has also other categories such as food and beverages, chemical products, textiles, footwear and apparel and paper products. The number of the products in the archive is also different: the IBU-EPD has 1613 building materials while the

Table 1 EPD programmes: construction products (processed data on 14 December 2020)

EPD programmes	Link	Number of EPDs	Country
FDES INIES	www.hqegbc.org/accueil/	3516	France
IBU-EPD	http://www.ibu-epd.com/	1613	Germany
International EPD® system	http://www.environdec.com	1087	Sweden
EPD-NORGE Norwegian EPD foundation	http://www.epd-norge.no	500	Norway
MRPI®	www.mrpi.nl/	135	Holland
EPDItaly	www.epditaly.it/	108	Italy
EPD Ireland	www.igbc.ie/epd-home/	101	Ireland
RTS EPD	http://www.epd.rts.fi/en/	88	Finland
BRE	www.greenbooklive.com	77	United Kingdom
DAP construcción®	www.csostenible.net/home/index?locale=es	73	Spain
EPD Danmark	www.epddanmark.dk/	31	Denmark
ZAG	www.zag.si/si/	14	Slovenia
DAP habitat system	http://www.daphabitat.pt/	13	Portugal
Bau EPD	www.bau-epd.at	8	Germany

International EPD® System 1087, even if the latter can be found in more countries (25 in Europe alone). Concerning the Construction Product Rules, it is worth mentioning that the International EPD® System has developed a PCR entitled Construction Products and Services (and 12 Sub-PCRs related to different material categories, valid at the access date of 29 April 2019), updated with the Product Category Rules (PCRs) for Construction Products, compliant with the European standard EN 15804: 2012 +A2:2019 (Sustainability of construction works-Environmental Product Declarations-Basic Product Category Rules for Construction Products), valid at the access date of 14 December 2020; the 8 c-PCRs currently existing, and to be used in addition to this document for the relevant product categories, shown in Table 2, and the 11 c-PCRs currently under development.

The IBU-EPD has developed 105 PCR divided into three large groups called 01) basic materials and precursors, 02) building products, 03) building services engineering, in which we can find some subgroups. For example, group 01 Basic materials and precursors includes 4 subgroups named 'Aggregates', 'Cement, building limes and other hydraulic binders', 'Other basic materials and precursors', and 'Products related to concrete, mortar and grout'. Within the subgroups, one finds the PCRs of the materials, as shown in Table 3. It should be noted that some materials with the same PCR are found in the same subgroups, which is a useful expedient to facilitate searching within the database; among others, we mention bulk granulate, both in the subgroups "Aggregate" and "Products related to concrete, mortar and grout".

5 The Case Study

Further analysis has been carried out on the LCA found in the EPDs of the material called autoclaved aerated concrete—of which, for confidentiality reasons, the manufacturing companies are not mentioned (see: www.ibu-epd.com; www.environdec.com)—developed by the two programmes (Tables 4, 5 and 6). Specifically, EPD developed by the International EPD® System based on the PCR '2012:01—Construction products and construction services 2.2', while the one developed by IBU-EPD is based on PCR 'Aerated concrete'. Both documents describe the field and objective of the LCA and establish the parameters for the assessment of the environmental performances necessary for the development of an EPD. What immediately seems clear is the different limit of the system used in the two cases. In the case of the EPD developed by the International EPD® System, the mentioned limit is cradle to gate with Options, related to Module A1–A3 (product stage) and Module C4 (Disposal) while in the case of the IBU-EPD the life cycle assessment refers to a cradle-to-gate analysis, therefore exclusively related to Module A1A3 (Product stage).

Table 2 PCR e sub-PCR developed by the International EPD® system (Access 14 December 2020)

PCR	Sub-PCR
PCR 2019: 14 construction products (EN 15804:A2)	c-PCR-001 cement and building lime
	c-PCR-002 ceramic tiles
	c-PCR-003 concrete and concrete elements
	c-PCR-004 resilient, textile and laminate floor coverings
	c-PCR-005 thermal Insulation products
	c-PCR-006 wood and wood-based products for use in construction
	c-PCR-007 windows and doors
	c-PCR-008 lifts (elevators)

Table 3 PCR and sub-PCR developed by IBU-EPD (Access 14 December 2020)

01 basic materials and precursors		PCR
	Aggregates	– Natural aggregates – Bulk granulate – Processed fly ash
	Cement, building limes and other hydraulic binders	– Cement
	Other basic materials and precursors	– Synthetic carpet yarns – Synthetic granulate – Rare earth oxides, metals, alloys and compounds
	Products related to concrete, mortar and grout	– Bulk granulate – Concrete admixtures

Table 4 Environmental impact of the autoclaved aerated concrete. EPDs developed by the International EPD® System and by the IBU-EPD (Access 14 July 2020)

Autoclaved aerated concrete Environmental impact						
Parameters	Unit	International EPD® system				IBU-EPD
		A1	A2	A3	C4	A1–A3
Global warming potential (GWP)	kg CO_2 eq	159	4.59	32.5	2.80	167.00
Acidification potential (AP)	kg SO_2 eq	332×10^{-3}	19.0×10^{-3}	90×10^{-3}	16.0×10^{-3}	1.83E−1
Ozone depletion potential (ODP)	kg CFC 11 eq	5.18×10^{-6}	851×10^{-9}	2.10×10^{-6}	716×10^{-9}	1.62E−10
Eutrophication potential (EP)	kg PO_4^{3-} eq	86.5×10^{-3}	4.32×10^{-3}	35.8×10^{-3}	5.99×10^{-3}	2.33E−2
Formation potential of tropospheric ozone (POCP)	kg C_2H_4 eq	19.6×10^{-3}	761×10^{-6}	6.52×10^{-3}	914×10^{-6}	1.68E−2
Abiotic depletion potential—elements	kg Sb eq	48×10^{-6}	8.12×10^{-6}	27.1×10^{-6}	2.52×10^{-6}	4.48E-4 ?
Abiotic depletion potential—fossil resources	MJ, netcalorific value	736	69.9	493	60.7	1.00E + 3
Water scarcity potential	m^3 eq	611	0.521	63.9	2.93	NOT PRESENT

Table 5 Use of resources of autoclaved aerated concrete. EPDs developed by international EPD® system and by the IBU-EPD (Access 14 July 2020)

Autoclaved aerated concrete Use of resources						
Parameters	UNIT	International EPD® system				IBU-EPD
		A1	A2	A3	C4	A1–A3
Primary energy resources—renewable TOT	MJ, net calorific value	51.3	0.907	588	1.60	3.63E+2
Use as energy carrier	MJ, net calorific value	51.3	0.907	588	1.60	1.98E+2
Used as raw materials	MJ, net calorific value	0	0	0	0	1.65E+2
Primary energy resources—Non-renewable TOT	MJ, net calorific value	839	75.4	560	65.5	1.09E+3
Use as energy carrier	MJ, net calorific value	839	75.4	560	65.5	1.02E+3
Used as raw materials	MJ, net calorific value	0	0	0	0	6.87E+1
Secondary material	kg	0	0	0	0	0.00
Renewable secondary fuels	MJ, net calorific value	0	0	0	0	0.00
Non-renewable secondary fuels	MJ, net calorific value	0	0	0	0	0.00
Net use of fresh water	m^3	207	4.26	112	3.11	2.47E−1

Table 6 Waste production and output flows of autoclaved aerated concrete. EPDs developed by the International EPD® system and by the IBU-EPD (Access 14 July 2020)

Autoclaved aerated concrete Waste production and output flows						
Parameters	Unit	International EPD® system				IBU-EPD
		A1	A2	A3	C4	A1–A3
Waste production						
Hazardous waste disposed	kg	–	–	0.092	–	4.02E−6
Non-hazardous waste disposed	kg	–	–	0.215	385	1.45E+1
Radioactive waste disposed	kg	–	–	–	–	3.21E−2
Output flows						
Components for reuse	kg	–	–	0	0	0.00
Material for recycling	kg	–	–	0	0	0.00
Materials for energy recovery	kg	–	–	0	0	0.00
Exported energy, electricity	MJ	–	–	0	0	0.00
Exported energy, thermal	MJ	–	–	0	0	0.00

6 Discussions

The case study allows underlining the difficulty of comparing materials of the same type, in this specific case autoclaved aerated concrete, highlighting the system boundaries used (also because the case study is not an exception). We can compare only the results of the modules common to both (for this comparison we refer to the following study, since in this paper we have focused only on the comparison of the system boundaries), without being able to examine the others used only by one of the EPDs, a limit that makes an exhaustive and well-founded comparative evaluation difficult.

One of the basic characteristics of the EPD is the possibility of comparing the environmental aspects of the evaluated products, as in the examination of the case studies. However, as it is highlighted in this paper, this can happen correctly only if the same system boundaries are taken into consideration. Despite the efforts dedicated to the standardization of the declaration rules, especially with the implementation of EN 15804, this aspect must be considered more carefully.

Therefore, EPDs can be an important starting point to develop a new approach towards the project. They can have a significant influence on the evolution of environmental awareness in the building industry, calling design to define technical-building choices with a more sustainable environmental and energy profile, but mostly giving, in every decision-making step, a central role to the environmental impact of the entire life cycle of materials, components and building systems. At the same time, the EPDs can be an excellent lever both for greater and more aware qualifications of designers and companies and for the economy of the sector in increasing the turnover thanks to eco-friendly investments. If this tool was already used in the design stage it would produce a paradigm shift—not immediate and rather complex due to the number of subjects involved and the large amount of information to find and to consider—to change the way of considering materials: from things (products and technologies) to systems (parts linked to each other and to the surrounding environment) conceived throughout their entire life cycle. In addition to the performance, technical and aesthetic characteristics, environmental parameters such as the embodied energy and the embodied carbon, environmental impacts (for example through EPDs) and the possible effects on public health (for example through the health product declarations) and, obviously, the economic effects (Arroyo et al., 2012) would support the decision-making stage. Everything would be organized systemically through software that finds valid operational support in the building information modelling (BIM). The challenge is on.

7 Conclusions

The current regulatory framework (Regulation 305/2011/UE on construction products, ZEB directive, CEN standards, GPP, etc.) and the many operational tools available (environmental assessment protocols, EPD certification, etc.), fostered by different bodies and born in different contexts and with different purposes, although worthy of having started a sustainability programme in building, they often appear in competition/conflict, generating confusion and disorientation amongst stakeholders. The environmental subject is often dealt with in a fragmentary way through the breakdown into sub-themes that leads to seek the optimization of some aspects to the detriment of others, without a systemic approach. One of the environmental certification tools that the building sector can use to address the environmental issue and to communicate the environmental performance of its materials and components but also to convey the process and product innovation, as reported in this paper, can be supplied by the environmental product declaration, regulated by ISO 14025 (2006).

The need to be more transparent on environmental information (Campioli & Lavagna, 2013) is mandatory and can no longer be avoided, and in this regard, the increase of EPDs published over the last few years underlines this aspect.

Amongst the subjects that can benefit from it, we include the producers, who can transparently declare the environmental performance of their products, the designers, who can select material also based on its environmental profile, and the users, whose purchases can be more informed and respectful of the environment. Since it is a voluntary-based tool, understanding the reasons that can push manufacturers to use it is important. According to a study published in 2016 (Ibáñez-Forés et al., 2016), 80% of producers state that the greatest limitation of the EPD lies in the fact that many users still do not know this tool, while its strength lies in the objectivity of the results reported and in the fact that using this tool improves the image of the company. Therefore, the dissemination of EPDs has not to be taken for granted because, on the one hand—to paraphrase Sinopoli and Tatano (2002)—the tool is taken in slowly before being able to change long-established practices, on the other, it must deal with limited knowledge or reticent professionals.

In Italy, an important boost to the spread of EPDs may come from the National Action Plan on Green Public Procurement (NAPGPP) with which the Ministerial Decree of 11 October 2017 made the Minimum Environmental Criteria (CAM) for public works operational (MATTM, 2017). These criteria establish the percentages of recovered or recycled materials, which can also be certified through a Type III Environmental Product Declaration. Actually, two recent surveys on the implementation of CAM—in the first (short) period of reference—do not show a prompt response from the Public Administration: the first survey reports that in 2017 out of a sample of 40 municipalities that organised 119 building-related tenders, only 6 (5%) included CAM among the project requirements (Punto 3 and Associazione dei Comuni Virtuosi, 2018); the second survey, conducted by Legambiente's Green Procurement Observatory on a sample of 54 administrative centres, reports a slightly higher percentage of 7. 1% (Nuova Ecologia, 2018). Both data are definitely not satisfactory, but we must take into account that we are in a transition stage in which, despite the contract law explicitly refers to the use of CAMs, their implementation clashes with the lack of preparation of the technical offices on handling them while evaluating the offers (since there are no official price lists or reference analysis) and with the necessary revision of the economic frameworks of the already planned works.

'Innovative', 'advanced', 'nanostructured' or 'resilient' are adjectives expressing a change referred to new ways to create and develop materials, components and building systems that today must necessarily be 'eco-friendly', for which the project can be a great driver, provided that its potential is understood and its effectiveness optimized (Lucarelli et al., 2012). The complexity of the building system—due to the relationships between the different sub-systems and between them and the whole building—requires accurate and detailed planning, capable of optimizing at the same time technologies that are very different (Campioli, 2011, p. 64) towards a single purpose: the sustainability of the 'building system'. Maria Chiara Torricelli has the same opinion (2017, p. 24), as she states: "The acceleration in technological innovations from other scientific and industrial areas has shifted the role of technological skills from those who systematize and design technology to those who know how to interpret it, finalize it, use it and make it work in the complex system of the design".

References

Arroyo, P., Tommelein, I., & Ballard, G. (2012). Deciding a sustainable alternative by 'choosing by advantage' in the AEC industry. In *20th Annual Conference of the International Group for Lean Construction* (pp. 1–10). Retrieved on February 17, 2019, from https://www.semanticscholar.org/paper/DECIDING-A-SUSTAINABLE-ALTERNATIVE-BY-%E2%80%98-CHOOSING-BY-Arroyo-Tommelein/74c79fc5395904b99c256464bdd22d385bc81977

Barucco, M. A., Verde, F., & Scalisi, F. (2016). Innovazione tecnologica di sistemi, componenti e materiali/Technological innovation of systems, components and materials. In M. T. Lucarelli, E. Mussinelli, & C. Trombetta (Eds.), *Cluster in progress. La Tecnologia dell'architettura in rete per l'innovazione/The Architectural technology network for innovation* (pp. 103–108). Maggioli Editore.

Bovea, M. D., Ibáñez-Forés, V., & Agustí-Juan, I. (2014). Environmental product declaration (EPD) labelling of construction and building materials. In Pacheco-Torgal, F., Cabeza, L. F., Labrincha, J., & de Magalhães, A., (Eds.), *Eco-efficient Construction and Building Materials. Life Cycle Assessment (LCA), Eco-Labelling and Case Studies* (pp. 125–150). Woodhead Publishing Limited.

Campioli, A. (2011). Architectural quality: innovation, technological research and design. *Techne, 1*, 62–69.

Campioli, A., Dalla Valle, A., Ganassali, S., & Giorgi, S. (2018). Designing the life cycle of materials: new trends in environmental perspective. *Techne, 16*, 86–95. https://doi.org/10.13128/Techne-23016

Campioli, A., & Lavagna, M. (2013). Innovazione ambientale dei processi di trasformazione del costruito e ciclo di vita. *Techne, 5*, 66–73.

Del Borghi, A. (2013). LCA and communications: Environmental product declaration. *The International Journal of Life Cycle Assessment, 18*(2), 293–295. https://doi.org/10.1007/s11367-012-0513-9

DGNB. (2019). *Life Cycle Assessments. A Guide on Using the LCA*. https://www.dgnb.de/en/news/reports/LCA-guide/index.php
EN 15804 (2012). Sustainability of construction works—Environmental product declarations—Core rules for the product category of construction products.
European Commission. (2011a). *A Resource-Efficient Europe—Flagship Initiative Under the Europe 2020 Strategy*, 21 final. https://eur-lex.europa.eu/LexUriServ/LexUriServ.do?uri=COM:2011:0021:FIN:EN:PDF
European Commission. (2011b). *Roadmap to a Resource Efficient Europe*, 571 final. http://www.europarl.europa.eu/meetdocs/2009_2014/documents/com/com_com(2011b)0571_/com_com(2011)0571_en.pdf
Ganassali, S., Lavagna, M., & Campioli, A. (2016). LCA benchmarks in buildings environmental certification systems. In *Sustainability and Innovation for the Future, 41st IAHS World Congress, 13–16th September 2016 Albufeira, Algarve, Portugal* (pp. 1–10).
Gargari, C., Hamans, C., & Torricelli, M. C. (2013). L'impegno dell'industria delle costruzioni per promuovere la sostenibilità dei prodotti: un approccio comune europeo per le prestazioni ambientali di prodotto. *Techne, 5*, 101–109.
Gonzalez, M. J., & Navarro, J. G. (2006). Assessment of the decrease of CO_2 emissions in the construction field through the selection of materials: Practical case studies of three houses of low environmental impact. *Building and Environment, 41*(7), 902–909. https://doi.org/10.1016/j.buildenv.2005.04.006
Hunsager, E. A., Bach, M., & Breuer, L. (2014). An institutional analysis of EPD programs and a global PCR registry. *The International Journal of Life Cycle Assessment, 19*(4), 786–795. https://doi.org/10.1007/s11367-014-0711-8
Ibáñez-Forés, V., Pacheco-Blanco, B., Capuz-Rizo, S. F., & Bovea, M. (2016). Environmental product declarations: Exploring their evolution and the factors affecting their demand in Europe. *Journal of Cleaner Production, 116*, 157–169. https://doi.org/10.1016/j.jclepro.2015.12.078
International Energy Agency. (2019). *Material efficiency in clean energy transitions*. IEA Publications.
ISO 14025. (2006). *Environmental labels and declarations—Type III environmetal declarations—Principles and procedures*.
ISO 14040. (2006). *Environmental management—Life cycle assessment—Principles and Framework*.
ISO 14020. (2000). *Environmental labels and declarations—General principles*.
ISO 14024. (1999). *Environmental labels and declarations—Type I environmental labelling—Principles and procedures*.
ISO 14021. (1999). *Environmental labels and declarations—Self-declared environ- mental claims (Type II environmental labelling)*.
Lucarelli, M. T., Mandaglio, M., & Pennestrì, D. (2012). The building envelope between process and product innovation. In P. De Joanna, D. Francese, & A. Passaro (Eds.), *International Conference on Sustainable Environment in the Mediterranean Region: From Housing to Urban and Land Scale Construction, Naples, 12–14 February 2012* (pp. 196–206). Franco Angeli Editore.
Maskell, D., Heath, A., & Walker, P. (2015). Use of Metakaolin with stabilised extruded earth masonry units. *Construction and Building Materials, 78*, 172–180. https://doi.org/10.1016/j.conbuildmat.2015.01.041
MATTM – Ministero dell'Ambiente della Tutela del Territorio e del Mare. DM 11/10/2017—Affidamento di servizi di progettazione e lavori per la nuova costruzione, ristrutturazione e manutenzione di edifici pubblici. *Gazzetta Ufficiale*, Serie Generale n. 259 del 6 novembre 2017.
Nuova Ecologia. (2018). Appalti verdi, ecco i dati ufficiali di enti locali e imprese. *La Nuova Ecologia*. Retrieved on December 12, 2018, from https://www.lanuovaecologia.it/appalti-verdi-ecco-i-dati-ufficiali-di-enti-locali-e-imprese/
Onchiri, R., James, K., Sabuni, B., & Busieney, C. (2014). Use of sugarcane bagasse ash as a partial replacement for cement in stabilization of self-interlocking earth blocks. *International Journal of Civil Engineering and Technology, 5*(10), 124–130.
OpenLCA. (2019). *The world's leading, high performance, open source life cycle assessment software*. https://www.openlca.org
Pacheco-Torgal, F. (2014). Introduction to the environmental impact of construction and building materials. In F. Pacheco-Torgal, L. F. Cabeza, J. Labrincha, & A. de Magalhães (Eds.), *Eco-efficient Construction and Building Materials. Life Cycle Assessment (LCA), Eco-Labelling and Case Studies* (pp. 1–10). Woodhead Publishing Limited.
Pacheco-Torgal, F., & Jalali, S. (2012). Earth construction: Lessons from the past for future eco-efficient construction. *Construction and Building Materials, 29*, 512–519. https://doi.org/10.1016/j.conbuildmat.2011.10.054
Point 3, & Associazione dei Comuni Virtuosi. (2018). *Primo Report di monitoraggio sull'applicazione dei Criteri Ambientali Minimi nelle procedure di approvvigionamento dei Comuni appartenenti all'Associazione Comuni Virtuosi*. https://ita.calameo.com/read/003376107bb1b3f39a315?language=it&page=1
SimaPro. (2019). *LCA software for fact-based sustainability*. https://simapro.com
Sinopoli, N., & Tatano, V. (2002). *Sulle tracce dell'innovazione: tra tecniche e architettura*. Franco Angeli.
Sposito, C., & Scalisi, F. (2019). Innovazione dei materiali naturali. Terra e nanotube di argilla per una sfida sostenibile|Natural material innovation. Earth and halloysite nanoclay for a sustainable challenge. *Agathón|International Journal of Architecture Art and Design, 5*, 59–72. https://doi.org/10.19229/2464-9309/572019
Torricelli, M. C. (2017). Technological culture, theories and practice in architectural design. *Techne, 13*, 21–26. https://doi.org/10.13128/Techne-21128
Treloar, G. J., Love, P. E. D., & Holt, G. D. (2001). Using national input-output data for embodied energy analysis of individual residential buildings. *Construction Management Economics, 19*(1), 49–61. https://doi.org/10.1080/014461901452076
UN environment. (2018). *Global Status Report 2018: Towards a zero-emission, efficient, and resilient building and construction sector*. Global Alliance for Building and Construction, International Energy Agency. https://wedocs.unep.org/bitstream/handle/20.500.11822/27140/Global_Status_2018.pdf?sequence=1&isAllowed=y

Public–Private Partnerships in Romanian New Urban Developments—A Potential Valuable Instrument in the Privatization Era

Luiza O. Voinea

Abstract

The large metropolitan areas from Romania are facing a lack of public land in the context of an increased demographic pressure, especially in the areas that concentrate on multiple economic activities (the population relocates according to professional opportunities), increased pressure from the real estate developers that want to monetize their properties, and last but not least an increased community demand for public facilities within reasonable distance from their home (education centres, healthcare facilities, public squares, markets, parks, playgrounds, cultural facilities, recreational areas). In addition, unlike other European Union member states, Romania currently has the highest percentage of private properties (~98% of the dwellings stock), together with multiple restitution of property rights after 1990 and the transfer of many old industrial platforms to private companies. All these factors had gradually led to a drastic decrease in public land and to the difficulties for the local authorities to develop the minimum public and social facilities. The expropriation law may be used for the development of public facilities, but it involves a lot of time and financial resources that authorities usually do not have. For this reason, most of the urban facilities were developed before 1990, based on the Disposition of the Romanian Communist Party from 1972. The public–private partnership represents a relatively new instrument defined by Romanian law in 2010 and its possibility to be extended as a territorial planning instrument was not yet explored. Starting from the analysis of the legal framework regarding this aspect in France and Poland (where this concept has been implemented since 1955 in France and 2004 in Poland), the objective was to identify the ways in which these partnerships can support the balance between public and private interest in the new urban developments, can become a tool for urban planning and can develop the public facilities using the financial potential of the private sector.

Keywords

Public facilities • Urban development • Public land • Peri-urbanization phenomenon • Public–private partnership

1 Introduction

The paper aims at analysing the legal and institutional framework of two European Union member states with demonstrated experience in implementing public–private partnerships (France and Poland) and to determine the development premises of this collaboration model in the new urban areas in Romania.

France is recognized as a country that implemented this type of partnership in multiple projects (infrastructure, education and health facilities, prisons and courts), with a stable legislation that was improved several times. On the other hand, Poland is a country with a more recent regulatory framework of public–private partnership, but which in a short period of time managed to equal France's achievements in terms of the amount of public projects implemented through this type of contract compared to the total public investments (approx. 5%). These two case studies offer a wide range of perspectives on how this partnership can be legislated, stimulated and coordinated by the central or local authorities. In addition, Poland may be considered a representative country for Romania due to the past communist regime, as they faced similar economic, political and social challenges, whose transition from a centrally planned economy to a market economy and from nationalization to privatization era, should be mentioned.

L. O. Voinea (✉)
Urban Planner Certified By The Romanian Register of Urban Planners, Bucharest, Romania
e-mail: odette.ilisoi@yahoo.com

H. Altan et al. (eds.), *Advances in Architecture, Engineering and Technology*,
Advances in Science, Technology & Innovation, https://doi.org/10.1007/978-3-031-11232-4_3

There is an extensive phenomenon of peri-urbanization identified around the main urban growth poles of Romania due to the increased migration inside the territory (for professional reasons), increased demand for housing in more affordable areas and dynamic urban development, all these under the pressure of a limited public budget that fails to ensure access to the minimum facilities. According to the latest statistics, in Romania, we have territories whose population has increased two or three times in the last 10–15 years due to the new economic activities that were located there. In this context, the central and local authorities have to identify ways in which they can support this dynamic development with functional infrastructure and public facilities.

One possible solution that has not been sufficiently exploited until now is the public–private partnership that may give access to the financial, human resources, technology and know-how owned by the private entities. This instrument also benefits from an increasing support from the European Commission which gave access to European funds for this type of contract. This was a very important measure that consolidated the perspective upon public–private partnerships and made them more attractive to private entities.

The aim of this research is to analyze the positive and negative aspects identified in the two case studies (France and Poland), to determine the potential legal and institutional obstacles from Romania and to issue some proposals for the current approach in order to increase its popularity among public authorities.

This complex tool for financing and managing projects of public interest has specific regulations according to each country in which is implemented, but the objective of this paper is to identify the common aspects from the legal and institutional framework that facilitated the public–private partnerships in France and Poland. These common structures, procedures and standards may represent a first step in developing a coherent regulation for this instrument in Romania.

Although public–private partnerships were first mentioned in 2002 in Romania with the publication of Law no. 137/2002 (2002), until 2019 no contract was signed using this financial instrument. This fact may suggest several obstacles that public authorities may encounter when trying to use this specific tool, which supports an average of 5% of the total public projects from the European Union.

In order to study the Romanian context on urban dynamics and to highlight the urgent need to promote a financial alternative for public facilities, the present research paper focuses on four metropolitan areas (Cluj, Timisoara, Iasi and Bucharest). The case studies were selected based on the demographic evolution, built-up area evolution, supply and demand of the housing market, distribution of public and private land and the ratio between new urban developments and public facilities. The last aspect reveals the public authority's capacity to respond to the increasing real estate pressure.

The analysis of the two case studies focuses on understanding the legal and institutional framework (distribution of responsibilities related to initiating, concluding, standardizing and regulating public–private partnerships) from France and Poland, followed by studying the changes and improvements that were implemented over time in order to support public–private partnerships. Secondly, the paper presents an overall image of the partnerships implementation in each chosen country (number of projects financed through this instrument compared to the total number of public projects) together with a set of representative projects.

The recommendations for the Romanian legal and institutional framework were made based on the common aspects identified in the regulatory framework, but also the practical experience from France and Poland.

2 Romanian Context Regarding Public Facilities in New Urban Developments

The large metropolitan areas are nowadays confronted with a high demographic pressure due to the unbalanced distribution of economic activities in the territory. There is a clustering trend in some areas considered favourable in terms of accessibility, availability of highly qualified human resources, availability of land for investments and finally the fiscal incentives and the total costs generated by performing an economic activity.

Analyzing only a few representative case studies, we are able to see the dimension of population mobility phenomenon among regions which lead to an accelerated expansion of the cities beyond their current limits, covering the peri-urban areas and even managing to integrate the communes within the metropolitan area. This dynamic development brings a lot of challenges for the public authorities that need to provide access to the main public facilities. In order to highlight this issue, the study focused on four metropolitan areas from Romania (Cluj, Timișoara, Iași and Bucharest) with a close look on the following indicators and variables: number of residents in 1977, 1992, 2002, 2011, built-up area evolution for the communes included in the metropolitan area, supply and demand of the housing market and also the price variation, distribution of land according to its regime (public or private) and the status regarding new public facilities developed in the recently developed areas.

The Metropolitan Area of Cluj Napoca consists of the municipality of Cluj Napoca and 19 communes and had the highest population growth among the other urban centres defined on national level (Urbasofia, 2015). Between 2002 and 2012 the population increased by 6,8% (from 385.434 to 413.761) mainly due to the peri-urbanization of the first ring of communes. Among the areas that experienced the highest pressure, Florești is to be mentioned with an increase of 205,4% (between 2002 and 2011), Baciu with 26,4% and Apahida with 21,6% (Fig. 1). Although the population of Cluj Napoca remained relatively stable at around 300.000, the communes that border the municipality became residential districts.

In the communes with the largest population growth, the study revealed an extension of the built-up areas proposed through the masterplans, as follows: the municipality of Cluj Napoca + 60%, Apahida + 65%, Florești + 50%, Chinteni + 45%, Ciurila + 28.8%, based on a comparative analysis between the last two general urban plans (Urbasofia, 2015).

Regarding the housing market, the number of new dwellings between 2013 and 2018 in Cluj Napoca increased four times. The high demand for residential areas influenced the average price per square metre, while in 2015 the price was around 873 EUR/sqm, in 2016—1000 EUR/sqm, in 2017 it increased up to 1211 EUR/sqm (+39%) (Veridio, 2018). The total value of property transactions kept the same upward trend, from 200 mil. EUR (2013) to 594 mil. EUR (2017) (Veridio, 2018). The communes near Cluj Napoca had similar challenges as the growth rate of new housing stock in Florești between 2007 and 2009 exceeded even the municipality of Bucharest and Ilfov county.

In the context of the real estate market boom, the new peripheral urban developments highly depend on the urban centres or municipalities, regarding the access to public facilities (health and education services, commercial spaces, public markets, parks) and job offers. According to the analysis published by the local administration from Florești (City Hall of Florești, 2019), they currently have only two schools, which is insufficient considering that 20% of the population is under 14 years old. The local authority has planned two new projects for schools and kindergartens, but these facilities are not yet developed as the financial resources are limited. Other issues mentioned in this document is the lack of medical facilities, social services units for the

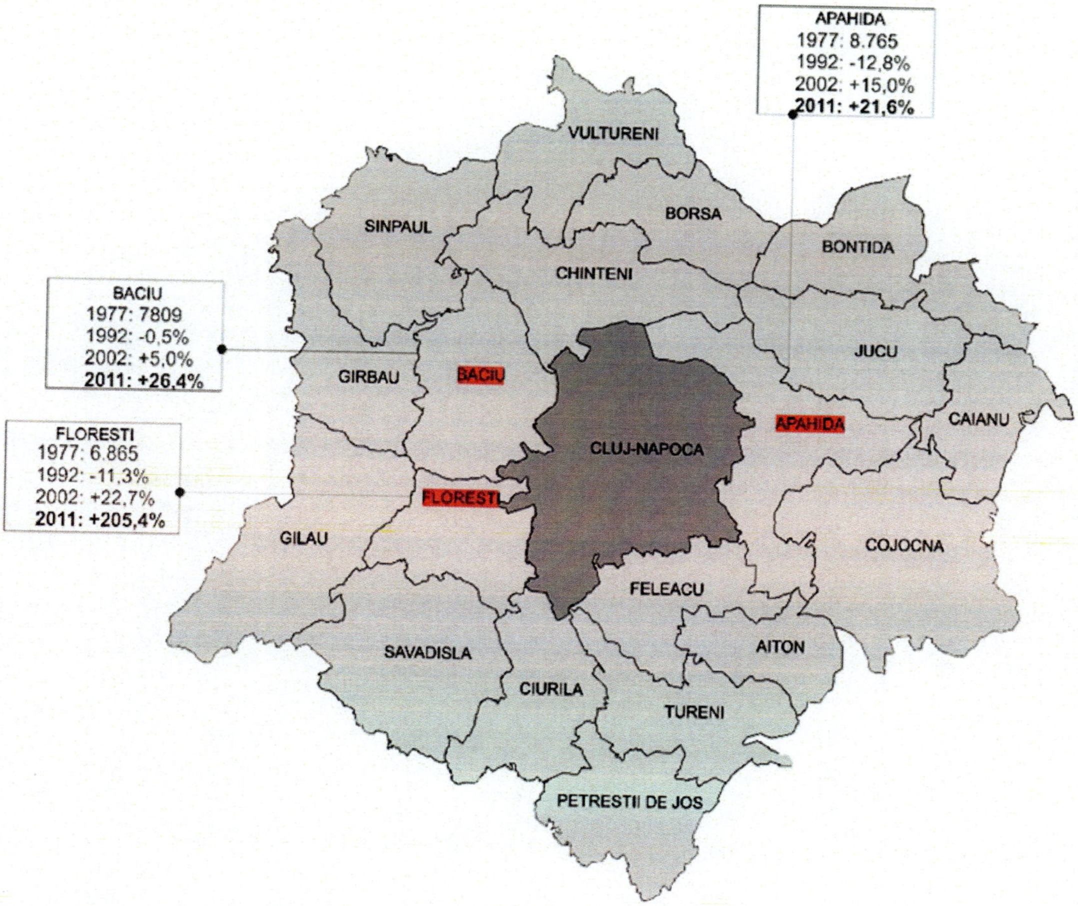

Fig. 1 Demographic evolution in the metropolitan area of Cluj Napoca (the communes marked with red had a population increase that exceeded 20%)

vulnerable people, as well as other facilities of public interest (the commune has no high school or nursery). All these essential investments are hard to be supported only from the local budget, as this commune is considered a residential district of Cluj Napoca with few economic activities that could contribute with taxes. In addition to all these difficulties, the local authorities usually do not own parcels on which they can develop these public facilities, so they first have to find optimal locations and purchase the properties.

The second metropolitan area relevant for this study is the one configured around the municipality of Timişoara, another urban growth pole due to its influence in the territory. Among its structure, there are seven communes that registered a demographic upward trend of over 20% during 2002–2011 (year of the last population census): Dumbrăviţa (+181.5%), Dudeştii Noi (+23.1%), Sânmihaiu Român (+39.2%), Giroc (+95.3%), Moşniţa Nouă (+61.5%), Ghiroda (+26.4%) and Giarmata (+20.3%) (National Institute of Statistics, 2003, 2012) (Fig. 2). The case is similar to Cluj Napoca as the urban population from the municipality of Timisoara remained constant, but the population from the peri-urban area increased in more than half of the communes.

The built-up area also increased significantly, for example in Dumbrăviţa (which also had the highest demographic growth) it was extended by 710.98 ha new lands proposed for different functional areas (AMH Mediu Expert, 2019). Some of the communes have not updated their masterplans, but instead, they approved several zonal urban plans for various functional areas: Dudeştii Noi—Zonal urban plan for leisure activities, sport, houses and services on a surface of 7.96 ha (MS-CAD S.R.L., 2017) and Giroc—Zonal urban plan for houses and public facilities on 4.1 ha (Tectonics house S.R.L., 2019). As for the housing market, the prices increased in the last ten years from 850 Euro/sqm to 1300 Euro/sqm (for new apartments) and from 770 Euro/sqm to 1350 Euro/sqm (for old apartments) especially due to a high demand for new residential areas (Imo Timisoara, 2020). In the first nine months of 2018, in the entire county of Timiş,

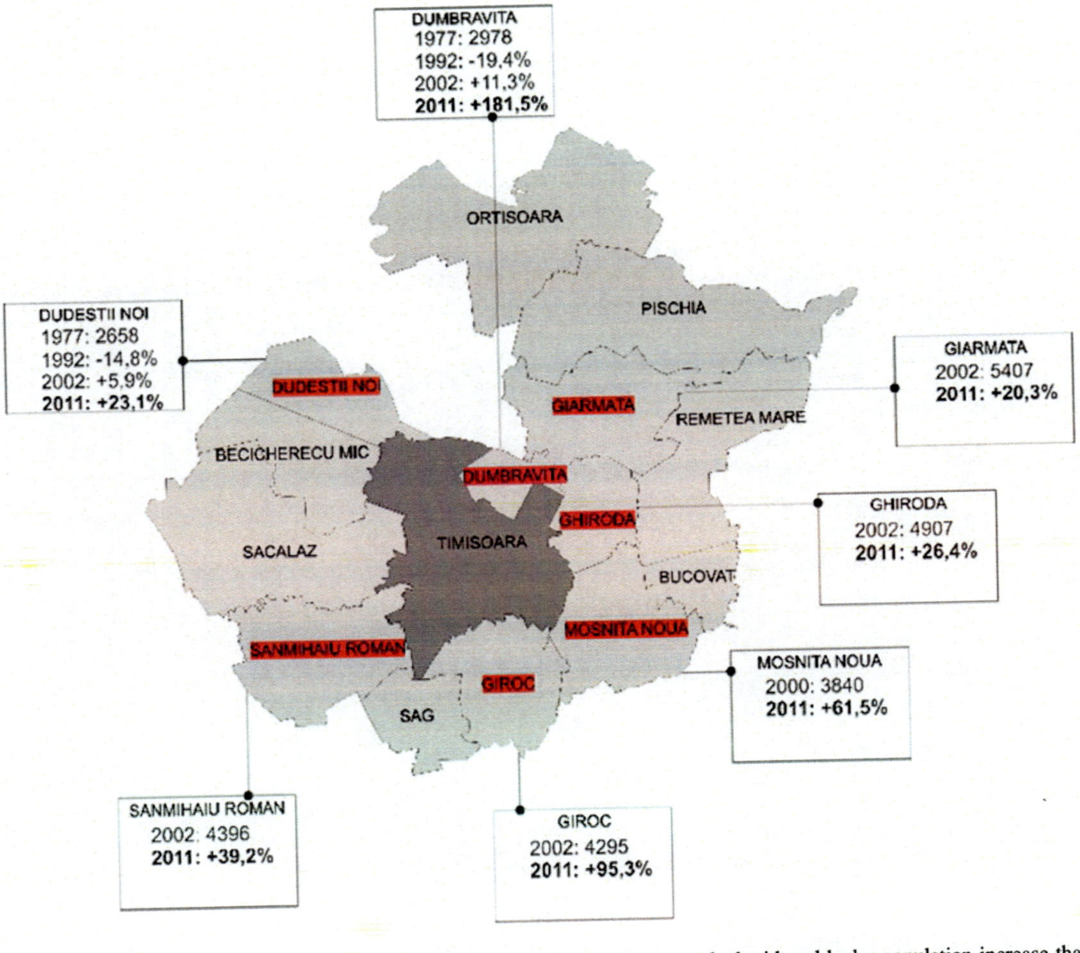

Fig. 2 Demographic evolution in the metropolitan area of Timişoara (the communes marked with red had a population increase that exceeded 20%)

the building permits were released for a total usable area of 623.992 sqm of new residential buildings. This number places Timiş county on the third place at national level, just below Bucharest and Ilfov, with an increase of 42% compared to the same period of 2017 (Timiş online, 2018). In this metropolitan area, the local authorities are facing the same problems as in the first case study, which is the limited financial resources compared to the requirements for new public facilities. For example, in the Local Strategy of Dumbrăviţa County they conclude that 'For the rehabilitation of the social infrastructure and for the development of new partnerships in the educational and cultural field, external resources are necessary: European or governmental funds or support from the County Council' (City Hall of Dumbrăviţa, 2014).

The metropolitan area of Iaşi is another example of how the peri-urbanization phenomenon affects the access to proper public facilities. The entire metropolitan area population increased by 9% from 2005 to 2014 (GEA Strategy & Consulting, 2015), but among all the communes there are four examples that stand out with a demographic growth starting from 30 to 60% between the last two censuses: Valea Lupului (+62%), Bîrnova (+46.9%), Miroslava (48.1%) and Rediu (+33.8%) (Fig. 3).

Regarding the urban sprawl tendency, after 1990, the communes bordering Iaşi municipality increased their built-up area significantly in order to allow the development of new residential districts: Valea Lupului by 76%, Miroslava by 56%, Rediu by 22%, Popricani and Bîrnova by 18% and Ciurea by 15% (GEA Strategy & Consulting, 2015). These new areas were mostly monofunctional, underestimating the need for public facilities. The housing market stock increased by average of 13% during 2009–2012, while there were communes in which the growth exceeded 50% (e.g. Communes of Miroslava and Bîrnova).

According to the strategy developed for the metropolitan area of Iaşi, the main public facilities are overworked, for example 'in the case of educational units, the demand is above national average in 10 out of 18 communes' (in Victoria there are 45 students per class) (GEA Strategy & Consulting, 2015). In addition, 'school laboratories, gymnasiums and sports fields are missing in many communes from the metropolitan area, as in Movileni, Schitu Duca, Valea Lupului' (GEA Strategy & Consulting, 2015). The analysis upon the health infrastructure reveals an even worse situation since all the communes depend entirely on the services offered by the Municipality of Iaşi. The only type of medical facility developed outside Iaşi is family doctor offices. Social services infrastructure is also missing in most of the communes.

The last case study that shows the urgent need to improve the access to public facilities is represented by the metropolitan area of Bucharest, where we find the highest demographic pressure and housing demand. According to the statistics, the population increased by over 20% in 22 out of 38 communes and cities in part of the metropolitan area

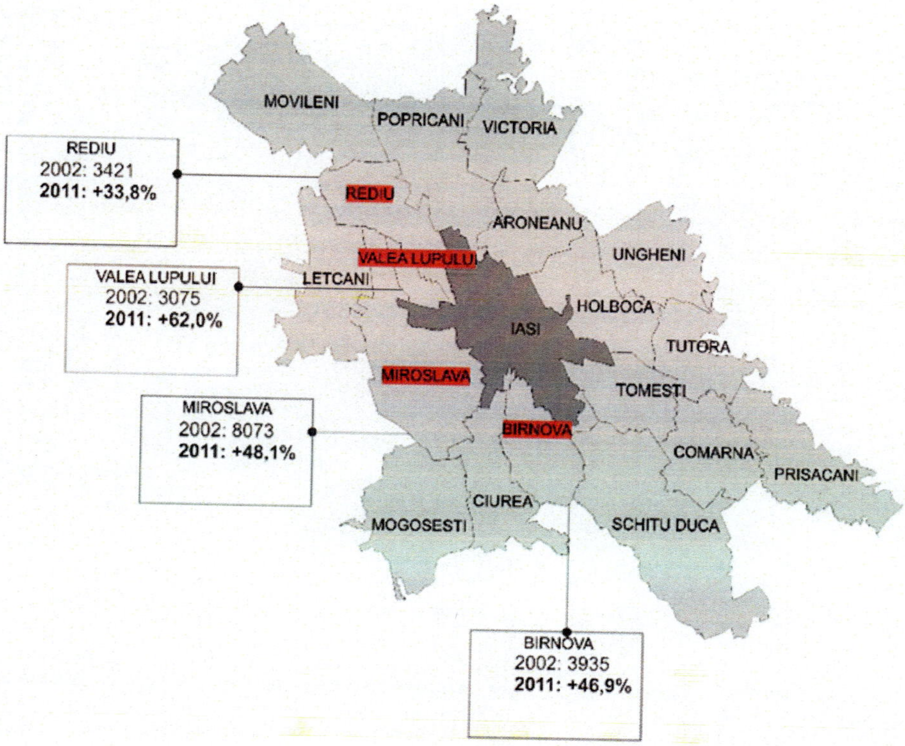

Fig. 3 Demographic evolution in the metropolitan area of Iaşi (the communes marked with red had a population increase that exceeded 20%)

(between 2002 and 2011). There are cases in which the growth was even more accelerated, as for example Bragadiru (+87.7%), Pantelimon (+59.8%) and Popeşti-Leordeni (+44.9%) (National Institute of Statistics, 2003, 2012) (Fig. 4).

The local administrations that have faced an important demographic pressure, were constrained to extend the built-up area through their masterplans, of which the following can be mentioned: Chiajna—1080.7 ha were added to the last area (+223.6%) (Geoid S.R.L., 2013), Domneşti—602.54 ha (+34%) (Amec Environment & Infrastructure S. R.L., 2014) and Corbeanca—1479.9 ha (+60,5%) (Mina-M-Com, 2008), Pantelimon—750.77 ha (+53,0%) (Plan.Co UrbisDesign S.R.L., 2013), Otopeni—459.2 ha (+16.9%) (Grigoraş PFA, 2013).

The dwellings stock in the metropolitan area of Bucharest also increased with 1739 more units in 2018, compared to 2017 (National Institute of Statistics, 2019). The number of new housing units built in Ilfov county reveals the impact that Bucharest has on its surrounding territories. In 2018, 5583 new units were built in Ilfov, a very close value to the one registered in Bucharest—5689 (these values are exceeded only by Cluj Napoca).

This peri-urbanization phenomenon has an important influence on the demand for new public facilities, starting from roads, public transport infrastructure, urban infrastructure systems (water, energy, information), education and health facilities, social services, parks, sport fields, playgrounds and other functional areas.

In the masterplans developed for the communes and cities inside the metropolitan area of Bucharest, the specialists mentioned about the issues regarding the access to public infrastructure: 'lack of health, education and commercial facilities', 'lack of social spaces', 'insufficient local financial resources to support all the investments' (Urban Architectural Management S.R.L., 2015a), 'deficit of cultural facilities', 'unbalanced distribution of public facilities such as kindergartens, school, dispensaries', 'insufficient funds for projects of national or local interest' (Europroiect, 2010), 'poor social infrastructure', 'insufficient sports and leisure

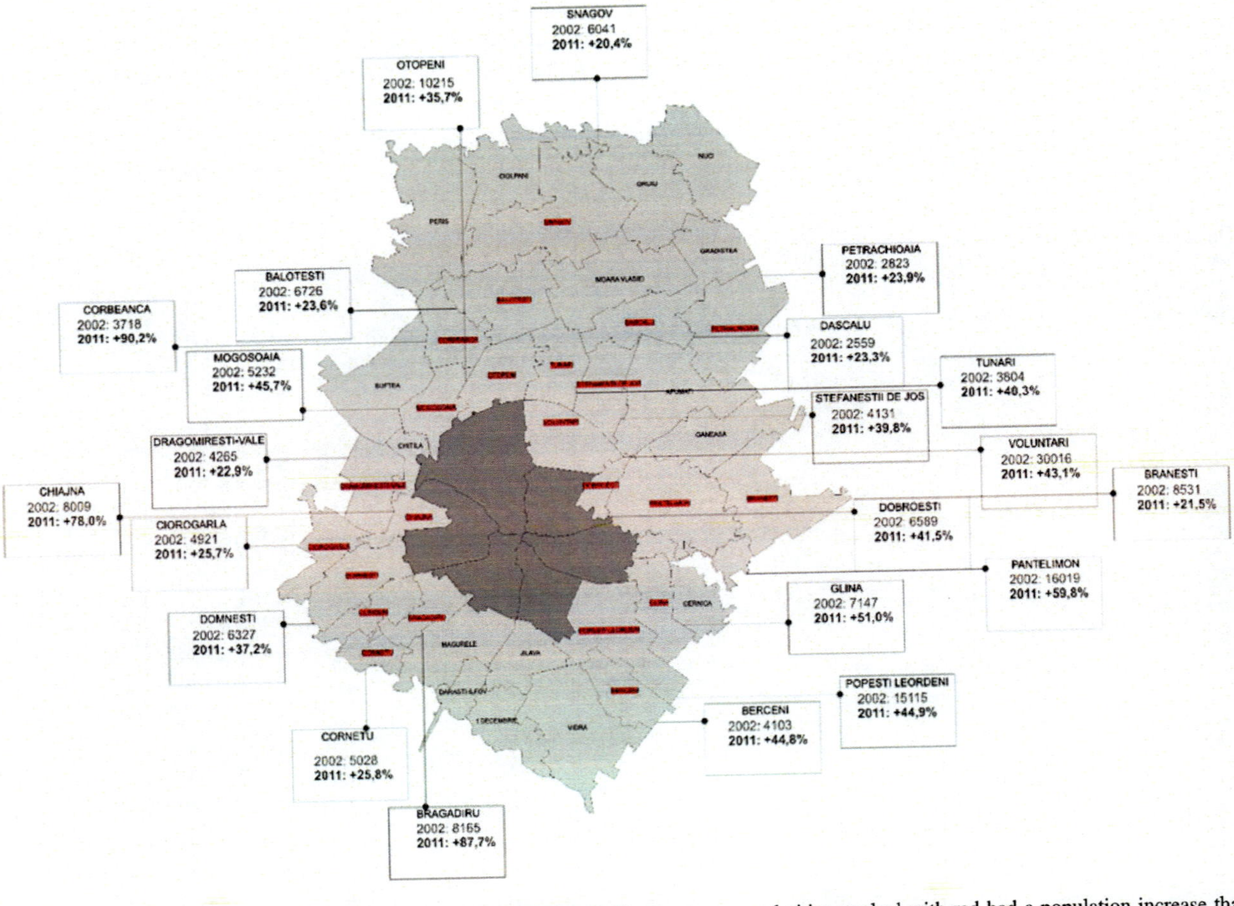

Fig. 4 Demographic evolution in the metropolitan area of Bucharest (the communes and cities marked with red had a population increase that exceeded 20%)

facilities', 'limited financial resources compared to the value of the mandatory investments in infrastructure, urban networks.' (Public research, 2017).

As regards to the property regime, the local authorities are facing a lack of public land that may be used for these important investments. For example, in Bragadiru, the local authority owns some parcels on the periphery, mostly used for weekly open markets or leisure activities, but not appropriate for other types of public facilities (Urban Architectural Management S.R.L., 2015b). In order to be able to develop educational, health or social services, the authority must have enough resources to acquire an appropriate land (considering the surface needed, location and accessibility), to contract the feasibility studies, technical projects and to finally build it, assuring periodically maintenance works.

Another obstacle is the fact that the local authorities in most cases do not have a clear overview upon the lands they own, an updated database with survey plans. In addition, considering the lack of land survey plans for most of the communes and cities inside the metropolitan area, the authorities are slowly progressing towards a complete land survey plan by making topographic measurements only for the lands of immediate investments (e.g. route of a future road), without considering the future investments on medium or long-term. Due to all these aspects, the masterplans are usually unable to make clear localization for new public facilities, as the authorities can be charged with restricting the owners' right to use their private property. As a result, masterplans usually propose mixt-use areas for both residential areas and public facilities, in order to allow these investments correlated to private entities' intentions.

According to Law no. 33/27.05.1994 (1994) related to the expropriation procedure for public utility, 'the buildings and lands necessary for investments in social housing and other public facilities of education, health, culture, sports, protection and social assistance, as well as public administration' are considered projects of public utility and they can follow this procedure, but the local authorities must have the financial resources to carry out all the necessary studies and projects.

Furthermore, 98% of the dwellings stock from Romania is owned by private entities and only 2% by public authorities (The World Bank, 2015). Regarding other functional buildings or lands, there are no public data available as there are no land survey plans that might provide an overview of the property regime. The significant reduction of public land was also influenced by the restitutions made after 1990. According to statistical data, 9026 buildings were reclaimed by the old private owners (including schools, kindergartens and hospitals), 340.011 ha of land inside and 709.305 ha outside the built-up area, while in 2015 there was still 300.000 ha for which the decisions were not yet issued (Finanțiștii, 2015). The old industrial platforms inside urban centres were important land resources that could provide support for public investments, but in many cases, they were returned to the initial owners and usually they changed their profile into residential, commercial areas or offices.

Another aspect that raises problems in developing new public facilities is the current norms and regulations that set a lot of requirements regarding the land surface and building capacity depending on the population density, demographic trends and district area. In areas of high building density, the local authorities have difficulties in finding lands considered appropriate for public facilities, according to the norms. For example, the plot of land necessary for a kindergarten is defined between 1000 and 2000 sqm in rural areas and 2000–4000 sqm in urban areas, while the construction must not exceed 25% of the land area. In addition, the building must have a minimum distance of 25 m to the street boundary and must not exceed two stories. Also, the acceptable walking distance for the kindergartens is 500 m (Ministry of Regional Development and Public Works, 1997). As for the nurseries, the requirements are even more difficult to meet. According to the General Urban Regulation of 1996, the maximum percentage of land use is 20% (Regulamentul General de Urbanism, 1996). The last report published by the Romanian Health Observatory in 2015 showed that there is an unbalanced distribution of health units and medical staff between counties and especially in rural areas and new facilities must be built in short-term (Romanian Health Observatory, 2015). For hospitals, the same regulation sets a minimum requirement of 10–15 sqm/patient of parks inside its plot, difficult to achieve given the fact that the functionality of the health facility is more important (without ignoring the benefits that green spaces can bring). The maximum percentage of land use, in this case, is also 20% (Regulamentul General de Urbanism, 1996). All these conditions made the local authorities' mission heavier as there is almost impossible to convert other existing buildings into public facilities. Because of all these reasons, most of the public facilities are built before 1990, later on being only rehabilitated or extended, while in only few cases new facilities were developed by public or private entities.

In conclusion, the large urban areas are confronting with high demographic pressure due to the increased population mobility and also with the pressure of the housing market. The new urban developments have started to expand beyond the urban centres limits, integrating the first ring of communes as residential districts. The demand for public facilities is increasing and the local authorities have two possible action plans: to stop all the private investments until they will have enough resources to develop the public infrastructure and facilities or to invest using the benefits of public–private partnerships.

3 Public–Private Partnerships Framework and Implementation in France and Poland

According to the statements made by the World Bank, 'Public–private partnerships (PPPs) can be a tool to get more quality infrastructure services to more people. When designed well and implemented in a balanced regulatory environment, PPPs can bring greater efficiency and sustainability to the provision of public services such as energy, transport, telecommunications, water, healthcare and education. PPPs can also allow for better allocation of risks between public and private entities.' (The World Bank, 2019).

One study entitled published by the Commonwealth Secretariat presents three main benefits of implementing public–private partnerships: possibility to make multiple investments without the constraints of a limited budget, more efficient process managed through one integrated contract and access to private companies' resources and know-how (Cambridge Economic Policy Associates Ltd, 2010).

Among the European Union member states that have an extensive experience in implementing public–private partnerships, this study will focus on the legal and institutional framework, projects and results from France and Poland.

3.1 France

In France, the first concession contract was signed in 1554 for the construction and maintenance of a canal over a period of 10 years (Bougrain, 2014). The concession law dates from 1955 and between 1956 and 1964 several companies with partially state-owned capital were created to build and operate part of the motorways system and receive the road taxes in return. Between 1980 and 1990 there was an attempt to extend this type of contract to public facilities (schools, penitentiaries), but due to the lack of transparency during the acquisition process, these contracts were restricted for 10 years and reintroduced in 2002 with a more specific regulation. After several updates and improvements, all forms of public acquisitions and procurements were finally defined in 2019 in the Public Procurement and Concession Agreements Code. The main purpose was to integrate the information from over 30 preceding laws and to present all the aspects related to these procedures. Multiple national authorities were created to coordinate and assist the public institutions and administrations when implementing public–private partnerships: Mission d'appui à la réalisation des contrats de partenariat public-privé, FIN-INFRA (responsible for evaluating the initial partnership proposals and the sustainability of the projects), Agence Publique pour l'Immobilier de la Justice (authority responsible for the projects of courts, penitentiaries and other similar buildings), L'Agence nationale d'appui à la performance des établissements de santé et médico-sociaux (Freire, 2020).

The main two forms of public–private partnerships are the concession contracts designed mainly for roads and network infrastructure and the partnership contracts for the rest of the projects in educational, health, cultural or social field (Fig. 5). The concession contract was defined as an agreement through which the state assigns to a private entity the right to implement a project or to manage a service. The return of investment is covered by the fees paid by the final users of the project or service. The partnership contract is an administrative act signed between a public and a private entity for a fixed period equivalent to the investment amortization period. The objective of this type of contract is to implement integrated projects (finance, design, build, maintenance works and management) and the investment is recovered by the monthly rents paid by the public entity.

The Public Procurement and Concession Agreements Code has defined three fundamental aspects to be followed in order to evaluate the sustainability of the projects. The first requirement is to have a preliminary assessment of the projects regarding the financial aspects, risks, maintenance costs and a comparative analysis of all the possible sources of finance. The public authorities should demonstrate that the public–private partnership is the best instrument for their investment. These evaluations are further analyzed and validated by the responsible central authority. The second requirement refers to the minimum information that should be included in the contract between the public and private entities: contract objectives and time frame, risk distribution, quality standards, financial obligations towards the private entities, obligations of the private partner regarding the use of the public infrastructure outside the project (if allowed), evaluation methods, sanctions and clauses and contract conclusion possibilities. The publishing of a model structure for all partnership contracts offers the support for the public entity to clarify and negotiate every important aspect regarding the investment right from the initial phase. The last requirement is related to the evaluation criteria of the received offers: financial value, objectives and time frame, the number of activities that the private entity cannot provide on its own and need to subcontract (French Government, 2018). An important condition for initiating a public–private partnership is to fulfil one of the following criteria: project of high complexity, urgent project (the investment must be made on short-term and the public entity does not have the necessary resources), increased economic efficiency using this type of partnership compared to other procurement alternatives (Freire, 2020). Otherwise, the public authorities are recommended to select another form of financing

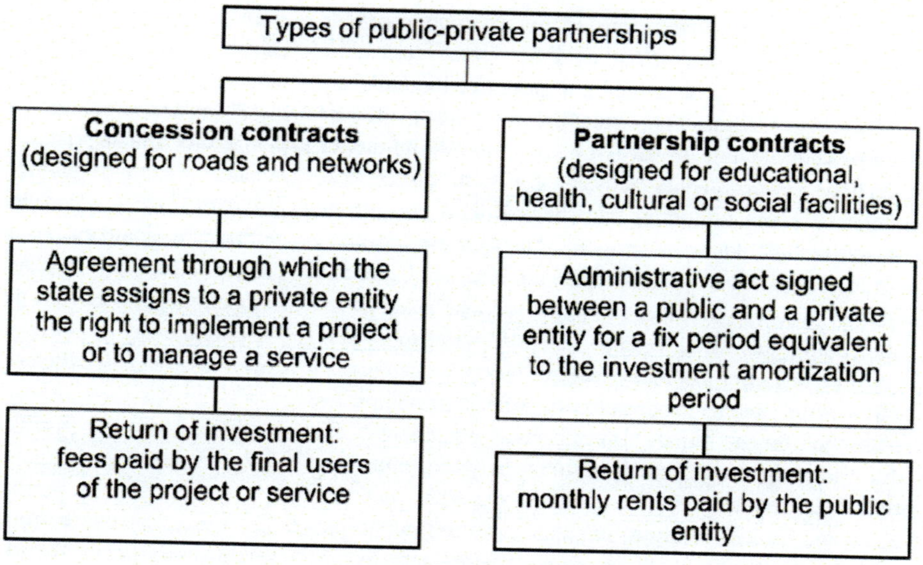

Fig. 5 Types of public–private partnerships

presented in the Public Procurement and Concession Agreements Code. At the end of the partnership contract, the construction becomes the property of the central or local public entity. The monthly or quarterly rates paid to the private partner include the works or services provided by the partner, tangible and intangible assets and maintenance costs (including the replacement of the equipment or consumables). The payment is not fixed, it can depend on multiple considerations, like the additional revenues obtained by the private entity from using the investment outside the project (if the contract allows it). The French state encourages this form of collaboration between public and private entities but does not guarantee the contracts (except from the projects of national importance). The local authorities on the other hand are allowed to guarantee the bank loans requested by the company responsible for implementing the project.

One national programme aimed for the development of the public health infrastructure is the hospital programme initiated in 2003 and improved in subsequent versions from 2007 and 2012. The public–private partnerships represent 12% of the total amount invested within this programme and the average contract period for health facilities is 15–35 years (most of them being signed for 30 years). The legal framework drafted three types of contracts for health facilities projects: BEH (Bail Emphytéotique Hospitalier), BEA (Bail Emphytéotique Administratif) and CP (Contrat de partenariat). BEA and BEH were designed only for local authorities, while CP can be signed by authorities from all levels of government. The adaptation of this investment instrument according to the project profile and the contractor has many advantages, as the French state can make specific recommendations or requirements.

To evaluate the results of public–private partnerships so far, the study will focus on the public facilities projects implemented through a CP (Contract de partenariat), as this contract is intended for all types of projects and public entities. Between 2004 and 2012, 200 contracts were signed and started (75% initiated by local authorities), totalling a value of 15 bil. EUR, to which are added the contracts using other public procurement models (BEA and BEH) with a total value of 3 bil. EUR (European PPP Expertise Centre, 2012). Due to the negative publicity, lack of transparency and the restrictions between 1990 and 2002, public–private partnerships represent today only 5% of the total public investments, although the studies have demonstrated that this alternative is both efficient and financially sustainable. ‘After ten years of implementation, the first available studies on the contribution of PPPs show clearly that the method works, both technically (facilities built to contract standards and delivered on time) and financially (rent budgets not exceeded) in over 90% of cases.’ (Bergere, 2016). The projects that benefited from the public–private partnership advantages cover a wide range of facilities: high-speed railways (e.g. Brittany-Pays de Loire and Nîmes-Montpellier high-speed rail links), educational infrastructure (universities, high schools, secondary schools—e.g.: 12 schools in Seine-St-Denis), sport and leisure facilities (e.g. stadiums in Lille, Marseille, Nice, Bordeaux, Dunkirk Arena), network infrastructure (e.g. internet in Auvergne), cultural and social facilities. According to the evaluation published by MAINH (Mission nationale d’appui à l’investissement hospitalier) in 2007 on 10 hospital projects developed in public–private partnerships, they concluded that the works were delivered on time, without exceeding the estimated budget and the authorities were satisfied with the collaboration and the final results. Questionnaires were also implemented to assess the community feedback regarding the services managed by the public/private authority or by a mixed regime and the results were positive for projects carried out by public–private partnerships (Bougrain, 2014). Public sector co-lending facilities for PPPs is a programme developed exclusively for education facilities. This solution was implemented by the French state due to the lack of predictability in the migration trends. People relocate for new professional opportunities and authorities must have a quick response to unanticipated demographic pressure. The education facilities built in public–private partnerships were functional after two and a half years in average, compared to those made by the public authorities which became functional after four years, according to Loiret Council (Bergere, 2016).

The main advantages of public–private partnerships are: access to private funds, management experience and advanced technology; risk distribution between public and private entities according to their competencies and efficiency and the possibility to implement several public projects simultaneously without being constrained by the limited public budget. This collaboration form also has a positive influence upon the local authority activity, as they are forced to plan their budget and long-term investments (without considering the term of a political mandate), to define a coherent and sustainable strategy on public facilities and to increase the project management competencies among its employees. These contracts, unlike other public procurement forms, are not so vulnerable to the economic instability, as they are usually planned for long periods. Another advantage would be that the proper maintenance of the buildings by the private entity is less expensive than periodic general rehabilitation that the public entities usually prefer. This is also a consequence of the several clauses or penalties included in the contract for the private entity in case the activity or service can no longer be performed in the best conditions. Public–private partnerships benefit from the private companies' experience in a specific field and the authority can sign a single contract for all stages of the investment (finance, design, build, maintenance and management). On the other side, the private entity is motivated to speed up the first stages in order to recover its investments and gain some profit, but also to provide a good quality service in order to reduce the maintenance costs.

The criticisms against this form of public procurement are: higher final value of the investments compared to using bank loans (although the value of the risk taken by the private entity cannot be quantified), possible opportunistic behaviour of the private entity, difficult project management that requires clear evaluation and control mechanisms as well as increased competencies among the project team. One of the main barriers of these partnerships is the difficulty to anticipate all the possible future situations or bottlenecks that may arise during the contract, as there are many factors that can influence the project on long-term (political, economical and social context, emerging technologies).

In conclusion, public–private partnerships represent a valuable instrument improved by the new procurement code, but not exploited at its full potential, as the percentage of these projects represents only 5% in France out of the total public investments. These partnerships can support a sustainable system of public facilities, without reducing the possibility for central or local authorities to invest in other sectors. The contracts have to be clearly defined, the procurement process must be transparent and encourage the competitiveness between the different private companies and the public entity has to ensure a stable assigned project team, regardless of political changes.

3.2 Poland

The legal framework from Poland on public–private partnerships consists of the following acts and regulations that address all public procurement forms: Communal Economy Act and Civil Code (1990), Act on public–private partnership (published in 2005 and revised in 2008), Act on Public procurement law (published in 2004 with several subsequently amendments), Act on concessions for public works or services (published in 2009, revised in 2010 and 2012), Act on Toll Motorways and the National Road Fund (1994) (Sîrghi, 2015).

As regard to the institutional framework, there is a distribution of different responsibilities in promoting, coordinating, validating and monitoring public–private partnerships between several entities: Public Procurement Office, Ministry of Infrastructure and Development—Public–private Partnership Platform, Ministry of Economy, Polish Agency for Enterprise Development, Ministry of Finance, Ministries for other sectors, other institutions. The department dedicated for this procurement model from the Ministry of Infrastructure and Development is responsible with developing the methodology, standards and reference documents, coordinating public entities who want to apply for a public–private partnership, monitoring projects workflow, recommending improvements to the current legal framework and afterwards assess the impact of these changes and promoting this concept among the public and private entities in the country. In order to facilitate and encourage the access of the public authorities to this partnership model, methodologies, practical guides and standard documents have been published: examples of risk analysis, questionnaire models for private entities, possible performance indicators for different projects categories (roads, network infrastructure, buildings) and contract models. In 2010, the Public Procurement Office in collaboration with the Institute for Public–Private Partnerships published the guide for public authorities on procurement and concession procedures. This guide came in response to the fact that under the

2005 law no public–private partnership contract was signed, the procedure being considered very difficult and confusing. In addition, the Polish Agency for Enterprise Development provided training sessions, courses and technical support to all entities involved. Besides the authorities created in order to facilitate this investment model, in Poland, there are also other mechanisms and public platforms meant to support cooperation and exchange of good practices among entities, for example: www.ppp.gov.pl—developed in 2011 and www.ppp.parp.gov.pl. The first online platform created by the Ministry of Infrastructure and Development also aims at presenting a database with all projects implemented through public–private partnerships that can serve as models to other public authorities. In order to stimulate the use of this procurement form, the Polish state introduced several incentives for the private entities, while for the public authorities, a new category of public debt was created exclusively for public–private partnerships, in order to avoid a distorted image of the public budget.

According to the legal framework, there are four types of projects that can be developed under public–private partnerships: construction or renovation of buildings, furniture and equipment replacement, and other categories of services (Fig. 6). The approach includes all project phases: finance, design, build and operate/maintain. Unlike the French legal framework, the buildings resulting from this partnership model can remain in the property of the private entities which, during the contract, will only receive the amount related to the renting and maintenance costs.

The law defined three risk categories for the public–private partnerships that need to be negotiated from the initial phases: risks within the construction process (e.g. non-compliance with technical requirements and time frame), risks within the building operation stage and the risks related to the flexible market demand for a service (Public–Private Partnership Platform, 2020). While the first two risks are assumed by the private entity, the last one lies on the public authorities as they pay the monthly costs regardless of the market demand. The private entity may also receive payments from the end users, as we usually find in the case of roads or urban network infrastructure. The evaluation criteria for the received offers are defined by the law as follows: distribution of income between the public and private entity, financial contribution, project management efficiency, compliance with project requirements (quality, technical parameters, maintenance), distribution of responsibilities and risks, contractual terms related to time frame and payments.

Due to the new hybrid contract model proposed by the European Commission, the entities involved in a public–private partnership can benefit from access to European funds. This fact increases the profit for investments that were not considered opportune by private entities or that could not be sustained only by private resources.

In terms of results, although Poland has a limited experience using public–private partnerships, between 2014 and 2018 they amounted 951 mil. EUR. Based on the law published in 2005, no partnership contracts were signed, mostly because of the lack of clarity on the procurement procedures. Since 2008 when the first revision of the law was published, there has been a significant increase in the number of projects implemented through public–private partnerships (2009—77 projects, 2014—211 projects). The projects developed in Poland are covering a wide range of public infrastructure: sports and leisure facilities (Mineral Swimming Pools Complex in Solec-Zdrój), social and student housing, parking areas, educational and health facilities, roads, urban infrastructure, waste management (Waste Management Facility System for the City of Poznań), urban revitalization projects (Development of the northern headland of Wyspa Spichrzów in Gdansk, Development of the area of the former tram depot in the Lower City of Gdansk, Revitalization of the Railway Station in Sopot), heritage protection projects (Interreg Central Europe, 2016). During 2013, the sectors that had benefited the most from public–private partnerships were: leisure, sports facilities and tourism infrastructure—417 mil. EUR, roads—848 mil. EUR and waste management—636 mil. EUR (Sîrghi P, 2015). In terms of contract value, 10% of the contracts had values less than 1 mil. PLN, while 70% were between 1 and 100 ml. PLN. Also, 95% of the projects were initiated by local authorities.

There was some uncertainty among the private entities as between 2009 and 2015 only 100 contracts were signed out of 300 listed projects. The possible reasons might be the lack of trust in this investment model, lack of information,

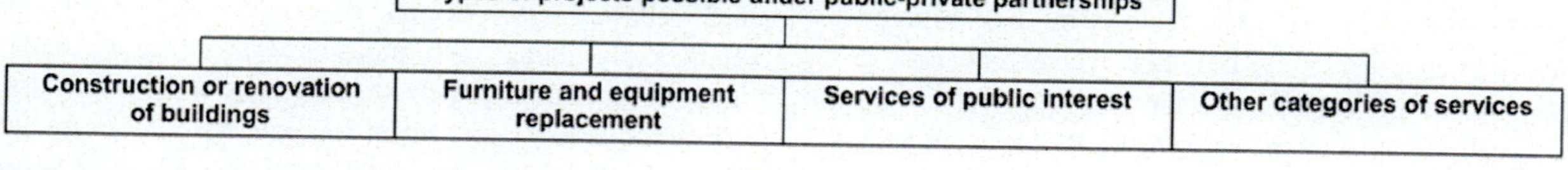

Fig. 6 Types of projects under public–private partnerships

unbalanced risk distribution between the public and private entity, low profitability, lack of research documentation or lack of financial and human resources. Despite these issues, the number of new public–private partnerships maintained an upward trend and in 2020 there were 100 more contracts signed than in the previous year, reaching 5% of the total public investments (similar percentage to the France achievements). The number of projects listed by public entities and the percentage of signed contracts compared to the number of listed projects also increased. In Poland, distinct from France, there are several regulations and norms that detail the public–private partnership, as they do not have a common procurement code. The responsibilities related to promoting partnerships, managing, validating and providing the necessary support are distributed among several public authorities. Furthermore, there is no national entity intending to correlate the activity of the multiple authorities engaged in this subject.

From my point of view, all the mechanisms aimed to promote public–private partnerships along with the access to the standardized documents have facilitated the expansion of this investment model, especially at the local level. In addition, the access to European funds has increased the attractiveness of public projects among private entities.

The European Commission also offers its support to improve the public–private partnership legislation and to promote this concept among member states through the Green Paper on public–private partnerships and community law on public contracts and concessions, published in 2004. A new authority has been created in collaboration with the European Investment Bank, called the European Public–private partnership Expertise Centre and it has multiple responsibilities in assisting public entities in the initial stages of the projects, exchanging information and good practices, publishing several reports and guides about this procurement model. The hybrid contract developed by the European Commission is another confirmation that public–private partnerships are encouraged throughout the entire European Union (Cieślak, 2016). A report made by the European Courts of Auditors in 2018 presents several recommendations for the current legal framework regarding public–private partnerships. First, they mention about the importance of transparency during the listing and acquisition procedure but also in the later stages. Another important aspect that all the authorities should consider is to have a comparative analysis of all the funding options in order to be convinced that the public–private partnership is the best solution for that particular project. The last recommendation is to increase the project management competencies among the authorities that want to implement this type of contract (European Court of Auditors, 2018).

4 Recommendations for Romanian Legal and Institutional Framework

In Romania, public–private partnerships have first been mentioned in 2002 with the publication of the Law no. 137/2002 regarding privatization procedures in which it is stated 'In order to attract new types of funding, public–private partnerships may be developed according to the dispositions provided by the general regulation'. Since 2002 the legal framework has been revised several times by Government Ordinances, Government Emergency Ordinances or other subsequent laws. This procurement form is currently defined by a series of acts, ordinances and methodological norms and this may be confusing for the interested authorities. The main legal documents are: Law no. 100/2016 regarding concessions of public works and services, GEO no. 39/2011, GEO no. 86/2011, GEO no. 96/2012, GEO no. 11/2014, Law no. 528/2004, Law no. 219/1998, GO no. 16/2002, GEO no. 34/20016, Law no. 293/2003 to which are added more acts that influence the public–private partnerships. The lack of a clear legal framework discourages the authorities that may be interested in this type of contract, as there is an increased risk of omitting some procedural details that are mentioned in additional acts or ordinances. These issues are also sustained by the fact that until 2019 no public–private partnership contract was signed based on the law from 2010. In 2018 by the Government Decision no. 357/2018, several strategic projects were assigned to be carried out with public–private partnerships (roads, railways, hospitals and airports), but until 2019 only feasibility studies were approved, the overall procedure lasts longer. After the feasibility study is approved by the government or other deliberative authorities, the contract assignment procedure is initiated, the best financial offer is accepted and the contract is signed. One of the main issues regarding the implementation mechanism is the lack of a section dedicated to these initiatives on the public platform SEAP (Electronic Public Procurement System), there is only one section dedicated to concessions (Vass Lawyers, 2012). This fact is reducing the visibility of the projects and discourages both private and public entities from initiating these procedures which are vulnerable precisely due to the lack of transparency. In addition, there are no platforms available for consultations or collaboration between the interested entities, as we have seen in Poland. The number of public–private partnership initiatives is very small, although the overall experience and results of countries promoting such procedures is positive. In the European Union the project value is around 5% out of the total public investments, but in UK and Australia is around 15%.

According to the Law no. 178/2010 (2010), 'centralized coordination and monitoring of public–private partnership projects is in the responsibility of the Department for Infrastructure projects, foreign investments, public–private partnership and export promotion', revised by GD no. 11/2014. This department transformed later in a General Division under the General Secretariat of the Government entitled 'National Commission of Strategy and Forecast' is responsible with developing the public–private partnerships database, but the online platform does not benefit from a very good visibility and the site design is not user friendly (National Commission of Strategy and Forecast, 2020).

Based on the comparative analysis of the legal and institutional framework from Poland and France, the current legislation from Romania on public–private partnership can benefit from important improvements in order to increase the visibility of this collaboration model and to facilitate the access to the resources and know-how owned by the private entities and finally to support the development of public facilities without the constraints of a limited public budget. The online platforms dedicated to public–private partnerships are a valuable instrument, but they must be updated periodically, must contain clear and structured information accessible to both private and public entities and must increase the transparency of these initiatives. Through these platforms, the assigned central authorities can provide support and both technical and financial consultancy to all the entities involved or interested in this type of investment. As to the legal framework, one country recently implemented a common procurement code (France) and one country for which the public–private partnership is detailed in several acts and norms (Poland), which makes it difficult to comprehend all the specifications and changes. Implementing a single legal framework has many advantages and ensures a correlation between the different regulations that were published over time. The frequent changes occurred to the law of public–private partnership through several Government Ordinances, Emergency Ordinances or other related laws affected the stability and reliability of this investment model and make it more vulnerable. Public–private partnerships are long-term commitments and the contracting procedure can last 1–2 years (depending on the complexity of the project) and for this reason, the legislative stability is mandatory, without discouraging the correction or improvement of certain aspects that proved inefficient.

As regards to the institutions and authorities with a consultative or deliberative role, the support from Poland may be considered more consistent as there are various entities meant to promote and facilitate this concept among different public or private entities, to offer consultancy or to develop methodologies, framework papers, guides or model projects and contracts. In France, these authorities are specialized on various project categories such as health facilities, courts and prisons. Depending on the complexity of the public investments and procedures, it may be better to have specialized departments or divisions, but this measure may be implemented at later stages when the volume of projects increases significantly. In Romania, it may be too early to propose this measure as there are few initiatives made by central authorities and not by local authorities. In Poland, the online platforms were developed as an important instrument to support the high increase in the number of public–private partnerships initiated by local authorities. In order to encourage and promote this concept, it is necessary to organize project management training for the local authorities so as to be able to follow up the project course and to properly evaluate the private entity's performance.

The procedure for public–private partnerships is not very different from the regular public procurement, but the later stages are more difficult for central and local authorities as they include project monitoring (time frame, objectives, performance indicators), risk management and periodic evaluations. The assistance of public entities must be continuous and should be supported through several means of communication: online platforms, consultations or working sessions. The Romanian government must also adopt measures to increase the attractiveness of these public investments among private entities, as they are not so popular nowadays (e.g. financial facilities, access to European funds, guaranteed loans by the state or local authorities, promoting public initiatives on SEAP platform and other dedicated online websites).

Regarding the new urban developments from Romania, there is a high demand for public facilities that are evenly distributed and at an acceptable walking distance from the residential areas. Public–private partnerships may be a valuable instrument that can improve the access to the most important facilities without the constraints of a limited public budget. First of all, the local authorities should undergo a study on the existing public facilities and the demand for new projects and to further identify the potential locations for them according to the surface needed and accessibility. The investments should then be prioritized based on the population density in those areas and the local authorities should further initiate the acquisition procedure using the expropriation law or public procurement law. After the parcels are registered in the public property inventory, the local authorities can start the feasibility study and list the project on the dedicated platforms in order to find a private entity interested in that investment. The local authorities may also consider making a reservation of a piece of land from the initial stages of a new urban development and buying it from the investor. Although this procedure involves certain acquisition costs, the resources needed are fewer than in the case of a project implemented solely by the local authority (parcel and building costs, maintenance, service management) and the same limited public budget would allow the development of more facilities simultaneously as they will have access to European funds and private resources. The risk distribution is also an important advantage in favour of the public–private partnerships as the demand for these facilities may fluctuate and on mid or long-term it may be more cost-efficient to conclude the contract and leave the building to the contractor, as the flexibility is higher.

Although this concept may seem to have more advantages to the public entities, it may also be seen as a win–win situation as the real estate developers can better evaluate their properties and housing units if there are new public facilities in the area. The buyers are starting to become more aware about the importance of having all the facilities at a reasonable walking distance in order to reduce the car dependency. The real estate developers will have two possible action plans, to sell the parcels designed for public facilities to the local authorities in order to allow them to make the necessary investments or to assume to build the facilities with their own resources, which may seriously affect their profit. In addition, the private companies that invest in public facilities may benefit from multiple advantages: more credibility in the market as they become a partner of a central or local authority, improved image as they show interest for the community needs, reliable fixed income on long-term for the public services they provide.

In conclusion, in order to improve the mechanism of public–private partnerships in Romania, the authorities should consider implementing a dedicated section on the public platform SEAP together with online platforms for consultation and collaborations between the interested entities, updated periodically with free access for both private and public actors. Furthermore, the central authorities should pursue to create one single and coherent legal framework that can ensure a correlation between the different regulations that were published over time and that highly discouraged the use of this instrument. The legislative stability is also an important aspect to be regarded, as the public–private partnerships are usually long-term collaborations, without discouraging the correction or improvement of some aspects that proved to be inefficient. The central authorities should assist all the public entities interested in using this instrument, through different means of communication: online platforms, consultations or workshops. Considering the different procedures specific to each type of public investment, it may be necessary to create specialized departments, but this measure is highly dependent to the volume of projects and the difficulties encountered during implementation.

5 Conclusions

The mobility of population and economic activity leads to a high and unpredictable demographic pressure on large urban areas and as a consequence the real estate developers are investing in new residential areas in the urban peripheries in order to meet the demand for new housing units. Furthermore, there is an extensive phenomenon of peri-urbanization of the first communes around the important urban centres which in some cases leads to their integration in the metropolitan area. The expansion with new residential districts brings a lot of problems to the local authorities that need to ensure access to the minimum public facilities (education and health facilities, sport fields, parks, social services and social housing), constrained by a limited public budget and no available public plots. The authorities need to find sustainable ways in which they can support this dynamic development without compromising the investments in other sectors.

In the context of promoting smart and adaptable cities and an increased complexity of urban developments, the concept of public–private partnership is a valuable alternative that is not affected by the limited financial capacity of the public authorities. Although the average value of these projects among European Union member states is just around 5% of the total public investments (considering that there are also states that have not signed any partnership commitment until now), the European Commission encourages these collaborations since 2004 when the Green Paper on public–private partnerships and community law on public contracts and concessions was published, but also with more recent papers and programmes. The main advantages that this investment model has are: risk distribution between the public and private entities according to their competences and efficiency in solving potential obstacles (market risks, development/planning risks, project risks, political risks, regulatory risks, financial risks), access to private resources and know-how, possibility to implement several public projects simultaneously without being constrained by a limited public budget, less vulnerability to the economic and political instability (long-term commitments), proper maintenance of the buildings, efficient management of contracts (integrated projects covering all the stages: finance, design, build, maintenance works and service management) and better access to public facilities on short-term. The potential issues of public–private partnerships are the need to increase the contract management competences of the project team (allocated by the public authority) and to implement clear project monitoring mechanisms (e.g. performance indicators, periodic evaluation). One major disadvantage is the multiple factors that may influence the project success rate on long-term as it is impossible to capture all the potential scenarios in the contract. This fact can be overcome by a good communication between the two entities, willingness to teamwork and full commitment to the project objectives. Both entities need to be flexible with the requirements mentioned in the contract as they can be adjusted during time according to the context. The legal and institutional framework from Poland and France revealed two perspectives upon the mechanisms that may encourage and support the initiatives for public–private partnership. Both states have created several national authorities responsible with developing standard documents, methodologies, guides and contract models and also online platforms for experience exchange and access to projects database. The legal framework is integrated in one procurement act in France and in multiple legislative papers dedicated for each type of procurement in Poland, both examples may be used as models for Romania, so as to achieve an integrated, correlated and stable legal framework.

In conclusion, public–private partnerships have many benefits to all the entities involved, starting from the property owner or developer that needs to better evaluate their new residential areas, the private entity that wants to invest in public facilities, the public authority interested to ensure the access to all the minimum facilities for the community and the final user who want to live in a sustainable and well-developed district.

References

Amec Environment & Infrastructure S.R.L. (2014). *Raportul de mediu pentru Planul Urbanistic General al comunei Domnești, județul Ilfov*.

AMH Mediu Expert. (2019). *Raportul de mediu pentru Planul Urbanistic General al comunei Dumbrăvița, județul Timiș*.

Bergere, F. (2016). Ten years of PPP: An initial assessment. *OECD Journal on Budgeting*. https://library.pppknowledgelab.org/documents/2851/download

Bougrain, F. (2014). *Public private partnerships in France: a significant and decreasing role in the delivery of public infrastructure and services*. RESEARCH GATE. https://www.researchgate.net/publication/334233867_Public_private_partnerships_in_France_a_significant_and_decreasing_role_in_the_delivery_of_public_infrastructure_and_services

Cambridge Economic Policy Associates Ltd, Commonwealth Secretariat. (2010). *Public-private partnerships Policy and Practice. A Reference Guide*. London: Commonwealth Secretariat.

Cieślak, R., & Marczewska, B. (2016). *Introduction to hybrid public private partnerships in Poland*. RESEARCH GATE. https://www.researchgate.net/publication/305041703_Introduction_to_Hybrid_Public_Private-Partnerships_in_Poland

City Hall of Florești. (2019). Analiză *S.W.O.T.* FLORESTICLUJ. https://floresticluj.ro/analiza-swot/

City Hall of Dumbrăvița. (2014). *Strategia de dezvoltare a comunei Dumbrăvița 2014–2020*. Dumbrăvița. PRIMARIA DUMBRAVITA. https://primaria-dumbravita.ro/wp-content/uploads/2019/05/Strategia-Comunei-Dumbravita.pdf

European Court of Auditors. (2018). *Public private partnerships in the EU: Widespread shortcomings and limited benefits*. ECA EUROPA. https://www.eca.europa.eu/Lists/ECADocuments/SR18_09/SR_PPP_EN.pdf

European PPP Expertise Centre. (2012). *France—PPP Units and Related Institutional Framework*. EIB. https://www.eib.org/attachments/epec/epec_france_ppp_unit_and_related_institutional_framework_en.pdf

Europroiect. (2010). *Planul Integrat de dezvoltare urbana a orașului Otopeni*.

Finanțiștii. (2015). *Afacerea restitutio in integrum la final! Statul nu va mai face retrocedari in natura catre cesionari*. FINANTISTII. https://www.finantistii.ro/afacerea-restitutio-in-integrum-la-final-statul-nu-va-mai-face-retrocedari-in-natura-catre-cesionari/

Freire, A. L., Santos Sombra, T. L., & Dos Santos, N. R. D. (2020). *The public-private partnership law review* (6th ed.). Law Business Research Ltd.

French Government. (2018). *The Public Procurement Code. Ordinance 2018–1074 and Decree 2018–1075*.

GEA Strategy & Consulting S.A. (2015). *Strategia integrată de dezvoltare urbană 2015–2030—zona metropolitană Iași*.

Geoid S.R.L. (2013). *Planul Urbanistic General al comunei Chiajna, județul Ilfov*.

Imo Timisoara. (2020). *House price index*. IMOBILIARE. https://www.imobiliare.ro/indicele-imobiliare-ro

Interreg Central Europe. (2016). *Country report on the Public-Private Partnership (PPP): Poland*. Version 1.

Law no. 33/27.05.1994 related to the expropriation procedure for public utility 1994. (Ro).

Law no. 137/2002 regarding privatization procedures 2002. (Ro).

Law no. 178/2010 on public-private partnerships 2010. (Ro).

Mina-M-Com. (2008). *Planul Urbanistic General al comunei Corbeanca, județul Ilfov*.

Ministry of Regional Development and Public Works. (1997). *Normativ privind proiectarea, realizarea și exploatarea construcțiilor pentru grădinițe de copii - NP 011–97*.

MS-CAD S.R.L. (2017). *Plan Urbanistic Zonal pentru dezvoltare zonă agrement, dotări, servicii și locuințe cu funcțiuni complementare*. Timis.

National Commission of Strategy and Forecast. (2020). CNP. http://www.cnp.ro/en/investitii_strategice_in_parteneriat_public-privat.

National Institute of Statistics. (2003). *Recensământul populației și al locuințelor, 18–27 March 2002*.

National Institute of Statistics. (2012). *Rezultatele definitive ale Recensământului Populației și al Locuințelor—2011*.

National Institute of Statistics. (2019). *Fondul de locuințe-anul 2019*. Bucharest.

Plan.Co Urbisdesign S.R.L. (2013). *Planul Urbanistic General al Orașului Pantelimon, județul Ilfov*.

Public-Private Partnership Platform. (2020). *Combining PPPs with EU funds*. PPP.GOV. https://www.ppp.gov.pl/combining-ppps-with-eu-funds

Public research. (2017). Strategia de dezvoltare urbana a orașului *Chitila, județul Ilfov 2014 - 2020*. Chitila.

Regulamentul General de Urbanism 27.06.1996. (Ro).

Romanian Health Observatory. (2015). *Hospitat 1.0. Activitatea sistemului public de sănătate din România*.

Sîrghi P. (2015). *Thesis: Public private partnership. Case of Poland*.

Tectonics house S.R.L. (2019). *Plan Urbanistic Zonal "Locuințe si funcțiuni complementare"*.
The World Bank. (2015). *Locuirea în România. Către o strategie națională în domeniul locuirii.*
The World Bank. (2019). *Overview*. WORLDBANK. https://www.worldbank.org/en/topic/publicprivatepartnerships/overview.
Timiș online. (2018). *Ce se întamplă cu vânzările de locuințe din județul Timiș?. TION.* https://www.tion.ro/stirile-judetului-timis/ce-se-intampla-cu-vanzarile-de-locuinte-din-judetul-timis-41237/
Urban Architectural Management S.R.L., Mina-M-Com S.R.L. (2015a). *Planul Urbanistic General al orașului Bragadiru, județul Ilfov.*
Urban Architectural Management S.R.L., Mina-M-Com S.R.L. (2015b). *Memoriu General. Planul Urbanistic General al orașului Bragadiru, județul Ilfov.*
Urbasofia S.R.L., MKBT Habitat S.R.L. (2015). *Studiu pentru actualizarea documentelor strategice pentru Polul de Creștere Cluj Napoca aferente perioadei de programare 2014 - 2020.* Cluj.
Grigoraș V. P.F.A. (2013). *Raport de mediu pentru Planul Urbanistic General al Orașului Otopeni, județul Ilfov.*
Vass Lawyers. (2012). *Public-private partnership the paradox of a state that harms even when it wishes to do good (an interview with the magazine Avocatnet).* VASSLAWYERS. https://www.vasslawyers.eu/public-private-partnership-the-paradox-of-a-state-that-harms-even-when-it-wishes-to-do-good-an-interview-with-the-magazine-avocatnet/
Veridio, Bândea, P., Bândea, A., Moiș, B., Teodorescu, P., & Becheș, L. (2018). *Ghidul tranzacțiilor imobiliare* (5th Ed.) Cluj.

Interaction of Prehistoric Heritage with Today's Landscape: The Case Study of Arslantepe Mound in Malatya, TURKEY

Aysun Tuna

Abstract

Like most mounds comprising the crucial cultural landscape information sources such as political, economic, religious, etc. dynamics, of their age, one of the most critical factors for Arslantepe Mound to survive to date and to host a number of civilizations are its location. It is thought that being located outside the flood plain borders of the Euphrates River, while also being located within an area surrounded with alluvial soil fed from the rivers, and accordingly bearing the characteristics to be a residential area from Late Chalcolithic Age to 5.000 Byzantine Period B.C. as a settlement area with the ability to govern the products and production process, including the ability to process the raw material in the region are recognized to be the most significant factors for Arslantepe to be named as the oldest city state known to world. On the other hand, it is also known that the location of mound above the environmental connections is of vital importance for the mound to have a powerful defense system. In the light of above-stated information, it is conferred that this area as listed within the Prehistoric World Heritage Candidates has managed to survive today thanks to its landscape characteristics. In this paper, it is aimed to investigate the interaction between the Arslantepe Mound from prehistoric age with today's modern landscape. Viewshed analysis method has been utilized for determining the interaction area of the mound. The natural and cultural landscape source values of the Arslantepe Mound interaction area were established within the boundary of the mound interaction boundary. As a result of the field studies carried out within the scope of the project numbered 217O290 supported by the Scientific and Technological Research Council of Turkey (TUBITAK) and the findings obtained from the excavations carried out to date, the landscape change has been revealed in today's context. Within the scope of field studies, natural landscape features such as topography, climate, hydrology, soil structure, geology, and flora and cultural landscape features such as settlement pattern, land uses, economic structure, and property status were examined in the interaction area of the mound. Within the scope of the obtained findings, our objective has been set to bring about the landscape changes of Arslantepe Mound within the historical development process.

Keywords

Prehistoric heritage • Landscape value • Archaeological landscape • Arslantepe Mound • Malatya

1 Introduction

Arslantepe Mound Archaeological Protected Area is situated in the territory of Orduzu quarter of the city of Battalgazi district, Malatya province (Fig. 1). The cultural filling of Arslantepe Mound is 30 m in height. As a result of excavation work from 1930 to our time, it has been considered to be a "mound" structure with its multi-layered artificial form. The systematic excavation work began in 1962 and it was identified as a result of excavation work from the sculpture of the lion, which was identified at the entrance of the palace at the beginning of the 1st millennium BC (Frangipane, 2012).

Arslantepe exhibits the characteristics of an uninterrupted settlement from the Late Chalcolithic Age to the Byzantine Period of 5000 BC as a settlement that can control its lands and process raw materials in the region due to its rich water resources and high agricultural potential, as a result of this, it is out of Euphrates River flood area boundary as a location choice (Bökönyi, 1993; Sadori et al., 2012). As a result of the excavations in the mound, a temple belonging to 3600–

A. Tuna (✉)
Department of City and Regional Planning, Faculty of Architecture, Bolu Abant Izzet Baysal University, Bolu, Turkey
e-mail: aysun.tuna@ibu.edu.tr

H. Altan et al. (eds.), *Advances in Architecture, Engineering and Technology*,
Advances in Science, Technology & Innovation, https://doi.org/10.1007/978-3-031-11232-4_4

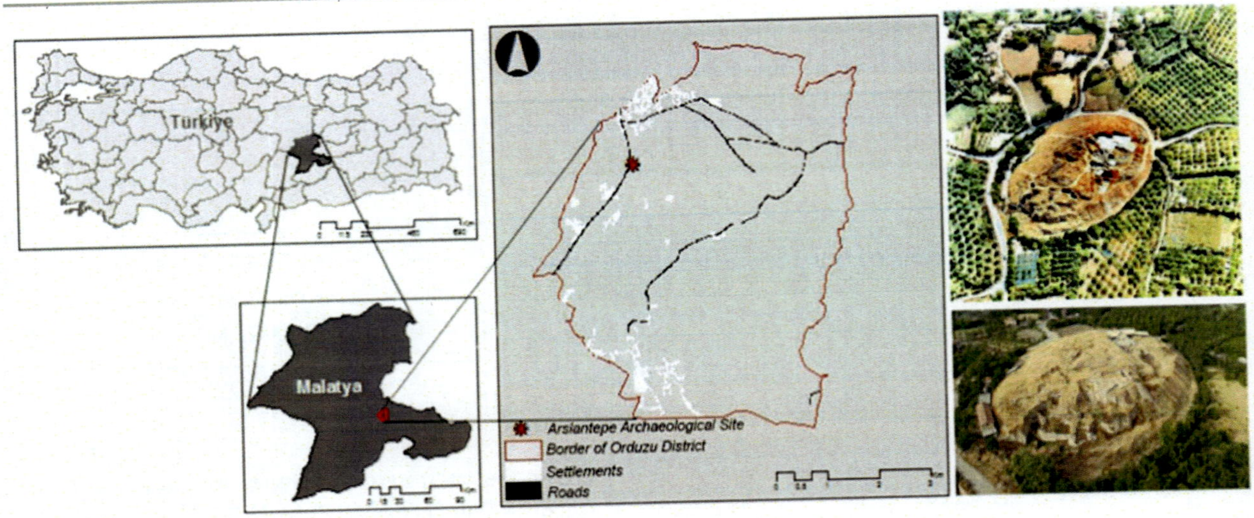

Fig. 1 Location of Arslantepe Mound (Tuna, 2019)

3500 BC, an adobe palace belonging to 3300–3000 BC, seal prints, and metal artifacts were found. According to the data obtained, Arslantepe is the official, religious and cultural center, in which aristocracy was born and the first form of state emerged (Frangipane, 2001; Liberotti et al., 2009).

The culinary structure and the containers (painted pottery belonging to Ubaid period) revealed in the excavations show that they belong to the Late Chalcolithic 1–2 period according to the radical carbon analysis (Ballossi Restelli, 2008). In the Late Chalcolithic 3–4 period, Temple C at the top of the mound and the side rooms in the temple, dishes, and seal prints in the rooms indicate that the temple had central authority at that time (Manuelli, 2010; Liberotti & Quaresimar, 2010). In line with the findings obtained, it is stated that the temple is not only a place where religious ceremonies take place but also a center where public and economic activities are carried out (Frangipane et al., 2001, 2003). Arslantepe Mound surface area expanded in 3350 BC (Late Chalcolithic 5) and a community of public buildings was built in this area. In the first known "public palace", it is not the place where the princes or manager lives, but the different public activities of central institutions (religious, economic, political, and administrative) are used as a structure. Among the findings obtained during excavation work in which two small-sized temples are located in the public palace and where the entry of people into the two temples is restricted. This shows that the realities of religion in the economic and political administration in society are not centralized (Frangipane et al. 2001). One of the buildings in the palace is the structure of warehouses. The warehouses inside the palace found that agricultural products were used for redistribution to a large number of people (Frangipane, 2000, 2001, 2008).

It is known that after the fires in the Late Phratry Period, the abandoned public space separated from the Syrian-Mesopotamian culture that continued its effects in Early Bronze I and a new culture based on Eastern-Anatolian Transcaucasian traditions that dominated from the beginning of Early Bronze II to Early Bronze III emerged. This finding is due to the construction of cities surrounded by walls as well as a settlement order in line with the urbanization tradition of Anatolia (Alvaro et al., 2008; Ardissone et al., 2008; Frangipane, 2003).

After 1700 BC, Arslantepe was used as a city called Melidia-Meliddu of the Hittite Empire, which expanded toward the Euphrates River. Melidia, the Hittite capital, was abandoned after it was captured by Assyrian king Sargın II in 712 BC. It was used as a Roman village between the fifth and sixth centuries AD and completed its settlement as a necropolis in the Byzantine period (Frangipane, 2012). Its mound today surrounds agricultural lands and apricot gardens. The first excavations in Arslantepe were carried out in the 1930s by the French team led by Louis Delaporte. Although deep boreholes were opened after World War II, continuous excavations were started in 1961 under the chairmanship of Alba Palmieri of La Sapienza University. Excavation works are underway under the presidency of Prof. Dr. Marcella Frangipane, from La Sapienza University, since 1990 to our time. In summary, the chronology of the mound is specified in Fig. 2 according to the excavation studies data in the Arslantepe Mound.

Arslantepe Mound is included in the Provisional List of UNESCO World Heritage Sites as of 2014 in the status of an outdoor museum. From the extraordinary universal value measures, which are a condition for the inclusion by the Arslantepe Mound World Heritage Committee on the World

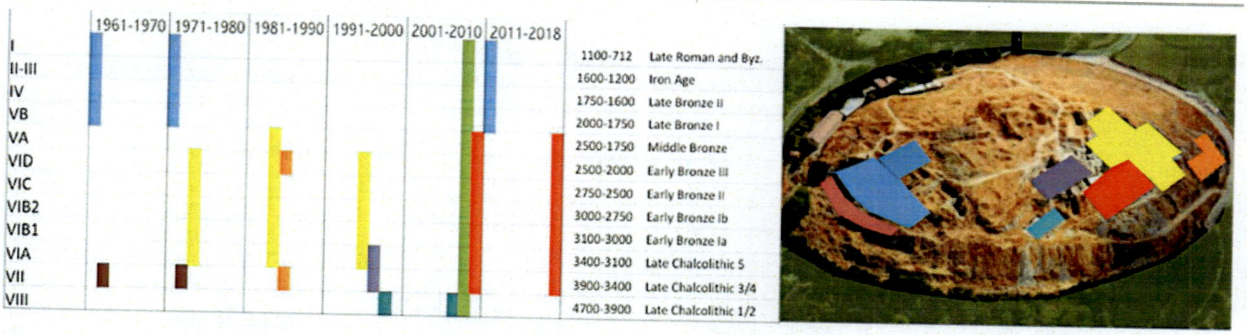

Fig. 2 Choronology of Arslantepe Mound (Balossi Restelli, 2019)

Heritage List, to 3 cultural criteria (criteria (ii, iii, iv)) as of 2019, Arslantepe Mound Archaeological Site has applied to ICOMOS as sole candidate of Turkey to become part of the UNESCO World Heritage List.

2 Material and Method

The objective of defining the communication with the landscape of the Arslantepe Mound, which dates from the prehistoric period, is to define the limit of the mound-environment interaction. As the boundary of interaction, the areas where Arslantepe Mound dominates in terms of visibility and the area where archaeological site and surface findings are dense have been determined.

Like many archaeological heritages, one of the most important factors in the ability of Arslantepe Mound to reach multi-layered cultural filling to date and to host many different civilizations is the location of the mound. It is thought to be a major factor in the fact that the river of the Euphrates is located outside the border of the floodwaters, that it is located in a river-fed area surrounded by alluvial lands having approximately 7000 years history and that it will be able to acquire the reputation of becoming the first known city state of the world. On the other hand, it is known that it is of great importance that the mound has a strong defense system and has a dominant position in environmental connections. From here, it is preferable to use the visibility (viewshed) analysis method to detect the mound interaction membrane. For visibility analysis, other heights have been taken into consideration from the point where Arslantepe Mound is located. The numerical height model of the area was created with a resolution of 3 m × 3 m according to the WGS84-UTM 37N coordinate system by photogrammetric documentation after fixed-wing drone flight and by means of ground control points established by RTK GPS method. The tools and materials used for photogrammetric documentation were obtained within the scope of the project "*Development of Archaeological Landscape Restoration and Management Strategy in Arslantepe Mound and Its Territory*" which was supported by the Scientific and Technological Research Council of Turkey (TUBITAK project number 217O290). Arslantepe Mound location was determined and transferred to open-access QGIS software. The Arslantepe Moorish location and movement-visibility analysis from elevation steps are utilized by the "Viewshed Analysis" module in the relevant software. After the visibility analysis, the zones visible from the Arslantepe Mound were detected. The regions that appear as a result of the analysis are overlapped with the region where the archaeological findings are intense and there are periodic similarities as a result of the surface research and Arslantepe Mound interaction area boundary is determined in Fig. 3.

The natural and cultural landscape source values of the Arslantepe Mound interaction area were established within the boundary of the mound interaction boundary. Within the scope of the findings obtained from the land works and excavation works, it is aimed to reveal the change of landscape in the context of today.

3 Results

3.1 Prehistoric Landscape of Arslantepe Mound and Its Surrounding Within the Scope of Archaeological Findings

Archaeological excavations in Arslantepe, which have been on for more than half a century, provide important data on the way the society is organized, the forms of power, and changes that have occurred over time, as well as in determining the pattern of settlement at that time. Although the Arslantepe site offers only a small amount of evidence regarding a very vast area—the Malatya Plain—the knowledge of archaeological history of Arslantepe allowed to establish a connection between the many events observed at the site and life in the surrounding plain, where a diverse set of landscapes was shaped over time by changes in population patterns and in the use of land. (Di Nocera, 2019). Surface investigations in the Plain of Malatya detected a

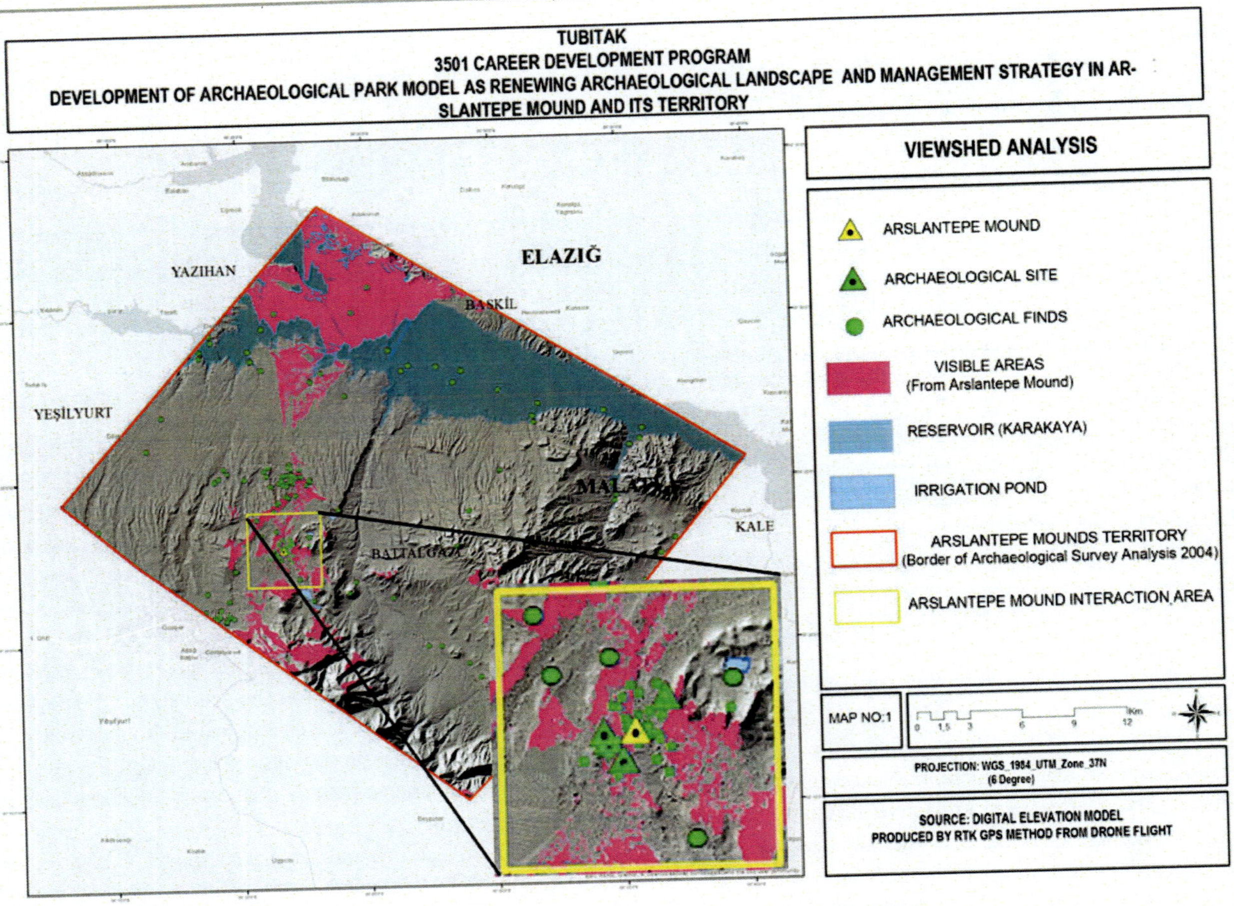

Fig. 3 Determining the border of Arslantepe Mound interaction area with viewshed analysis (Tuna, 2019)

very small number of settlements dating back to the same period as the Arslantepe Mound in Period VIII (Late Chalcolithic 1-2 (4700–3900 BC). The areas in which these settlement systems are located provide important data with the topography of that period. Two of sites are built on tells, one on a tabular summit and one on a glacis, that is, a level with a very slight slope connecting detrital sediments and the wide valley floor, strongly connected with water springs. None of these sites has a dominant character (Di Nocera, 2008, 2019).

In the Malatya Plain, sites dating back to the same age (Period VII, Late Chalcolithic 3-4, 3900–3400 BC) belong to different geomorphological categories, as in the previous period: 2 are built on *tells*, 1 on the *glacis*, 2 on *flat land* and 1 on a *complex of hills*. The impression is that, although some forms of administrative power started to emerge at Arslantepe, the economy of the site and the surrounding area was still family-based and enjoyed a certain amount of independence (Di Nocera, 2019; Di Nocera & Frangipane, 2012). Period VI_a 3400–3100 BC de Arslantepe reports that the settlements around the surrounding area have disappeared and the rural landscape has changed, and Arslantepe is the center of the Malatya Plain. Although livestock was predominantly at this time, the agricultural landscape started to develop in light of archeobotanical data. With agricultural production, the centralization layout is considered to have changed in relation to previous periods. In Early Bronz Age I-II, 3100–2500 BC several sites are distributed over the central plain and in the piedmont area, some of them along the Euphrates. 60% out of 38 sites is placed on natural mounds, most of them are characterized by single-phase settlements. In Early Bronze Age III (2500–2000 BC) the sites are distributed along the course of the Euphrates. 46% out of 45 sites is built on *tells*. The mounds started to form during the Early Bronze Age III and have a clearly stable nature, with several settlement phases. This tendency, bringing about a slight decrease in the number of settlements, continued without interruptions until the Middle Bronze Age (Arslantepe V-A) (Di Nocera, 2019).

The following periods, Hellenistic-Roman, Imperial Roman, Late Antiquity, Byzantine, and Islamic have been clearly recognized in the Arslantepe excavations; however, the site is not the only or dominant center in the area anymore (Di Nocera, 2019; Schneider, 1970). In these times, the concentration of settlement shifted from the center of power, Arslantepe Mound to the settlement of the Battalgazi (the Old Malatya), which is north of the mound. The historic Center of Battalgazi retained its central structure until the first half of the 20 century. It is thought that the use of mound as a military legion instead of a settlement in the post-Roman period was due to the devastation caused by the earthquakes in the north-west direction of the settlement to the Old Malatya (Battalgazi) region and as a result of the severe earthquake in Malatya in 1893 and 1905, it is believed that the settlement in Old Malatya moved to the region called today's Malatya City Center in the southwest direction.

Geographical investigation results describe the paleoenvironmental conditions in the periods when Arslantepe Mound was dated. The findings from the research were determined from the end of the last Molytic Center (B.C. 4200) an increase in erosion was discovered, but the reasons were not fully understood (Dreibrodt et al., 2014). Zooarcheological analyses have revealed the growth of the rabbit population in Arslantepe VI A period, while the month and deer species have decreased (Bartosiewicz, 2010). This indicates that the semi-open coniferous forest cover in the region has turned into an open meadow cover. This change was also monitored by the change in the tree varieties used in the structures in the mound (Alvaro, 2010; Bartosiewicz, 2010). Nevertheless, it is understood from the presence of dense hydrophilic plants that this period of settlement is under a rainier and more humid climate (Masi et al. 2012a, b). The presence of these plants has also documented the presence of a wetland near mound (Sadori & Masi, 2012). The presence of oak and pine seeds in a room (Period VIC) discovered in Arslantepe Mound as a result of archaeobotanical investigation proves that the humid climate prevails.

3.2 Natural Landscape Values of Arslantepe Mounds Surrounding (Today)

In defining the interaction, which is today's landscape of Arslantepe Mound and interaction area, the field studies carried out within the scope of TUBITAK project, the digital equipment provided and the information obtained from the

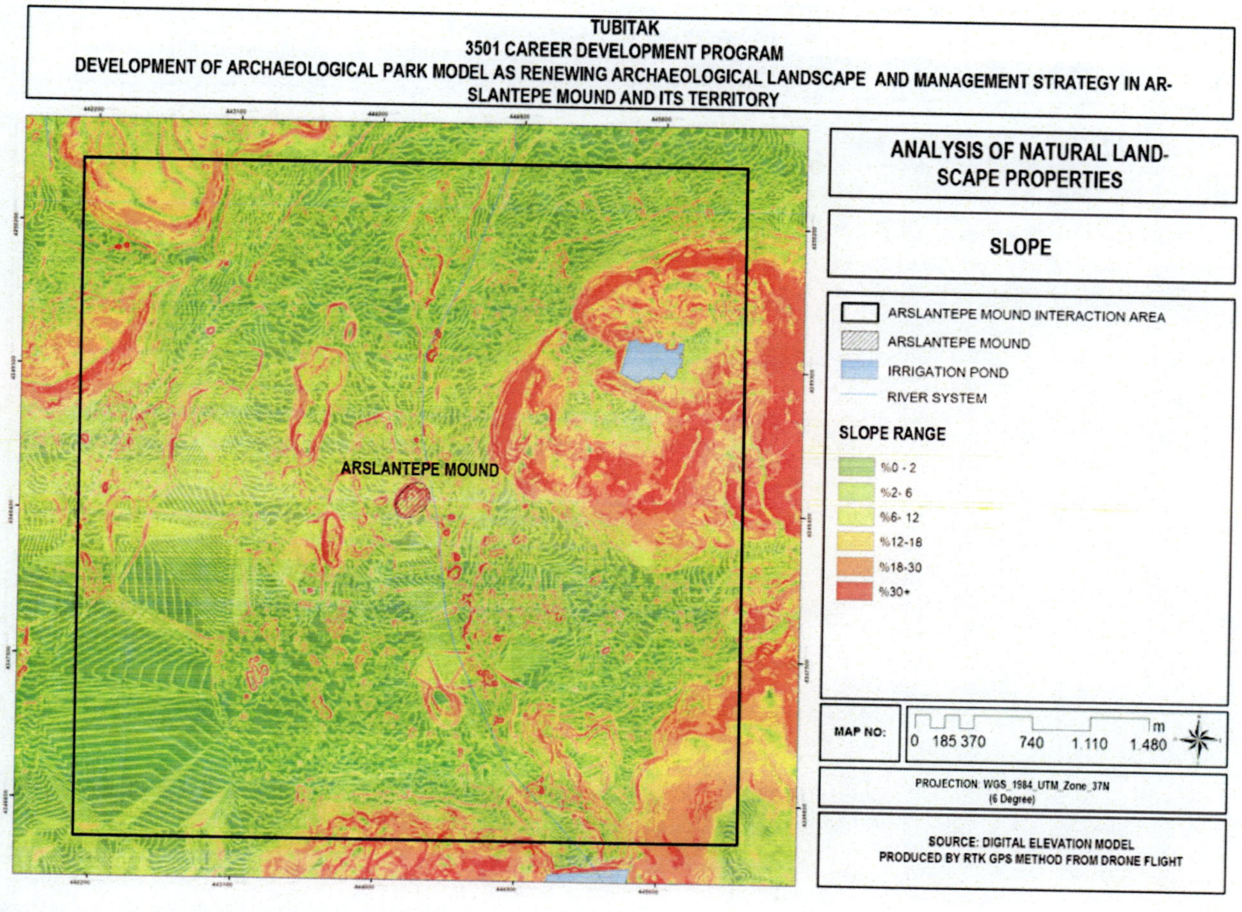

Fig. 4 Slope status of Arslantepe Mound interaction area (Tuna, 2019)

competent authorities (map, report, etc.) were utilized. Topography, climate, hydrology, soil structure, geology, and flora characteristics were examined within the scope of natural landscape features.

Topography: When the numerical height model of the mound interaction area is examined, it was determined that Arslantepe Mound environment is generally surrounded by flat areas and the mound interaction area of the dominant height group in the range of 1000–1200 m increases in the direction of south and south-east. When the geomorphological structure of Arslantepe Mound, which is determined to be 30 m high as a result of excavation studies, is evaluated, it is seen that mound wall is surrounded by plains and plateaus in northwest, southeast, and south of the area. It is known that it is in close interaction with Arslantepe Mound in terms of geological formations owned by Gelinciktepe and Yılancıktepe located on the large plateau located closest to the mound and archaeological findings detected in Gelinciktepe (Figs. 4 and 5).

When the slope status of the mound interaction area is examined, it is seen that the dominant slope group is 0–2 and 6–12, and the regions with high slope are concentrated in Gelinciktepe in the east of Arslantepe Mound.

Soil: When the soil groups within the border of the mound interaction area are examined, it is observed that the state soil group is surrounded by alluvial lands by 72.22% of the state territory group in the vicinity of the Arslantepe Mound. I. Class lands (arable land) constitute 66.67% of the mound interaction area.

Hydrology: The Arslantepe Mound shows that it is close to the water supply, the most important place selection criterion, as is the case in other mound settlements that have a significant historical history during the establishment period. As can be seen in Fig. 6, Orduzu Stream, which used to feed on Euphrates River today, takes its source from Karakaya Dam Lake and surrounds Arslantepe Mound in the east–west direction. A large number of wells and ponds identified by landfill observations with the 1/25,000-scale Land Use Map shows that agricultural production is dense. Çatlak and Üç Pınar on Orduzu Stream, which is the main water resource with its regular water regime, was an important water resource for Arslantepe Mound in the past and continue to be

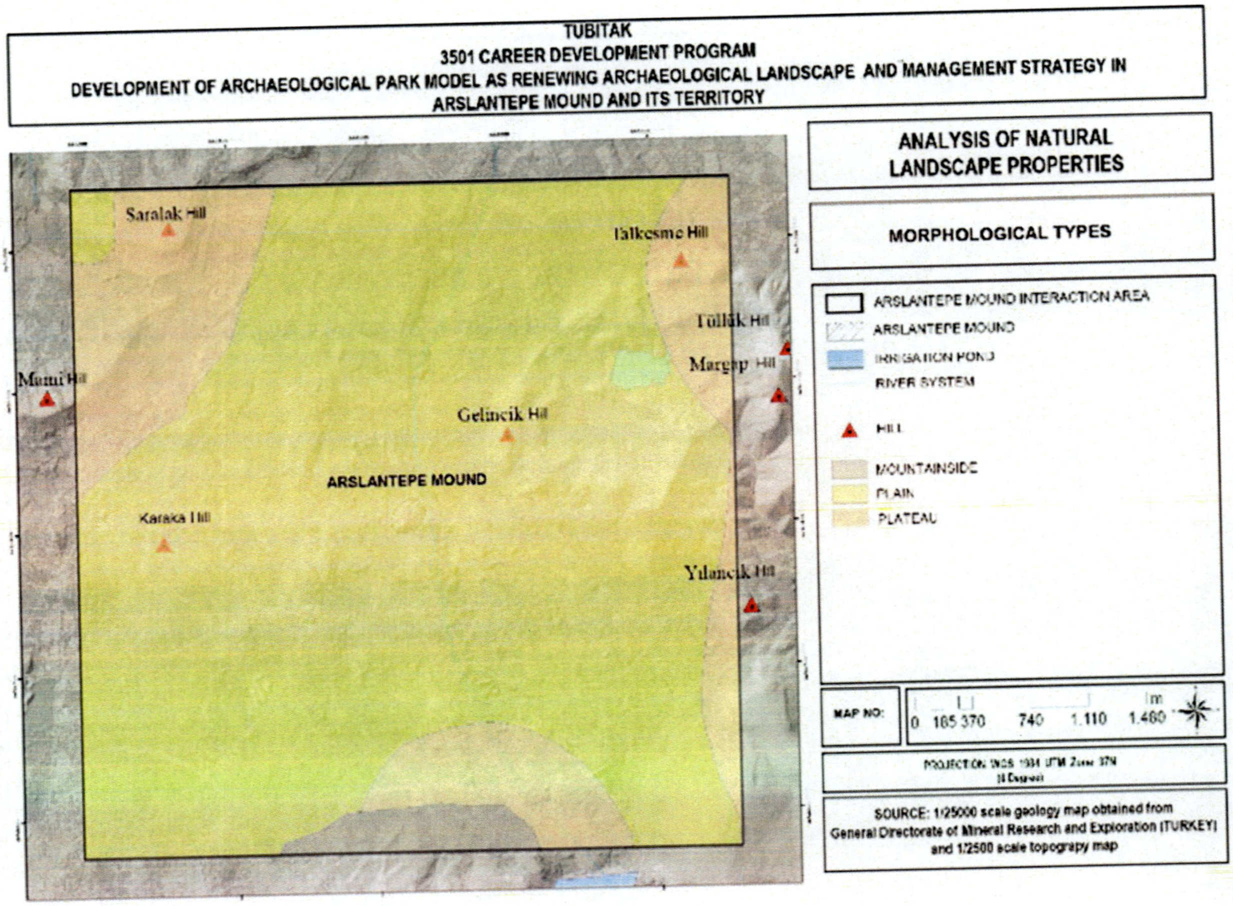

Fig. 5 Morphological types of Arslantepe Mound interaction area (Tuna, 2019)

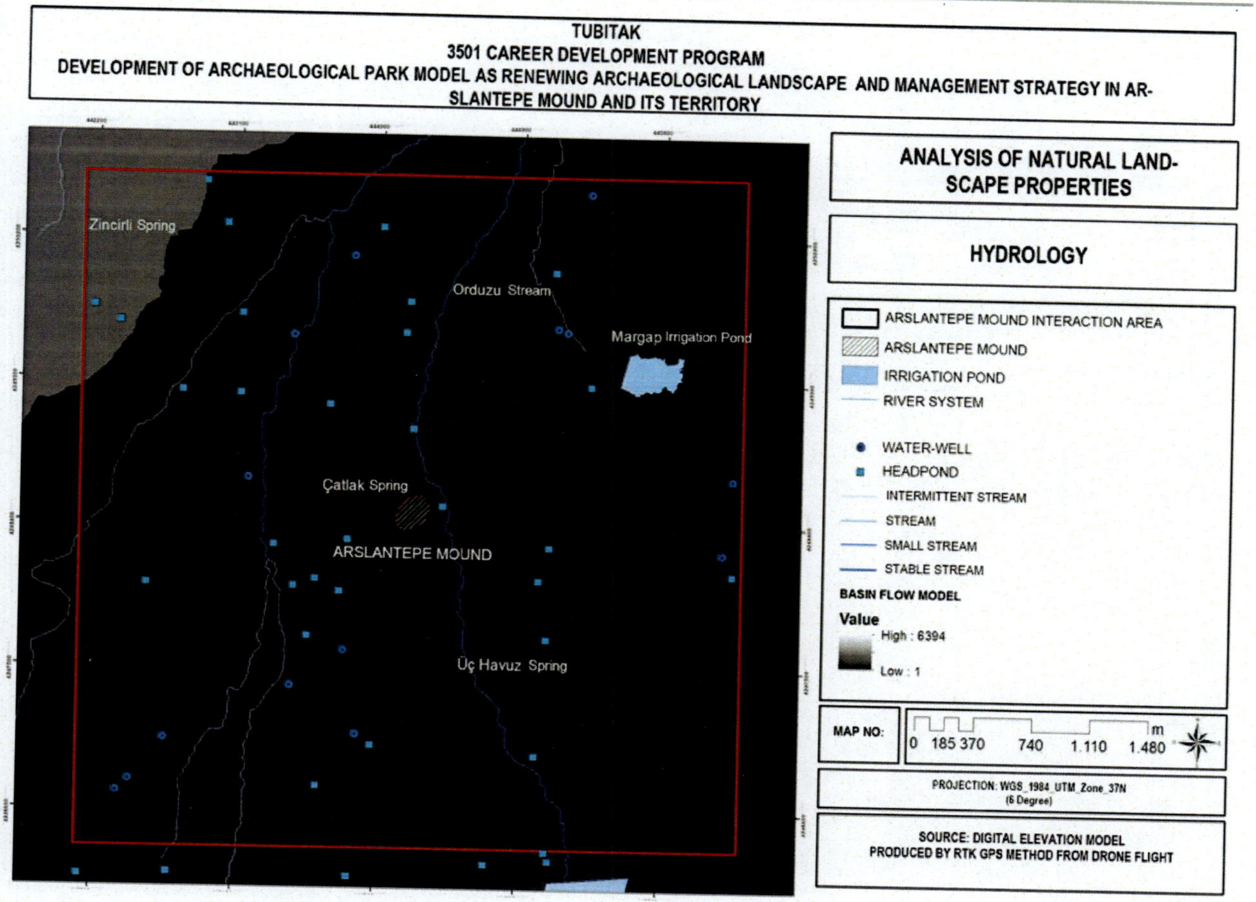

Fig. 6 Hydrology of Arslantepe Mound interaction area (Tuna, 2019)

important especially for agricultural production in Orduzu Neighborhood today. In particular, the irrigation channels fed from these streams form an important part of the settlement area (Fig. 6).

Geology: When we look at the geological formation map prepared for the mound interaction area, it is seen that predominantly quarternary aged alluviums in the immediate vicinity of Arslantepe Mound, tuff geological formation in the vicinity of Gelinciktepe-Tulluktepe, and pebble stone-sandstone-mudstone geological formations in the ages of lower Miocene and Pliocene in the north-west and south-east boundaries are concentrated (Fig. 7).

Flora: Since modern agricultural practices are intense in Arslantepe Mound and its close vicinity, it was observed that natural vegetation is significantly damaged. According to the information obtained within the scope of the field studies, it was determined that natural species are concentrated in Gelinciktepe in the east of Arslantepe Mound. In the vicinity of Gelinciktepe, *Achillea pseudoaleppica* Hausskn. ex Hub.-Mor., *Astragalus melitenensis* Boiss., *Cota wiedemanniana* (Fisch. & C.A. Mey.) Holub, *Elymus lazicus* (Boiss.) Melderis subsp., *Maarrubium globosum* Montbret and Aucher ex Benth. subsp. globosum, *Paronychia kurdica Boiss.* subsp. haussknechtii Chaudhri, *Salvia euphratica* Montbret and Aucher var. euphratica, Verbascum Euphraticum Benth., *Verbascum splendidum* Boiss, *Allium scabriflorum* Boiss., *Asphodeline damascena* (Boiss.) Baker subsp. rugosa E. Tuzlaci, *Gundelia tournefortii L.* var. armata Freyn & Sint., *Iris sari* Schott ex Baker, *Scorzonera tomentosa* L. Endemic plant species have been detected.

Using the orthophotos obtained from Land Registry Cadastre Malatya Provincial Directorate as an orthophotos base, the information obtained from Land Registry Cadastre General Directorate parcel inquiry open-access address and field studies and open-green area systems within the mound interaction area were identified in detail. As shown in Fig. 8, agricultural green areas are dense, and aqueous fields dominate the agricultural green areas. It was also observed on the lands and studies where the apricots, apples, mulberry, cherries, and walnut trees are planted, which are concentrated in aqueous fields. Another type of agricultural green area that has an important rate is wetland-garden-dry fields. These areas were observed to be used as seasonal dwellings, which are mostly described as

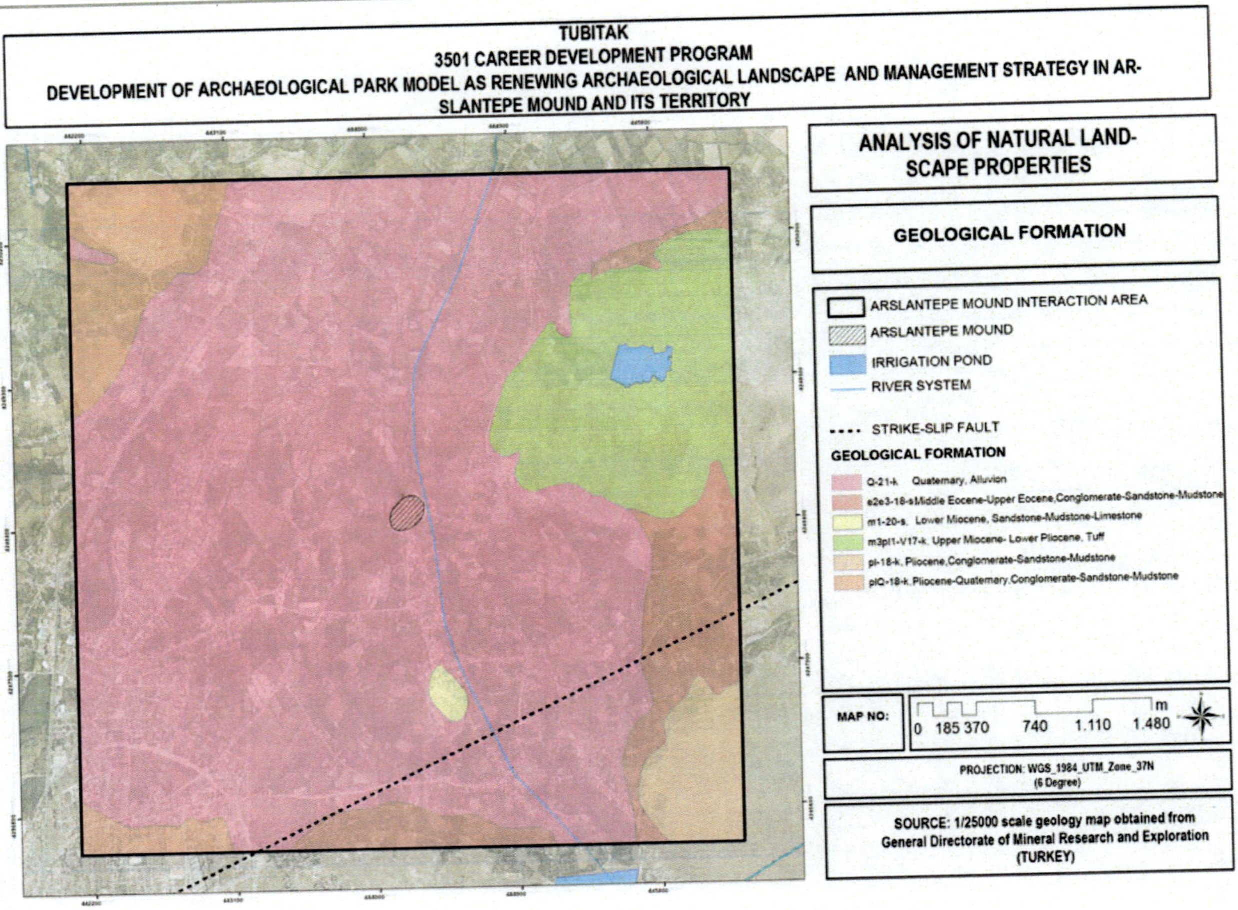

Fig. 7 Geology of Arslantepe Mound interaction area (Tuna, 2019)

"gardens" by Orduzu locals living in Malatya city center or other cities in summer months. It was determined in interviews with local people that poplar and willow fields along the stream bed are grown for commercial production purposes.

3.3 Cultural Landscape Value of Arslantepe Mounds Surrounding (Today)

Settlement Structure and Urban Space Utilizations: The field studies were conducted in 2018 and 2019 in order to determine the settlement pattern of the mound interaction area. Up-to-date orthophotos were used as underlays during field operations. Detected utilizations were transferred to the open-access QGIS software and a database was created. Analysis parameters such as building types, structural systems, physical properties of buildings, structure-floor heights, and traditional housing distributions were used in database design (Figs. 9 and 10).

When looking at the types of structures in the area of mound interaction, it is observed that 96.30% of housings are of high density, 1.42% is in housing-trade, 0.87% is stable-depot, 0.48%, and trade and educational structures are followed.

According to the data determined by land operations, the types of structures with a masonry (adobe) construction system within the mound interaction area are shown to be dense. In discussions with the local people, it was found that the use of adobe material continues and that the necessary property (white clay) is provided in the vicinity of Gelinciktepe and Yılancıkepe. It is observed that the density of reinforced concrete structures is 47.35%. The large percentage of reinforced concrete structures was found to have improved due to illegal construction.

When looking at the ply heights of the structures in the mound interaction area, it is observed that 42.53% of single-layered structures are dense. High-level (4 and above) settlement buildings were found to be concentrated along the southwest of the area of interaction of the mound, and are typically used as settlement areas for parcel applications.

When evaluating settlement typologies with traditional Turkish architecture, it was found that organic-developing

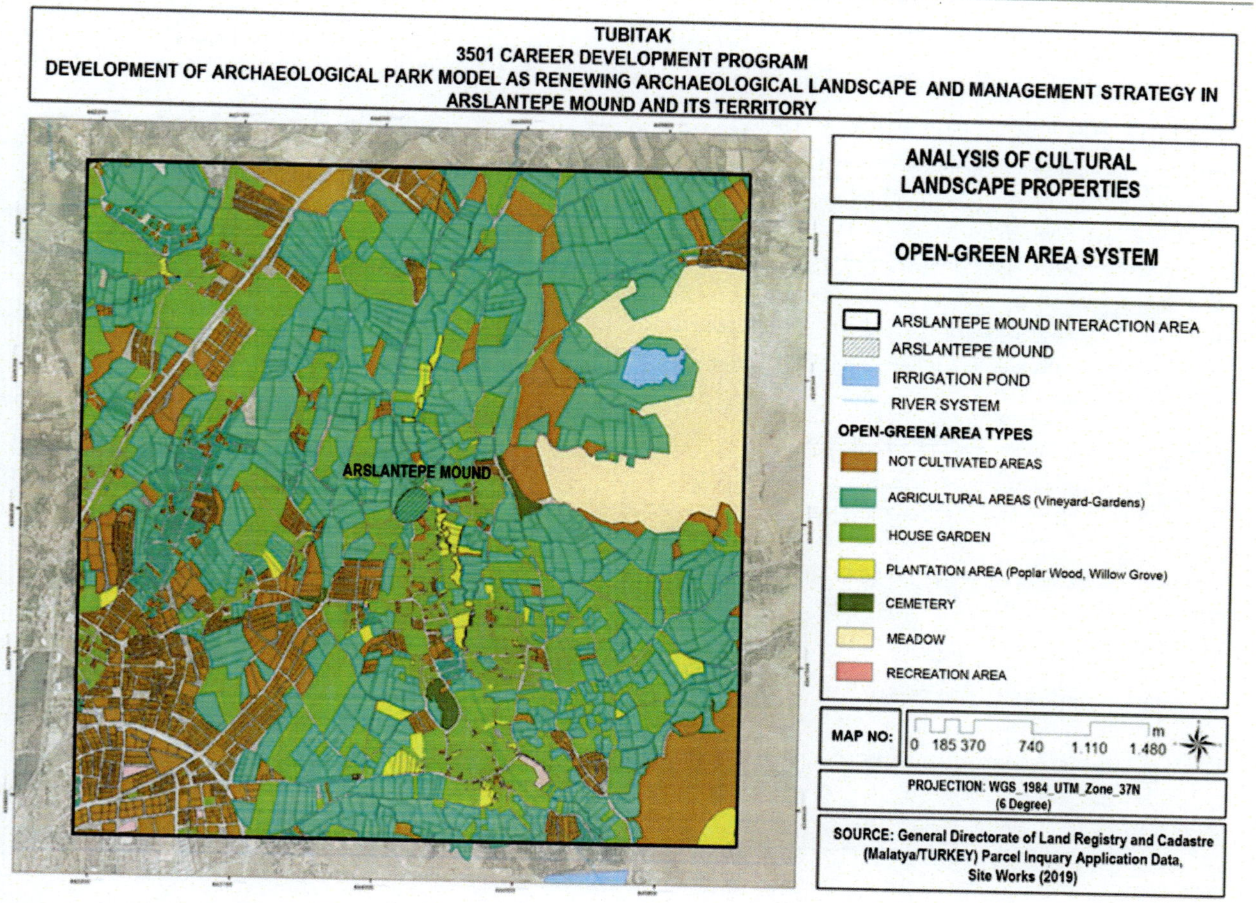

Fig. 8 Open-green system of Arslantepe Mound interaction area (Tuna, 2019)

settlements are mostly located in the attached buildings, mostly in the courtyard, and the floors of courtyard settlements are covered in the street, with the upper floors of the settlement and windows facing the street. The results of the land works indicate that the traditional street tissue was generally preserved on the close wall of the Arslantepe Mound, but it was found that the structure structures had serious destructive in the structural construction and the majority of which were unusable.

Ownership Status: Using the orthophotos obtained from Land Registry Cadastre Malatya Provincial Directorate as an orthophotos base, the information obtained from Land Registry Cadastre General Directorate parcel inquiry open-access address and field studies and land ownership status within the mound interaction area were identified in detail. According to the data obtained, there is a high concentration of private land. The fields belonging to the Municipality of Malatya, the Municipality of Battalgazi, the Ministry of National Education, and the Ministry of Treasury are other areas of public property.

Areas subject to Special Law: The areas subject to special law within the mound interaction area include agricultural protected areas and 1st and 3rd-degree archaeological protected areas. Interaction areas of Büyük Ova Protection Areas, which are the areas where land loss and land distortions are determined to develop rapidly due to various reasons such as high agricultural production potential, erosion, contamination, improper or misuse, and are therefore protected by the Council of Ministers Decision, cover 44.31% of the interaction area. The field boundaries that have gained agricultural protected area status under the scope of the Büyük Ova Protection area are provided by the Malatya Directorate of Agriculture and Forestry. The 3rd-degree archaeological protected area covers 3.5% of the mound interaction area and 1st-degree archaeological area covers 0.56% of the protected area (Fig. 11).

Economic Situation: The economy in Malatya is largely dependent on agriculture and the industry in which agricultural products are processed. 70% of the active population is involved in agriculture, livestock, fishing, forestry, and

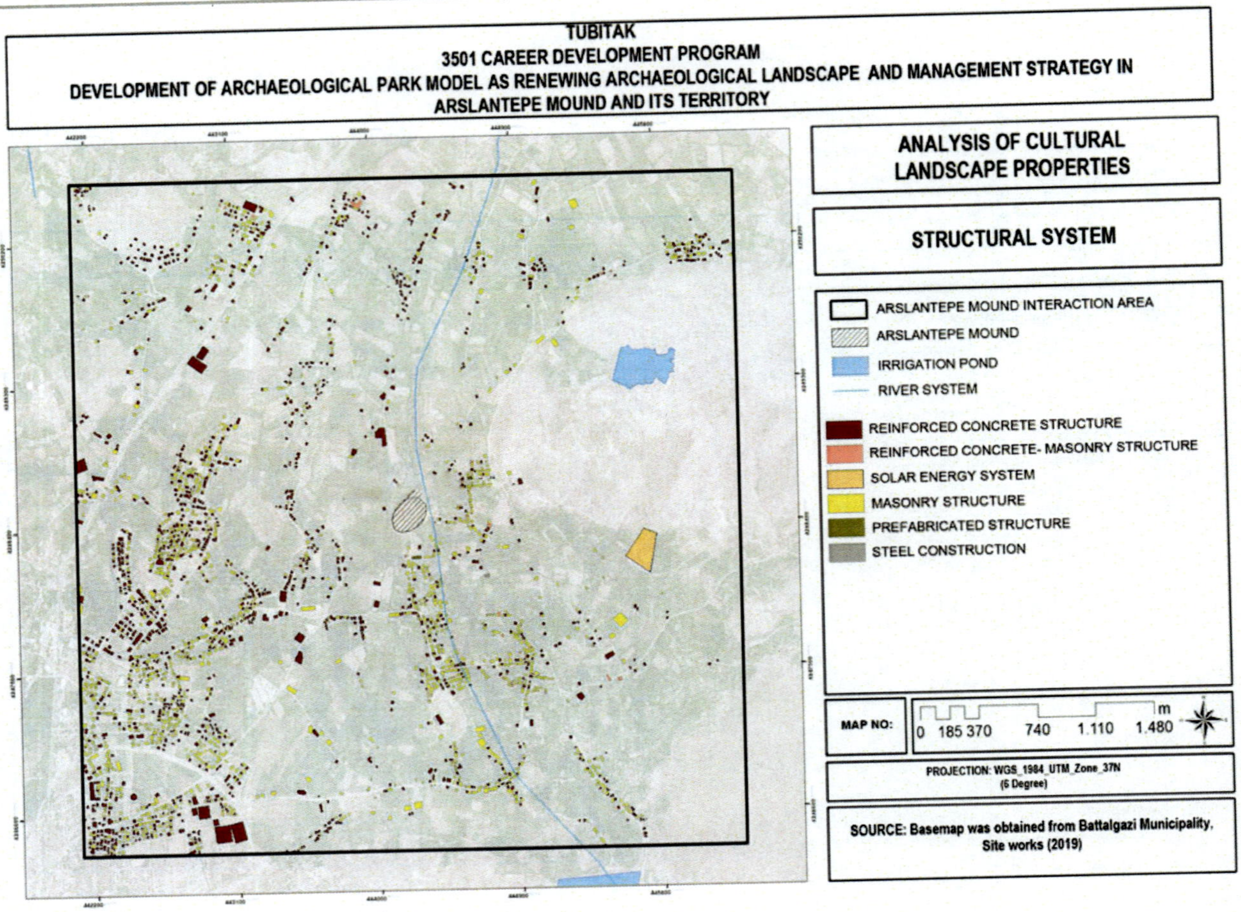

Fig. 9 Structural system of Arslantepe Mound interaction area (Tuna, 2019)

hunting. 35% of the annual gross state is provided from agriculture and 20% from industry and 12% from the service sector. Malatya meets approximately 53.4% of apricot production of Turkey and 85% of dry apricot exports of the world (URL 1), with an average yield of 329.655 tons of age.

Arslantepe Mound is situated in the territory of the Orduzu Quarter and here, the agricultural product pattern is varied substantially. Agriculture is, therefore, the main source of living. It contributes significantly on a city-wide basis in the production of, especially raw apricot. There are also 2 apricots-based after-market manufacturing factories in the quarter. The other economic source of input for the inhabitants of the quarter is excavation works of Arslantepe Mound. Excavations, which were carried out for many years (about 50 years) and which were attended by the people who provide seasonal employment and reside in Orduzu, are of high importance in terms of providing the first insurance entries of many citizens.

The most important change in the comparison of prehistoric landscape and the 21st-century landscape, which is tried to be defined in light of the data obtained from excavation and surface research, appears to be on climate. Climatic changes also bring about changes in the agricultural landscape and thus in manufacturing forms and settlement patterns. It is concluded that climate change can have an effect on the change of the central position of the Arslantepe Mound.

4 Conclusion

In this study, landscape changes from the date of Arslantepe Mound, which has proven to be a key center in the prehistoric period and where measures were taken toward becoming a UNESCO World Heritage with its superior universal values, were revealed up to this time.

Arslantepe mound, which was the center of the region where it was located in the prehistoric period, has lost its centralist position today. Today, the mound is located at the periphery of the current city center. It has been determined

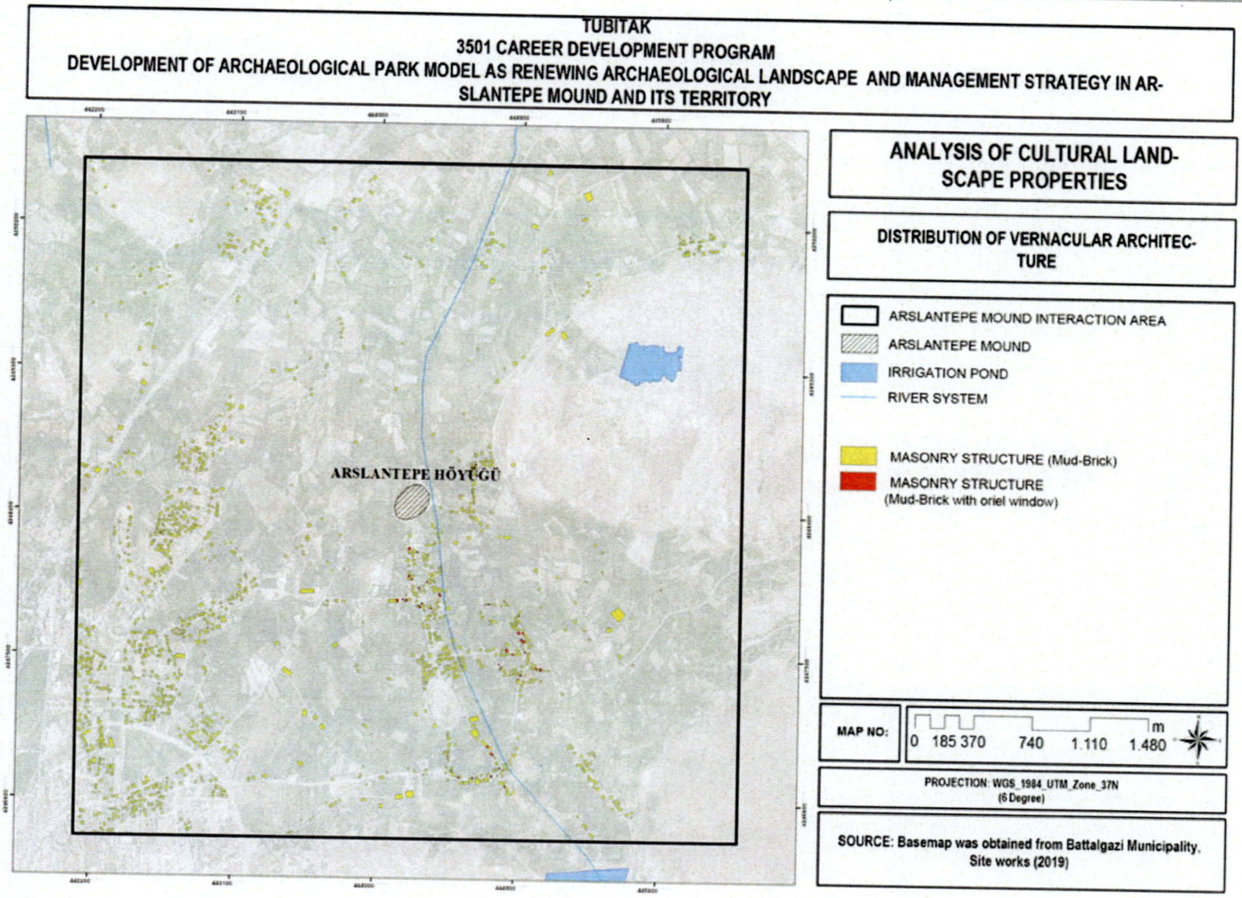

Fig. 10 The distribution of vernacular architecture of Arslantepe Mound interaction area (Tuna, 2019)

that agricultural production is dominant in the economic structure of the region from past to present.

When the green area system is examined, the intensity of the agricultural landscape character in the immediate environment draws attention. In the agricultural product pattern, it was determined that fruit growing (especially apricot cultivation) came to the fore. It is seen that there is a significant change in the green area character in the context of the past-present. According to the findings obtained from archaeobotanical and zooarchaeological studies, it has been determined that the species that can develop in humid climatic conditions have been replaced by flora and fauna species adapted to the terrestrial climate.

Considering the settlement pattern, mud brick, which is the characteristic building material of Arslantepe Mound, is also used in the houses in the immediate vicinity. However, as of the twentieth century, it was determined that the reinforced concrete structures were intensified and gradually reached the mound's immediate periphery. This situation negatively affects the original silhouette of the mound, which has the characteristics of world heritage.

The need to detect landscape changes is essential for ensuring the adaptation of archaeological sites, considered as background sources of information, to modern urban life and for today and to understand the past and shape the future for today. The investigation within this scope is also of utmost importance, especially in terms of the change in climate parameters, to provide the basis for the development of predictive models for future environmental conditions.

The detection of the change in environmental conditions will also contribute to the development of a historical environmental planning approach that will protect against the negative effects of urbanization activities in the immediate vicinity of the prehistoric Arslantepe Mound.

Acknowledgements This paper was supported by the Scientific and Technological Research Council of Turkey (TUBITAK project number 217O290).

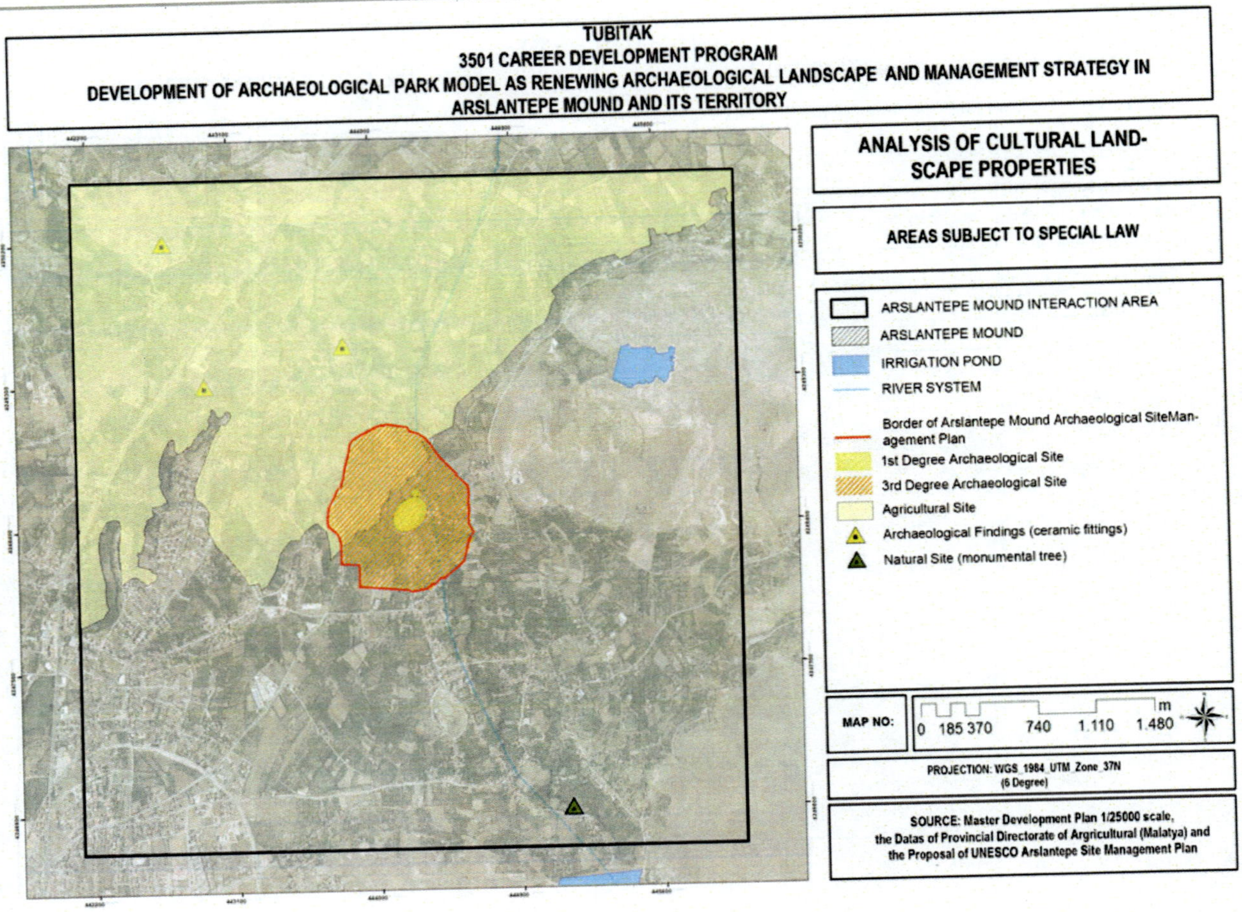

Fig. 11 Areas subject to special law of Arslantepe Mound interaction area (Tuna, 2019)

References

Alvaro, C. (2010). Architecture and the organisation of space, economic centralisation in formative states. The archaeological reconstruction of the economic system in 4th Millennium Arslantepe, Edited by Frangipane, M. *Studi Di Preistoria Orientale, 3*, 45–71.

Alvaro, C., Lemorini, C., Palumbi, G., & Piccione, G. (2008). From the analysis of the archaeological context to the life of a community. "Ethnographic" remarks on the Arslantepe VIB2 village. In *Proceedings of the 5th international congress on the archaeology of the ancient near east*, Madrid.

Ardissone, P., Balossi, F., Borgia, E., & Brienza, E. (2008). Web GIS solutions for the analysis and valorisation of archaeological sites in the Mediterranean basin. In A. Posluschny, K. Lambers, & I. Herzog (a cura di), Layers of Perception. In *Proceedings of the 35th computer applications and quantitative methods in archaeology conference,* Berlin, Germany, April 2–6, (*Kolloquien zur Vor- und Frühgeschichte, 10*). Bonn (Germany): Habelt.

Balossi Restelli, (2019). F. Arslantepe'nin Günümüz ve Gelecek Görünümü: Disiplinlerarası Araştırmalar, Yerleşimin Korunması, Müzeleştirilmesi ve İletişim, *I. International Arslantepe Archaeology Symposium,* 4–6 October Malatya, Turkey.

Balossi Restelli, F. (2008). Post Ubaid Occupation on the Upper Euphrates: Late Chalcolithic 1–2 at Arslantepe (Malatya, Turkey). In H. von Kühne, R. M. Czichon, & F. J. Kreppner (Eds.) *Proceedings of the 4th international congress of the archaeology of the ancient near east: The reconstruction of environment: Natural resources and human interrelations through time, art history, visual communication* (Vol. 2), Wiesbaden.

Bartosiewicz, L. (2010). Herding in period VI A. Development and changes from period VII, in economic centralisation in formative states. In M. Frangipane (Ed.) *The archaeological reconstruction of the economic system in 4th millennium Arslantepe. Studi di Preistoria Orientale 3,* Dip. di Scienze dell'Antichità, Sapienza Università di Roma.

Bökönyi, S. (1993). Late chalcolithic and early bronze I animal remains from Arslantepe (Malatya), Turkey: A Preliminary Report". *Origini 12*.2:581–597; Hunting in Arslantepe, Anatolia. In between the rivers and over the mountains. *Archaeologica Anatolica et Mesopotamica* Alba Palmieri Dedicata, Edited by, M. Frangipane, H. Hauptmann, M. Liverani, P. Matthiae,& M. Mellink, Università di Roma La Sapienza, Rome.

Di Nocera, G. M. (2019). Settlement patterns in the Malatya plain and metallurgical developments at Arslantepe: The role of surrounding communities. *I. International Arslantepe Archaeology Symposium,* 4–6 October 2018 Malatya, Turkey.

Di Nocera, G. M. (2008). Settlements, population and landscape on the Upper Euphrates between V and II millennium BC. Results of the Archaeological Survey Project 2003–2005 in the Malatya Plain. In J. M. Codoba, M. Molist, M. Pérez, C. I. Rubio, & S. Martinez (Eds.) *5th international congress on the archaeology of the ancient near east*, 3–8 April 2006, Madrid.

Di Nocera, G. M., & Frangipane, M. (2012). Discontinuous developments in settlement patterns and socio-economic/political relations on the Malatya Plain in the 4th and 3rd millennia BC. In F. Borrell Tena (Ed.), *Broadening horizons 3: Conference of Young researchers working in the ancient near east* (pp. 289–304). Barcelona.

Dreibrodt, S., Lubos, C., Lomax, J., Sipos, G., Schroedter, T., & Nelle, O. (2014). Holocene landscape dynamics at tell arslantepe, Malatya, Turkey-soil erosion, buried soils and settlement layers, slope and river activity in a middle euphrates catchment, Holocene.

Frangipane, M. (2000). The late chalcolithic/EB I sequence at Arslantepe. chronological and cultural remarks from a frontier site. In C. Marro & H. Hauptmann (Eds.), *Chronologies des pays du Caucase et de l'Euphrate aux IVe et IIIemillenaires* (pp. 439–447). Varia Anatolica XI, Paris.

Frangipane, M. (2001). The Transition between two opposing forms of power at Arslantepe (Malatya) at the Beginning of the 3rd Millennium, *TÜBA-AR,* 1–24.

Frangipane, M. (2003). Developments in fourth millennium public architecture in the Malatya Plain: From simple tripartite to complex and bipartite pattern. In M. Özdogan, H. Hauptmann, & N. Basgelen (Eds.) *From Villages to Cities, Studies presented to Ufuk Esin, Journal of Archaeology and Art* (pp. 147–169).

Frangipane, M. (2008). The Arslantepe Royal Tomb: new funerary customs and political changes in the upper Euphrates valley at the beginning of the third millennium BC. *The Arslantepe Royal Tomb: new funerary customs and political changes in the upper Euphrates valley at the beginning of the third millennium* BC., (pp. 169–194).

Frangipane, M. (2012). The evolution and role of administration in anatolia: A mirror of different degrees and models of centralisation. In *Archives, depots and storehouses goods storage and data recording in the hittite world: New evidence and new research* (pp. 107–122).

Frangipane, M., Di Nocera, G. M., Hauptmann, A., Morbidelli, P., Palmieri, A., Sadori, L., Schultz, M., & Schmidt-Schultz, T. (2001). New symbols of a new power in a "royal" tomb from 3000 BC Arslantepe, Malatya (Turkey), Paléorient, 27/2.

Liberotti G., & Quaresima R. (2010). Building materials in the 4th and early 3rd millennium monumental architecture at Arslantepe: mudbricks and plaster. In M. Frangipane (Ed.), *Economic centralisation in formative states. The archaeological reconstruction of the economic system the 4th millennium arslantepe, Studi di Preistoria Orientale 3,* Sapienza Università di Roma.

Liberotti, G., Alvaro, C., Frangipane, M., Giampaolo, C., Lo Mastro, S., Meloni, P., Quaresima, R.,& Volpe, R. (2009). Characterization of the 4th millennium mud-bricks of arslantepe, Malatya" (Turkey). In M. Achenza, M. Correia, & H. Guillaud (Eds.), *Proceedings of Mediterra 2009—The first mediterranean conference on earth architecture, Edicom Edizioni, Cagliari* (pp. 327–335).

Manuelli, F. (2010). Malatya-Melid: A new look at the twenty century's archaeological researches. Some remarks on the late bronze age pottery assemblages from arslantepe. In P. Matthiae, F. Pinnock, L. Nigro, & N. Marchetti (Eds.) *Proceeding of the 6th international congress on the archaeology of the ancient near east.* May, 5th–10th 2008, Sapienza Università di Roma, Wiesbaden.

Masi, A. (2012a). Stable isotope analysis of archaeological oak charcoal from eastern anatolia as a marker of mid-holocene climate changes. *Plant Biology,* 1–10.

Masi, A. (2012b). Climatic interpreattaion of carbon isotope content of mid-holocene archaeological charcoals from eastern Anatolia. *Quaternary International,* 1–9.

Sadori, L., & Masi, A. (2012). Archaeobotanical research at Arslantepe: Traditional approach and new challenges. *Origini, 34,* 433–446.

Schneider, E. (1970). Malatya-II. Rapporto preliminare delle campagne 1963–1968. Il livello romano-bizantino e le testimonianze islamiche, *Oriens Antiqui Collectio. X, Centro per le Antichità e la Storia dell'Arte del Vicino Oriente,* Roma.

Tuna, A. (2019). Project Report of the scientific and technological research council of Turkey (TUBITAK) (Project No: 217O290). In *Development of archaeological landscape restoration and management strategy in Arslantepe Mound and its territory.*

Hygromorphs: Exploring Morphologies of Architectural Adaptive Systems Using Hygroscopic Properties of Wood

Aly Ibrahim, Sherif Abdelmohsen, Akram Zayan, and Walid Omar

Abstract

The use of programmable materials for the passive actuation of low-tech adaptive architectural systems has received growing interest in the area of architectural computing, and particularly for responsive systems. Soft responsive systems that rely specifically on latent properties of natural materials as opposed to mechanical and rigid systems are becoming increasingly significant for achieving zero-energy adaptive systems. Several studies have attempted to capture the hygroscopic properties of wood and utilize its latent properties for passive actuation. These studies have partially addressed mechanisms for passive actuation but have not comprehensively defined all input parameters or resulting morphologies. This paper aims to develop an experimental methodology for identifying the different parameters affecting the actuation of natural materials like wood based on the induced flexible morphologies, where regulating different deformation parameters and motion behavior is utilized to respond to varying environmental conditions. We hypothesize that using natural programmable materials results in a variety of architectural morphologies for skins that react to external environmental stimuli, thus minimizing energy consumption and generating comfortable interior spaces. We classify specific components for examining hygromorphic material reactions under certain conditions and define methods of material tracking to document motion types and deformation angles using Kinovea video tracking and motion-sensing technology. We devise an experimental methodology that involves testing physical reactions of hygromorphic material samples under certain conditions and identifying parameters directly affecting material motion, thus demonstrating a variety of deflection patterns under three relative humidity conditions. We identify ten main parameters affecting morphologies of the resulting systems, including wood type, thickness, grain orientation, the slope of grain, artificial wood, laminated layering, dimensional proportions, temperature, perforations, and shape geometry. By embedding the combination and permutations of these parameters in the parametric environment, we define a palette of possible architectural morphologies for programmable passive actuators.

A. Ibrahim · S. Abdelmohsen (✉)
Department of Architecture, American University in Cairo, Cairo, Egypt
e-mail: sherifmorad@aucegypt.edu

A. Zayan · W. Omar
Faculty of Fine Arts, Alexandria University, Alexandria, Egypt

Keywords

Hygroscopy • Programmable materials • Passive actuation • Adaptive building skins • Building morphology

1 Introduction

Capturing the latent properties of programmable and natural materials has been of growing interest to develop zero-energy architectural adaptive systems using passive actuation. Wood, a typical anisotropic material, enables a variety of programmable kinetic configurations that can be attained through its hygroscopic nature and its response to humidity under varying climatic conditions. Recent research has demonstrated the ability to control and regulate the nature of motion, type of motion, and response speed, by accurately understanding and simulating a variety of hygroscopic design parameters.

Most studies that have attempted to investigate the hygroscopic behavior and properties of wood for adaptive building façade skins have addressed the physical experimentation for programming material response (Wood et al., 2016), and the numerical modeling of wood response to humidity based on its latent properties (Rüggeberg & Burgert, 2015; Abdelmohsen et al., 2019). Architecturally,

H. Altan et al. (eds.), *Advances in Architecture, Engineering and Technology*,
Advances in Science, Technology & Innovation, https://doi.org/10.1007/978-3-031-11232-4_5

several prototypes and full-scale models have been fabricated and tested using hygroscopic behavior as a catalyst for the low-cost and low-tech kinetic mechanism for enhancing the behavior of adaptive structures (Correa et al., 2015; Dierichs & Menges, 2016; Krieg et al., 2017).

Hygroscopic motion typically depends on the latent properties of wood related to shrinking and swelling based on the embedded moisture content, with no complex mechanism or additional energy required; a key factor to the sustainable utilization of such a natural material in adaptive façade building skins. Upon varying humidity levels, a set of parameters have been seen to affect the response behavior and speed of wood, including the type of wood, its thickness, its proportional aspect ratio, grain orientation, and lamination condition (Reichert et al., 2015; Holstov et al., 2016). Little has been done, however, to comprehensively enlist the key hygroscopic design parameters affecting the morphology of architectural adaptive façade prototypes. In this paper, we seek to identify these parameters and define a palette of possible architectural morphologies resulting from their permutations and combinations for programmable passive actuators. We hypothesize that using natural programmable materials such as wood results in a wide variety of morphologies for building façade skins that respond to external environmental stimuli, therefore, reducing energy consumption and generating comfortable interior spaces.

2 Hygroscopic Design Parameters

Wood as a natural material is known to respond to external stimuli based on its behavioral or hygroscopic design parameters. This section aims to extract the key parameters affecting morphology based on precedent studies. The first parameter identified from the literature is the *material type* used (hardwood or softwood). Botanically, softwoods are those woods that originate from gymnosperms (generally conifers) and hardwoods are trees that come from angiosperms (flowering crops). Hardwoods are generally porous, as they have wood pores with open ends that allow water molecules to flow through their walls causing swelling. Softwoods, however, do not contain these pores. This is significant in identifying active and passive layers of laminated composites in motion (Forest Products Laboratory, 2010).

The second identified parameter is the *cut direction and grain orientation* of wood. Due to its orthotropic nature, the mechanical properties of wood vary significantly when measured across its three different directions, which are typical in any timber cut; which are longitudinal direction, tangential direction, and radial direction. This quality is responsible for the variation in strength properties along all three directions. It was shown in the literature that three Young moduli, three shear moduli, and three Poisson's ratios (nine independent material parameters in total) are needed to describe the elastic response of wood (Cave, 1978; Hering et al., 2012).

The third identified parameter is the *wood thickness*, where different layer thicknesses and stiffness ratios demonstrate varied responses to humidity (Holstov et al., 2015). Having different thicknesses of wood veneer allows the use of specific material surfaces created to provide the highest possible reaction. The transverse rigidity of wood is usually much smaller than the thinner passive surface of most available passive surfaces to ensure optimal reaction. Thin rigid layers demonstrate greater flexural rigidity than those which are relatively thick but less rigid. This flexural rigidity is significant in terms of structural efficiency, where thicknesses varying between 0.05 mm and 0.5 mm are characterized by high rigidity and strength.

The fourth identified parameter is the *slope of grain*. Fine-grain and coarse-grain types are used to demonstrate different fiber directions, such as straight grain, spiral grain, and curly grain (Forest Products Laboratory, 2010). In some wood product applications, the direction of considerable stresses might not exactly fit the normal axes of wood fiber orientation. This may take place by choosing in design, by cutting the wood from the log or by grain defects that happened during tree growth. Elastic properties may be extracted from elastic theory in directions other than along the natural axes. The strength properties in paths across parallel to the perpendicular of the fibers can all be estimated with a method like Hankinson that shows the effect of the angle in motion (U.S. Department of Agriculture, 2007).

The fifth identified parameter is the *artificial wood*. Recent studies have explored 3D printing wood and bilayer composite materials (Correa et al., 2015; Vazquez et al., 2020). It was observed that 3D printed wood maintains the swelling and shrinking behavior based on the amount of absorption and loss of moisture. These motions are generally caused by the effect of environmental conditions such as humidity, radiation, and direct precipitation. Fused Deposition Modeling (FDM) wood filaments are formed by the combination of an appropriate 3D printed polymer and microwood fibers. As the wood fibers remain hygroscopically active, they also respond to changes in humidity, therefore, driving the material to transform. Laywood is an example of a commercially available wood composite consisting of strong cellulose co-polyester fiber (approximately 40%). While the 3D printed wood component reacts to humidity, other polymers are mixed with different substances to amplify the transformation, thus often responding more often to other stimuli rather than humidity (Correa et al., 2015).

The sixth identified parameter is the *multi-layer lamination*, where multi-layered laminated composites enable a component of regulation or control of morphology for a given wood sample, and a bending behavior that is not achievable solely by single layers. Typically, a hardwood layer would act as an active layer and expand due to an increase in moisture content, while a softwood layer would not experience such expansion (Brauns & Rocens, 1997). Both layers would need to be glued using a water-resistant adhesive for the bilayer system to work effectively with active and passive layer motion.

The seventh identified parameter is the *dimensional proportion* of a given tested sample. The extent and speed of bending of a given sample rely highly on its dimensional proportion. Most active components of a sample are constrained by the reaction based on available veneer sizes. The direction of veneer length is usually perpendicular to the grain and thus equivalents to the average width of the tree for quarter-cut wood material. This material configuration enables continuous curved curling (Reichert et al., 2015).

The eighth identified parameter is the *temperature* or thermal expansion. There is a limited body of research work in this area. Generally, the coefficient of thermal expansion is a function of the relative change of proportion caused by temperature variation. Coefficients of thermal expansion of fully dry wood are high across all directions, leading to expansion upon heat increase and vice versa. Significant temperature changes result in deformation changes due to an increase in the associated humidity by shrinking and swelling. Although thermal expansion is not usually significant, the influence of temperature on wood strength is crucial. High temperatures may result in both temporary and permanent decreases in timber strength. Also, it is hypothesized that bending is affected by factors such as type of wood, moisture content, heat source, intensity, and time interval (Hoadley, 2000).

The ninth identified parameter is the *perforation*. Moisture generally transfers faster through smaller thicknesses of veneer layers. The wood response tends to increase significantly when using perforated samples, especially when used as a passive layer. It was demonstrated in previous research that the rate of transformation in sample bending was 30% higher in perforated laminated samples than in solid samples, indicating a higher tendency for deflection and motion response speed (Holstov et al., 2016).

The tenth identified parameter is *shape geometry*. Studies interested in shape morphology and geometry for façade building systems reveal interesting results regarding the motion response of different geometrical shapes for a given sample, both in two dimensions and three dimensions. Efforts by Anis (2019) explored and analyzed different patterns of bending, speed, and transformation resulting from two-dimensional diamond-shaped samples and 3D tetrahedral modules under varying humidity conditions.

Although the previously identified parameters are based on precedents from the literature, very little has been discussed relating to the comprehensive inclusion of all ten parameters (or more) collectively in the study of the resulting morphology for adaptive building skins. In the study below, we devise an experimental methodology that involves testing physical reactions of hygromorphic material samples based on these identified parameters and defining their basic morphological transformations and deformation motion types.

3 Methodology and Procedures

For the purpose of this study, we classified specific components for examining hygromorphic material reactions under specific conditions to document motion types and deformation angles. Material tracking was conducted using Kinovea video tracking software, motion-sensing technology, and Grasshopper graphical algorithmic editor for parametric variation and documentation (Hisham et al., 2017; Adnan et al., 2018; Abdelmohsen et al., 2018; El-Dabaa et al., 2021). We used a digital camera on a tripod to record the motion of wood samples by means of time-lapse images and videos. The samples were marked with markers to evaluate the angle of deflection using the Kinovea; an image analysis and video tracking software that tracks motion via live analysis of linear and angular measurement per frame through defining fixed and movable markers. This was done by measuring the deflection angle of the sample for curved samples or by measuring both the angle and diameter of deflection for rolled samples.

We devised an experimental methodology for testing physical reactions of hygromorphic material samples under specific conditions. We then identified parameters that were seen to directly impact material motion and bending under three relative humidity conditions. Based on the ten identified parameters (wood type, thickness, grain orientation, slope of grain, artificial wood, laminated layering, dimensional proportions, temperature, perforations, and shape geometry), and their associated combination and permutations, we defined a palette of architectural morphologies for programmable passive actuators.

Figure 1 illustrates the basic setup for the experiment. We used a humidity chamber to capture the deformation readings of each of the wood veneer samples. We conducted ten experiments targeting each of the ten identified parameters to capture the resulting response behavior and motion type. The samples were first prepared with precise dimensions, cuts, grain orientation, wood types, lamination, perforation, etc., based on the experiment type. They were then placed inside

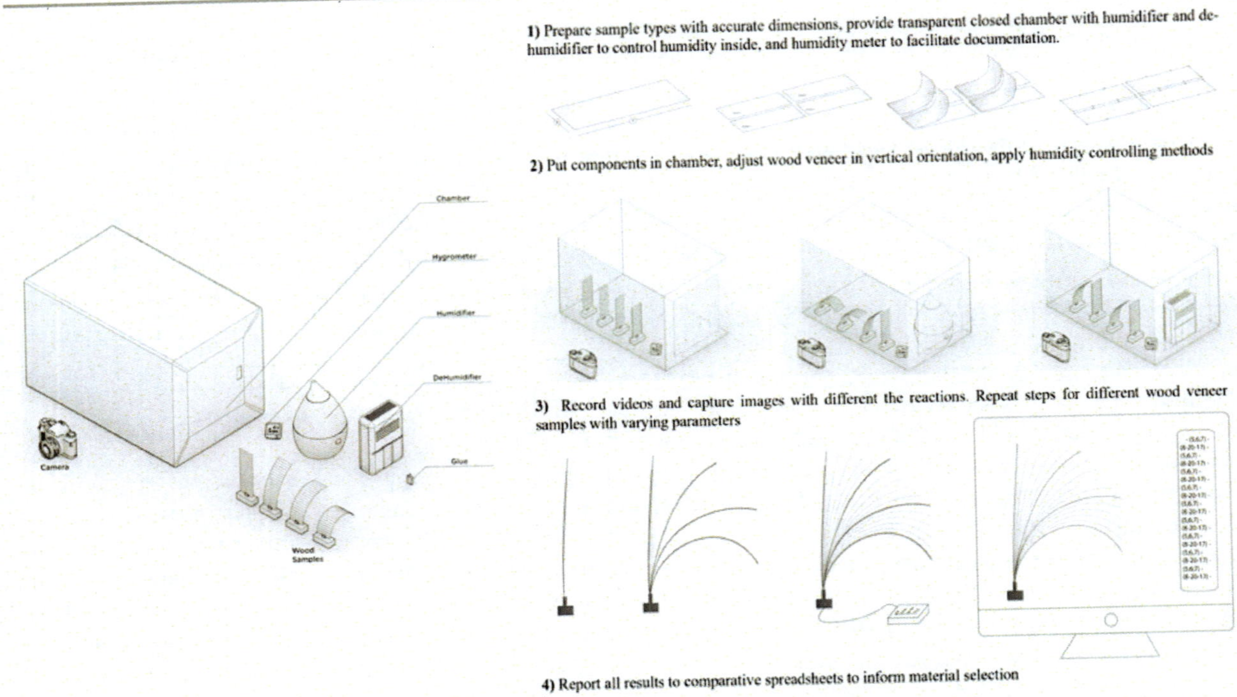

Fig. 1 Setup of the experiment (left: humidity chamber, hygrometer, dehumidifier, Flex sensor, Arduino kit, wood samples, and glue; right: testing steps and procedures: **a** sample preparation, **b** applying humidity to samples inside humidity chamber, **c** image and video capture using Flex sensor and Kinovea software, and **d** reporting and analysis of results)

the humidity chamber to provide a humidity-controlled environment.

For the purpose of the experiments, the wood veneer sample used as a base case and basis for comparison for all ten experiments exhibited the following properties: (1) material type: Beech (hardwood), (2) grain orientation: radial, (3) thickness: 0.5 mm, (4) slope of grain: 0°, (5) natural wood, (6) lamination: single-layer, (7) dimensional proportion: 700 × 1400 mm (ratio 1:2), (8) applied temperature: low temperature range (15–30°C), (9) perforation: opaque, and (10) shape geometry: rectangular. All samples were exposed to three conditions of relative humidity (low: 45–65%, medium: 65–80%, and high: 80–100%).

The wood samples were all fixed to metal clamps inside the chamber. A video camera fixed on a tripod was located at a specific fixed distance from all samples. For each experiment, readings were taken under three conditions of relative humidity. A hygrometer was used to accurately track humidity levels inside the chamber. The response behavior of each of the samples was measured and analyzed by comparing the maximum deflection angle with respect to time and using the Kinovea image analysis software for accurate monitoring and tracking. The maximum deflection was recorded in this case to indicate the peak value immediately before the wood veneer sample dries out and starts returning in reverse direction to its initial state (El-Dabaa et al., 2021). The deflection angle was measured between the location of the initial and final state points of the sample. Recorded videos from Kinovea were used to track the motion of wood as an input for the software, as shown in Figure 2. A flex sensor was also used to measure the rate of deflection caused by the sensor bending. The algorithm of the visual programming tool in Grasshopper, developed by Abdelmohsen et al. (2018), was used to automatically read the type of motion digitally. The script implements a computational closed loop that links a smart material interface—which records the deflection angle and speed—with a digital and tangible interface that consists of an Arduino microprocessor, Flex sensor, and the Firefly plugin. This allowed for the analysis, tracking, and documentation of the hygroscopic response by means of the continuous tracking of angular measurements in relation to time.

4 Results

This section demonstrates the readings and results of the ten experiments as per each of the identified parameters: (1) material type, (2) cut direction and grain orientation, (3) wood thickness, (4) slope of grain, (5) artificial wood, (6) multi-layer lamination, (7) dimensional proportions, (8) temperature, (9) perforation, and (10) shape geometry.

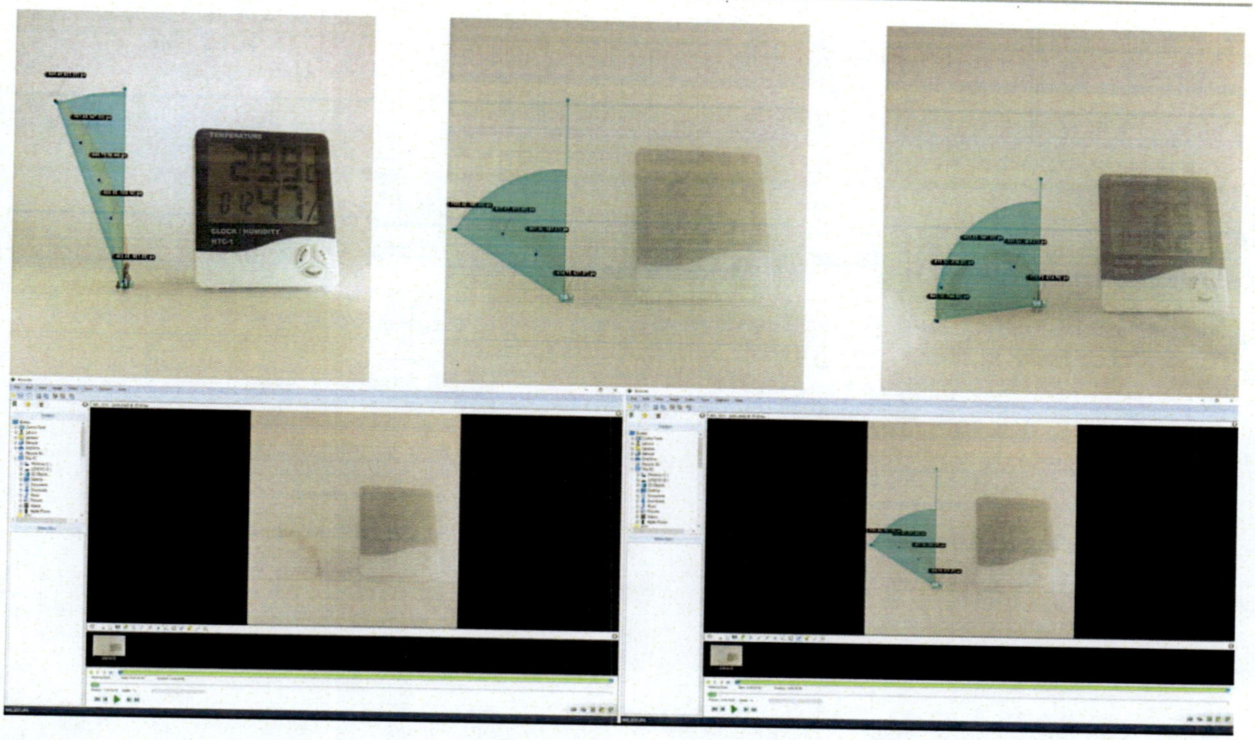

Fig. 2 Measuring and recording the deflection of a given sample (three average readings for consistency) using Kinovea image analysis software

4.1 Material Type

Four types of veneer were used in this experiment as representatives of hardwood and softwood that are commonly found in the area of study. These types were Beech (hardwood), Mahogany (hardwood), Oak (hardwood), and Pitch Pine (softwood). Upon exposing all four 700 × 1400 mm rectangular samples to the three conditions of relative humidity, it was observed that all samples exhibited a 2D bending reaction with varied angles of deflection. Beech veneer (base case) had the maximum deflection angle value (105°), followed by Pitch Pine (at 52°), then Oak (13°), and Mahogany (13°). Figure 3 illustrates the readings and maximum deflection values for each of the four types of materials.

4.2 Cut Direction and Grain Orientation

Two Beech veneer samples were used in this experiment with two different cut directions and grain orientations: a radial cut sample (base case) and a tangential cut sample. Upon exposing the two 700 × 1400 mm rectangular samples to the three conditions of relative humidity, it was observed that both samples exhibited a 2D bending reaction with varied angles of deflection. While the radial cut sample (base case) had a maximum deflection value of 105°, the tangential cut sample had a maximum deflection value of 123°. Figure 4 illustrates the readings and maximum deflection values for each of the two grain orientations.

4.3 Wood Thickness

Two Beech veneer samples were used in this experiment with two different sample thicknesses: 0.5 mm thickness (base case) and 1 mm thickness. Upon exposing the two 700 × 1400 mm rectangular samples to the three conditions of relative humidity, it was observed that both samples exhibited a 2D bending reaction with varied angles of deflection. While the 0.5 mm thickness sample (base case) had a maximum deflection value of 105°, the 1 mm thickness sample had a maximum deflection value of only 15°. Therefore, the results denoted a significant variation in terms of deflection response. Figure 5 illustrates the readings and maximum deflection values for each of the two sample thicknesses.

4.4 Slope of Grain

Five different Beech veneer samples with a varying slope of grain were used in this experiment: 0° slope of grain (base case), 30°, 45°, 60°, and 90°. Upon exposing all five 700 ×

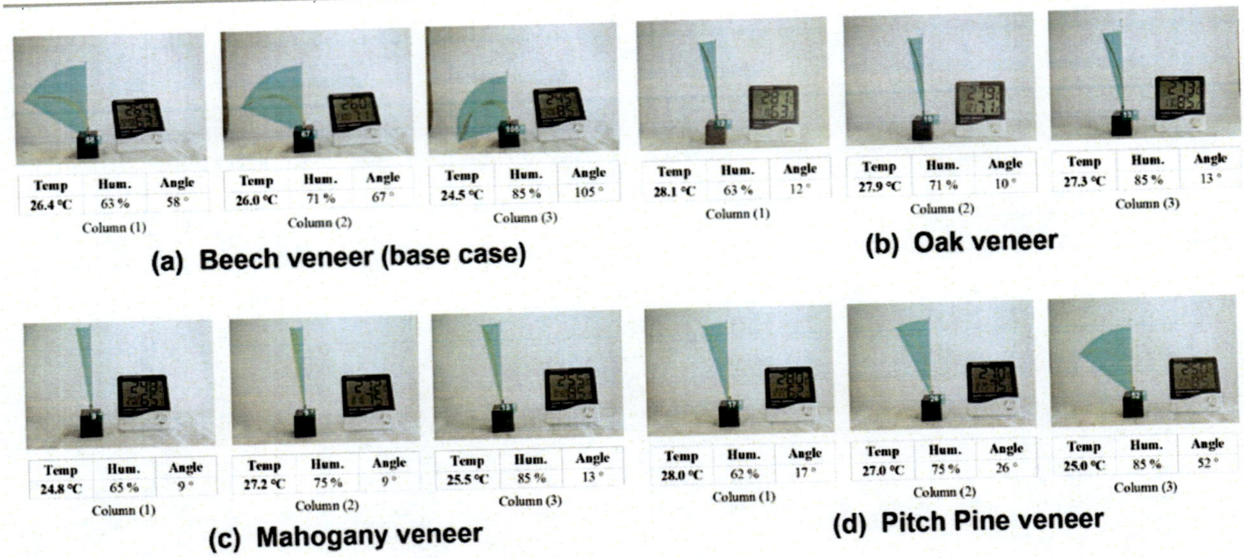

Fig. 3 Response of four wood veneer material types to three conditions of relative humidity

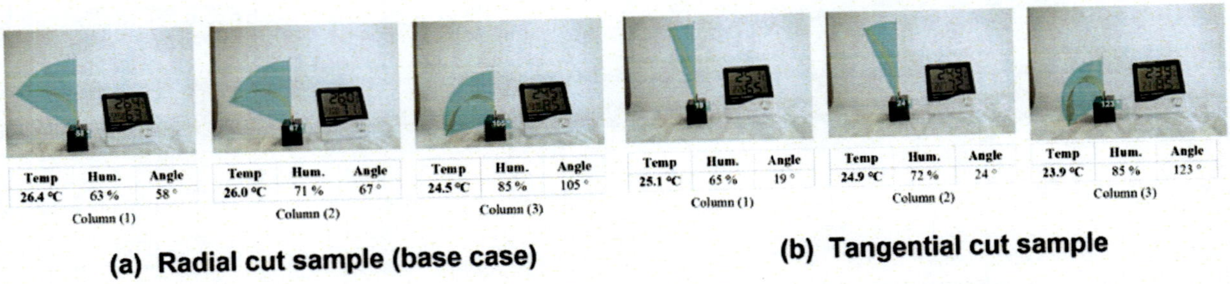

Fig. 4 Response of two Beech veneer samples with different grain orientations to three conditions of relative humidity

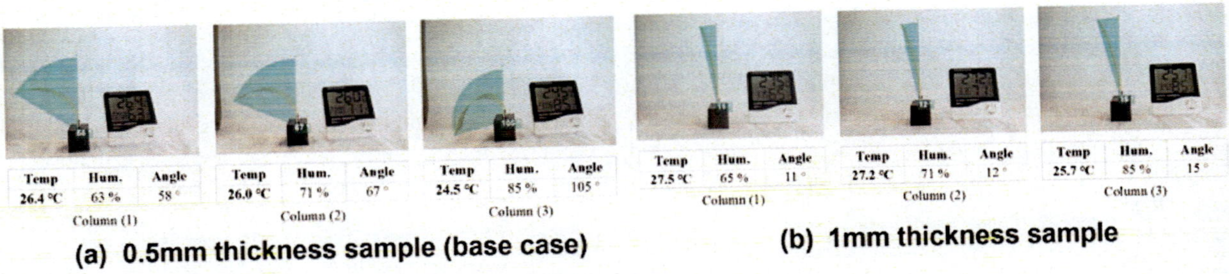

Fig. 5 Response of two Beech veneer samples with different thicknesses to three conditions of relative humidity

1400 mm rectangular samples to the three conditions of relative humidity, it was observed that the samples varied in their bending reaction. The 30° slope of grain sample had the highest maximum deflection value (126°), followed by the 0° slope of grain sample (105°), then the 45° slope of grain sample (86°), then the 60° slope of grain sample (14°), and finally, the 90° slope of grain sample (7°). While the 0° slope of grain and 90° slope of grain samples exhibited a 2D bending reaction, the 30°, 45°, and 60° slope of grain samples demonstrated a 3D bending reaction, where deflection occurred in two directions. Figure 6 illustrates the readings and maximum deflection values for each of the five samples.

4.5 Artificial Wood

For this experiment, the Beech veneer base case sample (natural wood) was compared to a 3D printed sample (artificial wood). Upon exposing the two 700 × 1400 mm

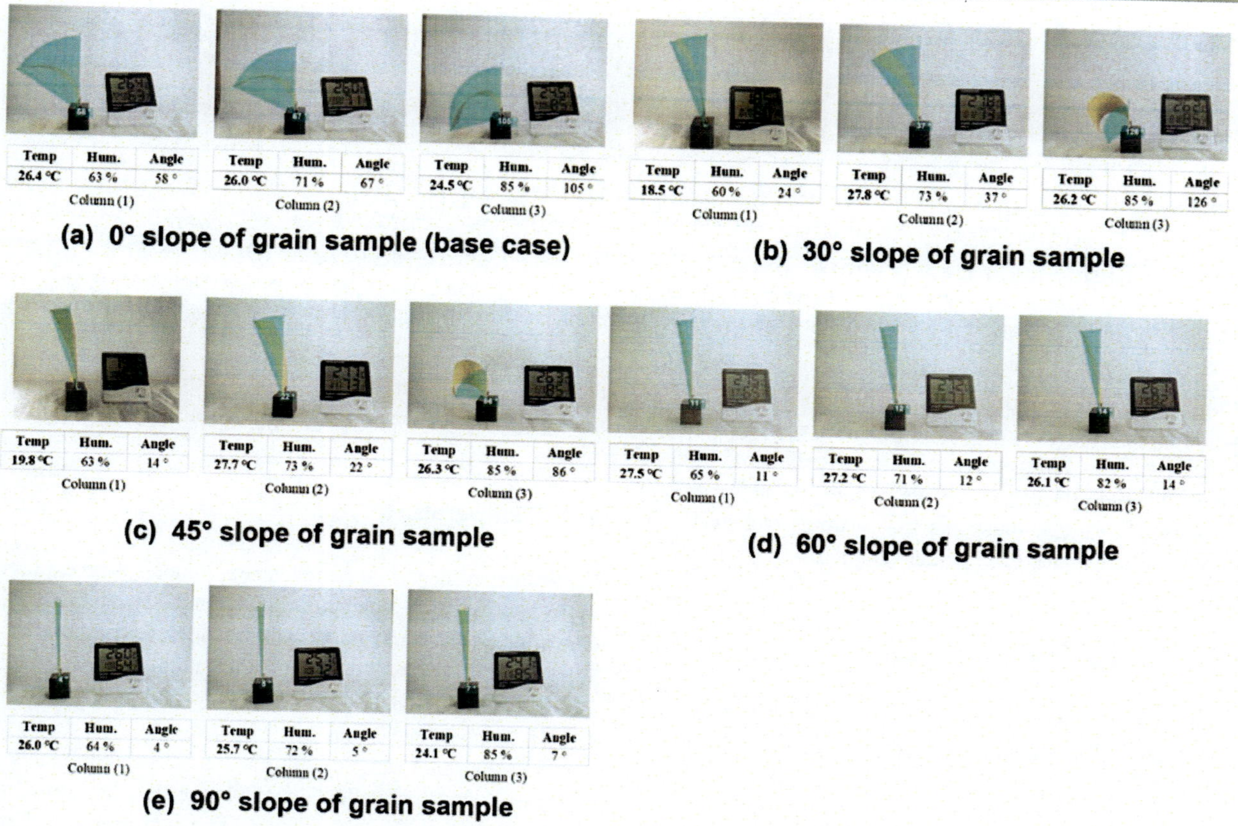

Fig. 6 Response of five Beech veneer samples with varying slope of grain to three conditions of relative humidity

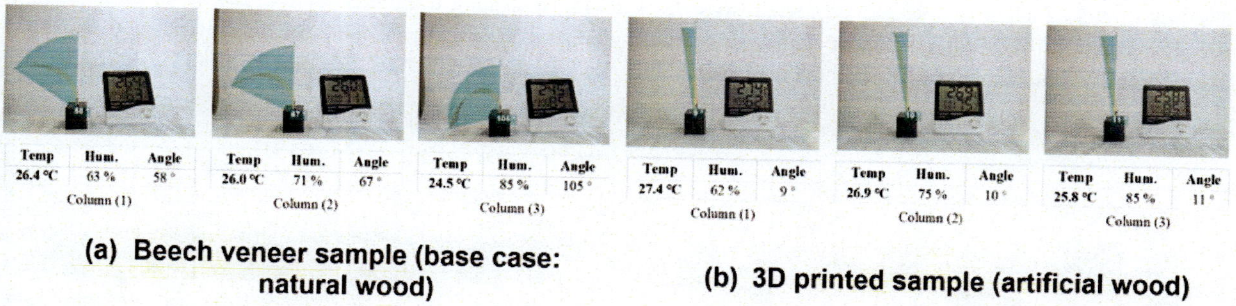

Fig. 7 Response of the Beech veneer base case sample (natural wood) and a 3D printed sample (artificial wood) to three conditions of relative humidity

rectangular samples to the three conditions of relative humidity, it was observed that both samples exhibited a 2D bending reaction with varied angles of deflection. While the natural sample (base case) had a maximum deflection value of 105°, the 3D printed artificial wood sample had a maximum deflection value of 11°, indicating a highly significant difference between both responses. Figure 7 illustrates the readings and maximum deflection values for each of the two samples.

4.6 Multi-layer Lamination

For this experiment, the base case single-layer Beech veneer sample was compared to a laminated veneer sample, with Beech (hardwood) as the active layer and Pitch Pine (softwood) as the passive layer. Upon exposing the two 700 × 1400 rectangular samples to the three conditions of relative humidity, it was observed that both samples exhibited a 2D bending reaction with varied angles of deflection. While the

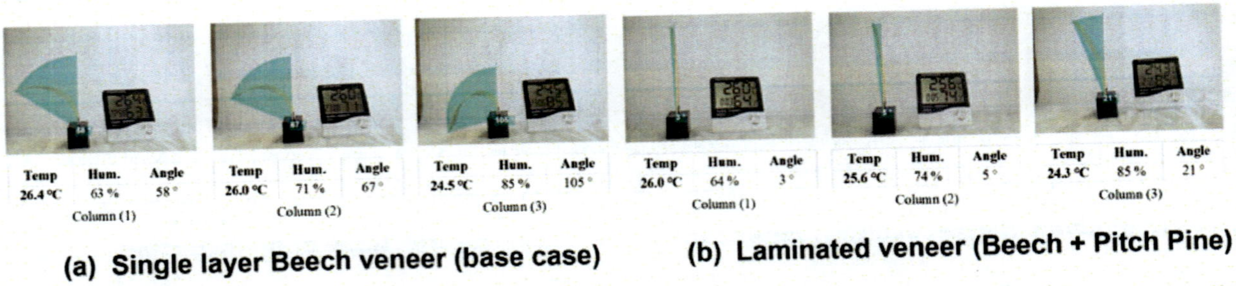

Fig. 8 Response of single-layer Beech veneer sample and laminated veneer sample (Beech + Pitch Pine) to three conditions of relative humidity

single-layer Beech veneer sample (base case) had a maximum deflection value of 105°, the laminated veneer sample (Beech + Pitch Pine) had a maximum deflection value of only 21°, indicating a significant difference between both responses. Figure 8 illustrates the readings and maximum deflection values for each of the samples.

4.7 Dimensional Proportions

Four Beech samples with varying proportional dimensions were used in this experiment as follows: 700 × 700 m (1:1), 700 × 1400 mm (1:2) (base case), 700 × 1750 mm (2:5), and 700 × 930 mm (3:4). Upon exposing the four samples to the three conditions of relative humidity, it was observed that all samples exhibited a 2D bending reaction with varied angles of deflection. While the 1:2 sample (base case) had a maximum deflection of 105°, the 2:5 sample had the highest maximum deflection value (114°), the 3:4 sample had a maximum deflection of 37°, and the 1:1 sample had the least maximum deflection (34°). Figure 9 illustrates the readings and maximum deflection values for each of the four samples.

4.8 Temperature

For this experiment, two Beech veneer samples were exposed to two different temperature conditions: (1) 24.5–26.4°C (classified as low temperature, falling within the range of 15–30°C) and (2) 34.7–43.1°C (classified as high temperature, falling within the range of 30–45°C). Upon exposing the two samples to the three conditions of relative humidity, it was observed that both samples exhibited a 2D bending reaction with varied angles of deflection. While the sample exposed to lower temperature range (base case) had a maximum deflection value of 105°, the sample exposed to the higher temperature range which had a maximum deflection value of 147°, indicating a significant effect caused by an increase in temperature. Figure 10 illustrates the readings and maximum deflection values for each of the samples.

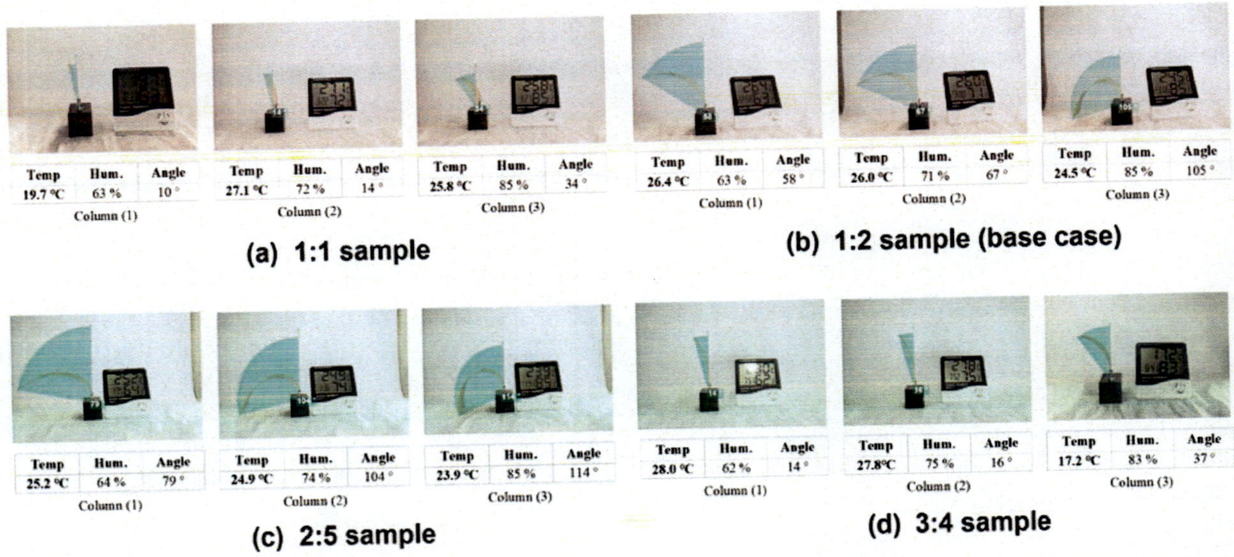

Fig. 9 Response of four Beech veneer samples with varying proportional dimensions to three conditions of relative humidity

4.9 Perforation

Three Beech veneer samples with varying perforation percentages were used for this experiment: Opaque (base case), 0.6 mm perforation, and 1 mm perforation. Upon exposing the three samples to the three conditions of relative humidity, it was observed that all three samples exhibited a 2D bending reaction with varied angles of deflection. While the opaque sample with no perforations (base case) had a maximum deflection value of 105°, the 1 mm perforated sample had the highest maximum deflection value (126°), and the 0.6 mm perforated sample had the least maximum deflection value (76°). Figure 11 illustrates the readings and maximum deflection values for each of the samples.

4.10 Shape Geometry

Four Beech samples with varying shape geometries were used in this experiment: rectangle (base case), triangle, rhombus, and hexagon. Upon exposing the four samples to the three conditions of relative humidity, it was observed that all samples exhibited a 2D bending reaction with varied angles of deflection. While the rectangular samples (base case) had a maximum deflection of 105°, the rhombus-shaped sample had the highest maximum deflection value (109°), the triangular sample had a maximum deflection value of 26°, and the hexagon-shaped sample had the least maximum deflection value (21°). Figure 12 illustrates the readings and maximum deflection values for each of the four samples.

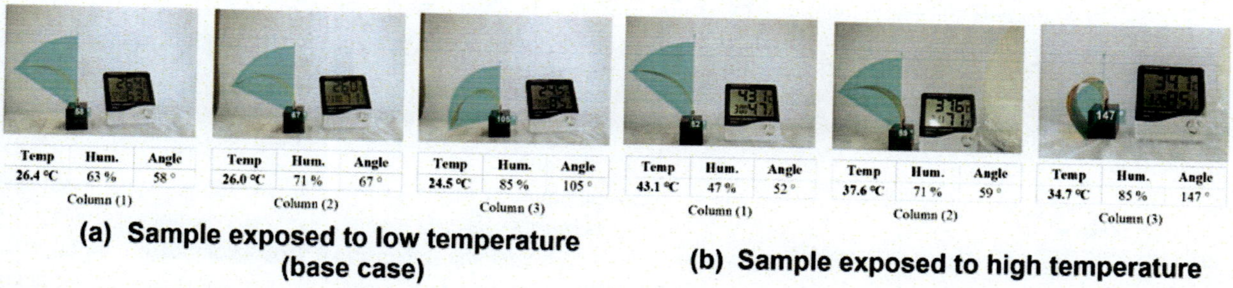

Fig. 10 Response of two veneer samples exposed to varying temperature conditions to three conditions of relative humidity

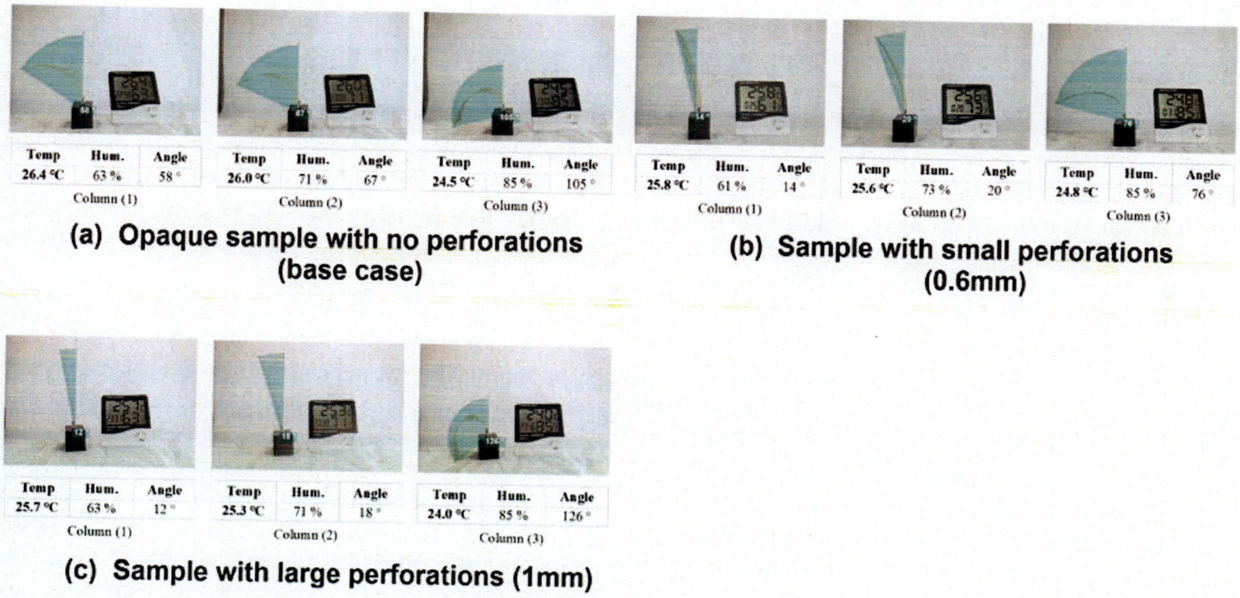

Fig. 11 Response of three Beech veneer samples with varying perforation percentages to three conditions of relative humidity

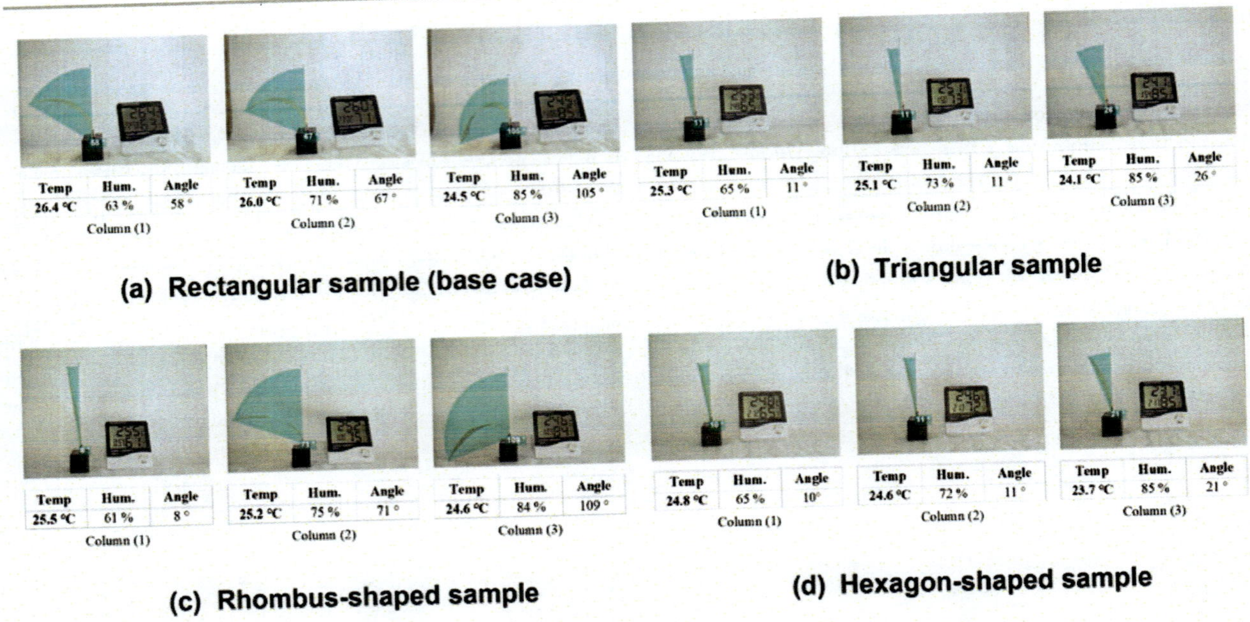

Fig. 12 Response of four Beech veneer samples with varying shape geometries to three conditions of relative humidity

5 Discussion

The experiment results demonstrate interesting findings related to the impact of each of the ten identified parameters on the wood response behavior and motion type. The following findings can be asserted for each of the parameters. For the material-type parameter (wood type), it was observed that the most significant factor affecting the resulting morphology was the physical and mechanical properties for each of the hardwood and softwood samples. It was also observed that the highest deflection occurs with tangential grain orientation. With respect to sample thickness, it was observed that deflection is significantly higher with thinner samples. Out of the ten experiments, the most distinctive motion type was observed with the slope of grain experiment, where the material was shown to afford both 2D and 3D bending based on the variations of grain angle.

The use of artificial (3D printed) wood as a parameter revealed interesting findings in terms of control and regulation. The similarity with the results of natural materials with comparable conditions indicates high possibilities of integrating artificial materials in such a process that could potentially overcome deficiencies in natural materials related to durability, stiffness, as well as enduring high temperatures due to the polymer content in the 3D printed wood filament. The concept of regulation and control was further accentuated by the use of cross-lamination, where different combinations of active and passive layers could be utilized with varying thicknesses (and possibly in conjunction with other parameters) to achieve the desired motion type speed and behavior.

Varying the dimensional proportions of the tested samples showed a significant difference between squarish ratios versus more rectangular or linear sample ratios, where deflection increases with higher ratios. The temperature was also observed to act as a catalyst together with increased humidity, as it was demonstrated to increase porous areas, therefore, allowing for more moisture gain and consequently enabling higher deflection and widening the solution space for more morphological transformations. The perforation parameter was also shown to affect the deflection angle due to the saturation of moisture propagation by virtue of the gaps created in between the wood fibers. Regarding shape geometry, it was shown that different geometries highly affected the nature of the deflection, where the narrower the angle, the faster the deflection.

The added value of these findings, however, does not lie solely in the independent results of each experiment, but rather in the collective and interrelations using combinations and permutations of the parameters altogether. As the main premise of this paper is addressing the potential of achieving a variety of shape morphologies for adaptive building façade skins, the merit of addressing the identified ten parameters and their relations in totality allows for an extensive dataset of morphological behavior. The main benefits of this dataset include: (1) an extended vocabulary of transformations, including 2D and 3D bending and (2) a higher ability to consciously control and regulate morphological behavior and deflection angle with a high degree of precision.

Future works based on these findings aim to address aspects related to the sustainable implementation of hygromorphic adaptive façades (Holstov et al., 2015) such as fatigue, deterioration, and upscaling, by means of the strategic combination of the identified parameters. For example, sample thickness introduces both an opportunity and a challenge, where upscaling of building façade prototypes becomes cumbersome with higher thicknesses. However, this can be mitigated by using thicker prototypes for stiffness and rigidity while introducing an adequate percentage of perforations for instance to counterbalance the fragile nature of thin samples while maintaining reasonable deflection. Likewise, artificial wood can be used in some cases where large-scale skins could possibly wear out easily were natural materials in place. Lamination could be used, given the adequate thickness for active and passive layers, to achieve specific angles of deflection, as opposed to single-layer prototypes. Using temperature as a catalyst for augmenting the capacity of the material to absorb more moisture content and further its capability of deflection is yet another example.

6 Conclusion

In this paper, we discussed the testing and identification of flexible morphologies induced by actuating wood as a natural programmable material that responds to varying environmental conditions, specifically its hygroscopic properties in response to variations in humidity levels. We conducted an experimental methodology to track and monitor the material behavior using physical testing, in terms of the nature of its motion response and deflection angles, using Kinovea video tracking and motion-sensing technology. We identified ten key parameters that were demonstrated to directly affect material motion and recorded the maximum deflection angles of different samples under three relative humidity conditions. These parameters included material type, wood thickness, cut direction and grain orientation, the slope of grain, artificial wood, multi-layered lamination, dimensional proportions, temperature, perforations, and shape geometry. Findings included identifying an extended vocabulary of transformations, including 2D and 3D bending, as a result of combining different parameters collectively, and a higher ability to consciously regulate morphological behavior and deflection angle. Future works will use combinations and permutations of these parameters in a parametric environment to identify a palette of architectural morphologies for programmable passive actuators that address the sustainable implementation of hygromorphic adaptive façades.

References

Abdelmohsen, S., Adriaenssens, S., El-Dabaa, R., Gabriele, S., Olivieri, L., & Teresi, L. (2019). A multi-physics approach for modeling hygroscopic behavior in wood low-tech architectural adaptive systems. *Computer-Aided Design, 106*, 43–53. https://doi.org/10.1016/j.cad.2018.07.005

Abdelmohsen, S., Massoud, P., El-Dabaa, R., Ibrahim, A., & Mokbel. T. (2018). A computational method for tracking the hygroscopic motion of wood to develop adaptive architectural skins. In A. Kępczyńska-Walczak & S. Białkowski (Eds.), *Computing for a better tomorrow—Proceedings of the 36th International Conference on Education and Research in Computer Aided Architectural Design in Europe* (1st ed., pp. 253–262). eCAADe (Education and Research in Computer Aided Architectural Design in Europe) and Faculty of Civil Engineering, Architecture and Environmental Engineering, Łódź University of Technology.

Adnan, N., Patar, M., Lee, H., Yamamoto, S.-I., Jong-Young, L., & Mahmud., J. (2018). Biomechanical analysis using Kinovea for sports application. In *IOP Conference Series: Materials Science and Engineering, 342*, IOP Publishing Ltd.

Anis, M. (2019). Responsive skin: Designing an adaptive building envelope for warm-humid climate with bamboo veneer as a hygroscopically active material. In C. Jarrett, P. Plowright & H. Rashed-Ali (Eds.), *The future of praxis: Applied research as a bridge between theory and practice—ARCC 2019 International Conference* (1st ed., pp. 338–347). https://www.arcc-journal.org/index.php/repository/issue/view/27

Brauns, J., & Rocens, K. (1997). Hygromechanical behaviour of wooden composites. *Wood Science and Technology, 31*, 193–204. https://doi.org/10.1007/bf00705885

Cave, I. (1978). Modelling moisture-related mechanical properties of wood Part I: Properties of the wood constituents. *Wood Science and Technology, 12*, 75–86. https://doi.org/10.1007/bf00390012

Correa, D., Papadopoulou, A., Guberan, C., Jhaveri, N., Reichert, S., Menges, A., & Tibbits, S. (2015). 3D-printed wood: Programming hygroscopic material transformations. *3D Printing and Additive Manufacturing, 2*(3), 106–116. https://doi.org/10.1089/3dp.2015.0022

Dierichs, K., & Menges, A. (2016). Towards an aggregate architecture: designed granular systems as programmable matter in architecture. *Granular Matter, 18*(25), 1–14. https://doi.org/10.1007/s10035-016-0631-3

El-Dabaa., R., Abdelmohsen, S., & Mansour, Y. (2021). Programmable passive actuation for adaptive building façade design using hygroscopic properties of wood. *Wood Material Science & Engineering, 16* (4), 246–259. https://doi.org/10.1080/17480272.2020.1713885

Forest Products Laboratory. (2010). *Wood handbook—Wood as an engineering material* (Centenntial ed.). U.S. Department of Agriculture, Forest Service, Forest Products Laboratory.

Hering, S., Keunecke, D., & Niemz, P. (2012). Moisture-dependent orthotropic elasticity of beech wood. *Wood Science and Technology, 46*(5), 927–938. https://doi.org/10.1007/s00226-011-0449-4

Hisham, N., Nazri, A., Madete, J., Herawati, L., & Mahmud, J. (2017). *Measuring ankle angle and analysis of walking gait using Kinovea.* International Medical Device and Technology Conference, pp. 247–250.

Hoadley, R. (2000). *Understanding wood: A craftsman's guide to wood Technology* (1st ed.). The Taunton Press.

Holstov, A., Morris, P., Farmer, G., & Bridgens, B. (2015). *Towards sustainable adaptive building skins with embedded hygromorphic responsiveness.* 10th Conference on Advanced Building Skins. Graz University of Technology. http://buildingskins.tugraz.at/home

Holstov, A., Farmer, G., & Bridgens, B. (2016). *Implementing hygromorphic wood composites into responsive building skins.*

11th Conference on Advanced Building Skins. Newcastle University. https://eprints.ncl.ac.uk/229088

Krieg, O., Christian, Z., Correa, D., Menges, A., Reichert, S., Rinderspacher, K., & Schwinn, T. (2017). HygroSkin—Meteorosensitive Pavilion. In S. Langenberg, F. Gramazio & M. Kohler (Eds.), *Fabricate 2014: Negotiating design & making* (1st ed., pp. 272–279). UCL Press. https://doi.org/10.2307/j.ctt1tp3c5w.37

Reichert, S., Menges, A., & Correa, D. (2015). Meteorosensitive architecture: Biomimetic building skins based on materially embedded and hygroscopically enabled responsiveness. *Computer-Aided Design, 60*, 50–69. https://doi.org/10.1016/j.cad.2014.02.010

Rüggeberg, M., & Burgert, I. (2015). Bio-inspired wooden actuators for large scale applications. *PLOS ONE, 10*(3), e0120718. https://doi.org/10.1371/journal.pone.0120718

U.S. Department of Agriculture. (2007). *The Encyclopedia of wood*. Skyhorse Publishing, Inc. https://www.skyhorsepublishing.com/9781602390577/the-encyclopedia-of-wood/

Vazquez, E., Gursoy, B., & Duarte, J. (2020). Formalizing shape-change: Three-dimensional printed shapes and hygroscopic material transformations. *International Journal of Architectural Computing, 18*(1), 67–83. https://doi.org/10.1177/1478077119895216

Wood, D., Correa, D., Krieg, O., & Menges, A. (2016). Material computation—4D timber construction: Towards building-scale hygroscopic actuated, self-constructing timber surfaces. *International Journal of Architectural Computing, 14*(1), 49–62. https://doi.org/10.1177/1478077115625522

Analyzing Public Transport System in Twin Cities of Rawalpindi, Islamabad

Abdul Basit Sheikh

Abstract

Cities are growing at alarming pace making huge burden on city planners and governments. Uncontrolled growth of cities is making urban land use unsustainable economically and environmentally. In transportation section, unprecedented growth of motorized traffic causes traffic congestion, time delays and pollution. Shift toward sustainable transportation system, i.e., improvement of public transport, development of Mass Transit Routes, cycle and walk trips-oriented road network is the need of the hour. Mostly public transport system constitutes a small portion of motorized traffic on city road network. In many developed and under developed cities, development of Mass Transit System is initiated to shift vehicular traffic to public transport. Mass Transit System only works when it is developed and operated as a whole and not as a part. Operations of one or two lines do not give benefits of modal shift, congestion relief and journey time saving in a city of more than 5 Million populations. Construction of complete Mass Transit System (MTS) including operations, integration of existing public transport routes and development of feeder routes is a gigantic task and in most of the cities could not be developed in single go. In this paper, MTS development in large cities from scratch is studied with following purpose and methodology.

- Purpose of the study is to analyze development, construction and operations of Mass Transit System (MTS) in old and large cities without MTS and to check whether this development is feasible and sustainable.
- Basic design of the study is to investigate different steps and phases involved in development of MTS in Twin Cities of Rawalpindi/Islamabad. For reference and comparison, development of MTS in cities of Lahore, Peshawar and Karachi is presented also.
- Major findings are that the development of Mass Transit System (MTS) in old and large cities which grew without MTS is a gigantic job requiring continuous Government will, allocation of funds and political stability. These things generally lack in most of under developed countries which in the long run jeopardizes the development of MTS projects in development phase.
- Instead the author recommends the idea of constructing new cities along peripheral of old cities which is more feasible and sustainable.

Keywords

Mass Transit • Ridership • Congestion relief • Public transport • Transit oriented • Mobility • Sustainable development • New cities • Social cost • Economic cost • Financial cost

1 Introduction

Increasing motorized traffic, in urban areas of Pakistan are calling for integrated transport planning and road network. Population growth and increase in vehicle ownership together with migration of people from rural areas to urban areas are posing serious challenges to traffic planners. The need for traffic planning of big cities is increasing world over. “In 1850, there were four cities in the world with more than 1 million people, and in 1950, there were about a hundred cities of this size. At present, there are about 400 cities of this magnitude” (Jotin & Kent Lall, 2002). In recent years, integrated public transport system along with Mass Transit Routes is being planned and executed in large cities of Pakistan to cope increasing traffic jams and delays on major roads of these cities. Cities, where Mass Transit System (MTS) is being initiated, are Lahore, Karachi, Peshawar,

A. B. Sheikh (✉)
NESPAK, Lahore, Pakistan
e-mail: abdulbasit_sheikh66@yahoo.com

H. Altan et al. (eds.), *Advances in Architecture, Engineering and Technology*,
Advances in Science, Technology & Innovation, https://doi.org/10.1007/978-3-031-11232-4_6

Twin cities of Rawalpindi-Islamabad and Multan. This study is aimed at analyzing different stages of development of MTS in Twin Cities of Rawalpindi and Islamabad and to see whether introduction of MTS in a belated stage when a city has grown is feasible and sustainable or not.

Rawalpindi and Islamabad are termed Twin cities as the two cities adjoin each other. Islamabad is the 9th largest city of Pakistan and is country capital. Islamabad is a modern and well planned city located in northeastern part of the country within Federal Territory. The city was built during 1960s as new capital of the country. Population of city has grown from 100,000 in 1951 to 1,009,832 in 2017 (PBS, 2018). Rawalpindi is located in the region of Punjab near Islamabad and is the 3rd most populous city of Punjab after Lahore and Faisalabad and is also the Military Headquarter of Pakistan Armed Forces. Population of city has grown to 2,875,516 in 2017 (PBS, 2018). Rawalpindi and Islamabad residents commute in both cities for work every day. Growing motorization coupled with inadequate planning and poor public transport has created acute traffic jams on almost all major roads during peak hours. City governments of twin cities have decided to introduce MTS as a sustainable transport solution to cope with severe traffic congestion in 2012–2013. Planning, designing and construction of Mass Transit System started simultaneously and are underway till now. Out of four priority corridors, one BRT line has been built and in operations since 2013. Construction of 2nd BRT line is in planning process but getting delayed due to other priority sectors under consideration by the Government.

Research Novelty Most of the researches in transportation planning, congestion and mobility is on issues concerning construction of mass transit infrastructure and transit oriented development, whereas this study opens new avenues of research for building new cities around old and unplanned cities for sustainable transportation solutions. Therefore, this research leads to a new approach of transport planning and solving mobility issues of big cities.

1.1 Problem Statement

Transit-oriented development (TOD) standard calls for urban development that enables high-quality car-free lifestyle with high-rise buildings within a walking distance of MTS stations to support use of public transport, walking and cycling as transport modes. Construction of MTS infrastructure in large and congested cities to achieve TOD goals at belated stage when cities already grew without a Mass Transit System is a gigantic job, involving huge public money, large-scale demolishing of built up areas, resettlement issues, construction of viaduct in the middle of city landscape and environment hazards.

1.2 Objective of the Study

This study was carried out to achieve following objectives.

- To analyze pros and cons of development of MTS in old big cities at belated stage when a city has already grown without a Mass Transit System.
- To carry out traffic impact and congestion relief analysis of MTS on city road network to check its feasibility and sustainability.
- To conduct a comparative study of development of different MTS in big cities of Pakistan.
- To asses and propose sustainable public transport system for city planners.

1.3 Methodology

For this research paper, development of MTS in Twin Cities of Rawalpindi/Islamabad is selected as a case study. The methodology adopted mainly consisted of milestones which have been briefed here under.

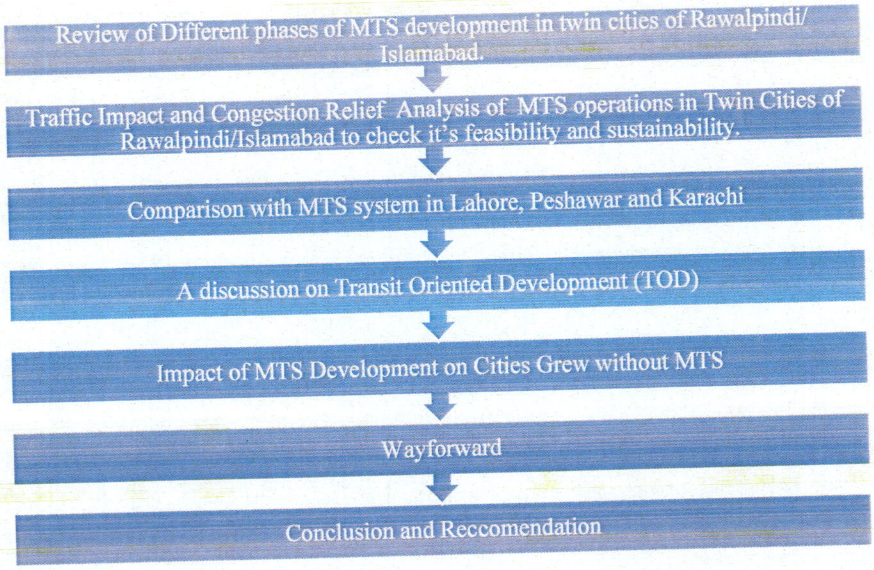

2 Different Phases of MTS in Twin Cities of Rawalpindi/Islamabad

2.1 Planning Phase

Planning of MTS in twin cities of Rawalpindi/Islamabad started in 2012 with pre-feasibility and feasibility studies. The original project of pre-feasibility study (PFS) funded by "Cities Development Initiative for Asia" (CDIA-ADB) was Islamabad focused. Rawalpindi was not in the original scope. However, when the consultant team was examining demand and travel patterns, etc., it became obvious that a large commuting market from Rawalpindi to Islamabad could be captured and improve the overall financial position of the project. The relevant representations were made to the Government of Punjab in Rawalpindi for the route to be extended into Rawalpindi. Later in a study (PMA, 2015) conducted by Punjab Metro Authority (PMA), four Bus Rapid Transit (BRT) lines were envisaged. Corridors of these lines are presented in (Fig. 1). Transit demand indicators, i.e., ridership, passenger kilometer per day and passenger kilometer per bus kilometer are given in Table 1.

2.2 Construction Phase

First priority line, i.e., Red Line was built in 2015 with a construction cost of Rs. 44 Billion and is in operation since then. Status of other lines is presented in Table 2.

2.3 Impact of Feeder Routes and Fare on BRT Ridership

Traffic impact and congestion relief analysis of MTS developed as a whole are carried out to evaluate benefits which this system gives to the city. Indicators presented in

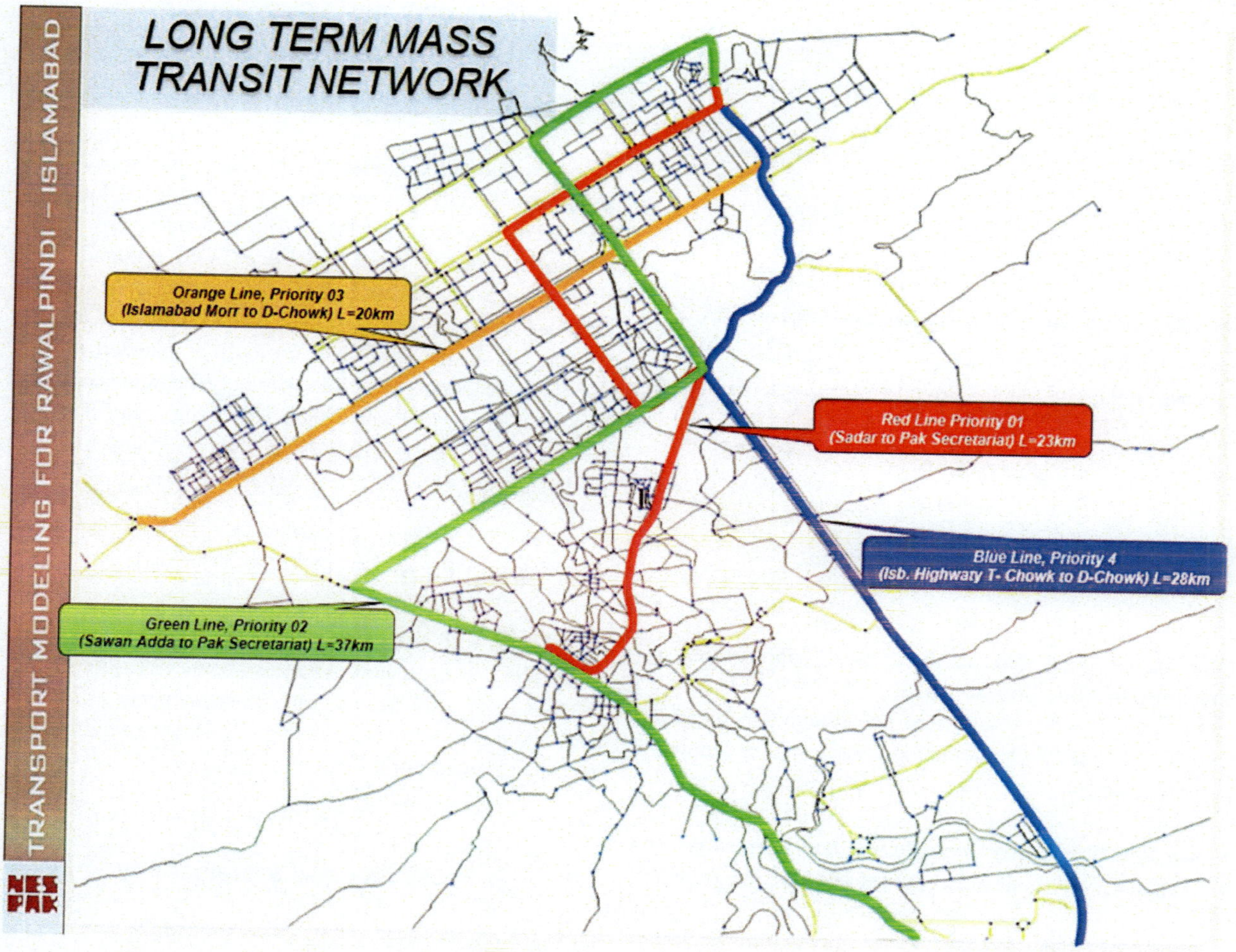

Fig. 1 Proposed Mass transit network (PMA, 2015) in Twin cities of Rawalpindi/Islamabad

Table 1 Long-term mass transit network demand indicators (PMA, 2015)

Priority	Line	Length	Ridership (pass/day) year 2014	Pass km/day (year 2014)	Pass km/ Bus km
1	PMA-Scenario: 2-BRT-01 (Red)	23	195,452	1,440,319	65
2	PMA-Scenario: 2-BRT-2 (Green)	37	126,876	886,953	25
4	PMA-Scenario: 2-BRT-3 (Blue)	28	71,731	533,847	20
3	PMA-Scenario: 2-BRT-4 (Orange)	20	87,993	470,527	25

Table 2 MTS priority lines' present status

BRT lines	Length (km)	Estimated cost Rs. (Million)	Actual cost Rs. (Million)	Status
PMA-Scenario: 2-BRT-01 (Red)	23	40,000	44,000	Operative since 2015
PMA-Scenario: 2-BRT-2 (Green)	37	58,081	–	Not started yet
PMA-Scenario: 2-BRT-4 (Orange)	20	24,419	–	Under construction
PMA-Scenario: 2-BRT-3 (Blue)	28	41,920	–	Not started yet

Table 3 Red Line Ridership with and without Feeder Routes (PMA, 2015)

Red line ridership	Year 2014–2015
Ridership without feeder routes/ fare integration	1,35,005 passengers/day
Ridership with feeder routes/ fare integration	2,19,500 passengers/day
Load	5,743 pphpd

Table 4 Impact of fare increase or decrease on ridership (PMA, 2015)

Impact of fare increase on ridership	
Fare rate (in Rs)	Percentage decrease in ridership (%)
20	0.0
30	8.03
40	10.68
50	13.51
100	28.34

following sections also highlight the feasibility and sustainability of Mass Transit System.

Impact of Feeder Routes and fare Integration on Ridership of Red Line (In Operations) was evaluated in detail with two Scenarios.

- Scenario: 1-With feeder routes and fare integration
- Scenario: 2-Without feeder routes and fare integration.

Impact of Feeder Routes and fare Integration on Ridership in passengers/day is presented in Table 3. Impact of fare increase or decrease on ridership is given in Table 4.

3 Traffic Impact and Congestion Relief Analysis

3.1 Volume Capacity Ratio Analysis

In the transportation model (Sheikh, 2015) developed by the author for the city government, all the proposed four BRT lines were modeled and analyzed for traffic congestion relief and improvement in Level of Service (LOS). Figure 2 shows road network of year 2035 with volume capacity ratios. Red bars are showing roads with near to capacity or over capacity. If all the BRT lines are built as planned till year 2035, traffic congestion will definitely be reduced on all major roads of the Twin Cities. This traffic relief is shown graphically in Fig. 3 with less red bar roads and bluer bar roads showing good Level of Service (LOS).

3.2 Modal Shift

Modal shift with MTS when all BRT lines are in operations (Year 2035) is estimated. Analysis shows 13% motorcycles (Sheikh, 2015), 8% cars and 80% bus passengers will shift to BRT, giving relief to traffic congestion on mix traffic lanes and environment pollution.

3.3 MTS Feasibility and Sustainability

Sections 2.3, 3.1 and 3.2 show that if all the four MTS line are built, it will give congestion relief on all major roads with a modal shift of motorized trips to public transport. This

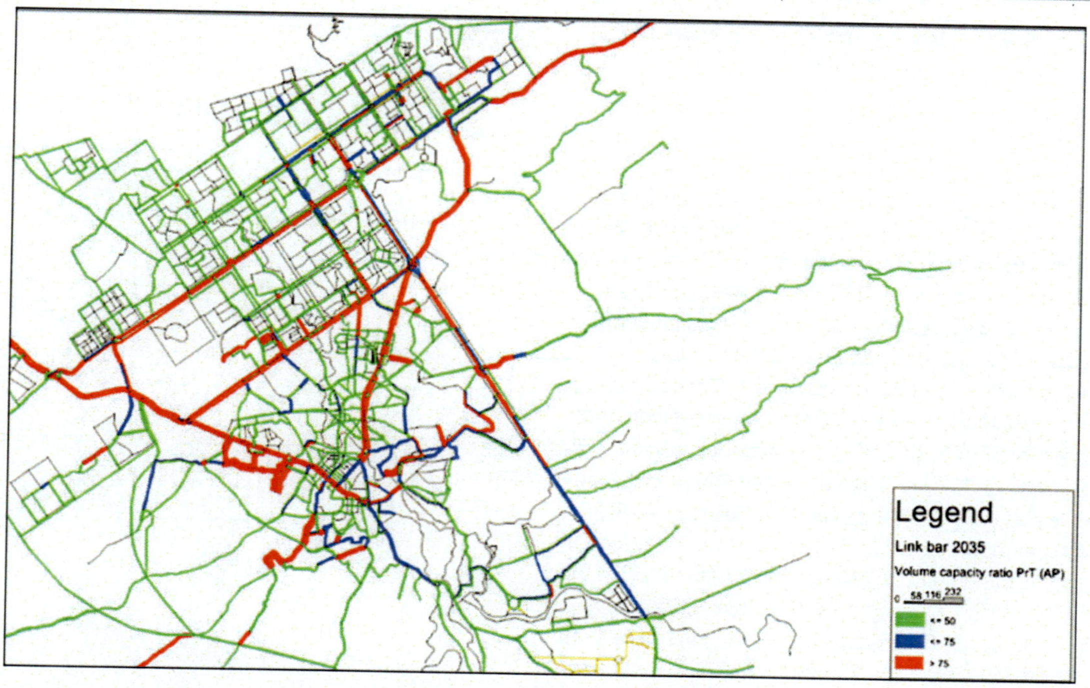

Fig. 2 Volume capacity ratios of city road network without MTS (without BRT Lines)

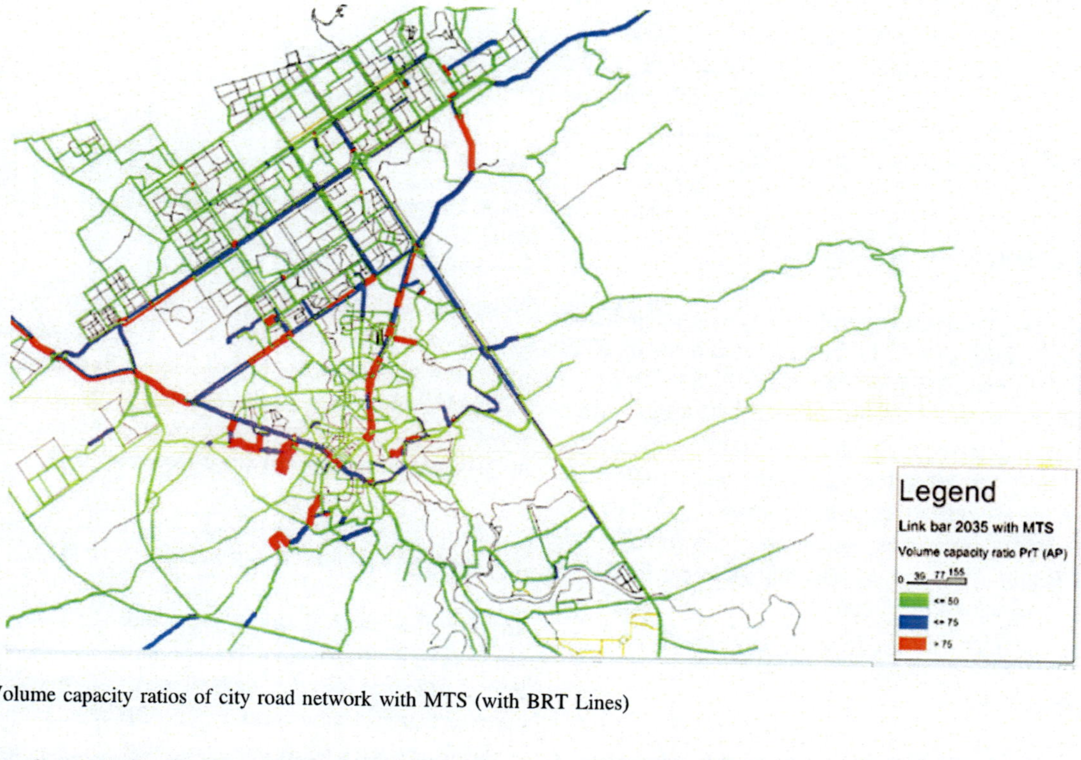

Fig. 3 Volume capacity ratios of city road network with MTS (with BRT Lines)

reduction of motorized trips in return reduces traffic jams, air pollution and noise. Saving in vehicle operating cost (VOC) and saving in value of time (VOT) cost makes all MTS projects economically feasible and sustainable. This saving is only when MTS is "**built as a whole**" and not in parts. Building of one or two lines will not give that ridership and congestion relief which is estimated for a whole integrated Mass Transit System.

4 Comparison with Under-Developing MTS in Lahore, Peshawar and Karachi

4.1 Lahore

Lahore is the 2nd largest city of Pakistan after Karachi with a population of 11,126,285 (PBS, 2018). In 1991, to cope with future public transport demand, JICA (1991) proposed a light rail transit system to city Government with more comfort, capacity and speed than the current bus service. The proposed LRT was of medium capacity Mass Transit System with elevated track structure. The project could not be materialized because of lack of funds, Government initiatives and political will. In 2012, transport department initiated the study of Lahore Urban Master Plan. The study proposed eight priority Mass Transit corridors shown in Fig. 4 with completion plan till 2020. Based on recommendations of Lahore Urban Master Plan (JICA, 2012b) out of proposed eight (08) priority corridors, Green Line project was initiated in 2012. Green Line, it was decided to be built as BRT and not as LRT to save money. BRT Green Line was built with infrastructure cost of Rs. 30 Billion and initial ridership of around 140,000 passengers/day. Despite demand of other corridors, Government could not initiate other MTS. In 2014, 2nd MTS priority corridor of Orange Line was initiated when China's Exim Bank agreed to provide a soft loan of $1.55 Billion. In Oct 2020 recently Light Rail Orange Line project is completed with an infrastructure cost of Rs. 187 Billion. Orange Line has started its operations with estimated ridership of 250,000 passengers/day.

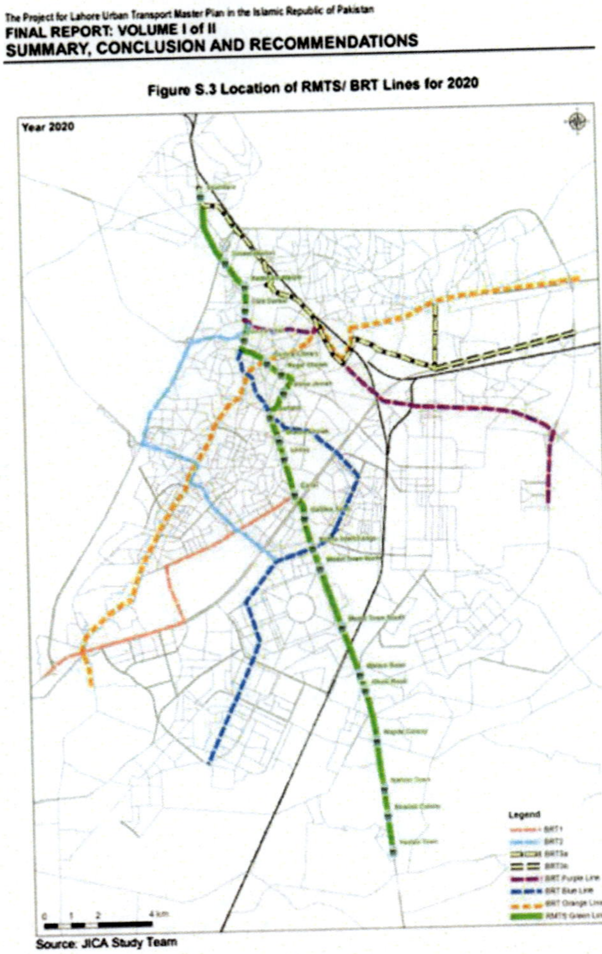

Fig. 4 Proposed mass transit lines for city of Lahore. *Source* JICA (2012b)

4.2 Peshawar

Peshawar is the 6th largest city of Pakistan with a population of 1,970,042 (PBS, 2018). In 2103, Government of KPK initiated to develop MTS in the city. In urban transport pre-feasibility study, (CDIA-ADB, 2014) seven Mass Transit Corridors were proposed. These corridors are shown in Fig. 5. In 2020, recently three integrated corridors are constructed and operative with Ridership of 110,000 passengers/day. Construction cost of these lines comes out to be Rs. 71 Billion. It is expected that other four corridors will be built in near future.

4.3 Karachi

Karachi is the largest city of Pakistan with a population of 14,916,456 (PBS, 2018). Karachi Metropolitan Corporation initiated a study "Karachi Transportation Improvement Project" in 2012. The study was aimed to revive Karachi Circular Rail (not operational since 1999) as Mass Transit Route and development of Mass Transit System in the city. Eight MTS lines were proposed in the study (JICA, 2012a) which are shown in Fig. 6. Green Line project was initiated in 2014 and is operational since December 2021.

5 Transit-Oriented Development (TOD)

The transit-oriented development standard (ITDP, 2013) is an effort by the Institute for Transportation and Development Policy (ITDP) for the development of transit-oriented car-free transport network. Gist of TOD standard published by ITDP is presented in following text:

"The TOD standard is based on ITDP's principles of transport in urban life:

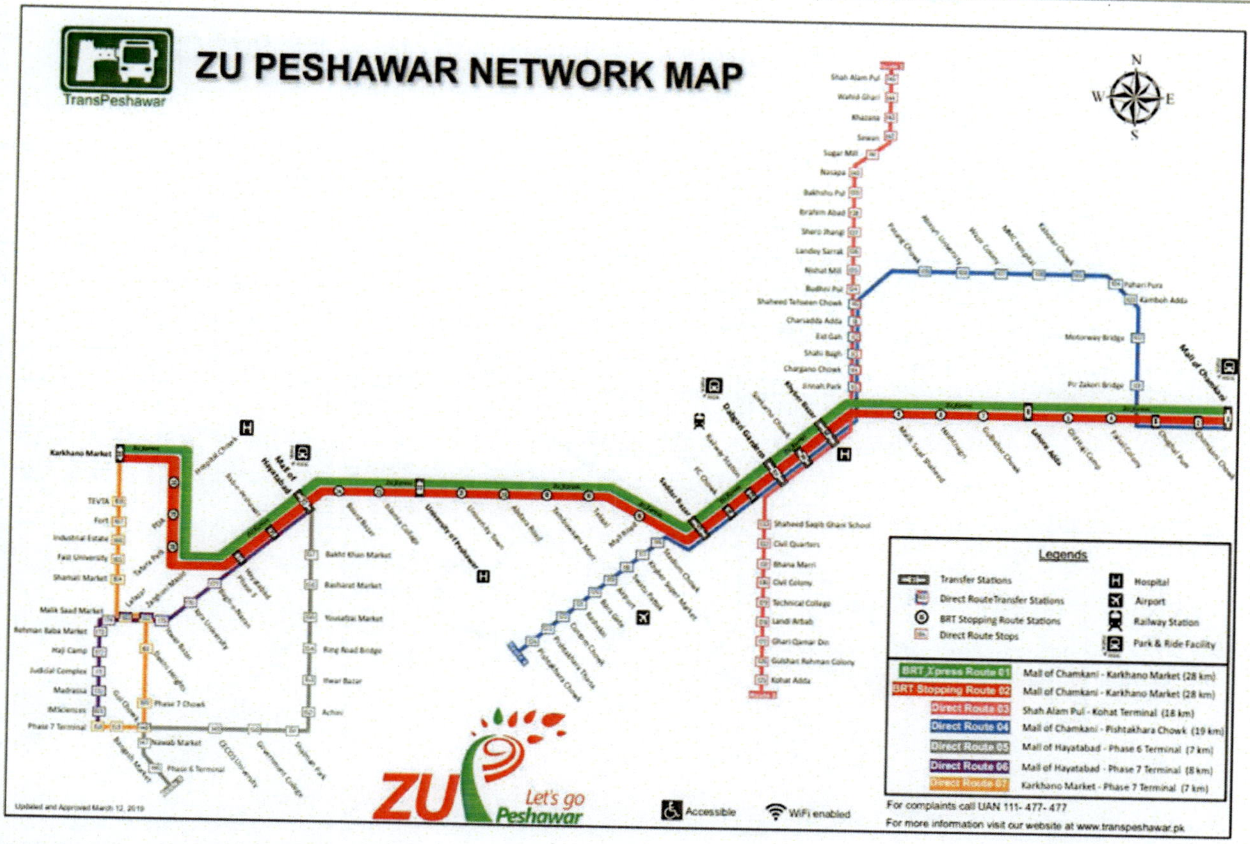

Fig. 5 Proposed mass transit lines for Peshawar. *Source* CDIA-ADB (2014)

1. Develop neighborhoods that promote walking **[walk]**
2. Prioritize non-motorized transport networks **[cycle]**
3. Create dense networks of streets and paths **[connect]**
4. Locate development near high-quality public transport **[transit]**
5. Plan for mixed use **[mix]**
6. Optimize density and transit capacity **[densify]**
7. Create regions with short commutes **[compact]**
8. Increase mobility by regulating parking and road use **[shift]**"
   (ITDP, 2013).

An overview of TOD principals highlights the need of transit oriented development of transport network. The standard calls for urban development that enable high-quality car-free lifestyle with high-rise buildings within a walking distance of MTS stations, to support use of public transport, walking and cycling as transport modes. All the eight (08) principals, i.e., walk, cycle, connect, transit, mix, densify, compact and shift mentioned above seem possible for small and new planned cities with planned zones and transport networks. For old and large cities, it is extremely difficult to achieve these goals without large-scale demolishing, resettlement and infrastructure cost.

6 Development of MTS in Large Cities

In the above Sects. 2, 3, and 4, the author presented the case of development of MTS in Twin Cities of Rawalpindi/Islamabad in detail and a brief history of MTS development phases in Lahore, Peshawar and Karachi. In Sect. 5, summary of TOD standard is presented which calls for car-free lifestyles. The above discourse clearly exhibits that despite 20–30 years of struggle to build MTS in aged big cities of Pakistan, only one or two lines could be built. Building of one or two lines does not give that ridership and mobility which is forecasted in an integrated system of 7 or 8 lines. Building of seven or eight lines is a continuous effort which requires political will, money and continuity of government policies. The other point is whether it is cost effective and environmental friendly to change landscape of a city as old as 100 years, almost at 180 degree turn as TOD envisages to

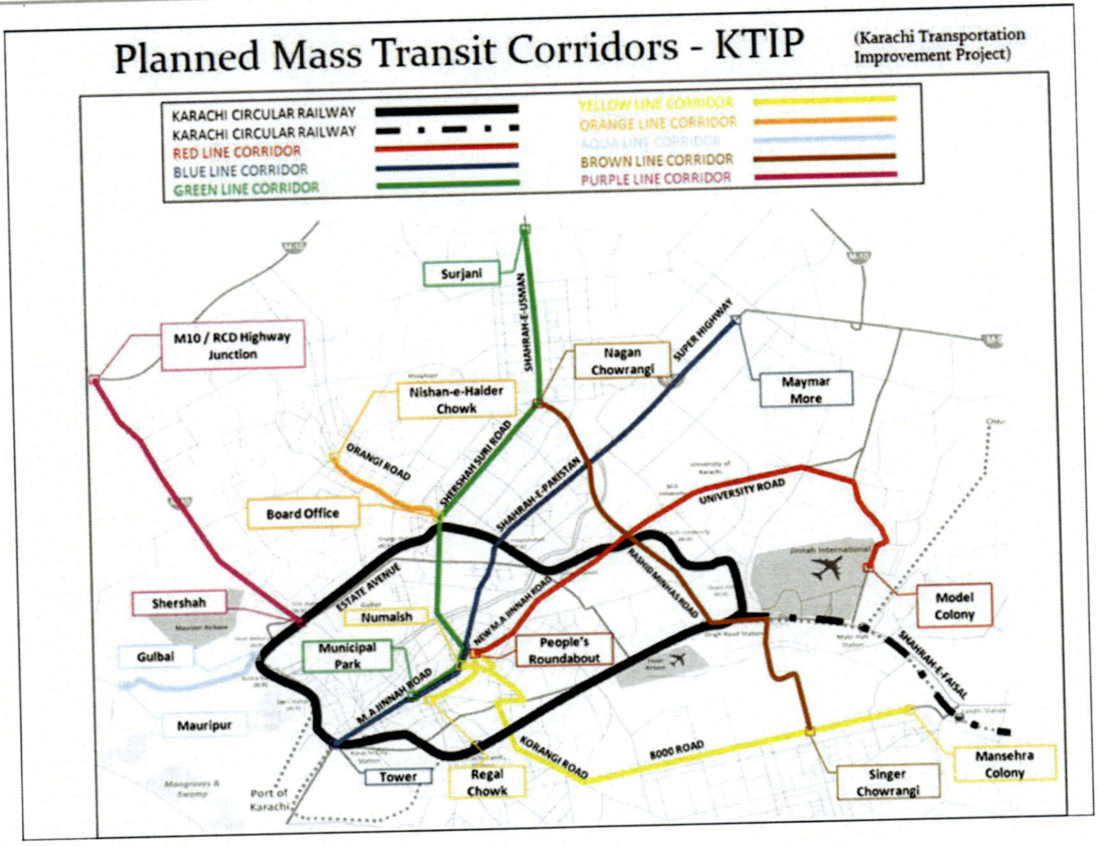

Fig. 6 Proposed mass transit lines for Karachi. *Source* JICA (2012a)

build high-rise buildings and MTS corridors on congested streets. The examples of MTS development in different cities mentioned above clearly show that in spite of Government will MTS could not be developed. Only one or two lines started operations in almost 20 years of planning and design. The reasons behind this slow development can be summarized as:

- To develop a complete MTS in an aged city from scratch, needs 30–40 years.
- In most of the cases, Government will and availability of fund is not possible for such a long time.
- Rich countries can only afford high infrastructure and operational costs, not possible for poor countries.
- Environmental hazards, as lot of buildings along MTS corridor and on stations need to be demolished to clear right of way.
- Construction of viaducts, elevated structures and tunnels in congested streets take many years to construct, hampering city economic activity and sometimes gives ugly face to city land scape like in Rawalpindi, Lahore, Karachi and Peshawar.
- Heavy reinforced structure closes the doors of future improvements and development.
- Successful operation of MTS requires Government commitment and continuation of policies regarding MTS.

6.1 Land Use Planning

Regarding the land use planning, this was largely not in the minds of Government until the 50s and 60s. The inputs by the architect and town planner Doxiadis were the basis of Islamabad design, and some of the areas of north west Karachi city. Prior to that, the lack of congestion and no specific land use planning skills by Government; city planners did not provide road networks based on future forecasts. Whereas, the transport modeling undertaken today for new cities is focused on the future and based on state-of-the-art tools.

7 Way Forward

Construction of MTS in aged cities grew without MTS is a gigantic task and generally if MTS is developed in 20–30 years, rise of population in 30 years will counter the benefits of MTS. It is also not easy to change city culture from car oriented to car-free society and to build high-rise buildings to facilitate walk trips and cycle trips. Other ways and means such as (i) building new cities with planned MTS, (ii) transforming the already grown cities to MTS effective in a gradual way should be sought for alternate solutions. A brief description is given below.

7.1 Development of New Cities

Development of new cities by kings and ancient rulers gives the idea that new cities shall be planned and developed. Construction of big towns by Estate Builders and Property Tycoons gives the clue that if one man can build city like towns why government could not build new cities. Instead of demolishing and then building a city with huge cost of MTS network, why not new cities are build based on TOD standards and principals. These new cities can be planned least at 30 km distance away from old cities. Islamabad as a new city adjacent to Rawalpindi is a practical example. Had Islamabad not been built in 1960s as a planned city neighboring Rawalpindi, the demographic and transportation profile of Rawalpindi would have been in shambles, giving picture of 4 to 5 Million densely populated city. Development of a new planned city Islamabad adjacent to Rawalpindi gave relief to the residents of old city and its environment. The pros and cons of building new cities are a separate research topic, the author briefly presents benefits of building new cities as follows.

- No cost of demolishing, land acquisition in built up area and utility shifting for MTS corridors.
- No damage to environment and no litigation.
- State of the art planning, i.e., at grade signalized intersections, at grade MTS dedicated corridors with signal priority and cycle-oriented road networks.
- More rural area enters into urban sprawl.
- Cheap government and open land available in rural areas for new city.
- Will decrease population growth of parent city and traffic congestion.
- It almost takes 10–20 years to develop new city from scratch, easy task for government to embark upon.
- It is free from big city problems.

Development of new cities in green spaces is definitely an option. In Malaysia, the creation of Putra Jaya for the Government workforce and the current wish of President Joko Widowo of Indonesia to create a new Capital in the province of Kalimantan in an attempt to deal with the overburdens of Jakarta are examples of this.

7.2 Gradual Development of MTS in Old Cities

Despite our discourse regarding the challenges experienced when developing an MTS for an aged city with narrow roads is correct, however, for the residents of these cities, they deserve ongoing improvements as the decades pass and technology. This is why subways are sometimes implemented. The problem with subways is the cost. Very expensive and from a political point of view, they are largely out of sight—when compared with an elevated structure as per the LRT in Lahore and the BRT on Murree Road, Rawalpindi.

Although it is very difficult to develop gigantic MTS system in aged cities with narrow roads, but by experience it has been learnt that with proper methodology allocation of funds and will, MTS lines can be provided to inhabitants of old cities as well. It is appropriate to start with the lines for corridors with sufficient ROW less property acquisition to be demolished. This will create less environmental hazard and public unrest. The methodology should cater for development of high-rise buildings in commercial manner for residential purpose as well as for boosting up the trade along the MTS line. In the beginning, some displacement of residents may be a question, but the same can be addressed by paying the market rates of land and property to affected inhabitants. It is observed that after providing MTS line to one part of the city, the inhabitants of other parts also like to have the same facility in other parts of the city. The comfort provided to locals will gradually overcome the public displacement effects. It is emphasized that the construction works should be carried out in well planned manner with available funds and using latest equipment and machinery without hampering the city environment and targeted schedule of completion.

8 Conclusion and Recommendations

The above discourse will help to develop a mindset of sustainable urban development practices into the future. If Government and the private sector can work together in a transparent manner and with the principles of sustainability of the urban centers, then we can look forward to improvements over the next decades. Based on the studies carried out following conclusions and recommendations have been accomplished.

8.1 Conclusions

Major results and findings

Major results and findings obtained from this study are summarized as:

- To develop a complete MTS system in a city from scratch needs 30 to 40 years. It is environmental hazardous as well as lot of buildings along MTS corridor need to be demolished which creates lot of unrest amongst the citizens.
- Construction of viaducts, elevated structures and tunnels in congested streets take many years to construct, hampering city economic activity and sometimes gives ugly face to land scape.
- Heavy reinforced structures closed the doors of future improvements and development.
- In many developed and under developed cities, development of Mass Transit System is initiated to shift vehicular traffic to public transport. Mass Transit System only works when it is developed and operated as a whole and not as a part. Operations of one or two lines do not give benefits of modal shift, congestion relief and journey time saving in a city of more than 5 Million populations. Construction of complete MTS including operations, integration of existing public transport routes and development of feeder routes is a gigantic task and in most of the cities could not be developed in single go.
- Huge infrastructure cost and operation cost of MTS is not bearable as for as poor countries are concerned.
- Development of MTS in old and large cities require continuous Government will, allocation of funds and political stability. These things generally lack in most of under developed countries which in the long run jeopardizes the development of MTS projects during development phase
- All the eight (08) principals of transit-oriented development (TOD), i.e., walk, cycle, connect, transit, mix, densify, compact and shift are possible for small and new planned cities with planned zones and transport networks. For old and large cities, it is extremely difficult to achieve these goals without large-scale demolishing, resettlement and infrastructure cost.
- It is not cost effective and environmental friendly to change landscape of a city as old as 100 years, almost at 180 degree turn as TOD envisages to build high-rise buildings and MTS corridors on congested streets
- Instead of demolishing and then building/converting cities with huge cost to TOD standards and with MTS network, new cities shall be built based on TOD standards and principals
- Development of a new planned city Islamabad adjacent to Rawalpindi gave relief to the residents of old city Rawalpindi and its environment.

Advantages of building new cities

Advantages of building new planned cities instead of converting existing cities to TOD standards are summarized as

- No land acquisition, resettlement and environment issues.
- More rural area enters into urban sprawl.
- Cheap government/private land available in rural areas for new cities.
- Attractive for investors
- Will decrease population growth of parent city and traffic congestion.
- It takes 10–20 years to develop new city from scratch, easy task for government.
- New city is free from big city problems.

8.2 Future Development of This Research Work

Most of the researches in transportation planning, traffic congestion and mobility is on issues concerning planning/construction of mass transit infrastructure and transit oriented development, whereas this study opens new avenues of research for building new cities around old cities for sustainable transportation solutions. Further research works are needed to elaborate:

(i) Issues, bottlenecks and fatal flaws in development of new cities around old cities
(ii) Economic and financial analysis in development of new cities around old cities with "Do something" and "Do nothing scenarios"
(iii) Case studies.

8.3 Recommendations

Construction of new cities in green spaces along the peripheral of aged cities should be sought for sustainable development. There are lot of benefits of constructing new cities. It involves no cost of demolishing, land acquisition in buildup areas and utility shifting. Construction of new cities is environmental friendly. Twentieth-century state-of-the-art ideas of city planning, transportation engineering, safe city concepts, ITS and TOD guidelines can be part of new city planning and design. At grade signalized intersections, at grade MTS dedicated corridors with signal priority and cycle-oriented road networks will ease burden on city planners. More of rural land will be added into urban area with increased mobility and comfort. Development of new cities will also automatically decrease population burden on parent city. It will create new jobs for local inhabitants, along with opportunities for traders, real estate managers and for people from all walk of life.

References

CDIA-ADB. (2014). *Urban transport pre-feasibility study in Peshawar, Pakistan-cities development initiative for Asia and Asian Development Bank (ADB).* Urban Policy Unit of the Government of KPK's Planning and Development Department (P&D).

ITDP. (2013). Tod Standards institute for transportation & development policy.

JICA. (1991). *Comprehansive study on transportation system in Lahore.* Traffic Engineering & Planning Agency (TEPA) Lahore.

JICA. (2012a). *Karachi Transport Improvement Project.* Karachi Metropolitan Corporation, The Islamic Republic of Pakistan.

JICA. (2012b). *Lahore urban transport master plan.* Transport Department, Government of the Punjab.

Jotin, C., & Kent Lall, B. (2002). *Transportation engineering.* Prentice-Hall of India.

PBS. (2018). *6th population & housing census-2017.* Pakistan Bureau of Statistics, Government of Pakistan.

PMA. (2015). *Transport modeling for mass transit system in Rawalpindi & Islamabad, Study Report.* Punjab Metro Authority.

Sheikh, A. B. (2015). *Ptv Visum transport model.* Punjab Metro Authority.

Assessing the Environmental Aspects of Road Network Resiliency

Behzad Bamdad Mehrabani, Luca Sgambi, and Negarsadat Madani

Abstract

Evaluating road networks' performance during and after a disruption and/or malfunction is of great importance. The performance of the road networks includes four concepts: reliability, vulnerability, robustness, and resilience. Among these concepts, the concept of resilience, which evaluates the road network's performance after a disruption/malfunction, is very significant. On the other hand, given that the road network is one of the primary sources of air pollution and plays a crucial role in urban sustainability, the amount of polluted emission should be considered in road network performance (resilience) analysis. The literature presents several measures such as travel time, queue length, recovery time, network's total cost, etc., to study road network resiliency. A review of previous studies demonstrates that the number of studies that considered environmental aspects in road network resiliency evaluation is scarce. Therefore, in this study, new network resilience measures that consider environmental factors are presented. These new measures show how the amount of polluted emission will change when a disruption occurs in the road network. After introducing and defining these new environmental resiliency measures, the Sioux Falls road network is simulated as the case study in Aimsun. The Sioux Falls road network (based on new resiliency measures) is evaluated when the speed of links (sections) is reduced randomly. The London Emission Model (LEM) is used for estimating the amount of polluted emission.

B. B. Mehrabani (✉) · L. Sgambi
Louvain Research Institute for Landscape, Architecture, Built Environment (LAB), Université Catholique de Louvain, Louvain-la-Neuve, Belgium
e-mail: behzad.bamdad@uclouvain.be

N. Madani
Local Environment Management & Analysis (LEMA), UEE Department, Université de Liège, Liège, Belgium

Keywords

Road network resiliency • Urban sustainability • Polluted emission • Traffic simulation • Environmental aspects • Aimsun

1 Introduction

Evaluation of transportation network performance under incident or disruption/malfunction conditions is of great importance. The transportation network's performance includes four concepts: reliability, vulnerability, robustness, and resilience (Calvert & Snelder, 2018). These concepts are very close to each other in such a way that reliability, vulnerability, and robustness are mentioned as resilience characteristics in previous studies (El Rashidy, 2014). The concept of resilience has attracted widespread interest in the last few years. This concept was born in the field of ecology and, at its origin, it was identified with an ecological system's resistance to change (MacArthur, 1955). Later on, it has found application in economics (Rose & Krausmann, 2013), engineering (Bruneau et al., 2003), and, lastly, transportation engineering (McDaniels et al., 2008).

Although resiliency concepts have been evolved for many years (since the 1950s), many researchers concluded that there is no unified definition of resiliency. So that, each researcher has given a specific definition to resilience according to the objectives of their project and the type of infrastructure being studied (Gauthier et al., 2018; Lhomme et al., 2013). Therefore, according to the objectives, resiliency in transportation systems is taken into concern in this article.

Resiliency in transportation systems is defined as "The ability to prepare for changing conditions and withstand, respond to, and recover rapidly from disruptions." (Scope, 2015). There are two types of incident or disruption (that leads to changing conditions) in the transportation system:

H. Altan et al. (eds.), *Advances in Architecture, Engineering and Technology*,
Advances in Science, Technology & Innovation, https://doi.org/10.1007/978-3-031-11232-4_7

(1) natural disaster (2) abnormal conditions. Natural disasters include earthquakes, floods, adverse weather conditions, etc., and result in network-wide failures. Usually, natural disaster leads to speed and capacity reduction at the network level. While abnormal conditions such as road maintenance sites, accidents, malfunction, etc., result in a road closure of specific links (Mehrabani et al., 2021). Usually, abnormal conditions occur in two ways: (1) incidents (for example, accident, road maintenance sites, man-made disasters); (2) road assets' malfunction, which leads to speed/capacity reduction (for example, pavement cracks). This article examines the road network's resiliency in abnormal conditions, which leads to speed/capacity reduction.

2 Literature Review

There are four different methods for the resiliency analysis of transportation networks: (1) topological models of resiliency (Gauthier et al., 2018; Lhomme et al., 2013; Zhang et al., 2015) (2) optimization models of resiliency (Kaviani et al., 2017; Omer et al., 2013; Patil & Bhavathrathan, 2016) (3) big data analysis (Liu & Song, 2020; Tympakianaki et al., 2018) (4) operational and simulation models of resiliency (Aghababaei et al., 2020; Bala et al., 2019; Ganin et al., 2017; Kamga et al., 2011; Sgambi et al., 2020). In what follows, some studies in each of the above groups are examined.

One of the studies using topological models for analyzing resiliency in natural disaster (flood) conditions is Lhomme et al., (2013). This study used Geographic Information System (GIS) to achieve network topological features. Using resistance capacity, absorption capacity, and recovery capacity, a new redundancy measure is defined. Zhang et al. (2015) evaluate the resiliency of various road network structures. The authors argue that different network structures (e.g., grid network, ring network, etc.) have different resiliency levels. Average degree and cyclicity metrics are used to indicate the redundancy level of the road network. The results suggest that the redundancies level and the resilience level have a direct relationship with each other. Another view of topological models of resiliency is presented in the study of Gauthler et al., (2018). The difference between this study and other topological models is that the traffic demand distribution and the road traffic dynamics are considered. This study concluded that various resiliency metrics lead to different importance levels for the network's links. The topological models of resiliency are able to analyze any system and require little time for their implementation. However, they are inefficient since they cannot capture the network's specific characteristics compared to other approaches (Sgambi et al., 2020).

The number of studies in which optimization models of resiliency are presented is less than the other approaches. One of these studies is the study of Omer et al., (2013). The model of this study tried to minimize the network travel time. This study used three measures for road network resiliency: (1) cost resiliency, (2) environmental resiliency, and (3) travel time resiliency. The environmental resiliency is captured by producing the CO_2 emission per fuel and the fuel consumed in the whole network. Although this study is among a few studies which consider environmental issues, the emission model (used for calculating the environmental resiliency measure) is at the macroscopic level, which does not account for vehicles characteristics in pollutant emission. A similar optimization model was reported by Patil and Bhavathrathan (2016). The transportation network's resiliency is evaluated using travel time optimization and different road network measures. This study presented a generalized index of resiliency using network cost. Besides, Kaviani et al. (2017) introduce a bi-level optimization model. This model finds the optimal location of guidance devices through the whole road network. The optimization model tries to minimize total travel time, which is an essential measure for road network resiliency. This paper's outcomes indicate that when the proposed optimization model optimizes roadside guidance devices' location, the recovery phase time is less than the usual condition.

Although topological and optimization methods of road network resiliency can be applied to any graph modeling, the following drawbacks still exist in these studies (Gauthier et al., 2018; Liu & Song, 2020): (1) they are usually demand-insensitive (2) less attention has been paid to recovery simulation (3) although, after a disaster, the entire capacity of a network does not get affected, these studies usually consider the complete removal of nodes or links. (4) different traffic modes cannot be viewed in graph theory, and (5) the specific characteristics of one system in relation to another are overlooked.

The studies that implement big data analysis for evaluating resiliency compare the traffic data before and after a disruption. For instance, Tympakianaki et al., (2018) employed link sensor counts, automated vehicle location, automated passenger count, automated fare collection, automated number plate recognition, taxi floating car data, and google floating car data to assess the effect of tunnel closure on multimodal transport. The results indicate that travel times are higher on closure days than on other days. This approach's disadvantage is that it requires a considerable amount of data; therefore, this approach is inefficient when no data is available.

In Kamga et al.'s (2011) study, the Chicago road network is simulated in incident conditions using VISTA software with a dynamic traffic assignment method. Network-wide total travel time is employed as a network performance measure. The results demonstrate that incidents affect both the incident location and the entire road network simultaneously (network-wide effects). Besides, the authors pointed out that the availability of information on the presence of incidence could help the incident management process. Ganin et al.'s (2017) study extracted the topological features from OpenStreetMap (OSM) data. This study's transportation model (a gravity-like model) uses the population of each zone, distances between zones, and distance factors as input and presents the flow between each origin–destination pair (origin–destination matrix) as output. Extra delays caused by 5% link disruption are employed as road network resiliency measures. The results show that many inefficient road links in normal conditions are resilient in an abnormal state. In contrast, some efficient links (in normal conditions) are not resilient in an abnormal conditions. This circumstance indicates the resiliency and the efficiency of the road network should be examined separately. Balal et al. (2019) aimed to compare different resiliency measures and employed the DynusT traffic simulator. The candidate resiliency measures in this study are detour road delay, upstream road delay, segment travel time, segment speed, and queue length. The finding of this study demonstrates that different resilience measures lead to different resiliency and efficiency levels of importance for links. Therefore, they recommend that each researcher should consider their measures based on their project objective. Using Aimsun mesoscopic simulator, Aghababaei et al., (2020) evaluate the New Zealand road network's performance after a natural disaster (earthquake). This study is one of the few studies that has examined the road network's performance at the country level (New Zealand). However, it should be pointed out that this study only considered the state highways and did not consider the main roads. Evaluated measures in this study include density, delay time, total traveled distance, travel time, traffic flow, and traffic count. The operational and simulation models of resiliency (used as the modeling approach in the current study) require much more work than topological models for their realization. However, they are more effective in analyzing network behavior (Sgambi et al., 2020).

On the one hand, the literature review demonstrates that many measures such as travel time, queue length, emission, time of recovery, peak distribution, network total cost, etc., (Murray-Tuite, 2006; Shang, 2016) have been employed as network performance and resiliency measures. Besides, some studies introduce new measures such as new redundancy indicators (Lhomme et al., 2013), generalized index of resilience (Patil & Bhavathrathan, 2016), and general link performance indicator for resilience (Calverta & Sneldera, 2018). Evaluating these measures suggests that most of the presented measures are based on capacity, delay, travel time, and network cost (traffic-related measures) (Balal et al., 2019; Kaviani et al., 2017; Murray-Tuite, 2006; Shang, 2016).

On the other hand, the transport sector accounts for 25% of greenhouse gas emissions, which is continuing to rise (Mehrabani et al., 2021). Besides, in previous studies, it has been stated that environmental issues are one of the principles of infrastructures' resiliency (Twumasi-Boakye & Sobanjo, 2018). However, despite the importance of environmental issues, fewer studies have considered environmental issues to measure network resiliency (Omer et al., 2013; Shang, 2016). Therefore, in this study, in addition to the existing road network resiliency measures, network resiliency measures in which environmental issues are taken into account have been studied. To achieve this goal, vehicle-based simulation (VBS) is used in the Aimsun environment.

In what follows, the network resiliency measures that have been examined in this study are introduced (Sect. 3). Then, in Sect. 4, the modeling process in Aimsun with some explanations about the case study will be presented. In Sect. 5, modeling outcomes are given. Finally, in Sect. 6, recommendations and suggestions for future studies are expressed.

3 Road Network Resiliency Measures

Road network resiliency measures can be grouped into three categories (Sun et al., 2018). The first group compares the functionality (loss) before and after a disruption. These measures are either traffic-related (travel time, throughput, capacity, etc.) or topology-based (e.g., connectivity and centrality). The second group uses measures that cover the recovery process after a malfunction. The second group's measures are usually calculated using the resilience triangle. The third group uses resilience measures to quantify an event's impact on society and the economy (environmental). In this study, four resiliency measures are considered, two of which are traffic-related and the two others are environmental-related.

3.1 Travel Time Resiliency

This measure compares normal conditions' travel time (for the entire network) with abnormal conditions' travel time. This measure's value is obtained by dividing the travel time in abnormal conditions (capacity or speed reduction in the network) by the travel time in normal conditions. Travel time resiliency is calculated by Eq. (1).

$$R_T = \frac{\text{Travel time (in abnormal condistion)}}{\text{Travel time (in normal condition)}} \quad (1)$$

In normal conditions (without a disruption), the travel time resiliency is equal to 1. A disruptive event raises travel time between two nodes. Therefore, the resiliency approaches values greater than 1. The more the value is near to 1, the more resilient the network is.

3.2 Mean Speed Resiliency

This measure compares the network mean speed before and after a disruptive event, given by Eq. (2).

$$R_S = \frac{\text{Mean Speed (in normal condition)}}{\text{Mean Speed (in abnormal condistion)}} \quad (2)$$

When an abnormal condition occurs in the network, the vehicles' speed usually decreases, so the higher the intensity of the abnormal condition in the network, the greater this measure's value. The more the value is near to 1, the more resilient the network is.

3.3 Environmental Resiliency

Environmental issues are one of the issues that have attracted much attention in recent years. After disruptive events in road networks, the vehicles' speed usually reduces, which is caused by capacity reduction. This speed reduction across the road network increases vehicle emissions. Therefore, in this study, the following measures are defined to examine whether the road network is environmentally resilient or not.

$$R_{E1} = \frac{\text{Emitted NO}x\text{ (in abnormal condistion)}}{\text{Emitted NO}x\text{ (in normal condition)}} \quad (3)$$

$$R_{E2} = \frac{\text{Emitted CO}_2\text{ (in abnormal condistion)}}{\text{Emitted CO}_2\text{ (in normal condition)}} \quad (4)$$

In this study, environmental resiliency is investigated by how the amount of polluted emission (by vehicles) changes before and after abnormal conditions. The amount of emitted pollution by vehicles is calculated using the London Emission Model (LEM), a mesoscopic traffic emission model. Please refer to Aimsun User's manual (2018) for more information about this model. It should be noted that the LEM is already embedded in Aimsun.

4 Traffic Simulation

The Sioux Falls road network is simulated in the Aimsun environment. The first step of traffic simulation is to import road network data (supply data) into the Aimsun. To do so, Sioux Falls road network information is extracted from OpenStreetMap. A code was implemented in the Overpass-Turbo environment to extract some specific types of roads. The types of roads extracted from OpenStreetMap and their features are illustrated in Table 1.

Table 1 Road network information

Type of road	Speed (km/hr)	Capacity (pcu/hr/ln)
Motorway	120	2500
Primary	100	2300
Secondary	90	2100
Tertiary	80	2000

The demand data (origin–destination matrix) is extracted from Leblanc et al.'s (1975) study, presenting 76 links and 24 nodes for the Sioux Falls road network. The origin–destination matrix is given in hundreds of vehicles per day. Therefore, the peak hour demand matrix is extracted using daily volume distribution (NCHRP, 2004). Finally, the morning peak hour (7:00–8:00) is simulated with 24 nodes and more than 2000 links (section) in Aimsun. The simulated network in the present study and the Sioux Falls graph (Leblanc et al., 1975) are shown in Fig. 1.

As mentioned earlier, the objective of the current paper is the network resiliency investigation under abnormal conditions. These abnormal conditions can be due to incidents such as car accidents or pavement failures. Therefore, it is clear that these abnormal conditions occur randomly in different road sections, and it is not possible to determine the place of speed reduction. As a result, in this study, ten speed reduction scenarios are considered for simulation. In the first scenario, the speed of 10% of the sections is reduced by 50%. In the second scenario, the speed of 20% of the sections is reduced by 50%. This process is done for 30, 40, …, 100% of sections. The candidate sections for speed reduction are chosen randomly using Aimsun scripting. The scenarios examined in this paper are given in Table 2. Since the size of the studied network in this study (Sioux Falls network) is on a city scale, microscopic simulation is very time-consuming. Therefore, mesoscopic simulation has been used. The traffic assignment method used in this simulation is dynamic user equilibrium (DUE) (Which considers 20 iterations for convergence).

5 Results

Ten scenarios were simulated in Aimsun, and for each of these scenarios, the values of road network resiliency measures (introduced in Sect. 3) were calculated. Simulation results and the calculated values of road network resiliency measures are given in Table 3. For better evaluation of the simulation results, the values of these measures are illustrated in Figs. 2, 3, 4, and 5.

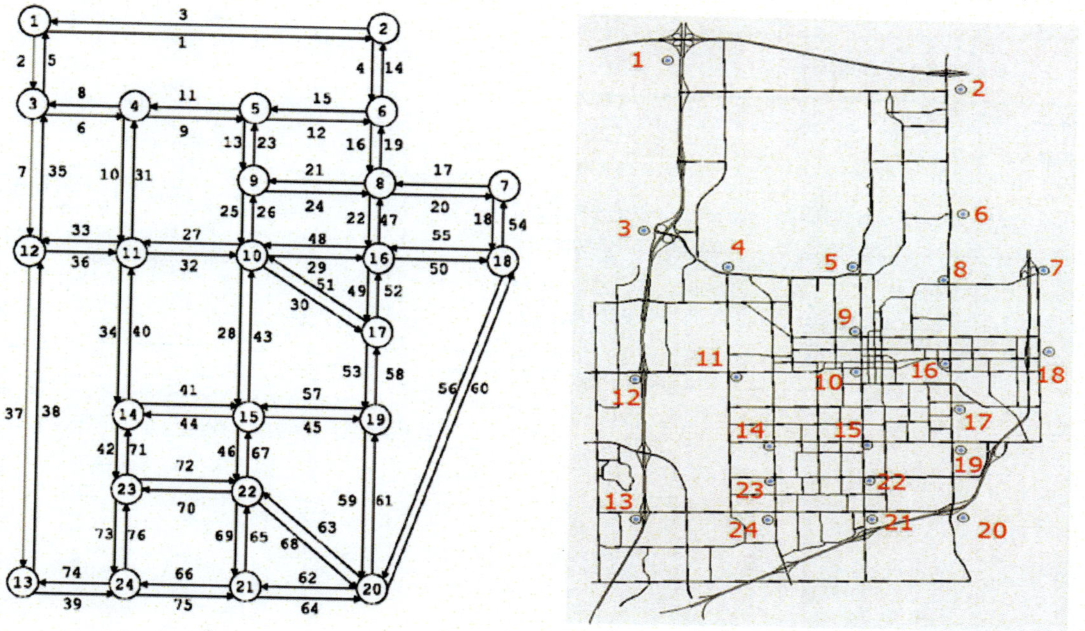

a b

Fig. 1 **a** Sioux falls network by Leblanc et al. (1975), **b** Sioux falls simulated network in Aimsun (present study)

Table 2 Simulation scenarios

No.	Speed reduction amount (%)	Section selection criteria	Percentage of sections in which speed reduction occurs (Percent of disruption)
1	50	Randomly	10
2			20
3			30
4			40
5			50
6			60
7			70
8			80
9			90
10			100

Figure 2 shows the values of the travel time resiliency measure (R_T). The line graph shows the value of R_T from 1 to 2 on the Y-axis against the percentage of disruption from 10 to 100 on the X-axis. As the percent of disruption increases, the value of R_T increases. In other words, there is a direct relationship between the number of sections affected by speed reduction and network travel time. Consequently, road network resiliency decreases with the increase in the number of section in which the speed is reduced. As can be seen, the variation rate in travel time resiliency changes nonlinearly. The rate of change in low and medium percentages of disruption (10–70) is very high, but it gets lower when we get closer to the end of the graph (percent of disruption 70–100). Therefore, it can be concluded that the network is less sensitive to speed reduction beyond the percent of disruption 70% (the breakpoint). This figure also shows that if the whole network's speed decreases by 50%, the total travel time will be less than doubled (R_T = 1.8). The mean speed resiliency measure is another measure studied in this paper. The values of this measure are shown in Fig. 3. As can be seen, with a 50% reduction in speed in all sections (scenario 10), the average speed in the entire network is increased by 80% (R_S = 1.8). Comparison of this measure with travel time resiliency measure (Fig. 2 versus Fig. 3) shows that the behavior of these two measures is very similar to each other. This circumstance is because speed and travel time have a direct relationship with each other.

Table 3 Simulation results

No	Percent of disruption	R_T	R_S	R_{E1}	R_{E2}
1	10	1.022	1.020	0.997	1.001
2	20	1.224	1.156	1.081	1.143
3	30	1.270	1.270	1.092	1.181
4	40	1.428	1.359	1.166	1.304
5	50	1.482	1.386	1.159	1.291
6	60	1.505	1.460	1.159	1.323
7	70	1.670	1.640	1.193	1.385
8	80	1.749	1.725	1.233	1.426
9	90	1.702	1.741	1.219	1.423
10	100	1.812	1.835	1.259	1.482

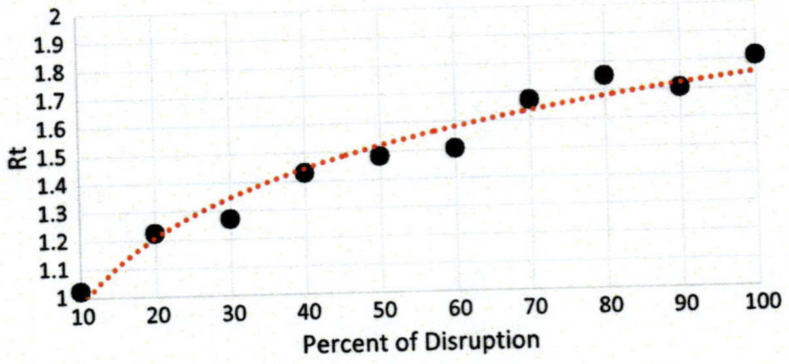

Fig. 2 Travel time resiliency measure in different scenarios

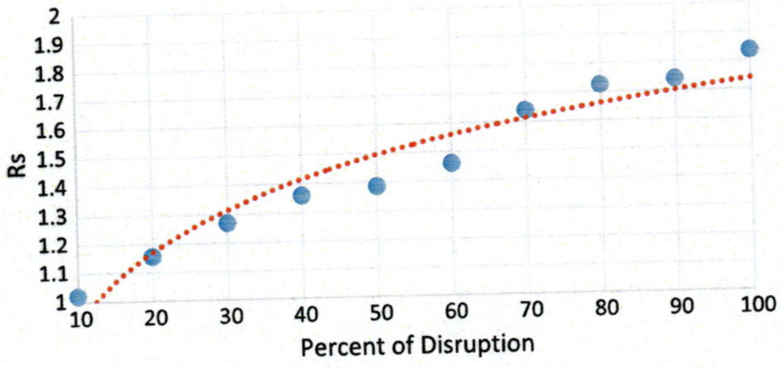

Fig. 3 Mean speed resiliency measure in different scenarios

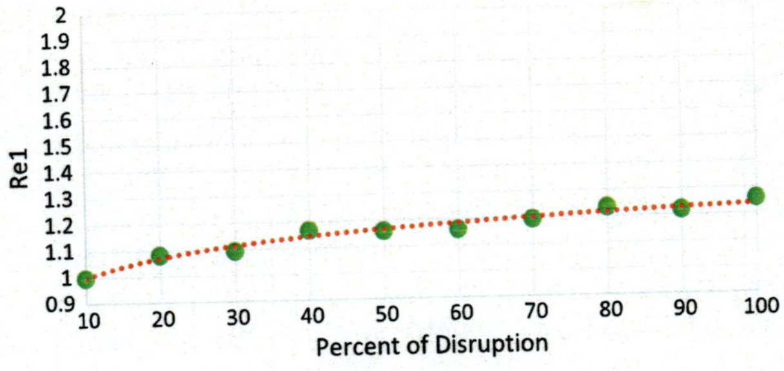

Fig. 4 Environmental resiliency measure (R_{E1}) in different scenarios

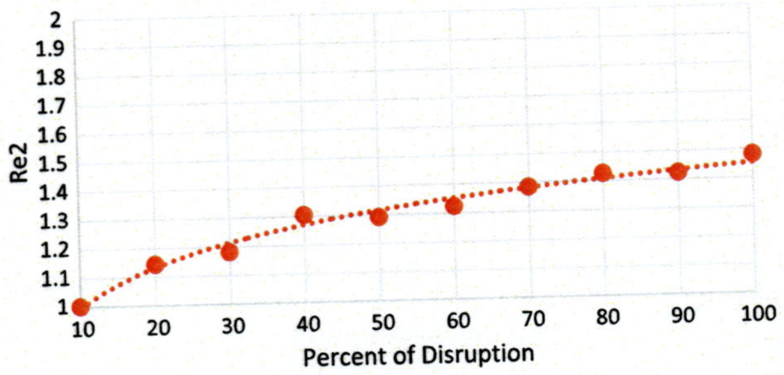

Fig. 5 Environmental resiliency measure (R_{E2}) in different scenarios

The last measures examined in this study are the ones related to environmental resiliency (R_{E1} and R_{E2}) for comparison of emitted pollutants under normal and abnormal conditions. The R_{E1} is used for the emittion of NOx, and the R_{E2} is defined for the amount of emitted CO_2. These measures are illustrated in Fig. 4 and Fig. 5, respectively. Examining these measures also shows that the amount of emitted NOx and CO_2 by vehicles increases with the percentage of disruption in the sections. So that if the speed of all sections is reduced by 50%, the amount of polluted emission will be approximately 1.5 times (R_{E1} = 1.259; R_{E2} = 1.482). It can be concluded that the amount of polluted emission by vehicles in abnormal conditions increases significantly. Therefore, environmental-related measures are highly important among traditional road network resiliency measures and should be considered in the decision-making process and evaluating road network performance. However, it should be noted that the rates of changes in environmental-related measures are lower than those of traffic-related measures. In other words, the network is less sensitive to emitted pollution than travel time and speed.

To compare the above measures better (R_T, R_S, R_{E1}, R_{E2}), the correlation matrix of these measures is shown in Table 4.

Table 4 Correlation matrix of resiliency measures

	R_T	R_S	R_{E1}	R_{E2}
R_T	1	0.988**	0.987**	0.991**
R_S	0.988**	1	0.965**	0.973**
R_{E1}	0.987**	0.965**	1	0.996**
R_{E2}	0.991**	0.973**	0.996**	1

**Correlation is statistically significant at 0.01 level

As can be seen, these four measures have a significant correlation with each other, which indicates that the disruption scenarios have almost the same effect on increasing travel time/emission and decreasing average speed. However, the rates of change are not the same. The rate of changes in R_T and R_S are very close to each other, while R_{E1} and R_{E2} has a lower rate of change in disruption scenarios.

6 Conclusion

The study of network resiliency in speed or capacity reduction scenarios (caused by natural disasters (e.g., earthquakes, floods, etc.) or abnormal conditions (e.g., vehicle accident, road maintenance sites, etc.)) is one of the issues that has attracted the attention of many researchers in recent years. Previous studies have introduced many measures for examining road network resiliency. Most of which are formed on traffic-related criteria (e.g., travel time, delay, etc.). A critical issue that has been overlooked in previous studies is the study of environmental issues in road network resiliency.

Therefore, in this study, four road network resilience measures were introduced, two of which are calculated using the total travel time and the average speed in the whole network (traffic-related measures). The other measures are based on polluted emissions by vehicles (NOx and CO_2). In the simulated scenarios, the speed decreases by 50% in a certain percentage of sections (which are selected randomly). For each scenario, the values of the above measures were calculated. These measures' values showed that the higher the percentage of speed deceleration in the sections (percent of disruption), the lower the network resiliency is.

The introduction of new network environmental resiliency measures showed that vehicles' amount of polluted emission increases under abnormal conditions. Although the four measures of R_T, R_S, R_{E1}, and R_{E2}, have almost the same behavior, the rates of change (when the network is in abnormal conditions) for the environmental-related measures (R_{E1} and R_{E2}) are less than those of the traffic-related measures (R_T and R_S).

Examining the resiliency measures' values shows that these measures vary nonlinearly in different failure scenarios. A noteworthy point that can be concluded from the measures' values is that the rates of change of the measures in low and medium percentages of disruption (10–70) are higher than high percentages of disruption (70–100). The results propose a nonlinear relationship between the percent of disruption and resiliency measures up to a breakpoint (percent of disruption = 70%). Beyond the breakpoint (percent of disruption = 70%), the changes' rates are meager and almost linear. This clarifies the need to present a nonlinear relationship between the percent of disruption and network resiliency measures in future studies. Besides, different traffic demand levels should be evaluated as the breakpoint is related to traffic demand.

Therefore, environmental-related indicators should be considered in the network resiliency studies and decision-making by decision-makers and policymakers. It can be concluded that a resilience network is a network in which not only traffic-related measures are considered but also other measures, including environmental-related ones, are taken into account. Moreover, the evaluation of these measures showed that they have a significant correlation. Therefore, future studies can focus on defining an indicator by combining the measures introduced in this study.

In the current study, the sections in which the speed reduction occurs were selected randomly, so it is suggested that the effect of speed reduction in pre-selected sections can be examined in future studies. So, the importance of each section and, consequently, the most critical sections can be specified. Also, this study considered a 50% reduction in speed. Other percentages of speed reduction should also be considered.

References

Aghababaei, M. T., Costello, S. B., & Ranjitkar, P. (2020). Assessing operational performance of New Zealand's South Island road network after the 2016 Kaikoura Earthquake. *International Journal of Disaster Risk Reduction*, 101553.

AIMSUN Version 8.4 User's Manual, (2018). *TSS-Transport Simulation Systems.*

Balal, E., Valdez, G., Miramontes, J., & Cheu, R. L. (2019). Comparative evaluation of measures for urban highway network resilience due to traffic incidents. *International Journal of Transportation Science and Technology, 8*(3), 304–317.

Mehrabani, B. B., Sgambi, L., Garavaglia, E., & Madani, N. (2021). Modeling methods for the assessment of the ecological impacts of road maintenance sites. In *Environmental Sustainability and Economy*, (pp. 171–193). Elsevier.

Bruneau, M., Chang, S. E., Eguchi, R. T., Lee, G. C., O'Rourke, T. D., Reinhorn, A. M., Shinozuka, M., Tierney, K., Wallace, W. A., & Von Winterfeldt, D. (2003). A framework to quantitatively assess and enhance the seismic resilience of communities. *Earthquake Spectra, 19*(4), 733–752.

Calvert, S. C., & Snelder, M. (2018). A methodology for road traffic resilience analysis and review of related concepts. *Transportmetrica a: Transport Science, 14*(1–2), 130–154.

El Rashidy, R. A. H. (2014). The resilience of road transport networks redundancy, vulnerability and mobility characteristics (Doctoral dissertation, University of Leeds).

Ganin, A. A., Kitsak, M., Marchese, D., Keisler, J. M., Seager, T., & Linkov, I. (2017). Resilience and efficiency in transportation networks. *Science Advances, 3*(12), e1701079.

Gauthier, P., Furno, A., & El Faouzi, N. E. (2018). Road network resilience: How to identify critical links subject to day-to-day disruptions. *Transportation Research Record, 2672*(1), 54–65.

Kamga, C. N., Mouskos, K. C., & Paaswell, R. E. (2011). A methodology to estimate travel time using dynamic traffic assignment (DTA) under incident conditions. *Transportation Research Part c: Emerging Technologies, 19*(6), 1215–1224.

Kaviani, A., Thompson, R. G., & Rajabifard, A. (2017). Improving regional road network resilience by optimised traffic guidance. *Transportmetrica a: Transport Science, 13*(9), 794–828.

LeBlanc, L. J., Morlok, E. K., & Pierskalla, W. P. (1975). An efficient approach to solving the road network equilibrium traffic assignment problem. *Transportation Research, 9*(5), 309–318.

Lhomme, S., Serre, D., Diab, Y., & Laganier, R. (2013). Analyzing resilience of urban networks: A preliminary step towards more flood resilient cities. *Natural Hazards and Earth System Sciences, 13*(2), 221.

Liu, W., & Song, Z. (2020). Review of studies on the resilience of urban critical infrastructure networks. *Reliability Engineering & System Safety, 193*, 106617.

MacArthur, R. (1955). Fluctuations of animal populations and a measure of community stability. *Ecology, 36*(3), 533–536.

McDaniels, T., Chang, S., Cole, D., Mikawoz, J., & Longstaff, H. (2008). Fostering resilience to extreme events within infrastructure systems: Characterizing decision contexts for mitigation and adaptation. *Global Environmental Change, 18*(2), 310–318.

Murray-Tuite, P. M. (2006). A comparison of transportation network resilience under simulated system optimum and user equilibrium conditions. In *Proceedings of the 2006 winter simulation conference*, (pp. 1398–1405). IEEE.

National Academies of Sciences, Engineering, and Medicine (NCHRP). (2004). Traffic data collection, analysis, and forecasting for mechanistic pavement design. Washington, DC: The National Academies Press. https://doi.org/10.17226/13781.

Omer, M., Mostashari, A., & Nilchiani, R. (2013). Assessing resilience in a regional road-based transportation network. *International Journal of Industrial and Systems Engineering, 13*(4), 389–408.

Patil, G. R., & Bhavathrathan, B. K. (2016). Effect of traffic demand variation on road network resilience. *Advances in Complex Systems* 19(01n02):1650003

Rose, A., & Krausmann, E. (2013). An economic framework for the development of a resilience index for business recovery. *International Journal of Disaster Risk Reduction, 5*, 73–83.

Scope, D. (2015). Transportation system resilience to extreme weather and climate change. *Federal Highway Administration (FHWA)*.

Sgambi, L., Jacquin, T., Basso, N., & Garavaglia, E., (2020). The robustness of infrastructure network assessed through a probabilistic flow model and a static traffic assignment algorithm—the case of the Belgian road network. In *Proceedings of the 10th international conference on bridge maintenance, safety and management (IABMAS 2020)* Hokkaido, Japan.

Shang, W. (2016). Robustness and resilience analysis of urban road networks, Imperial College London.

Sun, W., Bocchini, P., & Davison, B. D. (2018). Resilience metrics and measurement methods for transportation infrastructure: The state of the art. *Sustainable and Resilient Infrastructure, 5*(3), 168–199.

Twumasi-Boakye, R., & Sobanjo, J. (2019). Civil infrastructure resilience: State-of-the-art on transportation network systems. *Transportmetrica a: Transport Science, 15*(2), 455–484.

Tympakianaki, A., Koutsopoulos, H. N., Jenelius, E., & Cebecauer, M. (2018). Impact analysis of transport network disruptions using multimodal data: A case study for tunnel closures in Stockholm. *Case Studies on Transport Policy, 6*(2), 179–189.

Zhang, X., Miller-Hooks, E., & Denny, K. (2015). Assessing the role of network topology in transportation network resilience. *Journal of Transport Geography, 46*, 35–45.

Trends in ICT and Innovative Digital Technologies

Building Information Modeling on Construction Safety: A Literature Review

Orlean G. dela Cruz and Jason Maximino C. Ongpeng

Abstract

Building information modeling (BIM) technology has a significant contribution to construction safety, such as the automated detection of possible safety hazards and the prevention of potential risks likely to occur. The technology has been implemented in the architecture, engineering, and construction (AEC) industry for decades and has recently been introduced in construction safety. The technology can bring significant benefits to safety management in accident prevention and provide valuable reference material for rescue activities. Therefore, construction safety management must understand the technology, application, and challenges for better utilizing BIM as a safety tool. With these BIM safety tools, we should hope to see a shift in how safety is handled. To overcome this gap, concerning the challenges discussed, future research should consider: (1) generalization of the enhancement of BIM's technicality and functionality, (2) the cost–benefit analysis, and (3) construction practitioners' perspective on BIM applications on the results generated from different tools. It also recommends developing a database of all uncertain identified hazards and potential control measures subject to expert validation.

Keywords

Building information modeling (BIM) • Construction safety • Safety management • Safety tools • Latent Dirichlet Allocation (LDA)

1 Introduction

New techniques and construction methods introduced in the past years and up to now offer new ways to enhance construction safety management over the project's entire lifecycle. The main goal is to progress and not eliminate management-driven safety (Teizer et al., 2010). Building information modeling (BIM) is probably the most prosperous technology in the construction industry. This technology is a modern approach to the management of architecture, building, and infrastructure. A visual image of the construction process promotes the development process, knowledge sharing, and interoperability of information (Eastman et al., 2011).

Rajendran and Clarke (2011) specify which safety and health professionals can benefit from BIM technology in these areas: (a) Facility maintenance phase safety; (b) design for safety; (c) safety planning (job hazard analysis and pre-task planning); (d) accident investigations; and (e) worker safety training. BIM technology has allowed new technologies, collaboration opportunities, and practices to approach the site's safety facets in an appropriate way that can help facilitate high-quality site safety preparation even in a highly multinational ad competitive setting (Kiviniemi et al., 2011). BIM offers clear grounds for overcoming the fractured existence of the building industry. BIM's innovation allows 4D modeling, scheduling, and connecting to imagine the execution sequence to produce secure construction alternatives. The analysis carried out in a study (Wang & Chong, 2015) concludes that BIM must adapt to be wholly incorporated with other innovations, thus

O. G. dela Cruz (✉) · J. M. C. Ongpeng
Department of Civil Engineering, De La Salle University, 2401 Taft Avenue, 1004 Malate, Manila, Philippines
e-mail: ogdelacruz@pup.edu.ph

J. M. C. Ongpeng
e-mail: jason.ongpeng@dlsu.edu.ph

O. G. dela Cruz
Graduate School, Polytechnic University of the Philippines, G/F MH del Pilar Campus Valencia St. near R. Magsaysay Blvd. Sta. Mesa, 1016 Manila, Philippines

H. Altan et al. (eds.), *Advances in Architecture, Engineering and Technology*, Advances in Science, Technology & Innovation, https://doi.org/10.1007/978-3-031-11232-4_8

contributing to project success at different phases of the project's life cycle. While BIM is launched to contribute to the construction industry substantially, it is still in its early stages. The critical reasons for this are the need to incorporate BIM with other systems, which result in a lack of interoperability with a variety of specifications vying for data management (Martínez-Aires et al., 2018). Increased interest in research on the AEC industry has been using new BIM and BIM technologies to mitigate risk. The practical application of these innovations includes a thorough knowledge of the fundamentals, the necessary procedure, risks control approaches, and the interaction between modern and conventional methods (Zou et al., 2017). Moreover, the reliable sharing of data is essential to an efficient design process (Getuli et al., 2017).

Notwithstanding academic advancement, knowledge transfers to a physical manifestation of BIM, and incorporation of health and safety (H&S), there is still space for growth. Fast-changing technology in BIM application to construction is observed. The adaptation of BIM technology applied to safety management is limited, and inputs from industry practitioners as experts are not well recognized. With numerous BIM-based safety applications and other technologies, it is noticeable that the data generated requires expert validation. This paper's essential purpose is to study the applications, evolution, and BIM challenges as a construction safety management tool. Furthermore, the authors reviewed the current situation and proposed further proposals to promote and guide future research on BIM as a safety tool. By addressing these challenges, better utilization of technology is expected in the future.

2 Methodology

This section outlines a systematic overview to summarize and use the knowledge to recognize patterns and justify creativity in a new research initiative or gather precious findings that could be used in a sense other than that in which it was created (García-Holgado et al., 2020). A keyword-based source such as Scopus https://www.scopus.com was used to gather studies about interest and provides a comprehensive list of articles concerning all possible sources. Scopus supports Boolean syntax, a search that allows users to combine keywords with operators like AND. NOT and OR to further generate more accurate results. Scopus search was used in this paper because it tracks citation data for more journals and renders its journal impact measure (SNIP and SJR) available access all of the databases in the system (Bergman, 2012).

The aim was to classify research trends, formulate and apply the following questions for each publication:

RQ 1: What are the developments of BIM Technology in construction safety?
RQ 2: What are the functions of BIM as a construction safety tool?
RQ 3: How does the technology improve the quality of construction safety?
RQ 4: How does the technology affect the delivery of construction safety?

Figure 1 shows the following query string the "building information modeling,' and "construction safety,' which give 72 document results from 2011 to 2020. In the screening stage, 72 potentially relevant articles were identified. Followed by the screening stage, duplicate entry was excluded. Papers not related to the research question ($n = 6$) in the eligibility are also removed. A total of sixty-five (65) documents were identified.

Applied in this study is a Latent Dirichlet Allocation (LDA), a framework suitable for the study of high-level text relations is a series of probabilistic techniques called "topic models" (Roque et al., 2019). Topic modeling is a process that models each text as a mixture of themes and each subject as a mixture of terms (Roque et al., 2019). As defined by Osmani et al. (2020), the perplexity indicates the goodness-of-fit of the model describing a set of documents or simply that better models have lower perplexity. As shown in Fig. 2, a graph comparing the perplexity and appropriate time for a given number of topics. Furthermore, the number of topics selected is three (3).

LDA algorithm was used with three (3) topics; words cloud plots shown in Fig. 3 represent the LDA topic. As

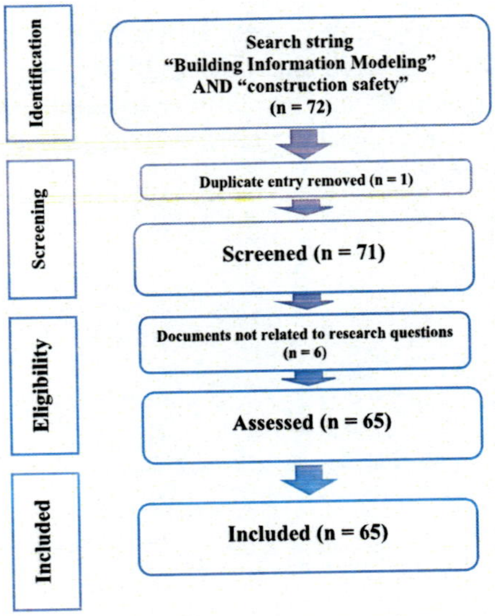

Fig. 1 Procedure of the systematic literature review

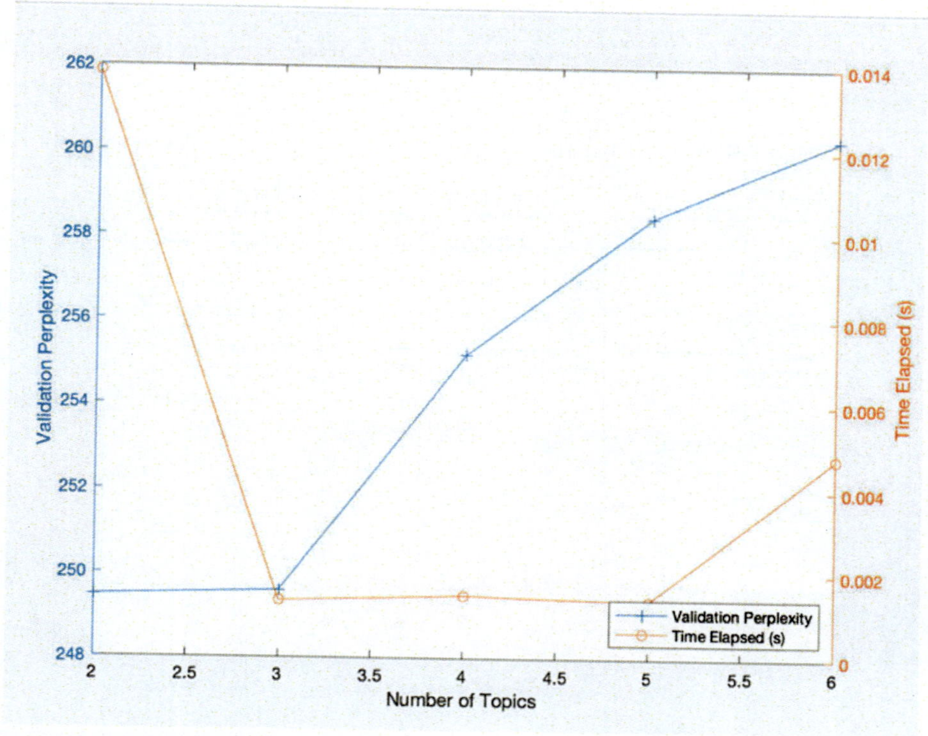

Fig. 2 Topics validation perplexity through MATLAB

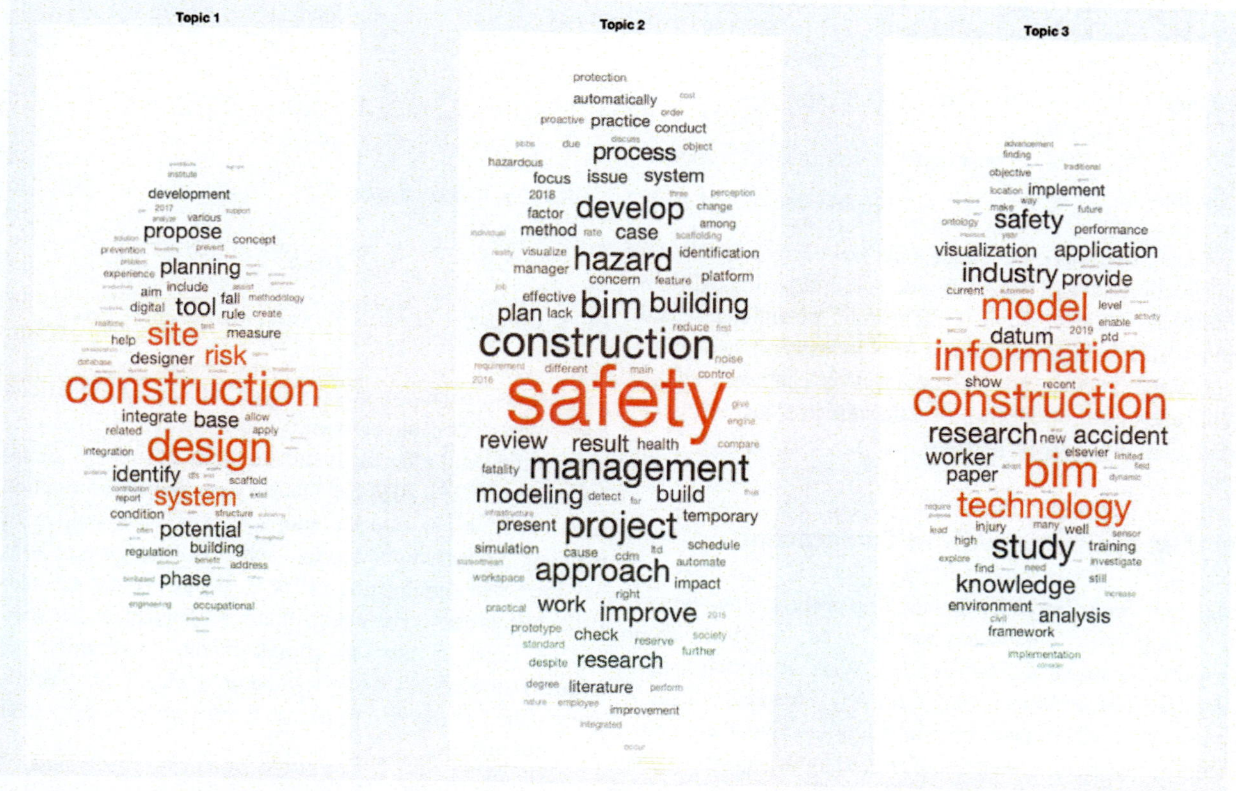

Fig. 3 Primary topics generated through MATLAB found in 65 included papers

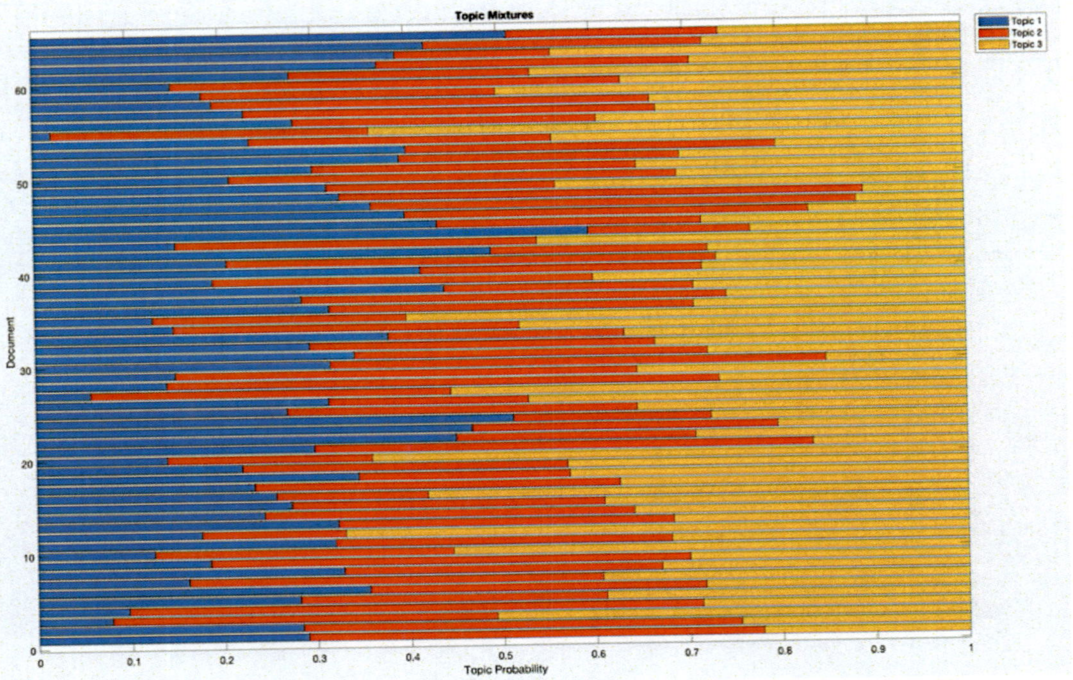

Fig. 4 Probabilities of topic mixtures generated in the 65 included articles using LDA

shown in Fig. 4, the topic mixture and probabilities of the 65 included articles. Primary topics from the 65 included articles have been listed and are as follows:

- Topic 1: construction, design, system, site, risk
- Topic 2: safety
- Topic 3: BIM, construction, information, model, technology

Topic 1 focuses on different tools and designs of BIM in construction sites. Topic 2 focuses on integrating BIM in construction safety. Topic 3 focuses on potential research model BIM technologies in the AEC industry. As shown in Fig. 5, the number of publications was characterized by publication year, indicating that the research interesting in BIM as safety tools increased.

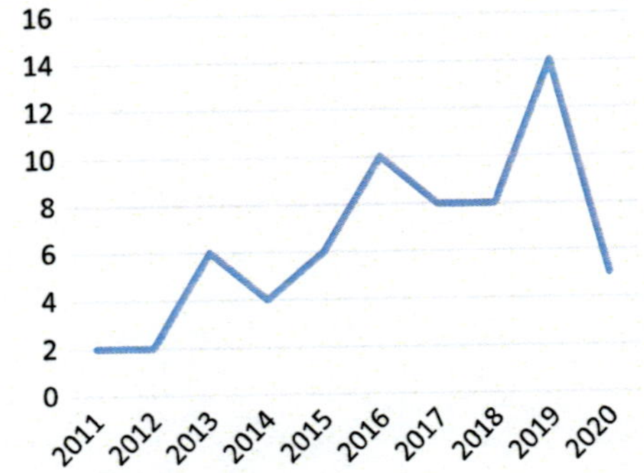

Fig. 5 Distribution of papers per year, published on the topic

3 BIM as Safety Tools in Construction

BIM technology improves the building process simulation to help employees understand safety risks in construction projects (Soemardi & Erwin, 2017). With all its rich visual characteristics, BIM technology, used as a broader potential construction safety platform, including its future commitment to improving protection (Teo, 2016), especially in the parametric modeling method (Yoo et al., 2016). To extend and utilize accessible resources and approaches, BIM provides a new and efficient foundation (Kasirossafar & Shahbodaghlou, 2013). BIM has made a remarkable contribution (Zou et al., 2017), and the adoption of BIM in support of planning, construction, service, and maintenance processes has increased significantly (Volk et al., 2014) and substantially affects the construction industry (Ramaji & Memari, 2015). Thus, an analysis illustrates the opportunities to incorporate BIM as a safety tool (Gambatese, 2004). Table 1 gives the comparison of different tools and their contributions.

Table 1 Critical areas of comparatives of BIM as safety tools

Authors	Tool	Design for safety tool	Visualization tool	Integrated BIM tool	Contribution in construction safety
Chunko and Benjaoran (2019)	Building information model for construction safety knowledge management (BIM-CSKM)	*			– It helps promote learning and make fair use of useful information to avoid future problems or injuries from repeating
Getuli et al. (2017)	BIM-based code checking	*			– Efficient execution of design practices by site safety preparation during the digital development process
Hongling et al. (2016)	BIM-based Design for Safety (DfS)	*			– Improving the safety performance of construction safety management and time and labor cost-reducing for safety measures
Jin et al. (2019)	4D BIM	*			– Contributes to the knowledge body by offering and illustrating the DAE approach for applying PtD quickly and reliably and addressing the risk management hierarchy's top level
Hossain and Ahmed (2019)	Automated safety checking system	*			– Reducing downside risks in the building industry
Hossain et al. (2018), Hossain et al. (2017, Zhou (2014)	Structured DfS rule-based knowledge library	*			– To provide safety knowledge to DAE's – Support to prevent any unexpected delay or expensive design late change to mitigate the risk
Qi et al. (2014)	Prevention Through Design (PTD)	*			– Helps DAE's to refine sketches to eliminate safety risks during construction
Riaz et al. (2017)	Confined space monitoring system or CoSMoS	*			– Continued surveillance of environmental and dangerous hazards impacting human health and safety
Qi et al. (2011)	Model checking software and constraint model/rule sets	*			– Automatically check fall hazards in the BIM and provide users with interface alternatives
Tixier et al. (2017)	Construction safety clash detection	*			– Provides valuable information about any technology-driven data and a range of safety preparation practices
Wei et al. (2017)	BIM-based workplace noise hazard prediction and visualization	*			– It is possible to incorporate wearable sensing data into BIM for noise hazard prediction in the construction workplace
Yuan et al. (2019)	Accident prevention through design (PtD)	*			– Automatically assesses the identified hazards or risks during the design phase
Zhang et al. (2016)	BIM-based risk identification expert system (B-RIES)	*			– Providing advice on risk evaluation and control of tunnel design would improve the possibility of successful projects in a dynamic environment

(continued)

Table 1 (continued)

Zhang et al. (2013)	BIM rule-based system	*			– Reduces the time and resources of safety personnel/engineers by way of an integrated safety code verification and simulation platform that assists in labor-intensive safety tests
Zhang et al. (2015a)	Construction safety ontology	*			– Helps avoid so-called housekeeping problems and circumstances due to human activity that lead to injuries
Kasirossafar and Shahbodaglou (2013)	BIM DfS		*		– Designing with safety software improves the designer's ability and awareness in identifying risks
Kim et al. (2016), Kim and Cho (2015), Feng and Lu (2017), Melzner et al. (2013), Liu et al. (2017), Yu et al. (2017)	BIM-based optimization / BIM-based scaffolding safety management model BIM 3D scaffolding BIM 2D scaffolding		*		– Identify several scaffolding plans and assist in the development of safer and more successful temporary structures/ scaffolding plans
Park and Kim (2013)	Novel safety management and visualization system (SMVS)		*	*	– Increase the recognition of safety threats and increase the capacity of employees to identify risks – Enhance real-time contact with administrators and staff
Shen et al. (2015)	Near miss information visualization tool		*		– Allows construction workers to view near misses during the entire construction process
Zhou et al. (2013a)	4D BIM		*		– An appropriate way to assess the extent of the danger and update item knowledge promptly depending on the metro's construction
Zhang et al. (2015b)	BIM-based fall hazards		*		– Supports human decision-makers in the examination process by avoiding all risks in the design and preparation processes and ensuring that protective equipment is procured and available for installation when necessary
Zhang (2015c)	Workforce location tracking		*	*	– Improves construction safety by strengthening the basis of how construction safety decisions and the future effect on a productive and unimpeded work environment are taken proactively
Alizandeh et al. (2018), Chen et al. (2019)	4D/BIM-UAV (unmanned aerial vehicles)-enabled safety model			*	– It helps safety managers to acquire, evaluate, and develop effective prevention techniques at construction sites
Arslan et al. (2018a)	Semantic trajectories using hidden Markov model (HMMs) and BIM			*	– Help deliver smarter construction services, increased automation, and information cohesion

(continued)

Table 1 (continued)

Bansal (2011)	GIS-based navigable 3D animation			*	– Manipulates the timetable, components, and sequences on one platform, thus ensuring that the secure sequence of construction quickly produces: and – Allows representatives of different files to communicate and collaborate
Cheung et al. (2018)	Real-time construction monitoring system for hazardous gas			*	– It reduces the risk of human exposure and increases social suitability in surveillance
(Getuli et al., 2020)	BIM and VR			*	– For safety preparation in conjunction with the site planning phase in real-world building projects
(Li et al., 2015)	Proactive behavior-based safety			*	– It offers a new direction for a continuous influence on building safety, to some degree
Tixier et al. (2016)	Natural language processing (NLP) tool			*	– Has a great deal of scope for convergence of specialized job packaging and construction information modeling applications
Yu et. al. (2017)	Integrated fall protection model based on a real-time locating system (RTLS) and BIM			*	– It improves the inconvenience of the conventional BIM-based fall security control (BIM-SCRS) method to assist construction engineers and managers in properly managing safety at work

3.1 BIM in Identifying Hazards as a Design for Safety (DfS) Tool

Construction hazards are the inherent features of an item that may interrupt the construction process and cause damage or injuries to persons. Identifying hazards is usually achieved by analyzing accident records and preparing the construction process while looking for potential instances of unforeseen dangers and harms. The management hierarchy routinely classifies all feasible steps for eliminating hazards; typically, as follows: (a) elimination, (b) substitution, (c) engineering controls, (d) administrative controls, and (e) personal protective equipment (PPE). Due to the nature of construction activities, some risks and hazards remain even in the safest designs (Yuan et al., 2019). However, it can be addressed with the Prevention through Design (PtD) concept, also known as Design for Safety (DfS), Design for Construction Safety (DfCS), and Construction Hazard Prevention through Design (CHPtD) (Yuan et al., 2019). Designers may recognize risks and hazardous situations using Design for Safety (DfS) tools. They can eliminate them (Gambatese et al., 2008). DfS also considered a practical approach to improving construction safety performance by considering safety problems during design (Hongling et al., 2016). Additionally, safety information may recommend suitable design improvements and adequately assess the method to remove building risks (M. A. Hossain et al., 2018).

Moreover, early detecting and addressing design hazards can reduce rework, yet it is far effective (Dewlaney & Hallowell, 2012). Preferably, PtD has three primary functions: (1) minimizing safety non-compliance during the design phase, (2) conducting safety non-compliance detection after the design works, and (3) correcting the safety non-compliance situation (Qi et al., 2014). The concept of PtD should be used to encourage sustainability through the design process, to ensure that staff, the public, and the community have the highest level of safety, energy, and environmental benefits (Kasirossafar et al., 2012). A promising idea in the building market was described as PtD to accomplish the zero-accident vision and encourage PtD implementation. Several researchers attempted to supply PtD resources to DAE and safety practitioners (Yuan et al., 2019).

Initial attempts include performing empiric experiments and employing best practices to build PtD-based checklist methods: such as DFCS toolbox (Gambatese et al., 2005; Gambatese & Hinze, 1999; Yuan et al., 2019). While implementation and correction of automatic hazard identification during building design and planning, BIM has demonstrated benefits in approaching safety management at the design phase of construction projects, e.g., compliance/rule check (Hongling et al., 2016) and hazard identification (Zhang et al., 2015b). It is a BIM tool changing how the construction approach analyzes building models to define threats and recommend preventive measures (Zhang et al., 2013). The merger of BIM and ESs was created to promote information and collaboration between distributed clients and experts during the safety risk recognition process. BIM-based risk identification expert system (B-RIES) was used to systematize explicit and implicit scattered information on tunnel construction (Zhang et al., 2016). Later, clash detection (Tixier et al., 2017), also known as BIM-based code checking (Getuli et al., 2017; Hongling et al., 2016; Hossain & Ahmed, 2019), was introduced for the validation and informative material review built into the framework. We know that the design phase and safety plan of construction projects are a core element of an automated operation.

Furthermore, to formalize expertise in safety management, a construction safety ontology was proposed. It is a new approach that organizes, stores, and re-uses the construction safety knowledge and provides a reasoning opportunity which supports safer and healthier execution of a construction project (Zhang et al., 2015a). With the motivation to develop a BIM-sensor based that improves construction workers' health and safety, particularly in confine spaced, a prototype system entitled confined space monitoring system or CoSMoS was introduced (Riaz et al., 2017).

A study suggests that safety regulations, documents, and best practices were adopted to extract safety knowledge to strengthen the PtD knowledge base (Yuan et al., 2019). Moreover, the 4D PtD tool was introduced to provide DAE's tool for assessing construction risks of multi-story building projects in a 4D environment (Jin et al., 2019). However, the construction company can share safety knowledge with other people by applying BIM-CSKM (Chunko & Benjaoran, 2019).

3.2 BIM as a Safety Tool for Visualization of Construction Projects

The ever more complicated architecture and design method also made those drawings challenging and less accurate. Besides, only minimal visual details in geometry, measurements, and maybe some information about the quality of the topic under examination can be provided by typical 2D or 3d sketches and technical specifications (Soemardi & Erwin, 2017). Visualization may be used to quantify blind spots to incorporate safety measures into the construction plan to define safety criteria for high-risk operations and associate temporary structural and design part safety measures (Zhou et al., 2012). Besides, the use of visualization in construction safety includes risk and hazards evaluation (Ding et al., 2014; Olugboyega & Windapo, 2019), integration of safety into construction site layout (Zhang, Sulankivi, et al., 2015b), the sequence of construction review (Bansal, 2011), and provision of information on the boundary conditions of risk elements (Park & Kim, 2013). Also, visualization technology is an essential strategy to help diverse site workers address obstacles and cultural and language barriers (Clevenger et al., 2011). Moreover, a case study expressed that visualization technologies were very useful in training and casual environments on building safety (Azhar, 2017).

The viewing of a schedule enhances the planning capacity and analyzes safety, safety management, communication, and better morale (Kiviniemi, 2011). Regarding the application, 3D simulated site simulation was suggested that reflects building circumstances such as machinery, temporary installation, inventories, and workforce (Li et al., 2003; Park & Kim, 2013). Decades ago, a study demonstrated by 4D BIM of the rule-based safety management framework for incidents involving safety guards and railing construction schedules (Sulankivi et al., 2010; Zhang et al., 2015b), BIM platform for detecting potential workspace conflict between the construction workers and lifting equipment or materials (Zhang et al., 2015c) and a system that incorporates temporary structures into the automatic plan and optimization of protection (Kim et al., 2016). However, adding OSHA safety rules on fall protection in 3D BIM models is feasible (Volk et al., 2014).

Despite certain studies concerning the optimization of the architecture and management of construction safety, still, safety regulations in metro construction cannot be implemented. Nevertheless, a 4D visualization for safety management was proposed (Zhou et al., 2013a). The findings demonstrate that 4D visualization can be easily used to identify safety hazards before and during the construction process that enables preventive measures. A proposed dynamic protection risk and visual monitoring method focused on BBS to achieve proper construction safety (Lee et al., 2019) should be applied.

3.3 BIM and BIM Technology Incorporation Integration as a Safety Tool

An increasing research focus has been using BIM and BIM technology to handle risks (Zou et al., 2017) and the most flourishing technology (Getuli et al., 2020) in the AEC industry. It presents a modern way of planning, construction and facilities management, which utilizes digital representations of a building process to promote knowledge sharing and interoperability (Eastman et al., 2011; Getuli et al., 2016). As an efficacious approach for avoiding incidents and increasing protection, several researchers have sought to incorporate emerging technology into safety activities (Alizandeh et al., 2018).

For example, GIS effectively predicts hazards were increasingly possible in a construction project because BIM and 4D modeling cannot edit attributes, including 3D elements, geospatial analysis, topographic modeling, and the generation and actualization schedules, and GIS has been successfully applied (Bansal, 2011). Also, to provide a piece of important reference information in the rescue task, the integration of BIM and WSN will be much help. Using this method, a building site may monitor the safety state of the site visually using a spatial, colored interface and automatically remove any harmful gas (Cheung et al., 2018). The use of unmanned aerial vehicles (UAVs) to monitor the workplace on a regular basis can also improve safety procedures. Construction site safety managers can now collect and analyze data using BIM and unmanned aerial vehicles (UAVs), giving them the information they need to detect potential dangers and devise mitigation methods at various stages of the project (Alizandeh et al., 2018). Further studies have shown that a combined schedule, BIM, and simulation may be utilized to forecast hazards and prevent workplace conflict, as well as an active scheduling tool. (Kim & Teizer, 2014; Moon et al., 2014).

Workers' movements on building sites are often very unpredictable, as their actions can vary due to dynamic conditions from predefined planning (Arslan et al., 2018b). However, we must define particular places in a building where they remained longer before extracting multiple consumer gestures. These shifts to the positions of the platform inevitably contribute to varying patterns of user mobility. For capturing the semantic of a dynamic environment, a system was proposed based on semantic trajectories; a framework based on semantic and hidden Markov model (HMM) and implemented in BIM was submitted for the semantics of a complex environment (Arslan et al., 2018a).

Although safety management practices or tools are related to incident avoidance and can represent a critical first step in preventing accidents, safety efficiency may also rely on workers' focused cognitive and emotional intervention. Therefore, businesses should also be concerned with winning over workers' minds and hearts through an efficient safety management scheme to mitigate/avoid injury-related problems (Wachter & Yorio, 2014). With the long-term application of the management, staff who can build a shared sense of secure ways to conduct all building tasks can become aware of safety attitudes (Li et al., 2015).

4 Challenges and Research Gap

Building information modeling (BIM) is one of the newest innovations implemented for construction safety worldwide (Liu et al., 2019; Muñoz-La Rivera et al., 2020). As a thriving and emerging approach in the construction industry's advancement (Martínez-Aires et al., 2018), BIM's application still imposes several limitations. It may offer impressive performance today and in the future, but BIM-based risk management remains to inflict challenges that are yet to be overcome. A significant dispute on its utilization in construction safety is the practitioners' unfamiliarity with its nature as an advanced technology (Eadie et al., 2013). Besides, it is indicated that BIM's necessity to be unified with other technologies places it farther from the spotlight, resulting in its inability to operate in different avenues and inadequate proficiency of the concerned organizations (Martínez-Aires et al., 2018). Their study further specified that the 3D model production, identifying the construction site environmental information (i.e., hazards, risks, and near misses), and the regular update of the 3D models depending on the existing condition of the site "as the construction process moves forward." Those are additions to the challenges that come with BIM's utilization that most construction practitioners are concern about. It also stated that in using BIM, safety practices components integrated with the selected modeling software have yet to be developed in the project (Zhou et al., 2013b). They indicated that site layout and safety planning components, which should naturally be required considering BIM's functionality, are missing and still need to be created with the contractor's aid. Moreover, they also pointed out the "limited internet functions for health and safety (H&S) rule browsing" as BIM's disadvantage and its single function for maintenance phase only and unfavorable inclination to collaborations.

Through BIM's data efficiency has already been proved under different circumstances, the data exchange and its accessibility gradation are still disputed (Mzyece et al., 2019). Moreover, it indicated that a BIM dimension intended for safety implementations, integrated with sensor-based systems, "can significantly improve the effectiveness of safety performance" (Asadzadeh et al., 2020); with this, by filling the gap imposed by traditional safety management. Additionally, the cost of availing BIM software and technical difficulties of operating has also raised challenges in applying the said innovation. Human behavior, which technologies have no control over at all, is also recognized as one of the significant drawbacks of BIM (Azhar, 2017). A study (Forsythe, 2014) also agreed that human behavior indeed poses an inevitable challenge when applying BIM technologies since it can be unpredictable and irrational, which he further noted as a gap that remains unsolved. Concerning, he also recommended human factor testing for future research endeavors.

The BIM application challenges in construction are also enumerated in a review conducted on the said subject: training, contractor selection, technical logistics, and on-site implementation (Muzafar, 2019). However, it further indicated that these could be confounded through the continuous development of visualization technology. Meanwhile, it concluded on a ground theory review on BIM construction safety that "visualization, simulation, animation, digital fabrication, and virtual reality' are the key features for BIM-enabled construction safety culture should have (Olugboyega & Windapo, 2019). Moreover, a study also supported these claims by stating that failure to adopt BIM is also caused by a lack of experts and skilled workers on visualization technologies, low demand from clients, and the software's high price (Marefat et al., 2019). While introducing revolutionary technology to BIM has been developed and applied, a need exists to envision safety data and interpret details (specifically near miss). Then, the near miss visualization tool was introduced (Shen et al., 2015), a tool intended to enhance decision-making for safety management and other building project stakeholders through the access and interpretation of lost information.

As the PtD concept has shown great potential for improvements in construction safety, various barriers have been identified. PtD's implementation has been hampered by a lack of effective tools and expertise, the absence of DAE's, separation of construction projects within their life cycle, and the fear of liability exposure (Gambatese, 1998; Gambatese et al., 2005, 2017; Toole, 2005). There was a research undertaken to address the issues of poor safety knowledge and an inadequate mechanism to assist the designer in identifying risk (Hossain et al., 2018) as they introduced a structured DfS rule-based knowledge library. A BIM-integrated risk review method that aids the designer in ensuring the design is safe. The library's design characteristics are vital to reducing the risk to the responsible parties. A 3D model is used to categorize and visualize the associated idea elements, as well.

Focusing more on this review's scope, presented in Table 2 are some of the BIM application challenges that the research studies revealed in their literature. After reviewing, the author's specified and mentioned limitations of BIM application in their paper and coming up with categories to have a generalized interpretation.

A total of nine (9) categories were extracted upon observing the challenges mentioned in the papers. These can be marked as challenges that a relatively new technology still imposes. This observation also indicates an ideal space for improving BIM's functionality in terms of the categories mentioned above. Besides, users' unfamiliarity with technology leads to operation difficulty and inconsistent performance in different scenarios. In a survey conducted by Enhassi et al., it was revealed that a substantial part of their respondents has little to no knowledge at all about the concept of BIM and does not consider using it in their future safety training. These challenges can collectively direct scholars to several knowledge gaps, further resulting in a broader range of research opportunities.

Some of the gaps that can be investigated, concerning the challenges discussed beforehand, are as follows: (1) general enhancement of BIM's technicality and functionality, (2) cost–benefit analysis, and (3) construction practitioners' perspective on BIM applications on the results generated from different tools. The first gap is interpreted by the mutual challenges presented in Table 2. It clearly shows that BIM has a long way to go in terms of data accuracy and adequacy and its primary methodologies. Secondly, a cost–benefit analysis will also be a significant subject in this study because the construction industry is a sector that also considers financial aspects in implementing innovations. Accordingly, it also emphasized that economic impacts hold a considerable part in adopting any new technologies. With this, evaluating how construction practitioners perceive these technological advancements also increases the likelihood of coming up with more desirable products that they can employ on-site (Akram et al., 2019).

5 Conclusion

This study's main goal was to systematically review the literature on prospective factors and future building information modeling (BIM) developments on construction safety. Part of the successful use of such technology is comprehensive knowledge of the principles of construction management, general practice, safety management, and the relationship between new and traditional approaches. Besides, it is the most promising technology because of its flexibility and versatility in various AEC industry areas, particularly in construction safety. Despite the rapid development of BIM and other enabling technologies in incorporating safety that offers the working environment and a visual interpretation of a worksite before the building process starts, it also facilitates a visual representation of site conditions (Martínez-Aires et al., 2018). However, there have been limited studies on authenticating the inputted data of advanced tool applications that require expert opinion and on-site validation. This paper recommends and challenges end-users and developers on developing web-based

Table 2 Specific challenges imposed by BIM tools in construction safety

BIM challenges	Author
Likelihood of conveying inaccurate data	Cheung et al. (2018)
	Getuli et al. (2017)
	Jin et al. (2019)
	Tixier et al. (2016)
Incompatibility to real-time environment	Arslan et al. 2018a)
	Qi et al. (2011)
	Zhang et al. (2013)
Subjective tendencies of the tool	Shen et al. (2015)
	Tixier et al. (2017)
Provision of inadequate information	Alizandeh et al. (2018)
	Hossain et al. (2018)
	Li et al. (2015)
	Yuan et al. (2019)
	Zhang et al. (2015b)
	Zhang et al. (2015a)
The complexity of defining tool functionalities, and general procedures	Boje et al. (2020)
	Qi et al. (2014)
	Riaz et al. (2017)
	Zhang et al. (2016)
The unfamiliarity of users to the technology leading to operation difficulty	Alizandeh et al. (2018)
	Park and Kim (2013)
	Soemardi and Erwin (2017)
Inconsistent performance to different scenarios	Liu et al. (2019)
	Zhang et al. (2015c)
Difficulty and unavailability	Chunko and Benjaoran (2019)
	Hossain and Ahmed (2019)
Financial disadvantage	Alizandeh et al. (2018)
	Soemardi and Erwin (2017)

knowledge for an expert in construction safety (open access). A database of all uncertain identified hazards and potential control measures that require expert validation—besides, an accessible database of all controlled safety measures of past incidents and accidents for the integration of the new technology.

To conclude, the AEC industry's implementation of BIM changes how safety is approached and verified. A zero-accident workplace can be achieved with corresponding prevention methods verified by an expert through the recommended web-based knowledge for an expert.

Acknowledgements The authors would like to extend appreciation to the PUP Engineering Science students' assistants for their untiring support in completing this research study.

References

Akram, R., Thaheem, M. J., Nasir, A. R., Ali, T. H., & Khan, S. (2019). Exploring the role of building information modeling in construction safety through science mapping. In *Safety Science* (Vol. 120, pp. 456–470). Elsevier B.V. https://doi.org/10.1016/j.ssci.2019.07.036

Alizandeh, S., Yitmen, I., Celik, T., & Arditi, Da. (2018). The effectiveness of an integrated BIM/UAV model in managing safety on construction sites. *International Journal of Occupational Safety and Ergonomics*.

Arslan, M., Cruz, C., & Ginhac, D. (2018a). *Exploiting semantic trajectories using HMMs and BIM for worker safety in dynamic environments*. 2018a International Conference on Computational Science and Computational Intelligence (CSCI). https://doi.org/10.1109/CSCI46756.2018.00107

Arslan, M., Cruz, C., & Ginhac, D. (2018b). *Understanding worker mobility within the stay locations using HMMs on semantic trajectories*. 2018b 14th International Conference on Emerging Technologies (ICET). https://doi.org/10.1109/ICET.2018
Asadzadeh, A., Arashpour, M., Li, H., Ngo, T., Bab-Hadiashar, A., & Rashidi, A. (2020). Sensor-based safety management. *Automation in Construction, 113*. https://doi.org/10.1016/j.autcon.2020.103128
Aslam, H. M., Abbott, E. L. S., Chua, D. K. H., Nguyen, T. Q., & Goh, Y. M. (2018). Design-for-Safety knowledge library for BIM-integrated safety risk reviews. *Automation in Construction, 94*(January), 290–302. https://doi.org/10.1016/j.autcon.2018.07.010
Azhar, S. (2017). Role of visualization technologies in safety planning and management at construction jobsites. *Procedia Engineering, 171*, 215–226. https://doi.org/10.1016/j.proeng.2017.01.329
Bansal, V. K. (2011). Application of geographic information systems in construction safety planning. *International Journal of Project Management, 29*(1), 66–77. https://doi.org/10.1016/j.ijproman.2010.01.007
Boje, C., Guerriero, A., Kubicki, S., & Rezgui, Y. (2020). Towards a semantic Construction Digital Twin: Directions for future research. In *Automation in construction* (Vol. 114). Elsevier B.V. https://doi.org/10.1016/j.autcon.2020.103179
Chen, Y., Zhang, J., & Min, B. (2019). *Applications of BIM and UAV to construction safety*. Proceedings, Annual Conference—Canadian Society for Civil Engineering
Cheung, W. F., Lin, T. H., & Lin, Y. C. (2018). A real-time construction safety monitoring system for hazardous gas integrating wireless sensor network and building information modeling technologies. *Sensors (Switzerland), 18*(2). https://doi.org/10.3390/s18020436
Chunko, K., & Benjaoran, V. (2019). Implementation of construction safety knowledge management via building information model. In *Lecture notes in civil engineering* (Vol. 9, pp. 431–437). Springer. https://doi.org/10.1007/978-981-10-8016-6_35
Clevenger, C., Lopez, C., & Puerto, D. (2011). *Using 3D visualization to train hispanic construction workers background and literature review*.
Dewlaney, K. S., & Hallowell, M. (2012). Prevention through design and construction safety management strategies for high performance sustainable building construction. *Construction Management and Economics, 30*(2), 165–177. https://doi.org/10.1080/01446193.2011.654232
Ding, L., Zhou, Y., & Akinci, B. (2014). Building Information Modeling (BIM) application framework: The process of expanding from 3D to computable nD. *Automation in Construction, 46*, 82–93. https://doi.org/10.1016/j.autcon.2014.04.009
Eadie, R., Browne, M., Odeyinka, H., McKeown, C., & McNiff, S. (2013). BIM implementation throughout the UK construction project lifecycle: An analysis. *Automation in Construction, 36*, 145–151. https://doi.org/10.1016/j.autcon.2013.09.001
Eastman, C., Eastman, C. M., Teicholz, P., & Sacks, R. (2011). *BIM handbook: A guide to building information modeling for owners, managers, designers, engineers and contractors*. John Wiley & Sons.
Feng, C.-W., & Lu, S.-W. (2017, July 1). *Using BIM to automate scaffolding planning for risk analysis at construction sites*. https://doi.org/10.22260/ISARC2017/0085
Forsythe, P. (2014). Proactive construction safety systems and the human factor. *Proceedings of Institution of Civil Engineers: Management, Procurement and Law, 167*(5), 242–252. https://doi.org/10.1680/mpal.13.00055
Gambatese, J. A. (1998). Liability in designing for construction worker safety. *Journal of Architecture Engineering, 4*(3). https://doi.org/10.1061/(ASCE)1076-0431(1998)4:3(107)
Gambatese, J. A., Asce, M., Behm, ; Michael, & Hinze, J. W. (2005). Viability of designing for construction worker safety. *Journal of Construction Engineering and Management, 131*(9). https://doi.org/10.1061/ASCE0733-93642005131:91029
Gambatese, J. A., Behm, M., & Rajendran, S. (2008). Design's role in construction accident causality and prevention: Perspectives from an expert panel. *Safety Science, 46*(4), 675–691. https://doi.org/10.1016/j.ssci.2007.06.010
Gambatese, J. A., Michael Toole, T., & Abowitz, D. A. (2017). Owner perceptions of barriers to prevention through design diffusion. *Journal of Construction Engineering and Management, 143*(7), 04017016. https://doi.org/10.1061/(asce)co.1943-7862.0001296
Gambatese, J., & Hinze, J. (1999). Addressing construction worker safety in the design phase Designing for construction worker safety. *Automation in Construction, 8*(6), 643–649. https://doi.org/10.1016/S0926-5805(98)00109-5
García-Holgado, A., Marcos-Pablos, S., & García-Peñalvo, F. (2020). Guidelines for performing Systematic Research Projects Reviews. *International Journal of Interactive Multimedia and Artificial Intelligence, 6*(2). https://doi.org/10.9781/ijimai.2020.05.005
Getuli, V., Capone, P., & Bruttini, A. (2020). Planning, management and administration of HS contents with BIM and VR in construction: An implementation protocol. *Engineering, Construction and Architectural Management.* https://doi.org/10.1108/ECAM-11-2019-0647
Getuli, V., Ventura, S. M., Capone, P., & Ciribini, A. L. C. (2017). BIM-based code checking for construction health and safety. *Procedia Engineering, 196*(June), 454–461. https://doi.org/10.1016/j.proeng.2017.07.224
Getuli, V., Ventura, S. M., Capone, P., & Ciribini, A. L. C. (2016). A BIM-based construction supply chain framework for monitoring progress and coordination of site activities. *Procedia Engineering, 164*, 542–549. https://doi.org/10.1016/j.proeng.2016.11.656
Hongling, G., Yantao, Y., Weisheng, Z., & Yan, L. (2016). BIM and safety rules based automated identification of unsafe design factors in construction. *Procedia Engineering, 164*(June), 467–472. https://doi.org/10.1016/j.proeng.2016.11.646
Hossain, M. M., & Ahmed, S. (2019). Developing an automated safety checking system using BIM: A case study in the Bangladeshi construction industry. *International Journal of Construction Management*. https://doi.org/10.1080/15623599.2019.1686833
Hossain, M.A., Abbott, E. L. S., & Chua, D. K. H. (2017). *Design for safety knowledge-based bim-integrated risk register system*. 9th International Structural Engineering and Construction Conference: Resilient Structures and Sustainable Construction.
Jin, Z., Gambatese, J., Liu, D., & Dharmapalan, V. (2019). Using 4D BIM to assess construction risks during the design phase. *Engineering, Construction and Architectural Management, 26*(11), 2637–2654. https://doi.org/10.1108/ECAM-09-2018-0379
Kasirossafar, M., Ardeshir, A., & Shahandashti, R. L. (2012). Developing the sustainable design with PtD using 3D/4D BIM tools. *Proceeding of World Environmental and Water Resources Congress*. https://doi.org/10.1061/9780784412312.279
Kasirossafar, M., & Shahbodaghlou, F. (2013). Application of visualization technologies to design for safety concept. *Proceedings of Sixth Congress Forensic Engineering*. https://doi.org/10.1061/9780784412640.040
Kim, K., & Teizer, J. (2014). Automatic design and planning of scaffolding systems using building information modeling. *Advanced Engineering Informatics, 28*(1), 66–80. https://doi.org/10.1016/j.aei.2013.12.002
Kim, K., & Cho, Y. (2015). BIM-based planning of temporary structures for construction safety. *Computing in Civil Engineering*, 436–444.

Kim, K., Yong, ;, Cho, K., & Kwak, Y. H. (2016). BIM-based optimization of scaffolding plans for safety. *Construction Research Congress 2016: Old and New Construction Technologies Converge in Historic San Juan*.

Kiviniemi, M., Sulankivi, K., Kähkönen, K., Mäkelä, T., & Merivirta, M. -L. (2011). *BIM-based safety management and communication for building construction*. VTT Technical Research Centre of Finland. VTT Tiedotteita - Research Notes No. 25978.

Lasda Bergman, E. M. (2012). Finding citations to social work literature: The relative benefits of using web of science, scopus, or Google Scholar. *The Journal of Academic Librarianship, 38*(6). https://doi.org/10.1016/j.acalib.2012.08.002

Lee, P. C., Wei, J., Ting, H. I., Lo, T. P., Long, D., & Chang, L. M. (2019). Dynamic analysis of construction safety risk and visual tracking of key factors based on behavior-based safety and building information modeling. *KSCE Journal of Civil Engineering, 23*(10), 4155–4167. https://doi.org/10.1007/s12205-019-0283-z

Li, H., Lu, M., Hsu, S. C., Gray, M., & Huang, T. (2015). Proactive behavior-based safety management for construction safety improvement. *Safety Science, 75*, 107–117. https://doi.org/10.1016/j.ssci.2015.01.013

Li, H., Ma, Z., Shen, Q., & Kong, S. (2003). Virtual experiment of innovative construction operations. *Automation in Construction, 12* (5 SPEC.), 561–575. https://doi.org/10.1016/S0926-5805(03)00019-0

Liu, K., Lee, Y., Shiau, Y., & Lin, C. (2017). Examining the lay-out of scaffolds and decks used in construction projects from the perspective of building information modelling. *ICIC Express Letters, Part B: Applications, 8*(2), 319–326.

Liu, Z., Lu, Y., & Peh, L. C. (2019). A review and scientometric analysis of global building information modeling (BIM) research in the architecture, engineering and construction (AEC) industry. In *Buildings* (Vol. 9, Issue 10). MDPI AG. https://doi.org/10.3390/buildings9100210

Marefat, A., Toosi, H., & Mahmoudi Hasankhanlo, R. (2019). A BIM approach for construction safety: Applications, barriers and solutions. *Engineering, Construction and Architectural Management, 26*(9), 1855–1877. https://doi.org/10.1108/ECAM-01-2017-0011

Martínez-Aires, M. D., López-Alonso, M., & Martínez-Rojas, M. (2018). Building information modeling and safety management: A systematic review. In Safety Science (Vol. 101, pp. 11–18). Elsevier B.V. https://doi.org/10.1016/j.ssci.2017.08.015

Melzner, J., Teizer, J., Zhang, S., & Bargstädt, H. J. (2013). Object-oriented safety planning of building construction by using Building Information Modeling. *Bauingenieur, 88*(November), 471–479.

Moon, H., Dawood, N., & Kang, L. (2014). Development of workspace conflict visualization system using 4D object of work schedule. *Advanced Engineering Informatics, 28*(1), 50–65. https://doi.org/10.1016/j.aei.2013.12.001

Muñoz-La Rivera, F., Mora-Serrano, J., Valero, I., & Oñate, E. (2020). Methodological-technological framework for construction 4.0. *Archives of Computational Methods in Engineering*. https://doi.org/10.1007/s11831-020-09455-9

Muzafar, M. (2019). Building information modelling to mitigate the health and safety risks associated with the construction industry: A review. *International Journal of Occupational Safety and Ergonomics*. https://doi.org/10.1080/10803548.2019.1689719

Mzyece, D., Ndekugri, I. E., & Ankrah, N. A. (2019). Building information modelling (BIM) and the CDM regulations interoperability framework. *Engineering, Construction and Architectural Management, 26*(11), 2682–2704. https://doi.org/10.1108/ECAM-10-2018-0429

Olugboyega, O., & Windapo, A. (2019). Building information modeling—enabled construction safety culture and maturity model: A grounded theory approach. *Frontiers in Built Environment, 5*. https://doi.org/10.3389/fbuil.2019.00035

Osmani, A., Mohasefi, J. B., & Gharehchopogh, F. S. (2020). Enriched latent dirichlet allocation for sentiment analysis expert systems. *37* August.

Park, C. S., & Kim, H. J. (2013). A framework for construction safety management and visualization system. *Automation in Construction, 33*, 95–103. https://doi.org/10.1016/j.autcon.2012.09.012

Qi, J., Issa, R. R. A., Hinze, J., & Olbina, S. (2011). Integration of safety in design through the use of building information modeling. *International Workshop on Computing in Civil Engineering*. https://doi.org/10.1061/41182(416)86

Qi, J., Issa, R. A., Asce, F., Olbina, S., Asce, A. M., Hinze, J., & Asce, M. (2014). *Use of building information modeling in design to prevent construction worker falls*. https://doi.org/10.1061/(ASCE)

Rajendran, S., & Clarke, B. (2011). Building Information Modeling: Safety benefits & opportunities. *Professional Safety, 56*(10), 44–51.

Ramaji, I. J., & Memari, A. M. (2015). *Information exchange standardization for BIM application to multi-story modular residential buildings*. AEI 2015: Birth and Life of the Integrated Building—Proceedings of the AEI Conference 2015, 13–24. https://doi.org/10.1061/9780784479070.002

Riaz, Z., Parn, E. A., Edwards, D. J., Arslan, M., Shen, C., & Pena-Mora, F. (2017). BIM and sensor-based data management system for construction safety monitoring. *Journal of Engineering, Design and Technology, 15*(6), 738–753. https://doi.org/10.1108/JEDT-03-2017-0017

Roque, C., Lourenço Cardoso, J., Connell, T., Schermers, G., & Weber, R. (2019). Topic analysis of Road safety inspections using latent dirichlet allocation: A case study of roadside safety in Irish main roads. *Accident Analysis and Prevention, 131*, 336–349. https://doi.org/10.1016/j.aap.2019.07.021

Shen, X., Asce, S. M., Marks, E., & Asce, A. M. (2015). *Near-miss information visualization tool in BIM for construction safety*. https://doi.org/10.1061/(ASCE)CO.1943-7862

Soemardi, B. W., & Erwin, R. G. (2017). *Using BIM as a tool to teach construction safety*. MATEC Web of Conferences, *138*. https://doi.org/10.1051/matecconf/201713805007

Sulankivi, K., Kähkönen, K., Mäkelä, T., & Kiviniemi, M. (2010). *4D-BIM for construction safety planning*. CIB 2010 World Congress Proceedings.

Teizer, J., Allread, B. S., Fullerton, C. E., & Hinze, J. (2010). Autonomous pro-active real-time construction worker and equipment operator proximity safety alert system. *Automation in Construction, 19*(5), 630–640. https://doi.org/10.1016/j.autcon.2010.02.009

Teo, E. A. L. (2016). Briefing: Determining productivity and safety indices using BIM. In *Proceedings of Institution of Civil Engineers: Management, Procurement and Law* (Vol. 169, Issue 4, pp. 143–144). https://doi.org/10.1680/jmapl.15.00054

Tixier, A. J. P., Hallowell, M. R., Rajagopalan, B., & Bowman, D. (2016). Application of machine learning to construction injury prediction. *Automation in Construction, 69*, 102–114. https://doi.org/10.1016/j.autcon.2016.05.016

Tixier, A. J. P., Hallowell, M. R., Rajagopalan, B., & Bowman, D. (2017). Construction safety clash detection: Identifying safety incompatibilities among fundamental attributes using data mining. *Automation in Construction, 74*, 39–54. https://doi.org/10.1016/j.autcon.2016.11.001

Toole, T. M. (2005.). *Increasing engineers' role in construction safety: Opportunities and barriers*. https://doi.org/10.1061/ASCE1052-39282005131:3199

Volk, R., Stengel, J., & Schultmann, F. (2014). Building Information Modeling (BIM) for existing buildings—Literature review and

future needs. In *Automation in construction* (Vol. 38, pp. 109–127). https://doi.org/10.1016/j.autcon.2013.10.023
Wachter, J. K., & Yorio, P. L. (2014). A system of safety management practices and worker engagement for reducing and preventing accidents: An empirical and theoretical investigation. *Accident Analysis and Prevention, 68*, 117–130. https://doi.org/10.1016/j.aap.2013.07.029
Wang, X., & Chong, H. Y. (2015). Setting new trends of integrated Building Information Modelling (BIM) for construction industry. In Construction innovation (Vol. 15, Issue 1, pp. 2–6). Emerald Group Publishing Ltd. https://doi.org/10.1108/CI-10-2014-0049
Wei, W., Wang, C., & Lee, Y. (2017, June 22). BIM-based construction noise hazard prediction and visualization for occupational safety and health awareness improvement. *Computing in Civil Engineering 2017*. https://doi.org/10.1061/9780784480823.032
Yoo, B., Yoon, H., Kim, Y., & Lee, K. M. (2016). Stepwise application of BIM-based parametric modeling to tapered slip-form system. *Procedia Engineering, 145*(512), 112–119. https://doi.org/10.1016/j.proeng.2016.04.028
Yu, W., Chang, H., & Cheng, S. (2017). An integrated fall-protection model based on real-time locating system and BIM. *Journal of the Chinese Institute of Civil and Hydraulic Engineering, 29*(4), 235–247. https://doi.org/10.6652/JoCICHE.201712_29(4).0003
Yuan, J., Li, X., Xiahou, X., Tymvios, N., Zhou, Z., & Li, Q. (2019). Accident prevention through design (PtD): Integration of building information modeling and PtD knowledge base. *Automation in Construction, 102*, 86–104. https://doi.org/10.1016/j.autcon.2019.02.015
Zhang, L., Wu, X., Ding, L., Skibniewski, M. J., & Lu, Y. (2016). Bim-Based Risk Identification System in tunnel construction. *Journal of Civil Engineering and Management, 22*(4), 529–539. https://doi.org/10.3846/13923730.2015.1023348
Zhang, S., Boukamp, F., & Teizer, J. (2015a). Ontology-based semantic modeling of construction safety knowledge: Towards automated safety planning for job hazard analysis (JHA). *Automation in Construction, 52*, 29–41. https://doi.org/10.1016/j.autcon.2015.02.005
Zhang, S., Sulankivi, K., Kiviniemi, M., Romo, I., Eastman, C. M., & Teizer, J. (2015b). BIM-based fall hazard identification and prevention in construction safety planning. *Safety Science, 72*, 31–45. https://doi.org/10.1016/j.ssci.2014.08.001
Zhang, S., Teizer, J., Pradhananga, N., & Eastman, C. M. (2015c). Workforce location tracking to model, visualize and analyze workspace requirements in building information models for construction safety planning. *Automation in Construction, 60*, 74–86. https://doi.org/10.1016/j.autcon.2015.09.009
Zhang, S., Teizer, J., Lee, J. K., Eastman, C. M., & Venugopal, M. (2013). Building information modeling (BIM) and safety: Automatic safety checking of construction models and schedules. *Automation in Construction, 29*, 183–195. https://doi.org/10.1016/j.autcon.2012.05.006
Zhou, W., Whyte, J., & Sacks, R. (2012). Construction safety and digital design: A review. *Automation in Construction, 22*, 102–111. https://doi.org/10.1016/j.autcon.2011.07.005
Zhou, Y., Ding, L. Y., & Chen, L. J. (2013a). Application of 4D visualization technology for safety management in metro construction. *Automation in Construction, 34*, 25–36. https://doi.org/10.1016/j.autcon.2012.10.011
Zhou, Z., Irizarry, J., & Li, Q. (2013b). Applying advanced technology to improve safety management in the construction industry: A literature review. *Construction Management and Economics, 31*(6), 606–622. https://doi.org/10.1080/01446193.2013.798423
Zou, Y., Kiviniemi, A., & Jones, S. W. (2017). A review of risk management through BIM and BIM-related technologies. *Safety Science, 97*, 88–98. https://doi.org/10.1016/j.ssci.2015.12.027

A Digital Twin-Based System for Smart Management of Office Spaces

Marco Marocco and Ilaria Garofolo

Abstract

The attention of the architecture, engineering, construction and operation (AECO) industry has been shifting from a great interest in the design and construction phases to the facility management (FM) and operational phase over the last decade. Disruptive technologies, such as information and communication technology (ICT), Internet of things (IoT) and building information modelling (BIM) have shown promising application to achieve a connected and effective management of buildings. Due to issues, such as COVID-19 and energy waste, governments have started promoting smart working to both private and public organisations. The expected benefits are twofold, namely social distancing in offices and better management of costs and spaces. This paper aims to define a digital twin-based system for smart management of office spaces. The system will help organisations to better manage their real estate and provide a basis for the development of a management platform.

Keywords

Building information modelling (BIM) • Internet of Things (IoT) • Digital twin (DT) • Architecture Engineering Construction and operation (AECO) • Facilities management (FM)

1 Introduction

According to the Official Journal of the European Communities, residential and tertiary buildings account for more than 40% of the final energy consumption in the European Union (Cox & Fisher Boel, 2003). On average, energy consumption inside office buildings can be mainly split in three parts: space conditioning accounts for almost 50% of the total energy consumption, lighting almost 20% and office equipment 20% (Stumpf Annette et al., 2011). As well as decreasing building energy consumption, diminishing greenhouse gas emission caused by travel, such as commuters' journey, is fundamental to environmental sustainability (Kumagai & Managi, 2019). Reducing the energy wastage and the carbon footprint of working practices, along with meeting the staff's aspiration of having an improved work-life balance are critical goals for organisations. Owing to these issues, smart working has gained increasing attention over the last few years (Bednar & Welch, 2020) and in particular, in the last year due to COVID-19-related restrictions (Deantonio et al., 2020). Smart working is the term used "to refer to the new ways of working made possible by advances in technology and made essential by economic, environmental and social pressures" (Andy Lake, 2015). This method includes mobility and flexibility in terms of time and location into the normal way of working (Imperial College London, 2019), resulting in less impact on the environment and improved effectiveness of work activities.

Among different facility management tasks, space management is the one which is mostly related to smart working, which provides adequate business service level and occupants satisfaction. With reference to offices, it is critical to manage how spaces, such as informal breakout spaces, meeting rooms, spaces for confidential one-to-ones, spaces for Skype calls and project rooms, are used and shared for various kinds of activities. Space management is regarded as an interdisciplinary effort that supports core business goals by integrating users, activities, space and technologies to manage working/living environment effectively (Li et al., 2017). Allocating and reallocating spaces are a continuous process that aims to increase productivity and employee's satisfaction and reduces costs (van der Voordt, 2004;

M. Marocco (✉) · I. Garofolo
Department of Civil Engineering and Architecture, University of Trieste, Piazzale Europa, 1, 34127 Trieste, TS, Italy
e-mail: marco.marocco2@phd.units.it

H. Altan et al. (eds.), *Advances in Architecture, Engineering and Technology*,
Advances in Science, Technology & Innovation, https://doi.org/10.1007/978-3-031-11232-4_9

Libecap et al., 2008). According to smart working principles, spaces are assigned to activities and not to individuals, therefore, it is critical to create office environments that facilitate collaboration and focus on outcomes rather than processes (Andy Lake, 2015). To this end, FM personnel need to have an evaluation tool for space utilisation, which periodically assesses whether the current allocation of space fulfils the needs and priorities of users (University of Michigan, 2012). Managing spaces according to their real usage and utility are critical to reduce direct and indirect costs. Space saving has been a primary concern since the beginning of the 2000s (Knight & Haslam, 2010). According to Space Management Group (2006), several organisations have tried to reduce the average size of enclosed offices, eliminate all solo offices and introduce "hotelling" and "hot-desking" models. The "hotelling" model allows employees to book their desk in advance, while the "hot-desking" one works on a first-come-first-served basis (Knight & Haslam, 2010). These models, along with teleworking, which means working from home using virtual technology, can improve the working environment and consequently increase employee productivity by almost 20% (Haynes, 2008). However, switching to a new way of working is a delicate matter that involves different stakeholders and necessitates the integration of performing technologies.

Recently, the concept of Digital Twin (DT) has gained increasing interest in the construction industry for dealing with issues related to FM activities. The Digital Twin is a "digital model, which is a dynamic representation of an asset and mimics its real-world behaviour" (Lu et al., 2020). The DT of a building mainly consists of two parts including the BIM model and the dynamic data of the building. BIM can be regarded as a platform storing building information, which will be used to improve decision-making processes during the asset life cycle (Nicał & Wodyński, 2016; Azhar, 2011). The dynamic data is provided by a system based on the Internet of Things (IoT) technology, which exploits a series of objects, which are provided with identifying, sensing, networking and processing capabilities (Jia et al., 2019). Digital Twin platforms can be the technology which enables smart working by providing an effective space management of buildings.

However, the implementation of this process is still in its infancy. Rethinking the office space is critical to drive organisations to shrink occupancy costs and enhance workers' productivity. For this reason, there is a need of innovative applications for the smart management of buildings. This paper presents a Digital Twin-based system to enable the smart management of office spaces. The system aims to book workspaces in organisations by avoiding repeated bookings or unused spaces and offers potential insights to new arrangements of spaces by using BIM technology and a post-evaluation analysis based on data provided by occupants and sensors. The rest of the paper is structured as follows: Sect. 2 proposes the DT-based system for space management, Sect. 3 discusses the application of the proposed system and Sect. 4 presents conclusions.

2 DT-Based System

This paper proposes a web-based decision support system based on four layers including presentation, application, data and perception layer, as shown in Fig. 1. This system integrates as-is BIM models, real time and historical information (e.g. historical booking records, comments and request of repair/adjustments) related to workspaces and monitoring data from sensors (e.g. presence or absence of occupants in office rooms and at workstations).

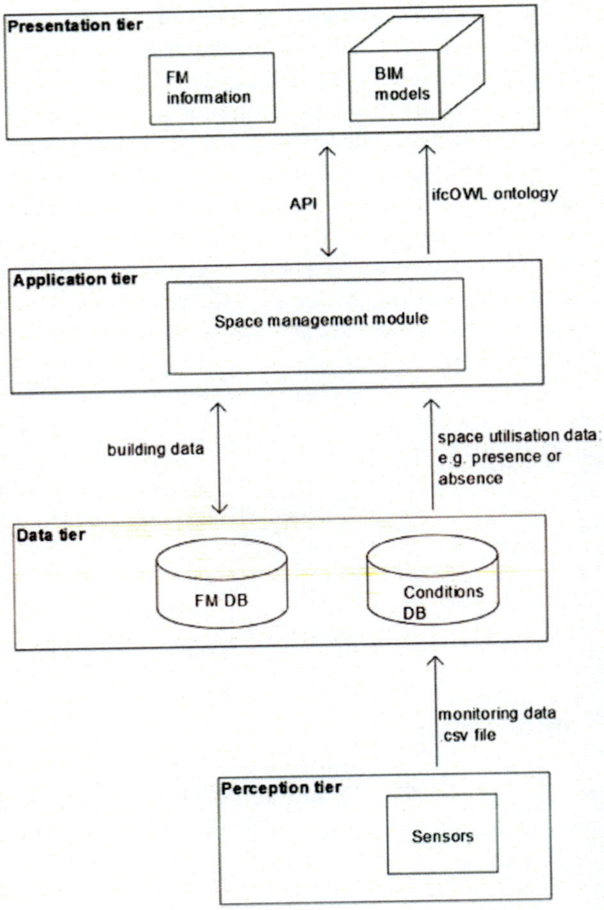

Fig. 1 Digital twin system

2.1 BIM

To support space management, reliable and updated information of facilities need to be available (GVA Grimley in association with Stellae, 2002). Information including location, space numbers, descriptions, boundaries, areas, volumes, intended use and actual status needs to be stored in a structured way (Burcin Becerik-Gerber et al., 2012). Traditionally, this task is conducted by using computer-aided design (CAD) tools. However, this way of managing internal building layouts usually encounters issues, such as inconsistent naming conventions and laborious attribute updates (Burcin Becerik-Gerber et al., 2012). On the contrary, BIM identifies elements with a specific Globally Unique Identifier (GUID), simplifying the process of identifying and tracking assets (Burcin Becerik-Gerber et al., 2012). BIM technology enables 3D modelling and interior navigation, which leads to huge potentiality of visualising spaces and their attributes in order to improve space analyses, and manages the allocation process and compares actual with planned space utilisation (Burcin Becerik-Gerber et al., 2012). Although BIM models can provide necessary information to perform space management (Kensek, 2015), continuous updates of models are needed. According to (Hardin & McCool, 2015), "maintaining a BIM model with regard to facility management information is similar to maintaining the actual facility. As components are replaced, repaired or removed, those changes will need to be reflected in the BIM file". To achieve the display of 3D BIM models on a web platform, the ifcOWL ontology can be used. This ontology is based on the translation of the IFC schema into the Ontology Web Language (OWL) (González et al., 2020). To this end, users do not have a limit to get access to the system and can view the 3D model online through any device anytime and anywhere.

2.2 Facility Management Software

Information is often stored in sheets of paper and/or file notes on a local computer, causing a time and space gap between facility location and where documents are stored (Lin et al., 2012). Manual data management is not able to generate a proper method for information management, leading to a lack of uniformity of data management storage (Aziz et al., 2016). To this end, several kinds of FM systems, such as computerised maintenance management system (CMMS), computer-aided facility management (CAFM) and integrated workplace management systems (IWMS), can help manage different data during the O&M phase (Bortolini et al., 2016). For instance, CAFM software enables FM managers to improve long-term planning by focusing on activities involved in space and workplace management, including asset management, reactive and preventive maintenance, operational facility services, room reservations and any other customer service request (Planon, 2020). Big organisations need a FM system able to manage all aspects of FM, but only part of those software solutions exploit the same database to manage more FM tasks. IWMS tools can integrate several application modules and use information stored in a unique database, which is in common among all application modules, for all FM tasks. This can bring several benefits to organisations, which can coordinate their diverse activities by avoiding misleading or scattered information flow. Eventually, FM tools often integrate a module to directly manage BIM models in their interface, leading to up-to-date models available in each module of the application.

2.3 IoT

Challenges arising from space management include lack of up-to-date and historical information regarding space utilisation. Having easily access to accurate information is paramount for facilities management decision-making processes (Oskouie et al., 2012; Akcamete et al., 2010). According to Qolomany et al. (2019), data acquisition should overcome issues, such as interoperability, reliability and integration, in order to serve smart buildings in the operational phase. Integrating different types of smart devices can provide accurate and reliable occupancy information (Labeodan et al., 2016). The Internet of Things (IoT) is defined as "interconnection of sensing and actuating devices providing the ability to share information across platforms through a unified framework, developing a common operating picture for enabling innovative applications" (Gubbi et al., 2013). A common IoT system is composed of five components including smart devices which capture data, networks, i.e. communication infrastructures, a data storing place, analytics and user interfaces (Jia et al., 2019). The architecture of an IoT technology can be summarised in a three-layer system including a perception layer, a network layer and an application layer. The perception layer is based on sensors and actuators, which generate an electronic signal from the environment and transform electronic signals collected by sensors into action, respectively (Jia et al., 2019). In this layer, the main utilised technologies consist of WSN, video cameras, RFID, QR codes and barcodes. The network layer deals with processing and sharing raw data captured in the perception layer. It usually transmits data through wireless and wired technologies such as Wi-Fi, Bluetooth, Zigbee, RFID and USB. The application layer is the user interface, where users can exploit data previously collected to perform analyses and enhance decision-making processes.

2.3.1 Sensors

Sensors are defined as "physical devices able to capture external signals and convert them into an analogue or digital voltage" (Maser, 1988). Sensors can be wireless or wired, where the former, such as wireless sensor networks (WSNs), is the most utilised technology due to its capability of flexibility and convenience (Xu et al., 2019). Sensors can detect the presence of people inside rooms by collecting data of CO2 levels, sound waves, emitted radiations, but also movements and activities (Mamidi et al., 2012; Tarzia et al., 2009). For instance, sensors embedded in chairs can conduct occupancy detection by checking strain, vibration and mechanical switch applied to chairs (Labeodan et al., 2016). Combining multiple detection systems can provide FM personnel with accurate and reliable occupancy building data. This enables a more efficient management of working spaces by providing the usage information, namely checking that spaces which are booked are also really used. Sensors have to be deployed in buildings in correspondence to office workstations and working spaces. As well as sensors, actuators need installing in correspondence to every workspace to reduce energy consumption whether not necessary. Data monitored is stored by using a csv format in a unique database (DB), which allows collecting information about spaces in a structured way.

2.4 Application Tier

To get access to the system, users have to log in with specific credentials to avoid the spread of information. Each type of user can see and use different applications according to the privileges of their account, as shown in Fig. 2. While employees have the possibility to book a workspace and provide feedback, managers dealing with space management have access to the modules of data analysis and space planning.

2.4.1 Booking and Feedback Modules

Traditional offices enable space booking only for meeting rooms. However, the advent of smart working has incentivised space management towards new ways of organising spaces. Thus, office spaces including informal breakout spaces, single office workstation, meeting rooms, spaces for confidential one-to-ones, spaces for Skype calls and project rooms can be booked.

The part of the system accessible to common users, such as employees, consists of three sections including a log in section, a searching section and a result section. Each result obtained from the searching section is linked to a page corresponding to the exact workspace searched. The page of a workspace contains all information regarding the workspace. Each workspace is structured following a certain code classification system including site, building, floor, room and workstation (if necessary). Additional information, such as technical equipment and furniture, is also added. A set of images are attached to the page in order to provide users with a better recognition of the space and understanding whether it can be suitable for their activities. Another attached source of information is the BIM model of the entire building with the workspace location highlighted. This allows users to navigate the building and immediately understand where the workspace is located and how to reach it. Finally, a digital calendar provides users with the exact period of time in which the selected workspace is or is not available and a system to book that space.

The application provides occupants with a clear vision of the space available in each building and allows selecting and

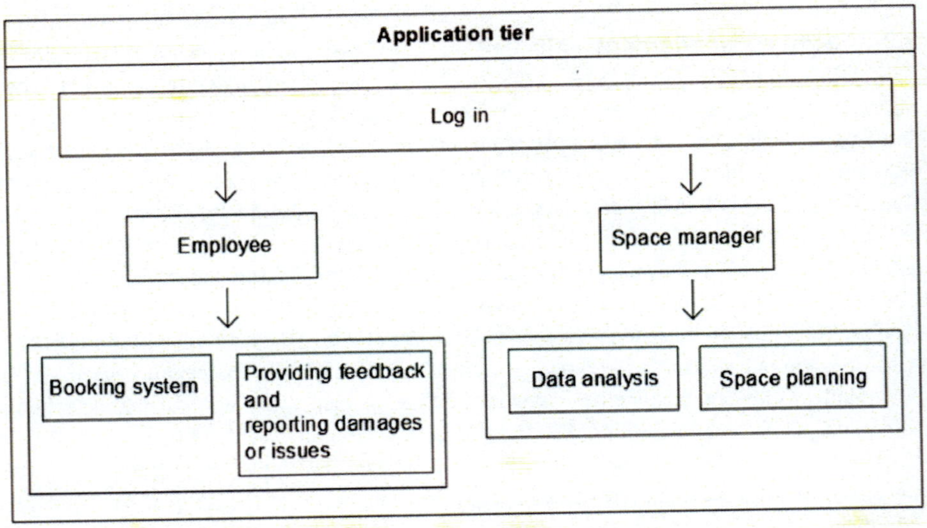

Fig. 2 Application tier

booking specific spaces. The booking procedure, as shown in Fig. 3, consists of three steps: (1) pinpointing the interested space, (2) checking time availability and (3) booking.

To locate the interested workspace and easily get access to the information of specific working spaces, QR code can be a potential solution due to its properties of security and high fault tolerance. Scanning tools are free and QR code can be scanned in many orientations without having trouble with problems of recognition. A QR code is applied to each office workstation and working space in order to exactly pinpoint the interested area. Scanning the QR code of workspaces or office workstations, users are linked to the page of the room/space searched. Otherwise, to solve the problem of searching a workspace suitable for specific activities, especially for users who do not have a complete knowledge of buildings, a smart query provides users with some filters to direct their search. The process of querying consists of three steps. First, users are asked to specify the site and the building where they want to book a space. Next, they are asked to select the type of space they are looking for, such as a meeting room. Then, furniture and equipment necessary to perform the activity are indicated if necessary. As a result, only workspaces which fit with the requirements previously selected are shown.

Booking a workspace for more time than actually necessary is one of the main issues related to space booking systems. Ensuring instant availability when spaces are not being used can be supported by sensors deployed around organisations, which check the presence of occupants. To this end, if a workspace is reserved, but sensors do not detect any occupants in that space for more than a certain threshold of time which try to avoid misleading situations, such as coffee break, the booking will be automatically deleted. Applications of IoT technology in energy management can provide the possibility to configure facility equipment according to specific requirements. For instance, actuators, which are connected to sensors, can avoid wastage of energy by performing specific tasks, such switching on/off lights, heating and air-conditioning, whether sensors do not detect the presence of occupants or according to the type of workspace and the preferences of occupants.

Users can provide feedback for the usage of space at the end of the booking. Inserting the code created by the booking system, the workspace reservation can be identified in terms of space and time. To better process a huge amount of information, a questionnaire based on a five-point score from one to five is asked to every user. The list of question is the following:

- Was the space suitable for the activity?
- Was the space clean?
- Was the space supplied with the equipment necessary?
- Was the level of noise acceptable?
- Was the level of light acceptable?
- Was the temperature acceptable?
- Was the space ergonomic?
- Was the space enough to work?
- Was the space comfortable overall?

For each question, additional free-form comment of the given score can be submitted to give the possibility to users to motivate the selected score and report possible damages or issues.

2.4.2 Data Analysis Module

After collecting information of space usage and comfort from sensors, bookings and feedback, data is organised and stored into a database. The data analysis module is divided into two parts: (1) evaluating the trend of occupant satisfaction by considering the feedback of space usage at the end of bookings; (2) analysing the trend of space utilisation by using several indicators according to the analysed workspace. Charts show the real usage of workspaces and their level of satisfaction by providing potential understanding of how, when and where work is carried out.

The shift to new office buildings and smart working models needs to be verified ex-post. An evaluation of space utilisation based on feedback provided by occupants is the critical step to implement and enhance space management effectively. This process helps to manage space planning according to a user-centred method and focuses on a user experience approach. Since the answers of the set of questions are based on a five-point score, statistical trends can be studied by using advanced techniques of analysis. As well as these structured answers, the additional free-form comments can strongly support the analysis. Although these results are less structured than the five-point score answers, they allow users to completely express their real evaluation.

As well as occupants' feedback, space utilisation indicators can provide potential insights to help decision-making processes of space planning. The historical and real-time monitoring of working space usage are a crucial task for avoiding wastage of organisations. The utilisation of workspaces is analysed by using a few indicators for each type of space. Since there are several types of workspaces, such as a single workstation in a co-working room and meeting room, there is a need to consider indicators for both the single workstation and the whole room. To gain additional insights from data, rooms are grouped according to type, departments and positions in buildings. The indicators which are taken into account to evaluate the usage of workspace are described in Tables 1, 2, 3 and 4.

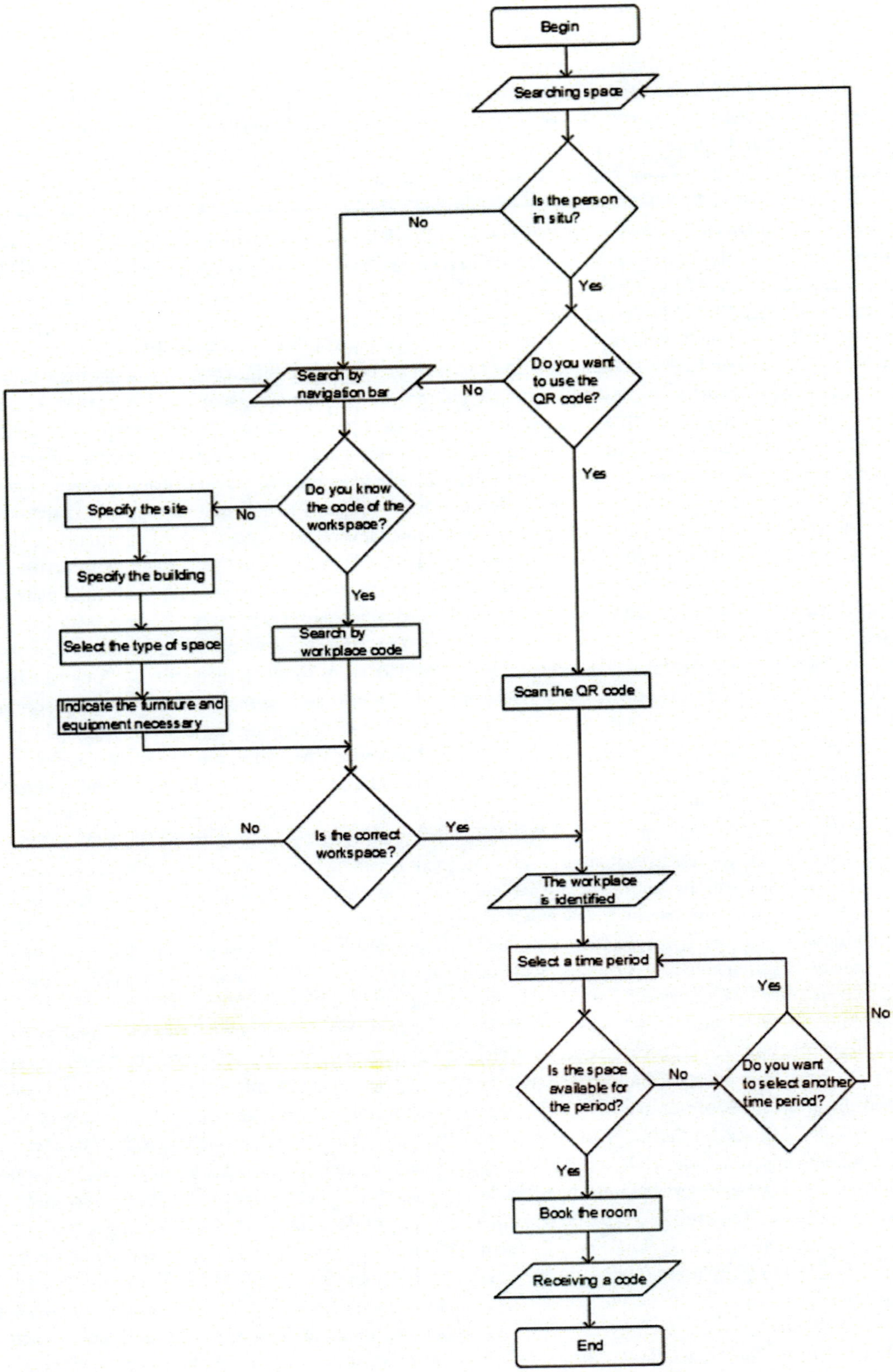

Fig. 3 Booking procedure

Table 1 Indicators for workstation

Indicators for workstation	Formulae
Workstation used [%]	$\frac{\sum \text{Time workstation is used}}{\sum \text{Working time}} \times 100$
Workstation used, but not booked [%]	$\frac{\sum \text{Time workstation is used, but is not booked}}{\sum \text{Time workstation is used}} \times 100$
Number of times a workstation is booked, but not used	Number of times a workstation is booked, but not used

Table 2 Indicators for room regarded as single workspace

Indicators for room regarded as single workspace	Formulae
Workspace used [%]	$\frac{\sum \text{Time workspace is used}}{\sum \text{Working time}} \times 100$
Workspace used, but not booked [%]	$\frac{\sum \text{Time workspace is used, but is not booked}}{\sum \text{Time workspace is used}} \times 100$
Number of times a workspace is booked, but not used	Number of times a workspace is booked, but not used
Square metres for worker [m^2/workstation]	$\frac{\sum \text{Net lettable room area}}{\sum \text{Number of worker}}$

Table 3 Indicators for room regarded as sum of workstations

Indicators for room regarded as sum of workstations	Formulae
Room used [%]	$\frac{\sum (\text{Workstation} \times \text{Time workstation in a room is used})}{\text{Room capacity} \times \text{Working time}} \times 100$
Room used, but not booked [%]	$\frac{\sum (\text{Workstation} \times \text{Time workstation in a room is used, but not booked})}{\sum (\text{workstation} \times \text{TIme a workstation in a room is used})} \times 100$
Square metres for workstation [m^2/workstation]	$\frac{\sum \text{Net lettable room area}}{\sum \text{Number of workstations}}$

Table 4 Indicators for type of room

Indicators per type of room	Formulae
Type of room used [%]	$\frac{\sum (\text{Room used}_{(i)} \times \text{Capcity room}_{(i)})}{\sum \text{Capcity room}_{(i)}} \times 100$
Type of room used, but not booked [%]	$\frac{\sum (\text{Room used,but not booked}_{(i)} \times \text{Capcity room}_{(i)})}{\sum \text{Capcity room}_{(i)}} \times 100$

The working time considers only workday, namely from Monday to Friday, and during each day, the boundaries of working time are from 8:00 to 18:00 o'clock.

Since a room can include more than one workstation, the room capacity is regarded as the number of workstations for a room.

2.4.3 Space Planning Module

To ensure the effective utilisation of workspaces, space should not be allocated on the basis of seniority, habit or personal preference, but considering the activities that have to be carried out and understanding what their space requirements are. For instance, activities that have to be conducted in informal meeting spaces should not take place at workstation desks. Having understood the amount and type of spaces needed in the office, a task force for space management has to establish strategies and targets for managing workspaces. Spaces that can be used in organisations are not unlimited, therefore, a precise planning and scheduling of the activities performed can improve the overall organisation. The results of space usage can highlight the need for more spaces of a certain type of workspace rather than others. Based on results provided by the data analysis module, re-allocation can be planned in the case of unsuitable space allocation. BIM can provide the task force with the clear view of rooms and up-to-date data, such as geometry and equipment of workspaces, therefore avoiding misleading information and potential controversy.

Establishing a method to continuously manage and improve processes is critical to efficiently run an organisation. With reference to space management, the method might consist in a preliminary space planning based on studies carried out inside the organisation and two review processes, which consider different periods of time, to validate the space management. Applying this method, issues regarding space utilisation priorities, space requests and occupant complaints can be regularly solved throughout the operational phase. Each of the two review processes consists of a build-measure-learn cycle that drives the optimisation approach after the first planning task. With reference to the cycle, the first step refers to the action of space planning and carrying out a specific space configuration, the second step regards the analysis of the upsides and downsides of the space configuration by using specific indicators and the third step concerns the recognition of potential mistakes and solutions to reallocate the space.

The first process is conducted after short periods, while the second one examines the management in the long term. In the short term, only data regarding feedback provided by occupants is analysed. This process consists of three steps, as shown in Fig. 4. The first one is the analysis of occupants' feedback. Using the answer to the employees' satisfaction questionnaire, it is possible to create a report which shows if there is any trouble with workspaces and which workspace is involved. If a workspace is reported, an inspection is carried out by going in situ and/or exploiting BIM technology. The last step concerns the adjustments of workspaces and little movements of workspaces if possible and necessary. This process allows improving workspace and ensuring more comfort for employees without requiring complex and expensive actions.

In the long term, data regarding indicators of space utilisation are considered, along with a statistical analysis of the occupants' feedback collected during the short periods. This case concerns major operations, such as reallocating activities, to better manage space. This process consists of three steps, as shown in Fig. 5. The first one regards an accurate analysis of space utilisation preferences by measuring how space is occupied and considering the feedback of the short period reviews throughout the fixed long period of time. If the indicators of space utilisation are under a certain threshold, workspaces related to those indicators need to be considered for improvements. Next, a list of unused or underused spaces is compiled by integrating data regarding space usage of different departments. Then, relationships and interactions between departments and teams of workers are mapped to have a better understanding of worker requirements. Lastly, space managers reallocate spaces into their organisations by exploiting BIM technology.

3 Discussion

The proposed system can provide space managers with an up-to-date schedule of when and how much time spaces are occupied and who the occupants are, leading to potential benefits to decision-making processes. Among the several types of information, users' feedback and space utilisation trends are fundamental to improve productivity and maximise the amount of time office buildings are used. If workspaces are used efficiently, costs related to wastage, which can derive from lighting, ventilation, heating and cooling systems, can be saved throughout the working time. As well as direct expenses for using a space, each space generates associated costs, such as cleaning costs. Taking into account, the space usage frequency and what kind of operation is performed inside, potential optimisation algorithms can maximise the savings by reducing useless and

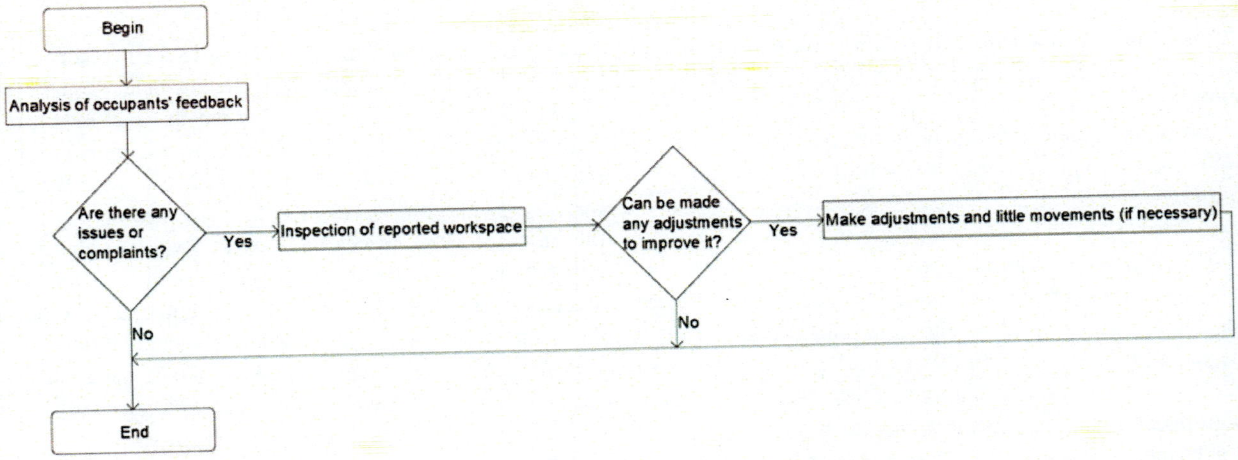

Fig. 4 Short period review

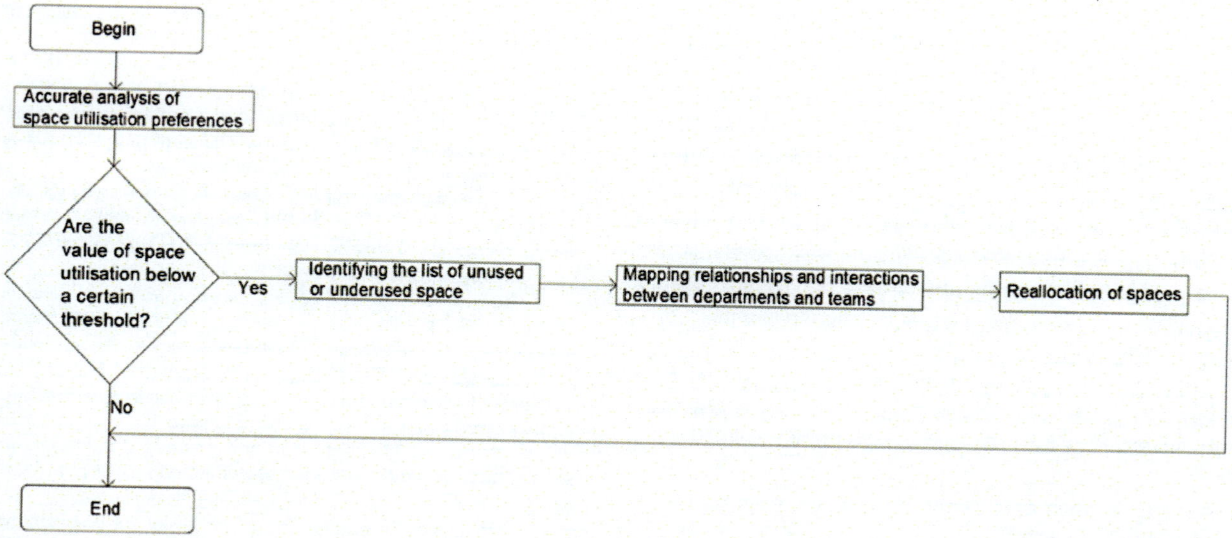

Fig. 5 Long period review

unnecessary cleaning operations and defining where and when to perform cleaning. To this end, tangible and intangible benefits and associated savings should be analysed in the next years to validate the proposed system. The system should be also expanded by developing more modules in the application tier, such as an energy management module and a maintenance management module, connected to the space management module in order to generate a IWMS for an integrated management of all activities regarding the facilities management.

Nevertheless, developing a tool is not sufficient to have the effective management of spaces inside organisations. The efficiency of processes depends on the overall management of organisations. Integrating the Digital Twin-based system with smart working processes can help to break down organisational silos, encourage cross-team working and improve the overall management of spaces. However, a resistance to change from enclosed offices to open spaces and desk-sharing configurations can be encountered among employees. As well as this hindrance, organisations still focus on management by presence rather than emphasising management by results. Thus, more effort on internal processes and responsibility issues are needed to change the way people work and organise their work by exploiting smart working. This will improve the relationship between work, the rest of life and support workers' autonomy. To this end, managers of organisations should focus on the cultural space transition by urging workers to think beyond the classical workplace and supporting management by outcomes rather than by presence.

4 Conclusion

Growing attention to environmental sustainability has led to substantial changes in the way buildings are managed. In particular, consumption and waste of energy are pivotal topics that have gained momentum for improving FM. On these lines, governments have started promoting smart working to both private and public organisations. This method can provide employers and employees with potential benefits but needs organised spaces to support work activities. Getting reliable information of how building spaces are managed can enable space managers to support decision-making processes regarding space allocation and re-allocation, along with boosting the overall utilisation and related savings. To reduce wastage in terms of energy and spaces, a promising starting point includes integrating disruptive technologies, such as BIM and IoT. This paper presents a Digital Twin-based system for smart management of office spaces. The proposed system allows two types of users, such as employees and space managers, to have access to different application modules. Employees can book workspaces and provide feedback, whereas managers dealing with space management can analyse space utilisation data and conduct space planning. Potential benefits of the proposed system include: (i) leveraging the advantages of BIM for visualising the exact location of workspaces and planning space allocation under complex organisational environment, (ii) conducting user-centred post-evaluation analysis to better understand employees' needs and

(iii) increasing the percentage of space utilisation and reducing costs associated with wastage of space and energy. The main limitation of this paper is that only office buildings are considered. However, office buildings are only part of the number of buildings that are under organisations control, especially in the case of public administrations. Future research should focus on extending the space management system beyond office buildings. Finally, it is hoped that this paper will contribute to improving the space management of real estate and providing a basis for the development of a management platform.

Acknowledgements This research did not receive any specific grant from funding agencies in the public, commercial, or not-for-profit sectors.

Declaration of Interest Statement No potential competing interest was reported by the authors.

Funding Source(s) This research did not receive any specific grant from funding agencies in the public, commercial, or not-for-profit sectors.

References

Akcamete, B. A., & Garrett, J. H. (2010, June). *Potential utilization of building information models for planning maintenance activities*. Proceedings of the International Conference on Computing in Civil and Building Engineering, pp. 151–157. https://www.researchgate.net/publication/260056325_Potential_utilization_of_building_information_models_for_planning_maintenance_activities

Annette, S., Hyunjoo, K., & Elisabeth, J. (2011). Early design energy analysis using BIMs (Building Information Models). *Building a Sustainable Future*, 426–436. https://doi.org/10.1061/41020(339)44

Azhar, S. (2011). Building information modeling (BIM): Trends, benefits, risks, and challenges for the AEC industry. *Leadership and Management in Engineering, 11*(3), 241–252. https://doi.org/10.1061/(ASCE)LM.1943-5630.0000127

Aziz, N. D., Nawawi, A. H., & Ariff, N. R. M. (2016). Building information modelling (BIM) in facilities management: Opportunities to be considered by facility managers. *Procedia—Social and Behavioral Sciences, 234*, 353–362. https://doi.org/10.1016/j.sbspro.2016.10.252

Bednar, P. M., & Welch, C. (2020). Socio-Technical Perspectives on smart working: Creating meaningful and sustainable systems. *Information Systems Frontiers, 22*(2), 281–298. https://doi.org/10.1007/s10796-019-09921-1

Bortolini, R., Forcada, N., & Macarulla, M. (2016). BIM for the integration of building maintenance management: A case study of a university campus. In *EWork and eBusiness in architecture, engineering and construction* (pp. 427–434). ISBN: 9781315386904.

Becerik-Gerber, B., Jazizadeh, F., Li, N., & Calis, G. (2012). Application areas and data requirements for BIM-enabled facilities management. *Journal of Construction Engineering and Management, 138*(3), 431–442. https://doi.org/10.1061/(ASCE)CO.1943-7862.0000433

Cox, P., & Fisher Boel, M. (2003). Directive 2002/91/ec of the European Parliament and of the council of 16 December 2002 on the energy performance of buildings. *Official Journal of the European Communities*. https://eur-lex.europa.eu/legal-content/en/ALL/?uri=CELEX:32002L0091

Deantonio, L., Bosetti, D., Cima, S., Martucci, F., Borgonovo, G., Di Bella, G., Pesce, G. A., Valli, M., & Richetti, A. (2020). #Stayathome: Smart working for radiation oncologists during the corona pandemic. *Strahlentherapie Und Onkologie, 196*(12), 1094–1095. https://doi.org/10.1007/s00066-020-01683-z

González, E., Piñeiro, J. D., Toledo, J., Arnay, R., & Acosta, L. (2020). An approach based on the ifcOWL ontology to support indoor navigation. *Egyptian Informatics Journal*, S1110866520301122. https://doi.org/10.1016/j.eij.2020.02.008

Gubbi, J., Buyya, R., Marusic, S., & Palaniswami, M. (2013). Internet of Things (IoT): A vision, architectural elements, and future directions. *Future Generation Computer Systems, 29*(7), 1645–1660. https://doi.org/10.1016/j.future.2013.01.010

GVA Grimley in association with Stellae. (2002). *Space management: A good practice guide*.

Hardin, B., & McCool, D. (2015). *BIM and construction management: Proven tools, methods, and workflows* (2nd ed.). Sybex, a Wiley brand.

Haynes, B. P. (2008). The impact of office layout on productivity. *Journal of Facilities Management, 6*(3), 189–201. https://doi.org/10.1108/14725960810885961

Imperial College London. (2019). *Smart working at imperial*. https://www.imperial.ac.uk/

Jia, M., Komeily, A., Wang, Y., & Srinivasan, R. S. (2019). Adopting Internet of Things for the development of smart buildings: A review of enabling technologies and applications. *Automation in Construction, 101*, 111–126. https://doi.org/10.1016/j.autcon.2019.01.023

Kensek, K. (2015). BIM guidelines inform facilities management databases: A case study over time. *Buildings, 5*(3), 899–916. https://doi.org/10.3390/buildings5030899

Knight, C., & Haslam, S. A. (2010). The relative merits of lean, enriched, and empowered offices: An experimental examination of the impact of workspace management strategies on well-being and productivity. *Journal of Experimental Psychology: Applied, 16*(2), 158–172. https://doi.org/10.1037/a0019292

Kumagai, J., & Managi, S. (2019). *Environmental behaviour and choice of sustainable travel mode in urban areas: Comparative evidence from commuters in Asian cities*. Munich Personal RePEc Archive. https://mpra.ub.uni-muenchen.de/97493/

Labeodan, T., Aduda, K., Zeiler, W., & Hoving, F. (2016). Experimental evaluation of the performance of chair sensors in an office space for occupancy detection and occupancy-driven control. *Energy and Buildings, 111*, 195–206. https://doi.org/10.1016/j.enbuild.2015.11.054

Lake, A. (2015). *The smart working handbook*. http://www.flexibility.co.uk/SmartWorkHandbook/index.asp

Li, L., Yuan, J., Ning, Y., Shao, Q., & Zhang, J. (2017). Exploring space management goals in institutional care facilities in China. *Journal of Healthcare Engineering, 2017*, 1–15. https://doi.org/10.1155/2017/6307976

Libecap, A., Wormsley, S., Cress, A., Matthews, M., Souza, A., & Joiner, K. A. (2008). A Comprehensive Space Management Model for Facilitating Programmatic Research: *Academic Medicine, 83*(3), 207–216. https://doi.org/10.1097/ACM.0b013e3181636f54

Lin, Y.-C., Su, Y.-C., & Chen, Y.-P. (2012). *Mobile 2D barcode/BIM-based facilities maintaining management system*. 2nd International Conference on Strategy Management and Research, Singapore, p. 5. https://www.semanticscholar.org/paper/Mobile-2-D-Barcode-%2F-BIM-based-Facilities-System-Lin-Su/34050f89edffbea993f41e0f91c9a2b0462d479d

Lu, Q., Xie, X., Heaton, J., Parlikad, A. K., & Schooling, J. (2020). From BIM towards digital twin: Strategy and future development for smart asset management. In *Service oriented, holonic and multi-agent manufacturing systems for industry of the future* (Vol. 853, pp. 392–404). Springer International Publishing. https://doi.org/10.1007/978-3-030-27477-1_30

Mamidi, S., Chang, Y.-H., & Maheswaran, R. (2012). *Adaptive learning agents for sustainable building energy management*. https://www.researchgate.net/publication/286192745_Adaptive_learning_agents_for_sustainable_building_energy_management

Maser, K. R. (1988). Sensors for infrastructure assessment. *Journal of Performance of Constructed Facilities, 2*(4), 226–241. https://doi.org/10.1061/(ASCE)0887-3828(1988)2:4(226)

Nicał, A. K., & Wodyński, W. (2016). Enhancing facility management through BIM 6D. *Procedia Engineering, 164*, 299–306. https://doi.org/10.1016/j.proeng.2016.11.623

Oskouie, P., Gerber, D. J., Alves, T., & Becerik-Gerber, B. (2012). *Extending the interaction of building information modeling and lean construction*. Proceedings for the 20th Annual Conference of the International Group for Lean Construction. 20th Annual Conference of the International Group for Lean Construction. https://www.researchgate.net/publication/278667245_Extending_the_interaction_of_building_information_modeling_and_lean_construction

Planon. (2020). *CAFM|Computer aided facility management*. https://planonsoftware.com/us/glossary/cafm/

Qolomany, B., Al-Fuqaha, A., Gupta, A., Benhaddou, D., Alwajidi, S., Qadir, J., & Fong, A. C. (2019). Leveraging machine learning and big data for smart buildings: A comprehensive survey. *IEEE Access, 7*, 90316–90356. https://doi.org/10.1109/ACCESS.2019.2926642

Space Management Group. (2006). *Promoting space efficiency in building design*. UK Higher Education Space Management Project. http://www.smg.ac.uk/documents/PromotingSpaceEfficiency.pdf

Tarzia, S. P., Dick, R. P., Dinda, P. A., & Memik, G. (2009). *Sonar-based measurement of user presence and attention*. Proceedings of the 11th International Conference on Ubiquitous Computing, pp. 89–92. https://doi.org/10.1145/1620545.1620559

University of Michigan. (2012). *Research space guidelines*. https://www.provost.umich.edu/space/other/ResearchSpaceGuidelines.pdf

van der Voordt, T. J. M. (2004). Productivity and employee satisfaction in flexible workplaces. *Journal of Corporate Real Estate, 6*(2), 133–148. https://doi.org/10.1108/14630010410812306

Xu, J., Chen, K., Zetkulic, A. E., Xue, F., Lu, W., & Niu, Y. (2019). Pervasive sensing technologies for facility management: A critical review. *Facilities, 38*(1/2), 161–180. https://doi.org/10.1108/F-02-2019-0024

Application of Virtual Reality to Enhance the Interpretation Dong Ho Folk Paintings in Museums of Fine Arts in Ho Chi Minh City

Ngoc Quynh Giao Pham, Minh Hieu Nguyen, Trieu Hoang Anh Le, Daniel Edem Adzovie, and Petr Stanicky

Abstract

In many works related to architecture, the museum is the only place to store documents and display artifacts related to the historical process, reflecting the culture of a city, a nation, or a country. However, the strength of the technological age and the consequences of war occurred as the storm destroyed the cultural identity layers. The city has also faced the abandon of past achievements in the modernization context. The relationship between conservation and development is, therefore, a controversial issue in the process of reservation art folk painting and art museums. It is still a painful problem that has not yet ended in Vietnam. An undeniable fact is that the city-county has not appropriately preserved museums and art folk paintings in Vietnam. This fact has led to the young generation in the city are not interested in visiting the museum. Among the folk paintings in Vietnam, Dong Ho paintings belong to the line of paintings printed on wood carving planks, created, produced by the villagers of Dong Ho village, and developed into craft villages. This is a line of painting that attaches and vividly shows the traditional Vietnamese agricultural society, the working life of a traditional farmer, and the daily life of the Vietnamese people. Currently, Dong Ho folk paintings are in danger of dying out due to the impact of the market economy, changes in people's aesthetic needs, and difficulties in the output of paintings. Besides, according to a number of painters, Dong Ho paintings are no longer as innocent, simple, "pure Vietnamese" as before, but are gradually being commercialized, with no rich colors like ancient paintings. Dong Ho painting profession today exists weakly, only a few families maintain. According to recent statistics, the number of artisans is only three people, the number of practitioners is about 20, the number of artists who are still capable of teaching is only two people (Mr. Nguyen Huu Sam and Mr. Nguyen Dang Che) are all elderly (Dung 2013). With the power of Virtual Reality, we can transfer not only the architectural aspect but also revive the cultural values hidden within art folk painting to the community as the best way to preserve the culture for all next generations without sacrificing the development potential of the country. Researching on "Application of Virtual Reality to enhance the interpretation, Dong Ho folk painting of Museum of Fine Arts in Ho Chi Minh city" was chosen for that purpose.

Keywords

Virtual reality • Museum of fine arts • Interpretation • Technology innovation • Dong Ho folk painting

1 Introduction

Museums are known to tend to display static exhibits. This is also understandable because the outside target is to provide knowledge information, and this is a place to store and secure historical arrays. Nonetheless, with the Virtual Reality museum in use, it is possible to dynamically change this feature.

N. Q. G. Pham · T. H. A. Le · D. E. Adzovie · P. Stanicky
Faculty of Multimedia Communications at Tomas Bata University, Zlín, Czech Republic
e-mail: hle@utb.cz

D. E. Adzovie
e-mail: adzovie@utb.cz

P. Stanicky
e-mail: stanicky@utb.cz

M. H. Nguyen
International Program - Architectural Heritage Management and Tourism at Silpakorn University, Bangkok, Thailand
e-mail: ngminhieu1207@gmail.com

N. Q. G. Pham (✉)
School of Media Design, UEH University of Economics Ho Chi Minh City, Ho Chi Minh City, Vietnam
e-mail: giaopham2910@gmail.com

H. Altan et al. (eds.), *Advances in Architecture, Engineering and Technology*, Advances in Science, Technology & Innovation, https://doi.org/10.1007/978-3-031-11232-4_10

Malreaux (1978) was one of the first people that introduced the concept of a museum which was a different kind of environment without walls for viewing and performing art. Lately, the concept of a virtual museum and technology was coined by Tsichritziz and Gibbs (1991) They named it the term "Virtual Museum" and it has emerged until nowadays.

The purpose of museum exhibitions is to disseminate precise messages from the objects to their visitors, which are being noticeable through the form and display of exhibited objects within the museum background. By using multimedia with virtual reality and providing multisensory experiences, scientific and cultural information communicated by museum exhibitions can be augmented (Burdea et al., 1996; Wickens et al., 1983).

With augmented reality features, museums in Ho Chi Minh City (HCMC) can bring art to life. Scientific concepts, architectural details, and other elements that visitors are granted by revealing it from a different and extensive perspective. This not only makes it easy for visitors to answer questions but also partially reflects the story behind an era style, an influence, and allows the user to have a sense of authenticity during the discovery process (Hoa, 2019).

The article Applying Virtual Reality (VR) to Enhance the Interpretation of The Museum of Art (TMOA) in Ho Chi Minh City is part of the project Dong Ho painting restoration project used in magic exhibitions at TMOA in Ho Chi Minh City, and has been taken in two years from 2020 to 2022. The project has focused on the propose of a Virtual Environment (VE) that would enable museum visitors to view and manipulate 3D exhibits. We will renovate the classic Dong Ho paintings, which most popular and famous in Vietnam culture from 2D painting to 3D modeling. In the next step, the project also redesigns the interior space of TMOA focus to serve the new 3D environment experiment. Furthermore, the necessity of numerous experimental evaluations defines for determining the best structure of VR devices. The quality of the configuration would enable users to navigate within the VE parallel control exhibits in an effective and perceptive manner. The number of people participating in museums required the diversity of their age, sex, occupation, and level social. They will attend through the VE by using the Oculus glass and fill in multiple-choice questions to explain their communication feelings in the end.

In this paper, we present the first part of the project's process of researching and building a modeling of Dong Ho folk painting characters. The first section of the paper discusses the status of TMOA in HCMC and introduces the virtual reality application in an exhibition in the context of the real museum. The second section presents some case studies of an exhibition in TMOA and the interpretation model for an art exhibition by VR technology. Finally, the recommended uses of a model experimental for manipulating exhibits are briefly mentioned.

2 The Current State of Art Museums and the Museum of Fine Arts in Ho Chi Minh City

2.1 The Current State of Art Museums in Ho Chi Minh City

Many social history and heritage places are particularly adapted for visitor attractions since they represent distinctive characteristics that express the history, culture, or environment (McKercher, 2002). These places also reflect the vibrant scene of a region's customs, cultural identity, and surroundings. Although the critical goal of museums is to serve as custodians of the heritage and cultures of areas and nations by preserving and restoring traditional historical objects connected with those places (Cho, 2013), this task often includes the propagation of information regarding and exhibit of their objects (Capriotti et al., 2012). Over more than a century, these features affected museums to be accessible to visitor arrivals for many purposes.

Museums are often being questioned to validate their viability, and this explanation is also started demanding economically challenging. Museums might play a critical role, such as adding regional or national income production, offering jobs or education, and minimizing cultural issues in the macroeconomic social structure. The audiences' absences at museums or art galleries in Ho Chi Minh City are progressively common in real life. The challenge lies perhaps not in the massive change from colonial archeological collections to 20th-century museums and galleries, but also in the possibility that the community is not the primary target audience. Although the museum's definition, which has never been implemented and incorporated into the citizenry's public activities, local involvement declined significantly. Museums of art present a substantial obstacle to rationalizing and attracting visitors as arts and museums are unfamiliar and unique to South-East Asian communities in general and especially in Vietnam. E.g., regarding items and objects taken from their original contexts in monuments or monasteries and placed in a nearby anthropology museum, citizens might see significantly less relevance. They did not notice that art was distinct from crafts, which often serve in public experience and are linked to rituals and spirituality. The definition of art as an aesthetics and amusement artifact had little part in Vietnamese local experience. Recent art projects may have opened their doors to the establishment's culture, but most citizens also had no code to appreciate art pieces. One challenge was the art space—the museum as a touch zone—which did not occur in connection with Western modernists' emergence in the region. The art museum's shortage has become more problematic as modernism was confronted by contemporary post-modern and various art forms. The divide between sculpture, galleries, and community has appeared to deepen.

2.2 The Museum of Fine Arts in Ho Chi Minh City

The majority of Vietnamese museums are historical museums that reflect the country. The two largest Art Museums in Vietnam: the Vietnam Fine Arts Museum (VFAM) in Hanoi and the Ho Chi Minh Fine Arts Museum (HCMCMFA) reflect the nationwide historical establishments that uphold Vietnamese nationalist agendas. The VFAM honors the ethnic origins of the Vietnamese people throughout Asia, defiance to external invasion, and endurance to war hazards. The art specimens of the Museum are categorized into three cycles: from the earliest, ancient to the eleventh centuries, from 11th to the 19th, and up to the twentieth century. The dual scope of the anthropological and fine arts is the critical role for the museum to display its acquisitions, informing the audience about the status of artifacts in Vietnamese culture while showcasing their artistic elegance. Vietnamese artists integrate the advantages of global art while maintaining native personalities (VNFAM, 2011).

Concerning the VFAM, the HCMCMFA maintains a national archive of Vietnamese paintings, but it exposes the absence of works from the post-colonial southern artists of the former capital area of the South of Vietnam. After the Đổi Mới (Reform Era), Ho Chi Minh City became Vietnam's socioeconomic capital, and the museum became a central site for the society to associate with the local and national art heritage. Therefore, South Vietnamese arts' distinctive heritage has either been ignored or otherwise blurred by the government-supported cultural factual record from post-colonialism to reconciliation duration; many art experiences and artists in the southern Republic of Vietnam have been substituted by their northern communist equivalents (Corey, 2014). Whereas the philosophy of acquisition of the HCMCMFA retains conservative, the museum has expanded its spaces to showcase more creative Vietnamese contemporary art. The museum has been leased to numerous arts organizations since the 1990s to organize installations and activities. Yet, this partnership is intensive; the museum controls exclusively the kinds of events that it approves, provides limited to none resource abilities to project administrators, and demands the planners to conquer many administrative challenges. The museum will terminate programs—as those it considers excessively controversial or inflammatory—at a certain period if the activity does not satisfy such criteria (Fig. 1).

The museum is planned to satisfy the requirements of the communities in the metropolitan area to embrace visual art and be a destination for tourists to discover and explore Vietnamese art historians. Over a preparation process for facilities and objects, the museum was launched to the community in late May 1989. At present, the museum has over 21,000 objects grouped into valuable collections. The museum artworks are subdivided into two critical areas: Classical art—native crafts and contemporary art with notable acquisitions such as rebellion paintings, artworks from Indochinese and Gia Dinh college artists. Other authors are Nguyen Gia Tri, Nguyen Sang, Diep Minh Chau, Kim Bach, Dinh Ru, Quach Phong, etc. They represent the core characteristics of the Vietnamese art community and fine art in the southern regions.

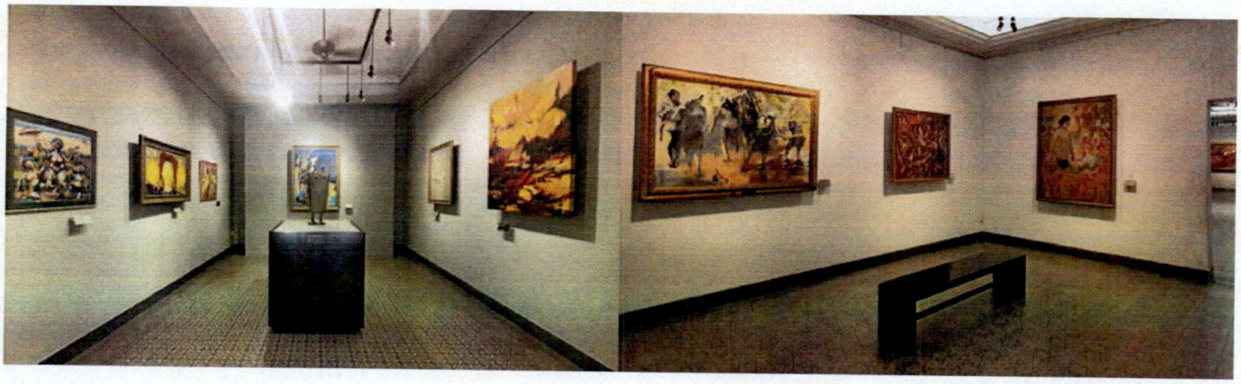

Fig. 1 Inside the Ho Chi Minh City museum of fine arts

3 Dong Ho Folk Painting. Suggestions for Visual Reality (VR), Application of an Art Exhibition in the Art Museum in HCMC, and Project Workflow

3.1 Dong Ho Folk Painting

According to the history of Dong Ho villagers recorded, Dong Ho painting was born in the eleventh century in the Ly Dynasty at Mai village, (Dong Ho village nowadays) located on the southern bank of Duong River, now in Thuan Thanh district, Bac Ninh province, about 35 km from Hanoi (Nguyet & Van, 2019). Meanwhile, many researchers believe that craftsmen started to print the first paintings in Dong Ho village from the Le dynasty, King Le Kinh Tong (1600–1619) (Education, 2015). During this period, Dong Ho was one of the places that made famous wood carving traditional paintings alongside other lines of traditional paintings, Hang Trong, Kim Hoang. In essence, Dong Ho painting is made of only two colors in black and white from wood carving molds (Anh, 2019). However, from the fifteenth century, the craftsmen in the village created other colors for the paintings such as red, yellow, green, and blue. Like many other villages that make wood carvings (Fig. 2) and print pictures, most of the villagers are involved in the production of pictures from wood carving "Diep" paper, using natural colors (Fig. 3) for creating different painting themes (AmiA, 2017).

The peak period of the painting village was from the late nineteenth century to the 40s of the twentieth century. At that time, there were 17 families in the village, they all made paintings. At that time, every year around July or August, the whole village was busy preparing for and the Tet season, the village was full of colors. Traditionally, Dong Ho paintings are often an indispensable element in Vietnamese families every Lunar New Year season. Dong Ho painting also has other names such as Tet Paintings or Xuan Tranh because of its bright colors, showing an optimistic and positive spirit (Nguyet & Van, 2019). The paintings also show pictures of Vietnamese village landscapes, and many other metaphorical images such as animals and babies that mean luck and success to families in the New Year (Anh, 2019).

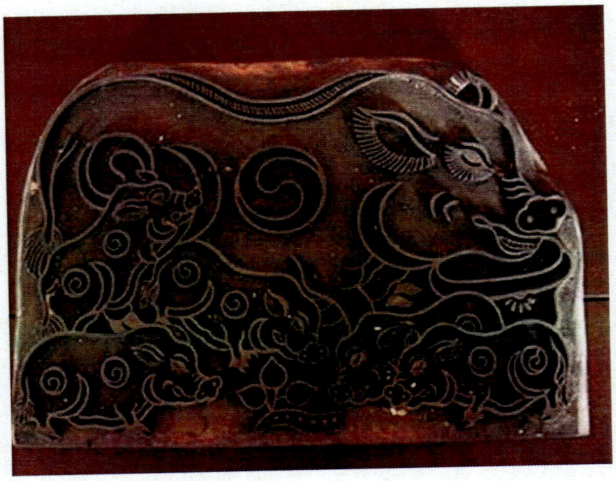

Fig. 2 Wood carving planks of Ying Yang Pig painting (Binh, 2015)

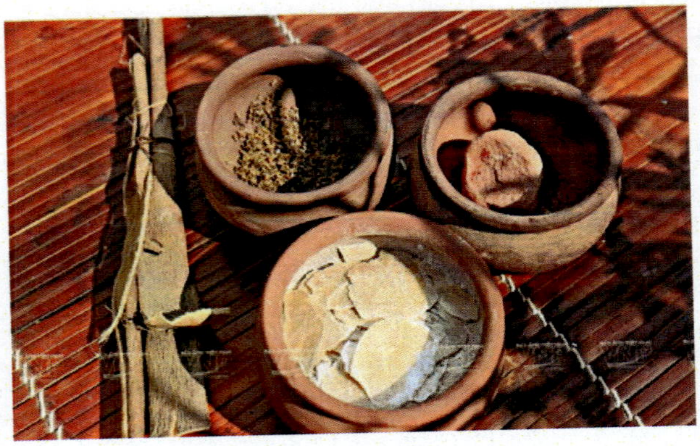

Fig. 3 The printing process and some of the main materials that make up the colors of Dong Ho's paintings are: yellow (Hoe flower), white and white (scallop shell), red (pebble, wine wood) (Dat, 2019)

The topics that the painter showed were very life images associated with the northern plain, from the images of animals attached to the house field, the lifestyle of people such as buffaloes, pigs, fish, mouse, flock of chickens, flock of pigs (Fig. 4), picture of a baby holding a Vinh hoa chicken, holding a precious duck. Topics of satirical paintings such as mouse wedding paintings (Fig. 4), jealousy paintings. Activities include stories: Wrestling, swinging, Dr. Vinh Quy, historical paintings such as Ba Trung, Phu Dong Thien Vuong, Ngo Quyen or Comic books like Kieu and Thach Sanh (Fig. 5). Besides, there are kinds of paintings such as bamboo, apricot, and four-quarter paintings.

Before 1945, Dong Ho village had more than 150 families making paintings. However, with the dominance of the wave of modern painting and the changing lives of Vietnamese people, Dong Ho paintings gradually disappeared from Vietnamese families on holidays and New Year. Most of the people who buy Dong Ho paintings today are tourists or as gifts for foreigners, who are still interested in traditional art (Nguyet & Van, 2019). For that reason, the villagers of Dong Ho village also quit the profession of painting to move to another job to earn additional income such as making joss paper and gold paper. At the same time, the tradition of printing old paintings by hand was also gradually replaced by a series printer, losing the folk character of Dong Ho's paintings (Anh, 2019).

In March 2013, Dong Ho folk painting profession was recognized as the National Intangible Cultural Heritage

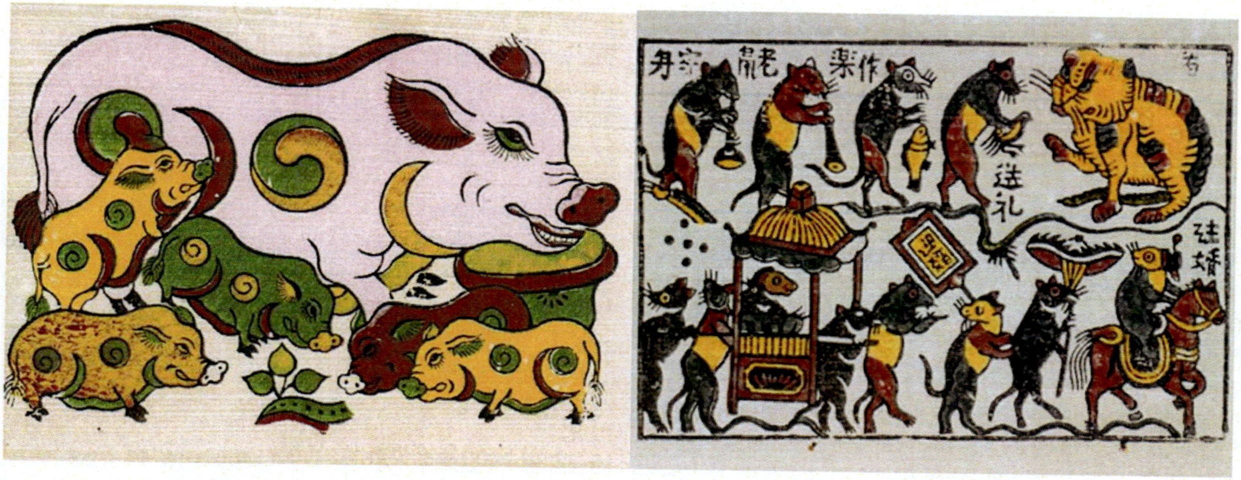

Fig. 4 Example of Dong Ho folk paintings: Ying Yang Pig painting (Left) (Sam)-mouse's wedding painting (Right) (Unknow)

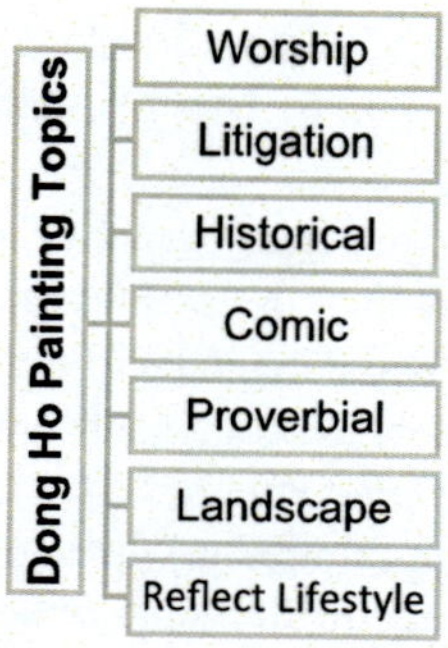

Fig. 5 Dong Ho painting topics (Author fig)

(Van, 2018). In 2018, the Ministry of Culture, Sports and Tourism (VHLSS) agreed with the proposals of Bac Ninh Provincial People's Committee on the development of Dong Ho Folk Painting Career Profile. The United Nations (UNESCO) lists the Intangible Cultural Heritage that may require world-level protection (Van, 2018).

3.2 Suggestion for Visual Reality (VR), Application of an Art Exhibition in the Art Museum in HCMC, and Project Workflow

Museums steadily embrace displays and support initiatives that prioritize human experience and collaboration (Abercrombie & Longhurst, 1998; Stylianou-Lambert, 2010). While conventional museological types consider tourists as inactive information users, the current museology strives to increase citizens' proactive participation via digital technology and customized personal interaction. It is suggested that this is more culturally integrated and encourages personal and interpersonal change (Stylianou-Lambert, 2010). O'Connor (2010) argues that digital technology's growth will introduce innovative ways of involvement and participation in the artistic field and encompass multiple influences and management approaches. Study results of experiences with culture and art facilitated by new technology, for example, find that those interested in museums and virtual art are equivalent or similar to people who attend and associate more frequently (e.g., Dean et al., 2010; Holdgaard, 2012).

The cultural objects displayed in a museum's physical surroundings are commonly seen in the display where minimal information is accessible. Museum items could be scanned and interpreted in an enjoyable digital environment at simulated museum exhibitions. A simulated museum might include specifics that traditional exhibits cannot present in a museum display. Thus, museum curators are granted a more satisfying experience due to a deep understanding of artifacts through the interactive context instead of the ones sealed in a display cabinet with a necessary explanation on a label. In such immersive experiences, consumers can interactively and flexibly navigate galleries. Digital museum exhibits offer virtual guests the opportunity to examine and interact with an item from all dimensions. Digital visitors will navigate themselves; they identify where the displays are located in the digital arena. The artifacts themselves will express their value when explored in combination with other area objects and with a narration that ties the artifacts to their history and helps bring to reality the possible cohesiveness of artifacts and their stories (*see also* Soren, 2004).

As described above, the number of young audiences attending at the art museum is decreasing since they can not pursue their specific motivation, tackling the issues of the Vietnamese Art Museum in general and the HCMCMFA in specific. The following levels of info about the artifacts are presented, as well as the museum's current approach to youth is indeed not exactly ideal. The VR approach will enable the HCMCMFA to handle the situation. Virtual museum-focused groups are museum curators and end-users. In the second group, three subcategories may be divided: experts, academics, and visitors (Bowen et al., 2004). Digital museum exhibits will provide a significant volume and level of experience, intended to expand insights, operate efficiently, and cultivate a broader understanding of each of the previously mentioned demographics of cyber guests. They might satisfy the demand for the essential and differentiated knowledge of ordinary tourists and need no additional support to decode the meanings and ideas within museum objects (Paternò & Mancini, 2000; Hooper-Greenhill, 1992).

Digital museums are often capable of presenting in-depth appropriated knowledge for different groups of visitors, thus bringing attention to not just one but several museums and art galleries, to various investigations, including the descriptive analysis criteria of experts and students. In comparison, artistic galleries can attract audiences who are rarely accessing museums or exhibitions and have no relevant familiarity or interest in the exhibition's subjects (Economou, 1998). Guests at digital art museums are offered chances to improve vital skills. A virtual museum will allow visitors to discover, practice individuality, and be prominent users, constructing their virtual tours and directions. Besides, interactive platforms can be used as intelligent systems that enable virtual tourists to surpass human experience limitations such as memory or decision-making weaknesses and achieve their expertise (Pea, 1985). A typical indication of this would be virtual museum tourists' potential to construct an individual multimedia showcase of digital artifacts, a display that suits their expectations and can exchange it with others. More educational opportunities arise in a simulated museum setting than in a traditional museum (Davallon, 1998; Mokre, 1998). Many digital museums have been constructed to reflect the functionalist concepts of educating by construction and experimentation, including engagement, experiencing, and learning together (Hein, 1991; Falk, 2000). In a simulated museum world, guests are not spectators but engage with learning artifacts and create information themselves. Moreover, interactive platforms are used as cognitive technology that enables simulated tourists to overcome human mind limitations such as memory or problem-solving limitations and create their expertise (Pea, 1985).

The restoration process we summarized in project workflow will be followed over the two-year project period. This process consists of three main steps: Planning, Modeling, and Programming. In the planning section, we will conduct research and select a concept story for the VR movie. We also conduct a selection of the most iconic and iconic paintings of Dong Ho folk painting. Then we will conceptualize the scenario and story setting. At the same time, we also created the storyboard for this period and corrected many times to complete the script. In the modeling section, we will continue to refine once more but more specifically the characters in the selected paintings. We use popular modeling and animation programs such as 3DMaya, 3DsMax, to reconstruct the character modeling story background. Texture mapping will use larger software such as UE4 Archive, Corona Archive to render 3D spatial scenes and can be used in VR. We also check, test, and fix errors in rendering and rendering. The Programming section links modeling environments and contexts to the same video. In this section, stages such as adding subtitles, annotations, sound, and animations are also edited and edited to be included in the final movie (Fig. 6).

4 Case Studies and the Structure of Space in the Virtual Museum for Exhibition, the Ho Chi Minh City Museum of Fine Arts

4.1 Case Studies of VR Exhibitions in the Museum in Europe

After many years of experimenting and applying VR technology in various types of exhibitions, many museums have turned to digital from 2016 to 2020. Especially, in the long-term covid-19 epidemic with restrictions on travel and sightseeing in countries around the world, which contribute to reinforcing the diversion of investment in museums, virtual reality is absolutely correct.

For example, in October 2019, a virtual reality experience explores Leonardo da Vince's classic Renaissance painting "Mona Lisa: Beyond the Glass" was released (Coates, 2020). Through interactive design, sound and animation help users discover details about the painting, such as the change in the background behind the Mona Lisa and the overall scenery as models (Fig. 7). Users through Oculus glasses can step inside the painting, experience another depth, and change of minds of the Mona Lisa.

In the UK, London's Tate Modern created a captivating VR exhibit alongside their Modigliani nostalgic from 2017 to 2018 (Coates, 2020). Visitors were capable to experience engagement in a 3D model of the artist's Paris studio. After a long researching period, the museum created a lively interior space of the artist's final studio as it has been compared 100 years ago to the actual studio space as a model. VR technology is a way of helping people link their emotions with an artist by communication of visuals. Nonetheless, various famous artists are deceased and difficult to comprehend their philosophy of art presented by paintings on flat walls, VR is a different way of absorbing that information and tells us the living story of art (Fig. 8).

In another case, The National Museum of Natural History turned to VR technology which intended to transform the concepts behind it more easily understandable. It launched

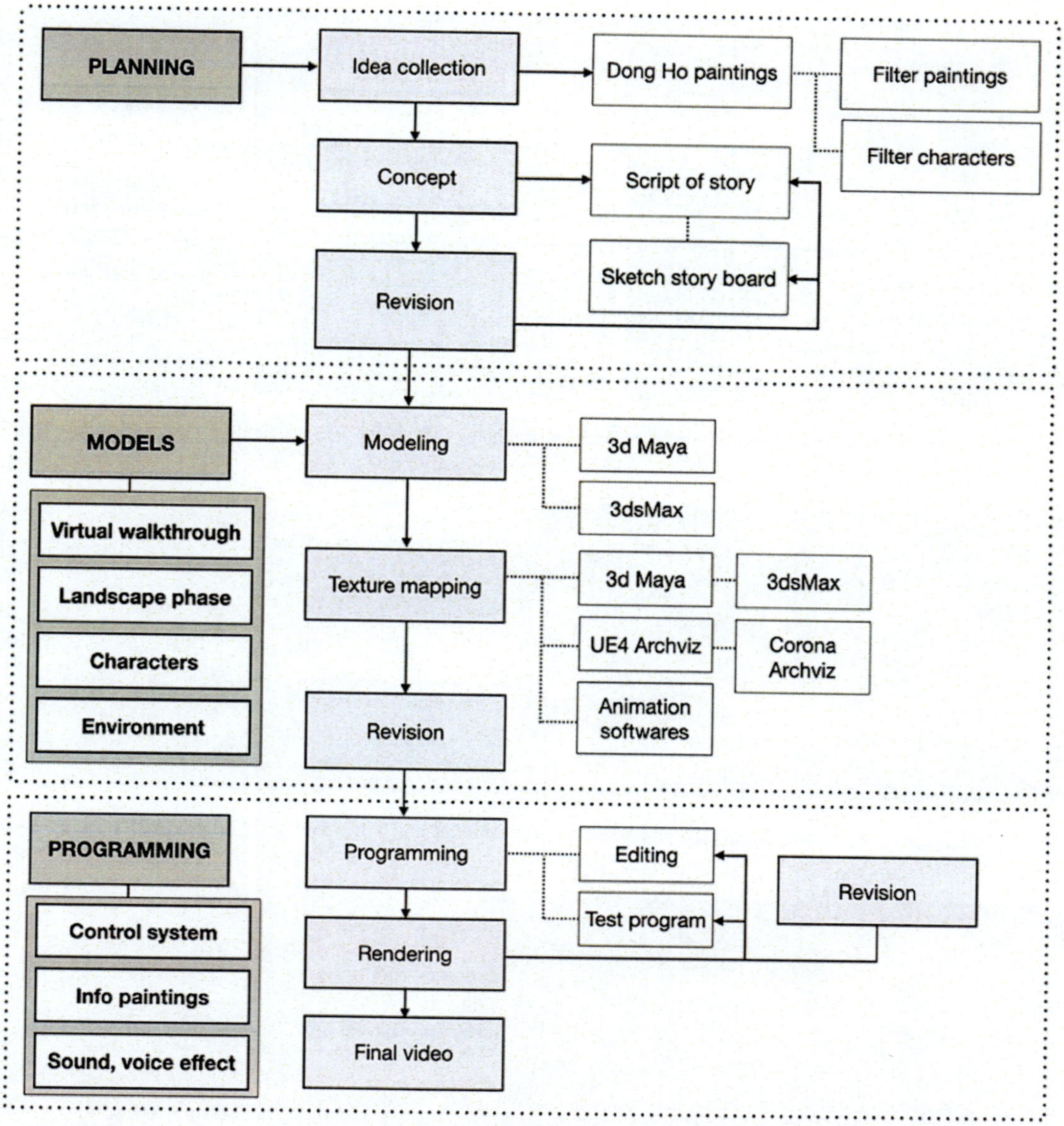

Fig. 6 Workflow for the development of the VR video—**a** PLAN (top), **b** MODELING (center) and **c** PROGRAMMING (bottom)

its first long-lasting VR exhibition in 2018 (Coates, 2020). When visitors enter the "Cabinet of Virtual Reality" with the VR headsets, they are entirely plunged into the journey of discovery evolution of history. They can travel between a million species and view detailed information such as a diversity of creatures up close (Fig. 9). Based on the positive feedback from visitors, the museum aims to develop and improve their permanent VR exhibition supplementary to prospects.

The last case is the Natural History Museum in collaboration with the broadcaster Sky developed Hole World interactive experience that lets you meet face to face with Sir David Attenborough, an English broadcaster and natural historian. The interactive capability brings you to London's Natural History Museum (Coates, 2020). The visitors were put in interactive with many rare specimens from its world-famous collection. It permits you to conduct and resize the objects when Sir Attenborough teaches you significant facts about the animal lived, ate, breathed, and furthermore.

4.2 Explaination of the Effort in Developing of VR Environment

Presently, Virtual Museum applications adjust from entire cave systems to partial multimedia presentations. VR systems bring a compelling sensory and affective experience that is certainly afforded by fully immersive might be the

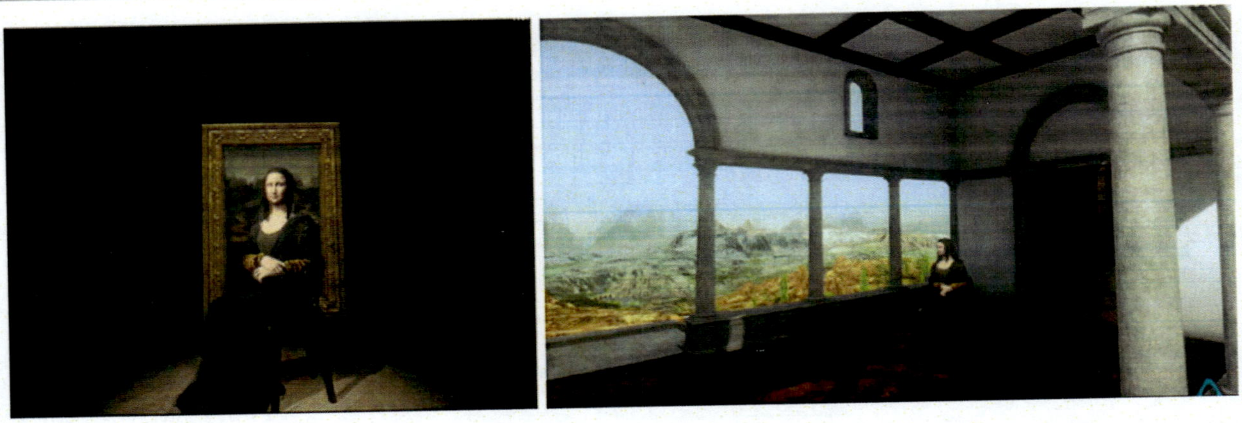

Fig. 7 Modeling of Mona Lisa and the landscape were created in VR technology (Emissive VR, the Louvre Museum, 2019)

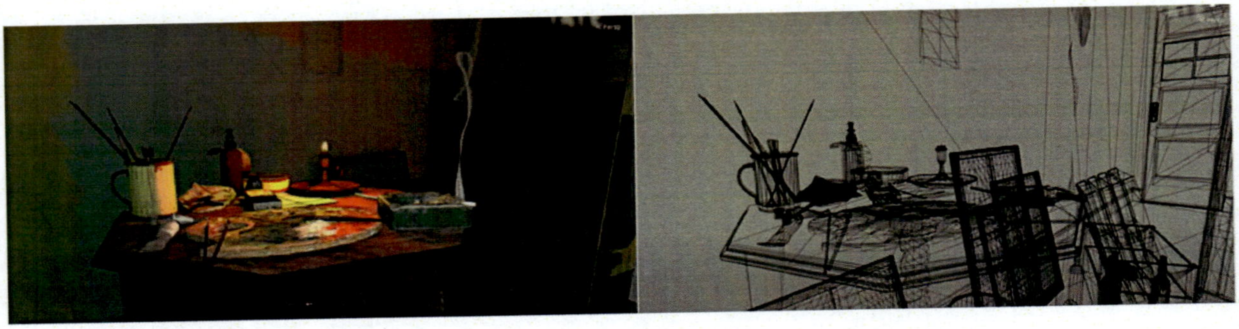

Fig. 8 3D model in grid with processing (Right) and after rendering with animation and lighting (Left) (TATE, 2018)

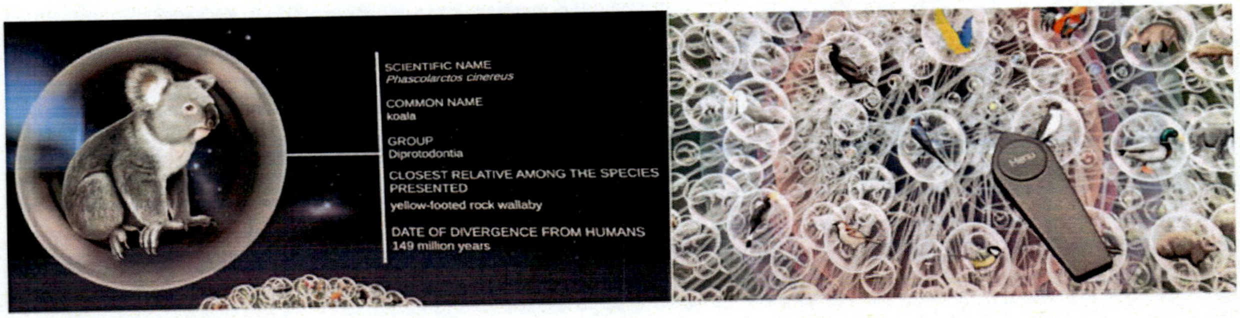

Fig. 9 Interactive screens select species-specific examples, Koala bears will display basic information next to their picture (French National Museum of Natural History, the Orange Foundation, 2017)

most effective system in an exhibition at the museum. The number of installations worldwide is limited because these systems use cutting-edge technology at an extremely high cost (Lepouras et al., 2004).

Changing from a classic model of an exhibition to create a VE may provide an enriching, intense, and entertaining presentation of assured exhibits to visitors. Furthermore, these previous examples of VR exhibitions in museums show several other reasons that may explain the exertion of improving such a system:

- Deficiency of space: Most museums display a small percentage of the exhibits they own since presentation space is frequently limited. Depending on the shape, size, and material of objects may be too breakable, cumbersome, or of high value to be exhibited. In this case, the stored objects can be successfully displayed by VR presentation within the environmental context of the real circumstance museum (Lepouras et al., 2004).
- Simulation of the exhibition environment: Visitors possibilities view a simulation of vital objects, species,

architectures, or environments by a VE system offering. These environments may either:

- No extended exist today
- be damaged and in the requirement of reconstruction renovation
- not be able to experience because they exist in a served weather condition or their conditions do not allow for their interior to travel through.
- Performance in an unsafe or isolated environment: Images within an ancient warrior, dinosaur, or on the mountains of Jupiter, which may be too complicated or too unsafe to physically visit. A VE system is the most guaranteed way of visiting this kind of environment.
- Ambulatory exhibition: A mobile VE system, which is a method where the digitized content of a museum is experienced in a practical manner, can be easily carried to any exhibition site or remote location. A wider audience may view significant exhibitions without the requisite of far-removed.

5 Selected Areas for V.R. Interpretation Experiment and Modeling of Characters from Dong Ho Folk Painting in the Ho Chi Minh City Museum of Fine Arts

5.1 Selected Areas for V.R. Interpretation Experiment in the Ho Chi Minh City Museum of Fine Arts

The building was designed by Mr. Rivea of a French architect in the late XIX century—early XX century harmoniously combined Eastern architecture in the exterior, decoration tiled roof and cladding, columns, pottery, and ceramic decorations on the roof (Vy, 2007).

The house was built not for a museum, but for housing and working. In 1987, the house was renovated and repaired into Ho Chi Minh City Museum of Fine Arts and officially opened to welcome visitors at the end of May 1989 up to now. The museum consists of three main buildings: the art gallery is located on Pho Duc Chinh street, the middle area, is the seasonal display. The building on Nguyen Thai Binh street displays ancient art and sculpture. The main area is a 3-story building with beautiful French-Chinese architecture, inside there is a large courtyard for ventilation purposes. Here, visitors can admire art before and after 1975, as well as contemporary works of art. More than 80% of the works on display are paintings. In particular, there is a display of national treasures, which is the lacquer painting of the North Central South Spring Garden by the artist Nguyen Gia Tri. The 3rd floor of the museum features modern paintings and shows the vision of a new generation of painters (Co, 2019).

In general, the structure of the museum primarily comprises three types of foyer. It depends on the allocation of movement within the museum compound:

- Entrance hall
- Museum-category foyer
- Museum foyer.

Classification of exhibits and museum management requirements which consequent organization of a particular space in each museum (Charitos et al., 2000). The form of each museum hall was defined by the following factors:

- The nature, size, and number of exhibits include.
- The specific demands for each exhibit category.
- The method is that the museum foyers are attached to the complete museum compound and subareas.
- The 3D direction technique is used in the VE.
- The approach of observing a set of exhibits.

In each exhibition, the way that these objects are organized will decide to position the exhibits within the virtual museum. The spatial constitution of an exhibition is determined by the method that occupations are organized inside each museum. In the point of fact, it depends on the purpose and objectives of each different exhibition. The creation of space within the museum environment has a target to support the participant in navigating within the VE, at the same time maintaining a sense of direction in an applicable design environment. By application of architectural comprehension, the participant's environmental knowledge was enhanced, and in precision viewer attention toward definite points within the exhibition space (Charitos et al., 2000).

The chosen spot for an experiment of V.R. interpretation is the 2nd-floor hall of the East Building in the whole museum complex—Currently selected as a place to display ancient and post-modern arts. Although it is adjacent to both Pho Duc Chinh and Nguyen Thai Binh Streets, this building is the furthest away from the Museum's main entrance, also the final destination in the tour, so the number of visitors here is the least in the entire museum. One reason to choose this site for an experiment was to see the most apparent progress in attracting visitors to the museum using modern interactive technology.

The chosen room is a square layout with 105 m^2, one of the Museum's largest rooms. Space morphology is still very luxurious thanks to the precise height of more than 8 m with the hexagonal ceiling surrounded by the fillet dome. The high elevation also allows many facades to decorate with

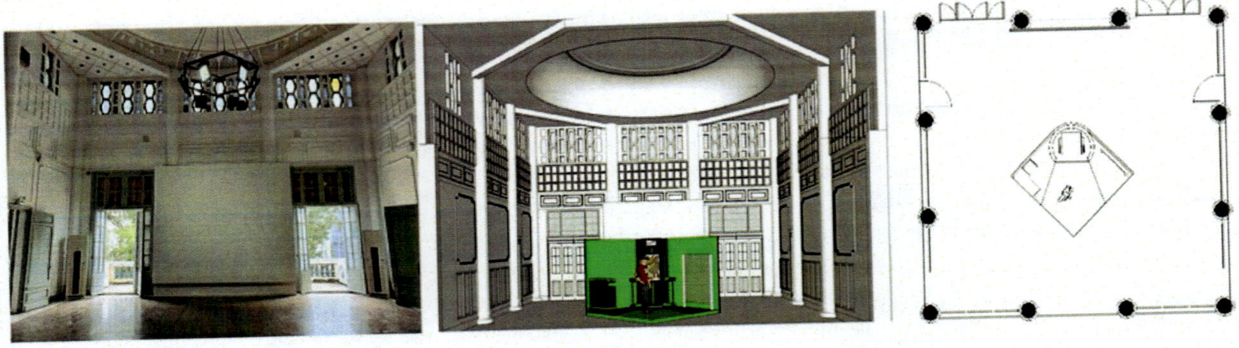

Fig. 10 Status of the pilot area (left), the 3D view of VR booth setup (middle), the plan of the VR booth (right)

columns, pilasters, floating ledges, tiles, and wooden louver doors, all typical Indochina architecture elements, and create a sense of luxury and art attraction. The white tone used for the entire interior space also differentiates itself from the Museum's signature yellow wall, making visitors feel ready for a new interactive experience. In terms of lighting, the whole room is calculated to combine both the artificial and natural light sources from the front and sides. This lighting method creates two bright areas of different intensity during daylight, depending on whether the viewer's position and direction of view are either forward or backlit. Therefore, the team chose to place the V.R. booth in the center to have a uniform artificial lighting effect, shading facing the sidewalls to eliminate its impact on the substance of V.R. Booth screening (Fig. 10).

5.2 Modeling of Characters from Dong Ho Folk Painting

The process we show in the framework of this article is the process of building a character model from 2D paintings into 3D modeling. After classifying the paintings based on more than 200 still preserved folk paintings, we classified the paintings into seven popular categories. Formerly, through another round of selection, we selected five typical characters in the picture and moved to the step of drawing 3D model frames for the characters. In the meshed frame, the models are added surface materials, particularly colors, and afterward, proceed to edit small details of the model. As a final point, there will be a rendering step to terminate the modeling (Fig. 11).

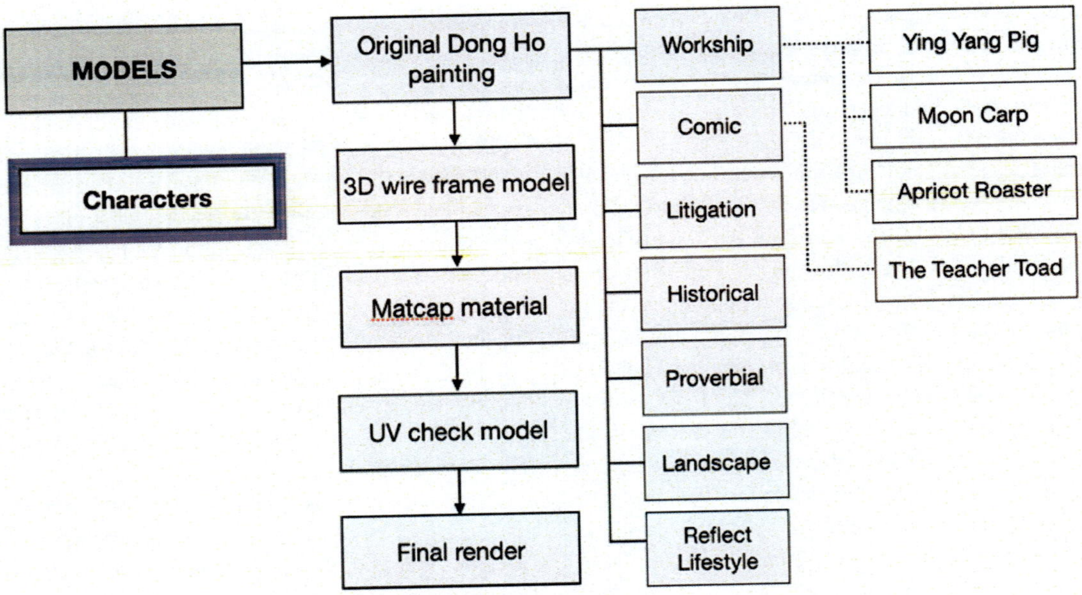

Fig. 11 Diagram of modeling characters in Dong Ho painting using a 3D VR model

In detail, as classified in the research section, we choose five types of paintings to restore by following the character modeling process from 2D paintings to 3D format. The restored characters are as follows: Yin-yang Pig, Apricot, Roaster, Toad Teacher, Moon Carp. How to restore the character model is shown through the four steps below for each character (Figs. 12, 13, 14 and 15). The characters are in default lighting environments in the 3DMax software. In the future, after testing and modifying more models, we will proceed to bring the model into the VR environment and continue with the next steps of the project.

6 Conclusions and Future Work

In this article, we have identified the overall ground for the VR installation in the Ho Chi Minh City art museum space. We found that there are a number of other factors affecting the VR testing space such as soundproof walls, light from windows, and interior design for the experience to be comfortable. These issues will be studied in detail and determined more in-depth solutions in the future paper.

We also initially conducted the selection of characters and animals in Dong Ho paintings to rebuild from a 2D version to 3D characters that can function in the final video. The next steps in the Dong Ho painting restoration process are to transform the landscape of Dong Ho village in the XV–XVI century into 3D format. It will then take on the characters, voices, sounds, and words of the native people. We in addition plan to complete a 5–6 min long video clip that brings together the most unique characters and scenes of Dong Ho's painted village. Users can experience the feeling of stepping inside the Dong Ho painting village space and admiring the characters in the paintings in a vivid form. As a final step, the footage will be shown in an actual exhibition in a suitable museum space. Viewers will be assessed and surveyed on the perception and effectiveness of visual communication after comparing between watching Dong Ho painting exhibition in a 2D version and experiencing the VR movie.

Moreover, we expect to complete this project within one and a half years and come up with a solution to preserve the traditional culture and art of Dong Ho folk painting, a line of painting that expresses the national identity of Vietnamese. Our aspiration is that the research will be completely

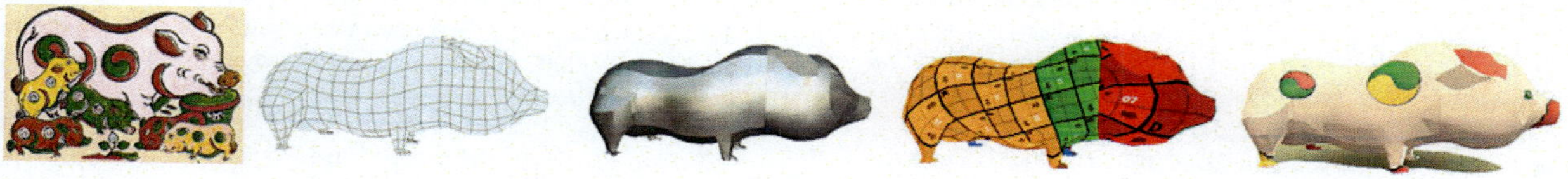

Fig. 12 Yin-yang Pig from Dong Ho painting with 3D VR model

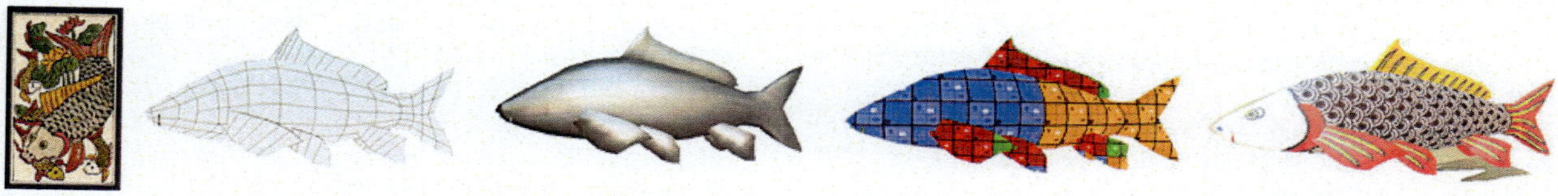

Fig. 13 Moon Carp from Dong Ho painting with 3D VR model

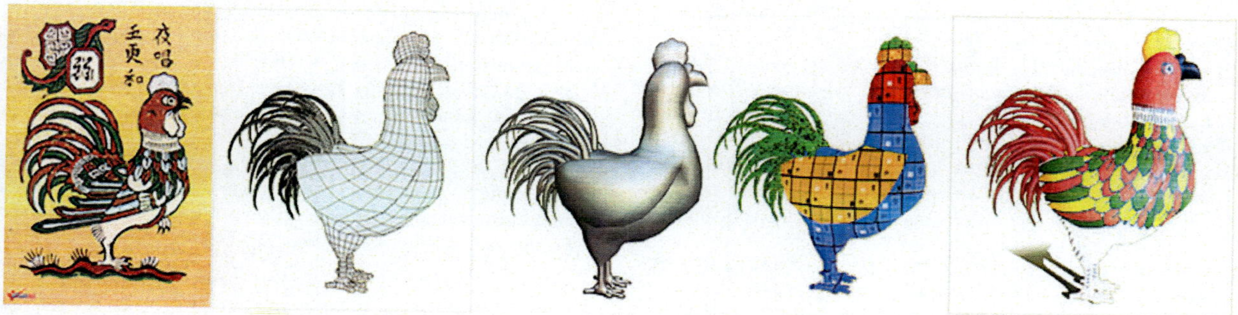

Fig. 14 Apricot Roaster from Dong Ho painting with 3D VR model

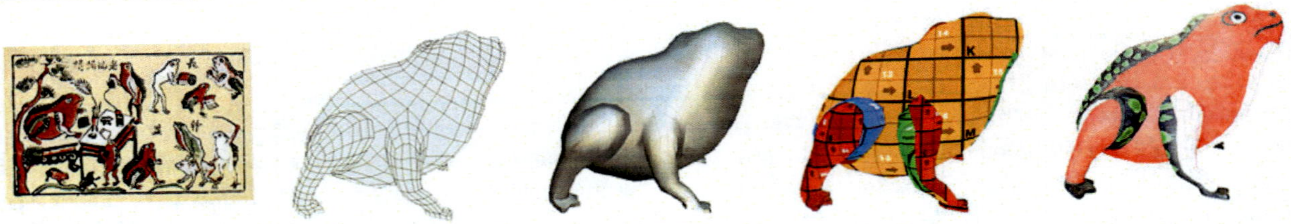

Fig. 15 Teacher Toad from Dong Ho painting with 3D VR model

developed. If it can be widely exhibited to all levels of the people in society, it will bring the popularization of the knowledge of traditional folk arts together with domestic copper and other countries in the world.

Acknowledgements This article was developed from the Ph.D thesis from Faculty of Multimedia Communications, Tomas Bata University, Czech Republic. I am particularly grateful to my advisors, colleagues, and friends for their help and support.

References

AmiA, X. (2017, January 26). *Khám phá lịch sử tranh dân gian Đông Hồ*. Retrieved from tranhtreophongkhachdep.com: https://tranhtreophongkhachdep.com/lich-su-tranh-dan-gian-dong-ho/

Anh, C. (2019, February 3). *Làng tranh Đông Hồ xưa và nay*. Retrieved from baodansinh.vn: https://baodansinh.vn/lang-tranh-dong-ho-xua-va-nay-89931.htm

Binh, A. (2015). *Làng Tranh Đông Hồ (Bắc Ninh)*. Retrieved from www.tranhdangiandongho.vn: http://www.tranhdangiandongho.vn/tin-tuc/lang-tranh-dong-ho-bac-ninh-7

Burdea, G., Richard, P., & Coiffet, P. (1996). Multimodal virtual reality: Input-output devices, system integration, and human factors. *International Journal of Human-Computer Interaction*, 5–24.

Capriotti, P., & Kuklinsk, H. P. (2012). Assessing dialogic communication through the internet in Spanish museums. *Public Relations Review*, 619–626.

Charitos, D., Vassilakis, K., Lepouras, G., Katifori, V., & Halatsi, L. (2000). An approach to designing and implementing virtual museums. *Engineering*.

Cho, H. (2013). Fermentation of intangible cultural heritage: Interpretation of kimchi in museums. *Museum Management and Curatorship*, 209–227.

Co, L. T. (2019, August 28). *BẢO TÀNG MỸ THUẬT THÀNH PHỐ HỒ CHÍ MINH*. Retrieved from lythanhco.com: https://lythanhco.com/2019/08/28/bao-tang-my-thuat-thanh-pho-ho-chi-minh-ho-chi-minh-city-museum-of-fine-arts/

Coates, C. (2020, July 17). *Virtual Reality is a big trend in museums, but what are the best examples of museums using VR?* Retrieved from www.museumnext.com: https://www.museumnext.com/article/how-museums-are-using-virtual-reality/

Corey, P. N. (2014, March 11). *A biography of a building: The Ho Chi Minh City museum of fine arts*. Retrieved from www.guggenheim.org: https://www.guggenheim.org/blogs/map/biography-building-ho-chi-minh-city-museum-fine-arts

Dat, C. (2019, October 14). *Tranh Đông Hồ - Hơi thở của làng Việt*. Retrieved from vietnam.vnanet.vn/: https://vietnam.vnanet.vn/vietnamese/tranh-dong-ho-hoi-tho-cua-lang-viet/427969.html

Dung, K. (2013). *Tranh dân gian Đông Hồ*. Retrieved from svh.gov.vn: http://dsvh.gov.vn/tranh-dan-gian-dong-ho-3151#::text=Vềnộidung20tranhdân,hộicôngbằngtốtđẹp

Education, A. B. (2015). *Lịch sử Tranh đông hồ* . Retrieved from www.tranhdangiandongho.vn: http://www.tranhdangiandongho.vn/tin-tuc/lich-su-tranh-dong-ho-3

Emissive VR, The Louvre Museum. (2019, October 21). Mona Lisa: Beyond the glass at the louvre I HTC VIVE ARTS. Paris, France.

French National Museum of Natural History, the Orange Foundation. (2017, December 20). *Journey into the heart of Evolution*. Retrieved from www.mnhn.fr: https://www.mnhn.fr/en/explore/virtual-reality/journey-into-the-heart-of-evolution

Hoa L. T. (2019, May 6). *5 Lý do tại sao các bảo tàng nên sử dụng công nghệ thực tế tăng cường AR*. Retrieved from tourzy.vn: https://tourzy.vn/tin-tuc-chung/Ung-dung-AR-vao-bao-tang

Lepouras, G., Katifori, A., Vassilakis, C., & Charitos, D. (2004). Real exhibitions in a virtual museum. *Virtual Reality Volume*, 120–128.

Malraux, A. (1978). *The voices of silence: Man and his Art*. (S. Gilbert, Ed.). United Kingdom: Princeton University Press.

McKercher, B. (2002). Towards a classification of cultural tourists. *International Journal of Tourism Research*, 29–38.

Minh Nguyet, Bich Van. (2019, October 14). *Tranh Đông Hồ - Hơi thở của làng Việt*. Retrieved from vietnam.vnanet.vn: https://vietnam.vnanet.vn/vietnamese/tranh-dong-ho-hoi-tho-cua-lang-viet/427969.html

Sam, N. H. (n.d.). Lon Am Duong–Ying Yang Pig. *Lon Dan–Dong Ho painting*. Bac Ninh. Retrieved from https://vi.wikipedia.org/wiki/Tập_tin:Tranh_Lợn.PNG

Tsichritzis, D., & Gibbs, S. (1991). Virtual museums and virtual realities. *Computer Science*, 17–25.

TATE. (2018, January 3). *The making of Modigliani VR: The Ochre Atelier*. Retrieved from www.tate.org.uk: https://www.tate.org.uk/whats-on/tate-modern/exhibition/modigliani/modigliani-vr-ochre-atelier

Unknown. (n.d.). Đám cưới chuột. http://www.booksite.ru/fulltext/1/001/008/071/524.htm. Bac Ninh.

Van, T. (2018, July 20). *Nghề làm tranh dân gian Đông Hồ: Đường đến di sản cần được bảo vệ khẩn cấp*. Retrieved from daidoanket.vn: http://daidoanket.vn/dan-toc/nghe-lam-tranh-dan-gian-dong-ho-duong-den-di-san-can-duoc-bao-ve-khan-cap-tintuc410020

VNFAM. (2011). *Vietnam national fine arts museum about us*. Retrieved from www.vnfam.vn: https://www.vnfam.vn/en/about

Vy, D. (2007, September 16). *Từ nhà chú Hỏa đến Bảo tàng Mỹ thuật*. Retrieved from tuoitre.vn: https://tuoitre.vn/tu-nha-chu-hoa-den-bao-tang-my-thuat-80015.htm

Wickens, C. D., Sandry, D. L., & Vidulich, M. (1983). Compatibility and resource competition between modalities of input, central processing, and output. *Human Factors: The Journal of the Human Factors and Ergonomics Society*, 227–248.

Design a Computer Program for Sustainable Materials to Apply in Egyptian Housing

Bishoy Magdy Tawfeeq, Hossam Eldin Hassan Othman Elborombaly, and Ali Kamal Ali Altawansy

Abstract

Public buildings are not without the use of energy and resources to operate and adapt them to the needs of their users. The close link between environment and development has led to the emergence of a concept of development called sustainable development to increase better opportunities for people in a society without a shortage to meet their basic needs for a long period of time. To achieve prosperity and happiness for society that the leading role of conscious architects aware of the importance of the application of energy conservation systems, rational consumption, and the use of renewable energy alternative to preserve the environmental and reduce pollution resulting from using of traditional systems of power generation. Many policies have been developed to address these growing problems associated with the nature of social and economic changes in Arab societies. The housing crisis is one of the most pressing crises the country has faced for decades. The high rate of population growth and the increase in migration from rural to urban areas have made it difficult to cope with the urban population and its demands for services and facilities. Since the mid-1970s, the state began to search for land for the work of housing projects and began to provide subsidized housing, especially low-income but did not provide the success of these projects to the problems and did not meet the needs of users. The problem is the lack of availability of computer programs in the Ministry of Housing, government bodies and special housing agencies that help designers to make the right decisions in choosing the right materials in each region that tolerate climatic conditions and adapted to nature. So, the data was collected like materials, specifications, and prices from those Institutions. Then some programs and tools were used to enter all information and data of materials, specifications, and prices by designing a computer program to apply them in economic housing. The research aims to reach modern methods to reach construction sustainability and to devise ways to measure the application of methods and determine the degree of sustainability and how to apply technology and organize management, sustainability and environmental conservation through the work of designing a computer program for sustainable materials for appropriate solutions in buildings in Egypt. The research presents using of new technique like design a computer program that consider the introduction of using local and sustainable materials to guide designers and companies in selecting appropriate materials from the results and graphs in each region to reduce costs and achieve sustainable development.

Keywords

Housing • Local materials • Construction • Computer program • Cost

1 Introduction

Throughout the ages, humans had developed tools and techniques that helped and provided their needs and improve production to live in well-being and comfort (Afifi, 2011). They began to learn and innovate to find the best solutions, especially in the construction of the house or shelter, where the age was characterized by the continuous development and acceleration and the rapid change of images and materials quickly social and economic data. Technology has begun to emerge through ideas and creativity to solve the environmental problems that have emerged in recent

B. M. Tawfeeq (✉) · H. E. H. O. Elborombaly
Architecture - Faculty of Engineering, Ain Shams University, Cairo, Egypt
e-mail: Eng.Bishoy-89@hotmail.com

A. K. A. Altawansy
Department of Architecture, Faculty of Engineering, October 6 University, 6th of October City, Egypt

H. Altan et al. (eds.), *Advances in Architecture, Engineering and Technology*, Advances in Science, Technology & Innovation, https://doi.org/10.1007/978-3-031-11232-4_11

periods, which led to reconsider the rebalancing of man and his environment and call for the development of technology to serve the human and improve the surrounding environment (Osman, 2013).

They had been begun to deal with natural materials before the emergence of industry and technology, where he began to form in the stones pharaonic civilization in the temples and tombs, also formed wood materials to get different shapes and sectors and then emerged porcelain and bricks in the Assyrian civilization, while Greece used marble and stones and added Roman Muazico and gypsum engraved and decree They used natural concrete, then iron was used in the eighteenth century, then steel appeared and was widely used in the 18th. Industrial cement emerged as it expanded significantly in the concrete industry and became the most prevalent materials. Also, the plastics and aluminum industry developed and the timber industry developed. Technology entered the material industry and innovated in modern materials and covers, which became modern styles in architecture, which made it difficult for the designer to choose the suitable building materials in each project by location (Osman, 2013) as shown Fig. 1.

The research is based on the following questions:

i. Are there other procedures and solutions to improve housing projects during design and operation?
ii. What are the benefits when applying the idea of design a computer program for materials that can be selected?
iii. Which is the way to control the project costs and provide the needs of users as much as possible?

2 Compare the Development of Computer Programs with the Current Program Designed

The computer has become one of the foundations of contemporary life, which has been reflected in architectural creativity in its various stages. Many computer programs appeared to reduce and solve complex problems and facilitate your approach to the architectural designer. In the sixties, the beginning was the drawing of design elements, and they still represent the basis in the two-dimensional programs used today. At the beginning of the seventies, programs began to develop in three dimensions, and at the end of the seventies and the beginning of the eighties.

The modeling process reached levels of development, then with the development of technology and the development of human thought in the nineties and twentieth century. Programs began development through companies and individuals programmed to design easier and reduce problems. These programs have evolved and gone through various stages until they have concluded that a set of programs with each other design or assistance in the design of the building, not only architectural but also in more than one major, as well as calculating the quantities of materials used in the project and estimating its cost and others of operations (Mahmoud, 2012).

Table 1 reviews some of the programs used, their types and their different uses, as it explains the history and development of the various programs that have emerged to assist architects in the building.

3 Research Objective

The research aims to reach appropriate methods of construction sustainability through designing a computer program for sustainable materials to choose appropriate solutions in buildings in Egypt. Research takes the following proceeding to achieve the goal.

Design a computer program by name (sustainable materials in economic housing) that assists in the introduction of local and sustainable materials to assist designers and companies in selecting appropriate materials from the results and graphs that are appropriate in each region to minimize costs and achieve sustainable development.

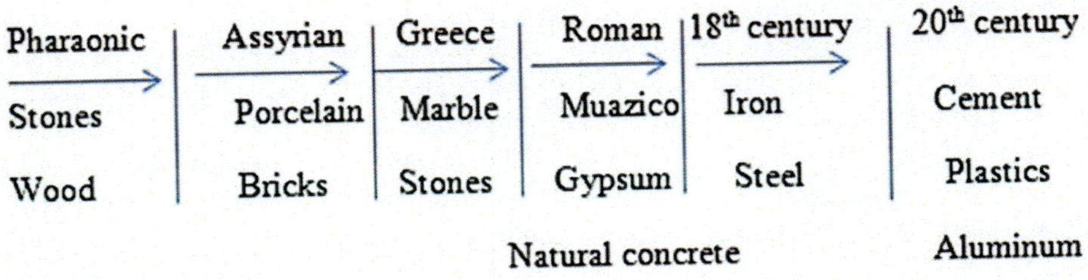

Fig. 1 Locations and availability of local building materials in different civilizations. *Source* Al Saeed & Jamal Abdul Ghani (2017)

Table 1 Some of the programs used, their types and their different uses, as it explains the history and development

	Problem analysis models (DP)		Building inference models for previous cases (CBR)		Productive design systems (GDS)		A number of different directions
	YASMIN	RBSaace	CADRE	MONEO	GAT	FWT	SEED
The producer of the program	Faculty of Engineering, Shubra, 1 University	Texas A&M	Swiss Federal Institute of Technology	A subscription between the University of Alexandria and the University of Kaiserslautern, German	Washington University'	.Alexandria University Faculty'of Engineering	Carnegie Aiellon
Program designer	Ashraf Abdel Moneim Jaafar	Scott Anthony Artin	Joint work between arclutects And computer scientists	Dina Alohamed Samel, Talia	HudaHuna-ouiu	IHosni.Ahmed.Al-	Robert 'Ulrich Si James Snyder' Coyne
Time period	1996	2004	1990	2006	2007	2010	1994
The goal of the program	A proposal to assist architects in the early design stages	A proposal to assist architects in the early design stages	It helps to reuse models Previous design in producing new designs	A tool usedin assisting the designer in he initial design stages of the process Design offers only similar solutions	A tool used m the production of designs Assist in the initial design stages For the design process	A tool used in the production of designs Assist in the initial design stages For the design process	Supports and benefits the designer in the first stage To design the building
Cases in the program	It is not limited to a specific type. Architectural projects	It is not limited to a specific type Architectural projects	Various cases divided into complex cases And simple cases according to its histoid sign	Low and medium housing units Costs	It is not limited to a specific type of Architectural projects	Tool for producing prototypes of waterfalls villa	Architectural projects of a recurring nature, as well as hospitals, housing units, and others
Design production	It depends on a mathematical model	It depends on a mathematical model	Creates designs by changing dimensions For previous models	The program displays cases similar to the functional relationship diagram, solving the horizontal projection of the residential unit 1	It produces a number of alternatives based on Genetic algorithms	It does not retrieve cases but rather produces alternatives emanating from Allla Waterfalls	Design cases are automatically retrieved after Conformities and requirements with The current state to be designed depending Object-Oriented on object
Output design processing	Design processing is primarily dependent on Designed Engineer	It allows the designer to interact with him to modify and sole the design problem	The user participates in the program in setting new design settings and the program develops a solution ora combination of cases	It does not treat design cases but Show it only	Design processing is primarily dependent on Designed Engineer	It does not handle cases, but only produces them according toasetofiules, then evaluated by the design engineer	Supports interactive correction of the problem

Source Mahmoud, A. R. M. (2012). *Evaluation of Selected Programs for Computer-Aided Architectural Design*. Faculty of Engineering in Shubra, Banha University, Egypt

4 Methodology

4.1 Theoretical Approach

Includes a comprehensive theoretical study of the relationship between building materials and architecture, defining building materials, types and conditions.

4.2 Applied Approach

- It deals with the extraction of indicators and elements of modern methods that achieve sustainability that benefit the designer and the community to minimize the negative effects through designing a computer program by name (sustainable Materials in Economic Housing) to evaluate the appropriate materials through the work of matrices to choose the best materials suitably, Specifications and prices for application in residential buildings.
- Also calculating the total price of the required quantity of each raw material per meter to choose appropriate specific materials to reach the objectives of the research and finally, the research finds a set of conclusions and recommendations.

They are some programs and tools used in design a computer program in the research:

(a) **Programs**:
1. Visual Studio 2013.
2. SQL server 2012
3. Team Foundation Server (TFS)

(b) **Libraries**:
1. Bootstrap
2. Kendo
3. Jquery
4. Entity framework
5. Microsoft Office

(c) **Tools**:
1. c#
2. OOP
3. MVC
4. Linq
5. HTML
6. Javascript
7. Jquery
8. CSS

5 The Relationship Between Building Materials and Architecture

The relationship between building materials and architecture remained simple until the industrial revolution, where construction materials were chosen either because of their availability or because of their external shape and comfortable (Hilal & Mahdi, 2014). Over the nineteenth century, architects adopted the choice of building materials on the form and function together and the availability of such materials in the local environment, where the role of building materials changed dramatically as the industrial revolution progressed (Wdeh et al., 2019).

Therefore, the concepts of sustainability and environmental architecture have emerged to reflect on the use of local building materials from renewable sources and low emissions and pollutants, characterized by durability, high resistance and recyclability which are the most important characteristics of sustainable materials (AbdulQawi & Walid, 2018).

6 Building Materials

Building materials are elements of architectural output in an era of previous eras with the development of possibilities, traditional building materials and increased knowledge of the possibilities and architectural and structural properties helped to develop architecture throughout the ages (Gamal Eldin, 2018).

Building materials include a huge boom in the construction industry and all stages from construction to internal and external finishing, including natural materials such as stone, clay, wood and sand, natural materials made of natural raw materials such as bricks, materials manufactured on natural materials such as concrete and steel and then industrial materials such as plastics, glass and paints. Construction materials are classified by the designer in the selection of type, specifications and price in the implementation and formation of buildings and output to achieve the sustainability of the building and longevity (Al Saeed & Jamal Abdul Ghani, 2017).

(a) **Natural materials**
(b) **Manufactured materials**
(c) **Blended materials.**
These materials and their properties are studied and the impact on the climate, environmental or social and economic aspects. It is the responsibility of the designer to choose the right materials to achieve environmental balance and user satisfaction. The materials were classified in three categories as shown in the Fig. 2.

Therefore, building materials must meet two basic conditions:

The materials should not affect high energy consumption whether in manufacturing, installation or even maintenance.

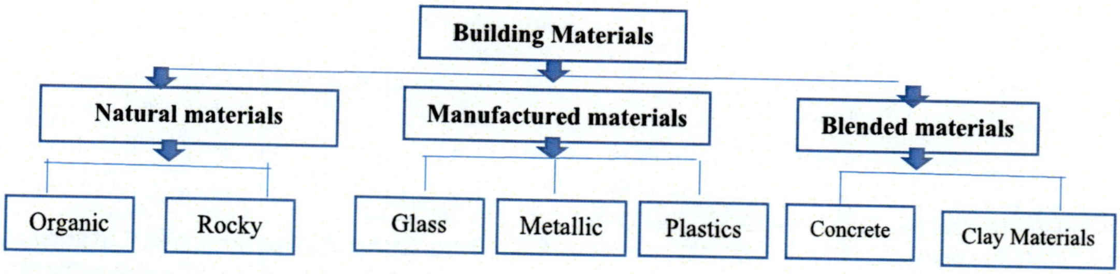

Fig. 2 General classification of the most important construction materials. *Source* Al Saeed and Jamal Abdul Ghani (2017)

It should not cause an increase in the internal or external pollution of the building. Therefore it is preferable to use natural building materials that bear the external environment factors and adapt to the environment (Arafa, 2019) as shown Fig. 3.

6.1 Conditions of Building Materials

First: Manufacturing Processes are carried out in manufacturing or operation on the raw materials in the site, condition of the soil and its characteristics include: specifications, chronological life, and composite materials consisting of the manufacturing material. For example, timber is a natural environmentally friendly manufactured materials and needs little energy to manufacture and is the most lightweight and environmentally friendly materials and this is what made developed countries use in Europe and America diffuse.

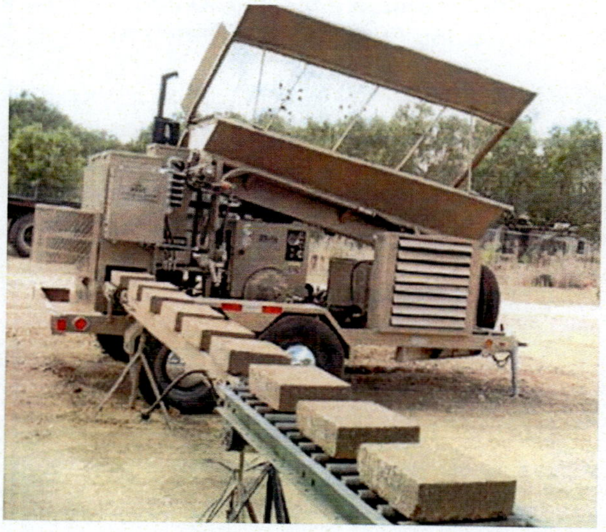

Fig. 3 Production and use of bricks from local materials at sites. *Source* Al Saeed and Jamal Abdul Ghani (2017)

Second: Insulation Materials must be characterized by good insulation properties of increased temperature and noise or any other environmental factors, which leads to save energy consumption and reduce pollutants.

Third: longevity life of the material in buildings that led the relationship is direct, the age of the material gives additional points for buildings to the criteria for sustainability. For example, stone is one of the most longevity materials used by the pharaohs in building the pyramids for thousands of years.

Fourth: Recycling: It is preferable to recycle demolition waste and constructions that reduce costs and continuous consumption whenever the material is recyclable, it is considered environmentally friendly as shown Fig. 4.

6.2 Construction Waste

Construction waste is defined as solid waste that is not hazardous and harmful to the environment except when collected, sorted and recycled from construction activities such as demolition, construction, development and restoration, and materials resulting from the work on-site include: asphalt, brick blocks, wood, glass, iron … etc. (Al Saeed, 2017).

The principle of recycling takes into account the reduction in the use of energy-consuming materials and the increase of pollutants so that buildings are designed and constructed in a way that makes buildings sustainable and longevity, so there are some materials and the proportion of waste as shown Table 2.

6.3 Recycling Building Materials

Recycling of materials contributes to the preservation of the environment and the reduction of pollutants through its role in achieving the following:

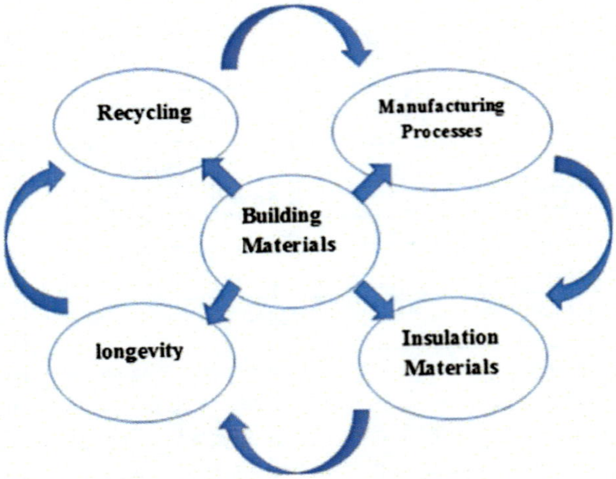

Fig. 4 Conditions that must be met in environmentally friendly building materials

1. Conservation of energy resources and renewable materials.
2. Reduce consumption by prolonging product life.
3. Reduce consumption through recycling.
4. Energy-saving by reducing production processes.
5. Produces high economic returns from the sale of recyclable products and manufacturing as shown Fig. 5.

6.4 The Life Cycle of Sustainable Building Materials

There are criteria for analyzing building materials during the three basic stages of the building life cycle:

(a) Pre-construction phase such as extraction of raw materials manufacturing of materials—packaging materials—transport to the construction site.
(b) Construction phase such as implementation, operation, and maintenance.
(c) Post-construction phase such as recycling–reuse as shown Fig. 6.

6.5 Plan to Apply Sustainability Standards and Use Environmentally Friendly Materials to Solve Building Problems in Egypt

Figure 7 reviews plan to apply sustainability standards and use environmentally friendly materials to solve building problems in Egypt.

7 Design a Computer Program

Significant advances in computer technology and information systems contributed effectively to the construction industry, architectural design in general and sustainable architecture in particular, which helped and solved problems on the emergence of forms. It became an interactive architecture that helped the architect consider the human and environmental requirements to choose the right materials and decisions (Shukr & Aboud, 2016).

7.1 Designing a Simulation Program for Building Materials and Choosing the Most Suitable for Application in Housing (Inputs of the Program)

The simulation program for building materials is designed to enter data and specifications of building materials, where general information about the regions, climate, temperature, and humidity, the average income of individuals and prices of building materials are entered.

Table 2 Some materials and the proportion of waste:

Material	Percentage of waste (%)
Concrete and stone mixture breaking	40–50
Wood	20–30
Cutters	5–15
Asphalt	1–10
Metals	1–5
Block Horde	1–5
Plastic	1–5

Source Al Saeed and Jamal Abdul Ghani (2017)

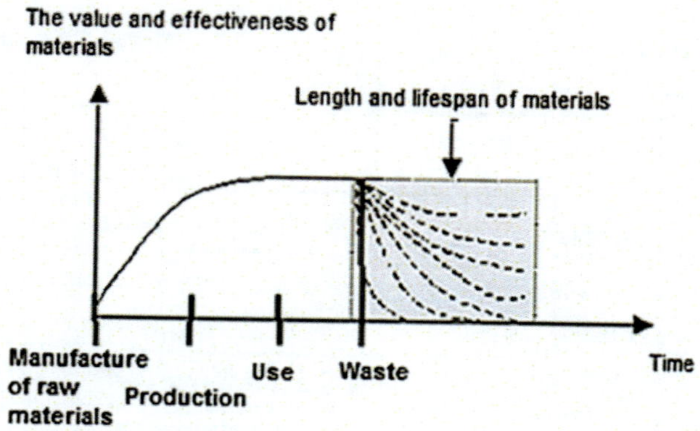

Fig. 5 Recycled building materials. *Source* Al Saeed and Jamal Abdul Ghani (2017)

Fig. 6 The life cycle of sustainable building materials

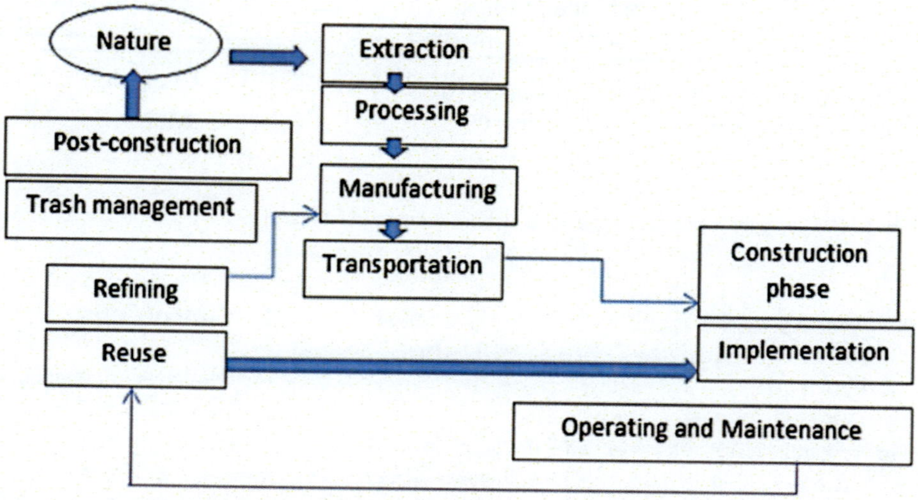

Through this, the regions were determined and thus determine the climate of temperature, humidity, materials to choose the appropriate materials, thus reducing the costs of transportation and utilization of local materials.

7.2 The Contents of the Program

1. Regions data in Egypt.
2. Data of climatic regions and their governorates through the Egyptian code and the annual statistical book of the General Authority for Mobilization and Statistics.
3. Data of local building materials available in each region, specifications and prices.

7.3 Characteristics of the Program

1. View the regions and provinces in Egypt.
2. View climate data in each region for possible identification.
3. View local materials were available in each region to be selected for application in housing projects, which reduces transportation costs and materials imported from abroad.

7.4 The Program Structure

See Fig. 8.

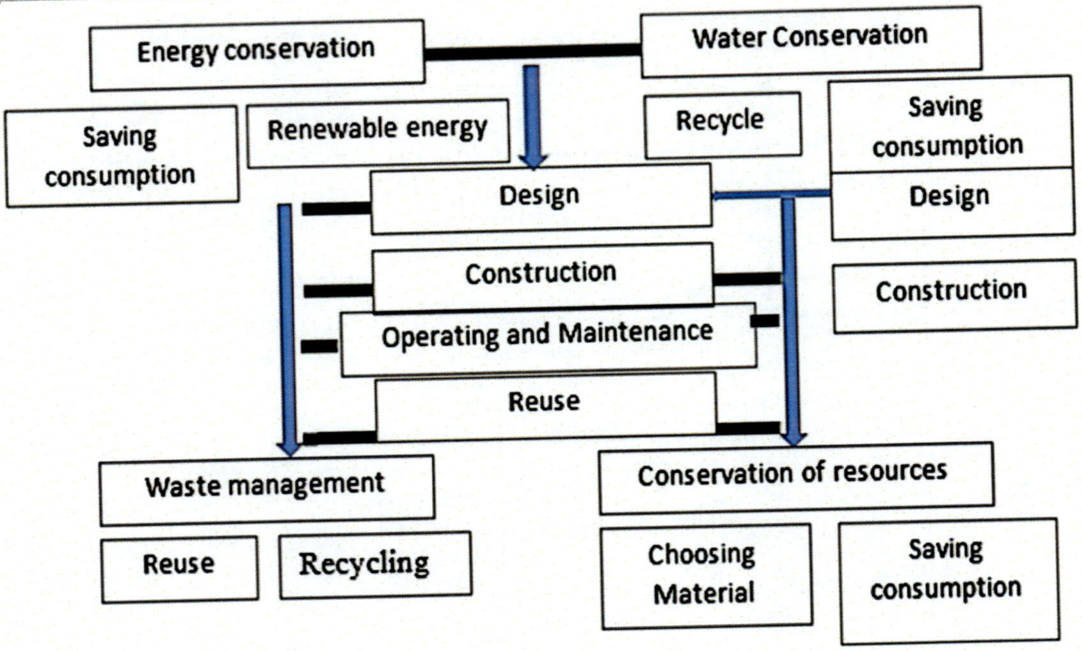

Fig. 7 Plan to apply sustainability standards and use environmental materials

8 Results and Discussion: The Computer Program Interface (Phases of Using Program)

The research results to work and develop an evaluation method for local materials through databases through the design of computer program by collecting data for materials, specifications and prices to reach the appropriate materials and achieve sustainability to reach the objectives of research through:

- Review of data and review of climate tables, regions, population, locally available materials and the average income of purchasing power of materials for housing construction at appropriate costs and identify them.
- Identify local building materials available in each region.
- Review and apply the Egyptian code to improve energy efficiency in residential buildings through climate schedules for each region and building materials to choose appropriate materials.
- Evaluate the appropriate materials through the work of matrices of those materials to choose the best materials, specifications, and prices in housing.
- Calculate the total price of the required quantity of each raw material per meter to choose appropriate specific materials from graphs to make analysis and housing report as shown Table 3.

9 Conclusions

The results were obtained through the study to examine this program that had reached building materials to apply in housing by using local building materials that adapt and withstand environmental conditions to take advantage of natural ventilation and lighting resulting in reduced use of cooling and heating devices. Materials should be recycling and using of building waste materials that meet the criteria of sustainability throughout their life cycle and do not adversely affect the environment in the stages of manufacturing and operation.

The selection of materials is carried out in the feasibility stages of the new buildings, especially in the new cities, where the architect must collect information and data for the specifications of the materials and choose them as available in the local market, which suits the cost and environmental sustainability. After that follow-up and continuous evaluation of new cities and communities to record experiences and take the views of the people and the government to follow-up and resolve defects and problems to work to correct the course in future projects.

This paper finds design a computer program that assists in the introduction of local and sustainable materials to assist designers and companies in selecting appropriate materials from the results and graphs that are appropriate in each region to minimize costs and achieve sustainable development.

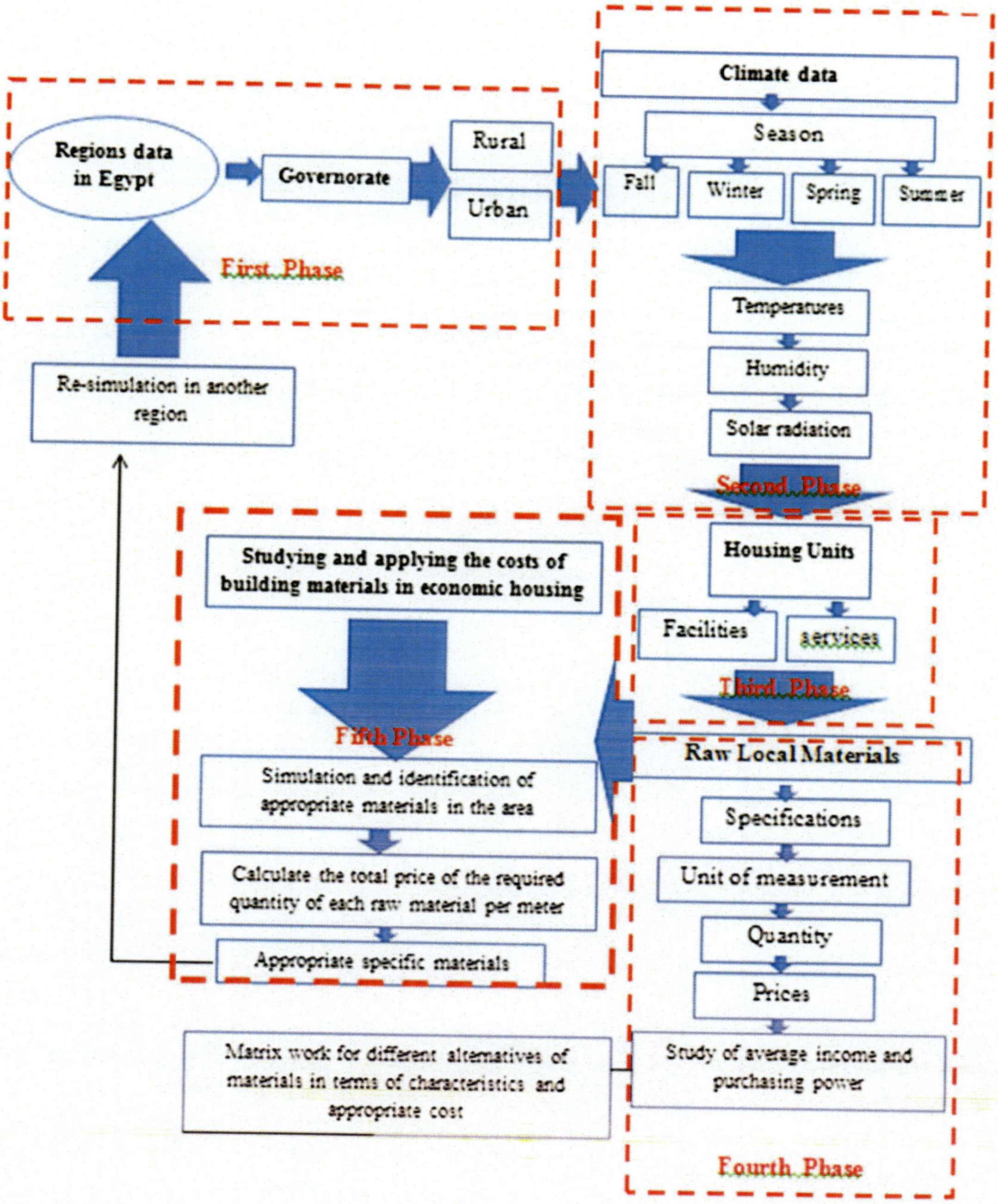

Fig. 8 The program structure. *Source* The Author

10 Recommendations

1. Encouraging companies and factories to utilize natural and local raw materials to produce sustainable materials, thus reducing the operation, consumption, and energy costs of the state.
2. Sustainability criteria must be adhered to periodically according to the environmental situation and development level.
3. The choice of the construction system must be compatible with the local capabilities of labor and equipment available besides attention and develop considerations of environmental problems to solve.

Table 3 The computer program interface (phases of using program)

Phases	Inputs	The computer program interface (Screen shot and outputs)
First phase: screen to add climate characteristics for each governorate:	– Code: governorate code appears automatic with the choice of the name of the governorate – Region: the region is selected from the list already prepared – Governorate: the name of the province is chosen from the list – Classification: select the governorate classification (rural/urban) – Season: season is selected (fall–winter–spring–summer) – Temperature: entering the degrees of great and minimum temperature – Relative humidity: the state of humidity is entered – Sunlight: enter the state of sunlight (weak–moderate–high)	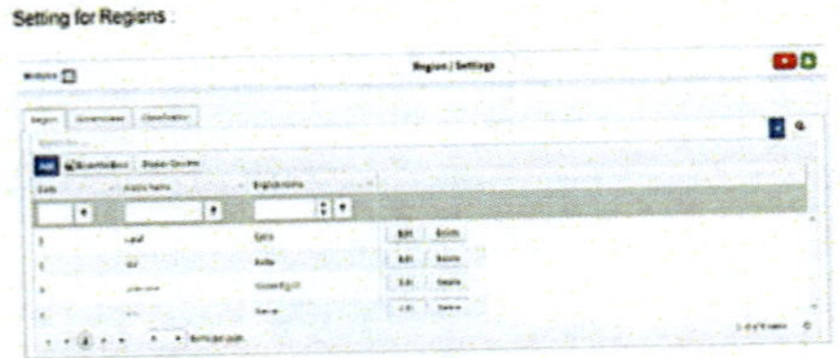 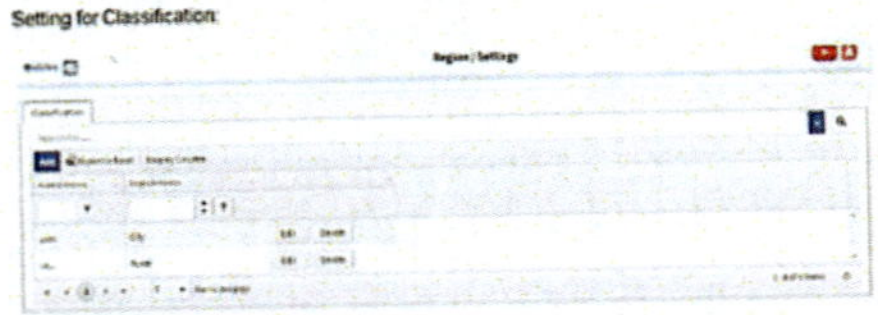 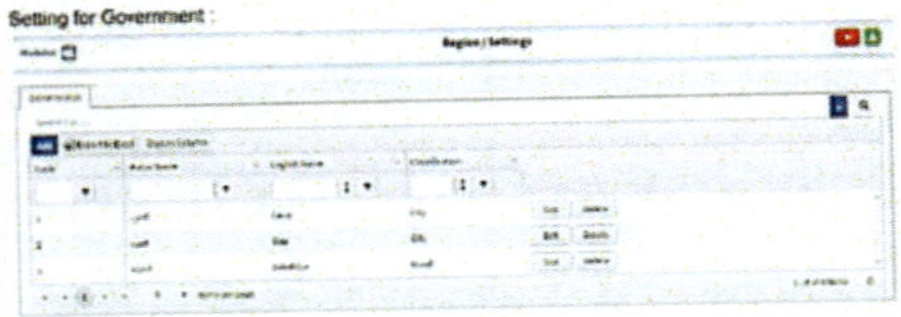 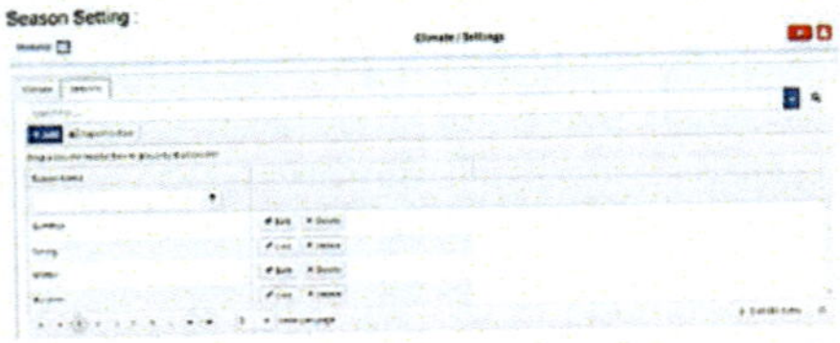 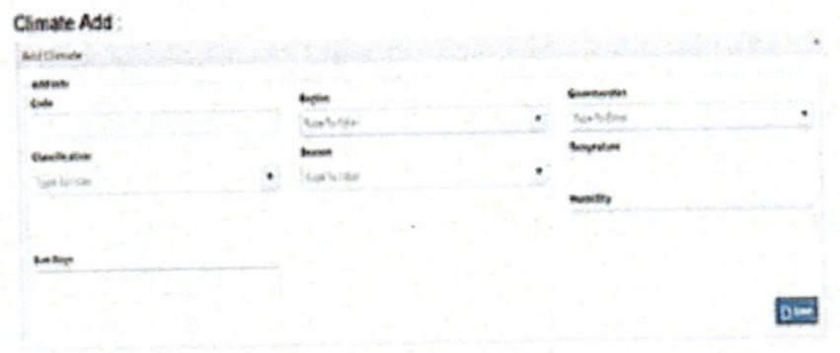
Second phase: screen to add residential units	– Classification: select the governorate classification (rural/urban) – Housing unit: enter the name or number of the housing unit to be added – Services: a tick will be checked if the housing unit is connected to facilities and services	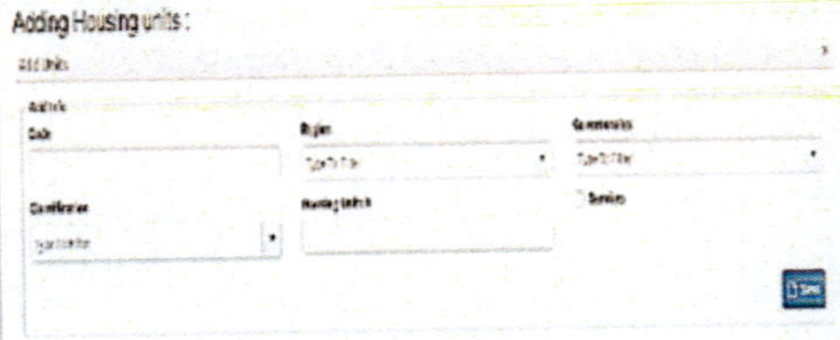 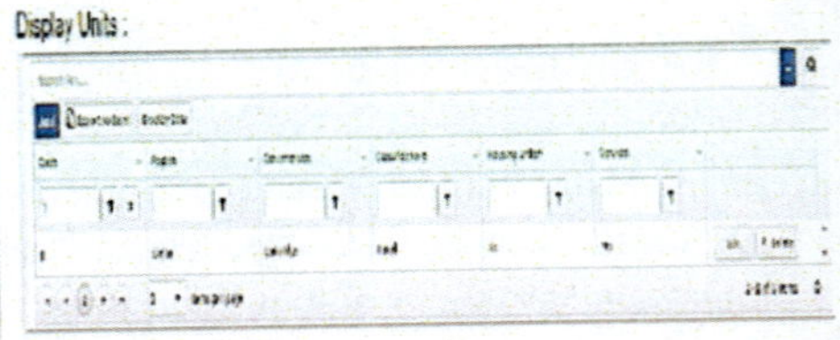

(continued)

Table 3 (continued)

Phases	Inputs	The computer program interface (Screen shot and outputs)
Third phase: screen to add raw materials that are shown and extracted from each governorate	The name of the material: write the raw material to be added and existing in the Governorate – Unit of measurement: the unit of measurement for the raw material is selected from the list – Price: enter the unit price of the raw material to be added	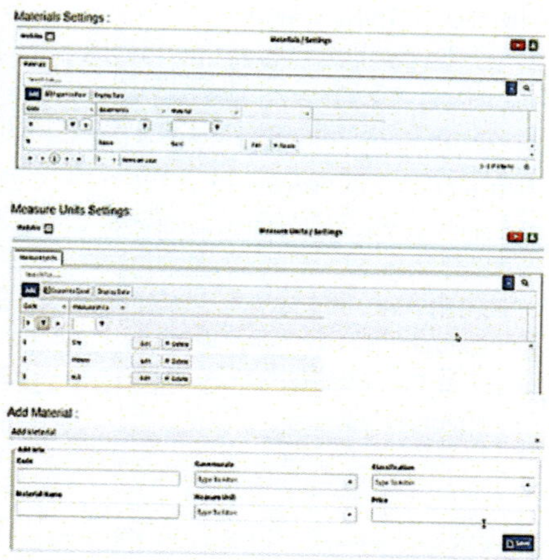
Fourth phase: screen to make a statistical raw materials used to build every 1 m in each governorate	– Area: enter an initial value of 1 square meter – Raw material code: all raw materials previously entered for each governorate are shown in the raw materials screen – Name of the material: show the raw material to be found in the province – Unit of measurement: the unit of measurement of the raw material is shown from the list – Quantity: the required quantity is written in raw materials – Price: enter the unit price of the raw material to be added	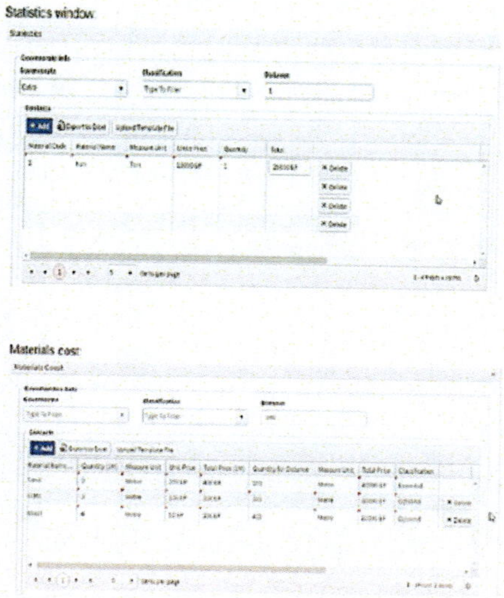
Fifth phase: screen to display the cost of the raw material used for construction according to the area of the building entered	– Total: shows the total price of the required quantity of each raw material, graphs to analysis make housing report	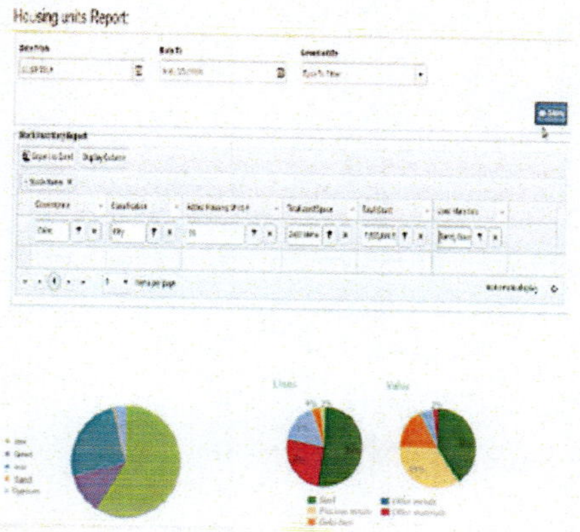

Source The Author

4. Work and develop training programs for engineers and workers and how to communicate to make decisions and choose materials with considerations of costs and risks associated with some production processes in the construction sector.
5. Work on the production and development of simulation programs for the performance of buildings so that they correspond to the regions of the climate in Egypt to facilitate communication with designers and decision makers to reach positive results.
6. The state of government should be communicating with international and local companies to produce design programs and the need to participate in the proposal of innovative evaluation systems to reach sustainable architectural design.

Acknowledgements First of all, I am thanking *my God* Almighty for accomplishing this modest work, hoping that holds the good for the scientific researchers and students locally and internationally. Deep respect and gratitude to my honorable *Professor Dr* . ***Hossam Eldin Hassan Othman Elborombaly*** for his valuable instructions and for being my ideal ethically and scientifically.I would like to acknowledge particular thanks to *Assoc. Prof, Dr* . ***Ali Kamal Ali Altwanasy*** under whose supervision this paper was accomplished. Finally, I also thank all the thanks, love and appreciation of Cloud Soft 5, which is headed by the most respectable engineer ***Khalid Khalifa*** and his assistants, whom welcomed all welcome and stood with me for the work of the design computer program in the research and technical support.

References

AbdulQawi, & Walid. A. (2018). *Policies of allocation and management of land and its impact on the sustainability of new cities*. Cairo University.

Afifi, A. M. (2011). *Towards the activation of an integrated strategy for the development of policies to provide and facilitate housing in the Arab World, an analytical study of the Egyptian experience*, Egypt.

Al Saeed, & Jamal Abdul Ghani. (2017). Building materials and their impact on green buildings. In *Third Arab development & investment conference in real estate and industry*, Fujairah, UAE.

Arafa, Z. M. (2019). *The environmental design of social housing building as a way to solve the problem of urbanization in the new cities*. Mansoura University.

Gamal Eldin, M. (2018). *Auditing standards and regulations for the selection of sustainable building and finishing materials for internal spaces of residential buildings*. Cairo University.

Hilal, M. M., & Mahdi, H. (2014). Sustainability in architecture: A study on the role of sustainable design strategies in reducing impacts on the urban environment. In *Paper published at the 13th Al-Azhar international conference*, Egypt.

Mahmoud, A. R. M. (2012). *Evaluation of selected programs for computer-aided architectural design*. Faculty of Engineering in Shubra, Banha University, Egypt.

Osman, F. O. (2013). *Selection of building materials for sustainability in architecture in developing countries (standards and mechanisms)*. Sohag University.

Shukr, A. A., & Aboud, B. A. (2016). Sustainable materials & construction techniques in Iraq. *Journal of Planned and Development, No. 33*, Iraq.

Wdeh, H. N., Roula, & Tanjour, K. (2019). The role of building materials in achieving sustainability of buildings according to global sustainability systems (Case study administrative buildings). *Tishreen University Journal for Research and Scientific Studies—Engineering Sciences Series, 41*(3), Syria.

Digital Methods and Tools for Mapping Public–Private Partnership Opportunities and Evaluating Proposals in the Regeneration Plans of Building Assets

Elisa Cacciaguerra, Ilaria Garofolo, and Nicola Strazza

Abstract

In recent years, Italy is experiencing a process of digital transformation and innovation intending to simplify the perspective it is a more efficient system within the public administrations. At the same time, there was a strong increase in the demand for urban redevelopment and sustainable development of the city, also thanks to the recent sustainable policies for the containment of land consumption. The research aims to analyse the potential of innovative methods, connected to information technologies (BIM and GIS) that public administrations can use to stimulate the interested parties to act on a specific territory through the presentation of proposals of public–private partnerships. The first phase of the proposed work focused on analysing the methods of making information available during a "traditional method" of making information available. In the perspective of a plan for the redevelopment of urban contexts, the research evaluates the potential for integration of various information, such as the results of relations in the area (connections, services, infrastructures, etc.) and details on the built heritage (state of conservation and history). Information can be managed by commonly used tools, such as point clouds, BIM models with an adequate level of details and GIS platforms. These platforms make a wide range of users able to access the information through the web interface, intending to define a common language between different actors, in the redevelopment process.

Keywords

BIM • GIS • Public–private partnership • Urban regeneration • Trieste

E. Cacciaguerra (✉) · I. Garofolo · N. Strazza
University of Trieste, Engineering and Architecture Department, Trieste, Italy
e-mail: elisa.cacciaguerra@dia.units.it

Nomenclature

GIS	Geographic Information System
BIM	Building Information Modelling
PPP	Public-Private Partnership
PMC	Preliminary Market Consultation
PA	Public Administration
IT	Information Technologies

1 Introduction

This contribution frames the progress of the research in progress, in particular, it takes up the state of the art on the subject and formulates the first proposals for the development of innovation processes of public–private partnerships. In-depth knowledge of the territory at all its scales is a fundamental requirement for effective resource management.

With the aim of a redevelopment plan of the city or parts of it, the ability to integrate various information such as the persistent relations in the area (connections, services, etc.) down to the details relating to the buildings and their history, is essential in the rationalizing of opportunities.

The aim of the research conducted is to probe the capacity of innovative methods, linked to Information Technologies, that public administrations can adopt to stimulate stakeholders to act on a specific territory through the presentation of proposals of public–private partnership.

In particular, the research starts from the study of an innovative workflow for a preliminary market consultation for the activation of a public initiative PPP operation, intending to define a standard to be replicated for the creation of an information platform, that stimulates the private entities, to present PPP proposals for private initiatives, with the consequent general revitalization of the city. In the Italian legal system, to start PPP operations, it is possible to use Preliminary Market Consultations, a preliminary phase,

H. Altan et al. (eds.), *Advances in Architecture, Engineering and Technology*,
Advances in Science, Technology & Innovation, https://doi.org/10.1007/978-3-031-11232-4_12

described in Public Contract Code, that makes information related to the operation available to potential investors. In detail, this work intends to compare the current method of making information available during a Preliminary Market Consultation and an alternative method related to the digitalisation of information through the use of GIS and BIM systems.

The proposed methodology is tested on the redevelopment of Trieste port warehouses, in particular on the investigation of a partnership proposal workflow for Warehouse 20 in the Old Port.

2 Public–Private Partnership

In recent years, the ability to "create value" through the enhancement of public real estate assets has been at the centre of a political, social and cultural debate regarding the appropriate ways to manage the "res publica" regarding new uses and the effects on the territory.

The real estate enhancement, seen as the maximization of the efficiency of real estate use, presupposes an in-depth study of the operation, both in regard to the efficiency of the recovery of the asset, and the urban context in which it is inserted, without neglecting a correct reading of the market needs and of the cultural and social vocation in which the operation is built.

In order to remedy the scarce availability of financial resources of public administrations, the intervention of private operators for the enhancement of public real estate assets has been used more and more frequently.

However, this area opens the way to a series of issues related to the danger of selling off the public asset, which has led many governments to seek alternative solutions to the transfer of ownership.

In Italy, the "enhancement concession" is placed within the range of public–private partnership (PPP) operational tools, i.e., characterized by the involvement of private operators in public initiatives (Provisions concerning privatization and enhancement of Public Real Estate Assets Legislative Decree (2001) art. 3-bis of No. 351) (Tajani et al., 2018; Vacca et al., 2018).

In particular, with regard to public buildings, the enhancement concession provides for the attribution to private entrepreneurs of the right to use public buildings for a specific period of time (generally a time of at least 25/30 years), in view of their functional conversion, extraordinary and ordinary requalification and maintenance. Private operators therefore become managers and non-owners, giving the public administration a share of the proceeds, in terms of financial burden and / or public works for the community (Ricchi, 2016).

It is clear that the ability to predict and analyse the behaviour of the building in all its phases, for a private investor, can only be a strong incentive to develop concrete proposals, and at the same time, the public administration can maintain transparent control over the correct management of the public good.

Once the period of concession of the asset has expired, the public administration falls within the full availability of the properties temporarily sold, with the acquisition of every transformation and improvements.

From a financial point of view, therefore, a private investor has the convenience to participate in a PPP procedure when the operation is sustainable, or if the initial investment capacity is able to generate a profit margin capable of remunerating the (market) risk of the initiative (Perino et al., 2018; Unità Tecnica Finanza di Progetto, 2010).

2.1 Public and Private Subjects, Roles, Risks and Critical Issues

Within these operations, it is essential to define and respect the roles between the different operating entities:

Public subjects have the task of identifying the public interests to be protected and guaranteeing the community the most appropriate tool to achieve them (Garilli, 2007).

As far as public actors are concerned, they have the task of identifying the public interests to be protected and guaranteeing the community the most appropriate instrument to archive them. Moreover, the role of control and protection of public good is central for the public administration, paying particular attention, where necessary, to safeguarding the context in which the projects are placed (EUBIM Taskgroup, 2018).

Private operators make their financial and technical-commercial capacities available, asking in exchange for the right to portray a specific utility (which can be represented through a fee recognized by the granting body or in the form of a direct income from the service management).

ANAC, the National Anti-Corruption Authority, has defined the guidelines to support the control activities of the administrations on the correct allocation of the risks associated with the PPP operations and the methods and terms for maintaining the risks incumbent on the private individual, the latter only for the entire duration of the contractual relationship (ANAC, 2018).

The 2018 report on PPP in Italy produced by Cresme Europa Servizi with data from the National PPP Observatory confirms a steadily growing trend of PPP tenders (Fig. 1). However, this growth does not correspond to the success of

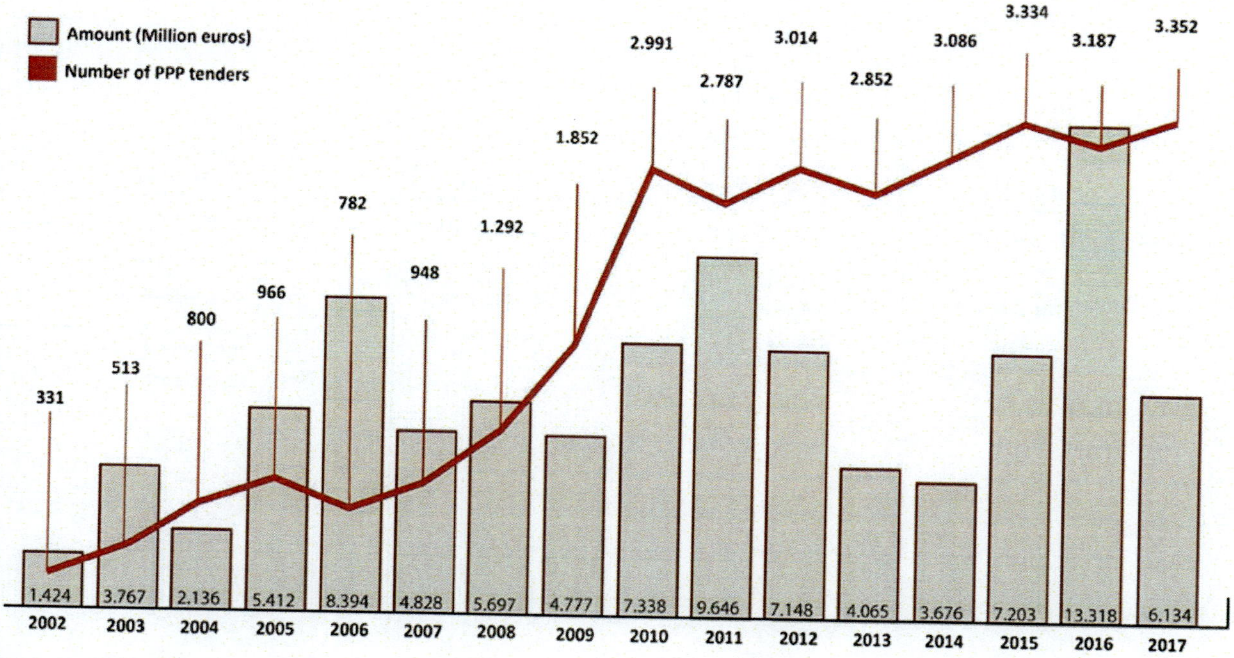

Fig. 1 Number of PPP tenders and amounts from 2002 to 2018, from the National PPP Observatory of 2018 (Cresme Europa Servizi, 2018)

all the operations: between 2002 and 2016, 13% of the procedures concerned "interrupted procedures" or cancelled tenders, deserted and not awarded tenders and revoked awards (Pasquini et al., 2019; Cresme Europa Servizi, 2018).

The difficulties encountered in these transactions were attributed to the limited preliminary and feasibility studies of the transaction, which meant that the economic assessments could not be verified, because they did not provide an updated picture of possible business risks, lengthened the time and it did not make it possible to find the right balance between ascertained risks and public guarantees (Siclari, 2019).

It is, therefore, clear that for the private operator it is essential to be fully aware of the amount of expenditure to be faced in the redevelopment of the work, in order to rationalize the hypotheses as much as possible and correctly evaluate the risk allocation, as well as the economic-financial balance of the operation.

2.2 Preliminary Market Consultation: Effectiveness and Effects Within the PPP Process

The preliminary market consultation is an instrument made available to public administrations (introduced by art. 66 of the Code of Contracts) which is part of a programmatic phase of the process (Fig. 2) which is not aimed at awarding a contract, but is a tool to initiate an informal dialogue with private operators experienced in the sector in order to verity the strategic nature of operation and to set up the best way of tendering for its success (ANAC, 2019).

The ability to gauge the market interest in a given transaction at the programming stage means that It can be built by ensuring maximum competitiveness in resources management, through the ability to attract the attention of a greater number of potential players who can provide their views before the administration formalises the request for the tender (Bardelloni et al., 2013).

The effect of this is that there will be a substantial gain in time in the conclusion of the races (i) a greater number of participants (with consequent improvement in quality of the project) (ii) and a lower cost (iii).

3 Traditional Method and IT Innovative Approach

In view of the discipline, still little used in practice, the National Anti-Corruption Authority (ANAC) considered it appropriate to adopt ad hoc Guidelines, with the explicit aim of encouraging the use of the institute and promoting its legitimate exercise, given that the tool in question allows public authorities to reduce information asymmetries on certain markets and a better knowledge of the technical aspects of the purchases envisaged by public administrations allows the market to produce offers more effectively aimed at satisfying public needs. From the point of view of the

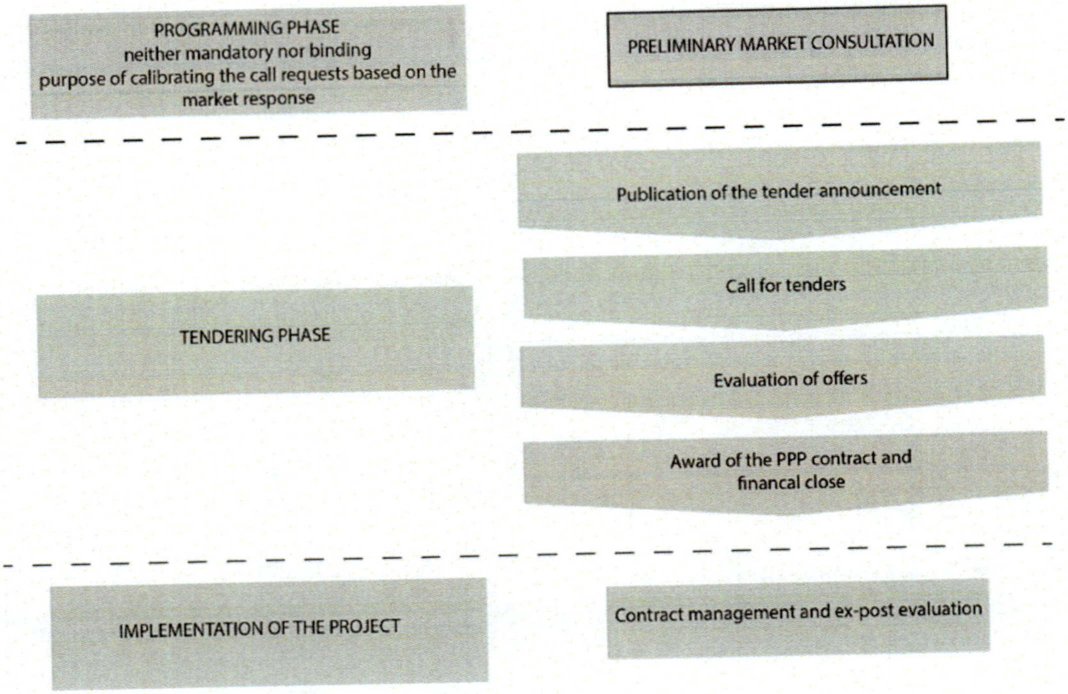

Fig. 2 Role of the PMC in the PPP operations, elaboration by the author

economy and efficiency of the administrative activity, the institution also reduces the risk of deserted tenders and represents an exercise of loyal collaboration between public and private sector (ANAC, 2019).

To date, a public administration that intends to launch a preliminary market consultation makes public a notice in which it makes available a file containing technical information related to the asset as well as a policy document in which it makes known its wishes (Fig. 3).

It is known that the modality with which the procedure will be set up will be able to determine its success, and so the choice of the information made available is of particular importance. In detail, the need for data reprocessing or the incomplete transparency of the actual state of the assets may lead to an incorrect assessment of the operation.

The possibility of associating the use of Information and Communication Technology (ICT) with the performance of the preliminary market consultation can significantly affect the strategy of the subsequent tender.

Currently, a contracting authority that intends to test the context of the market in which it would like to start a public–private partnership operation carries out a preliminary market

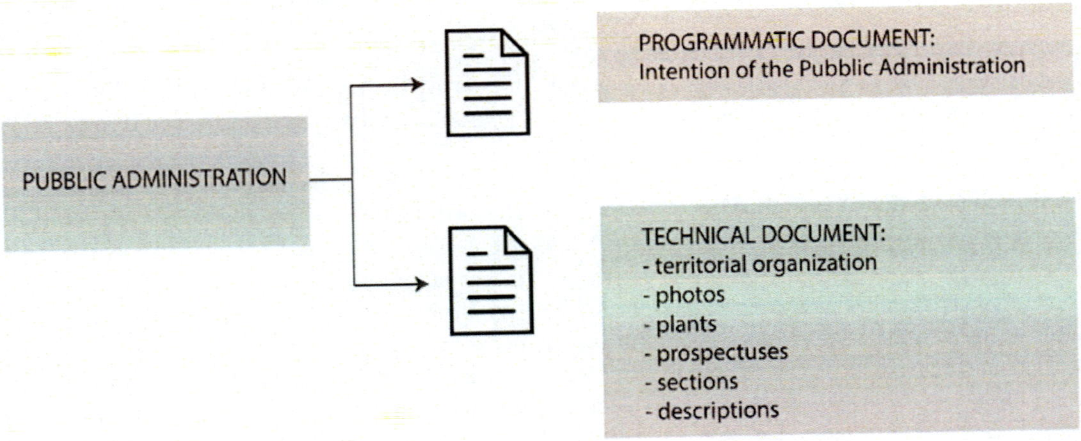

Fig. 3 Traditional PMC elements, elaboration by the author

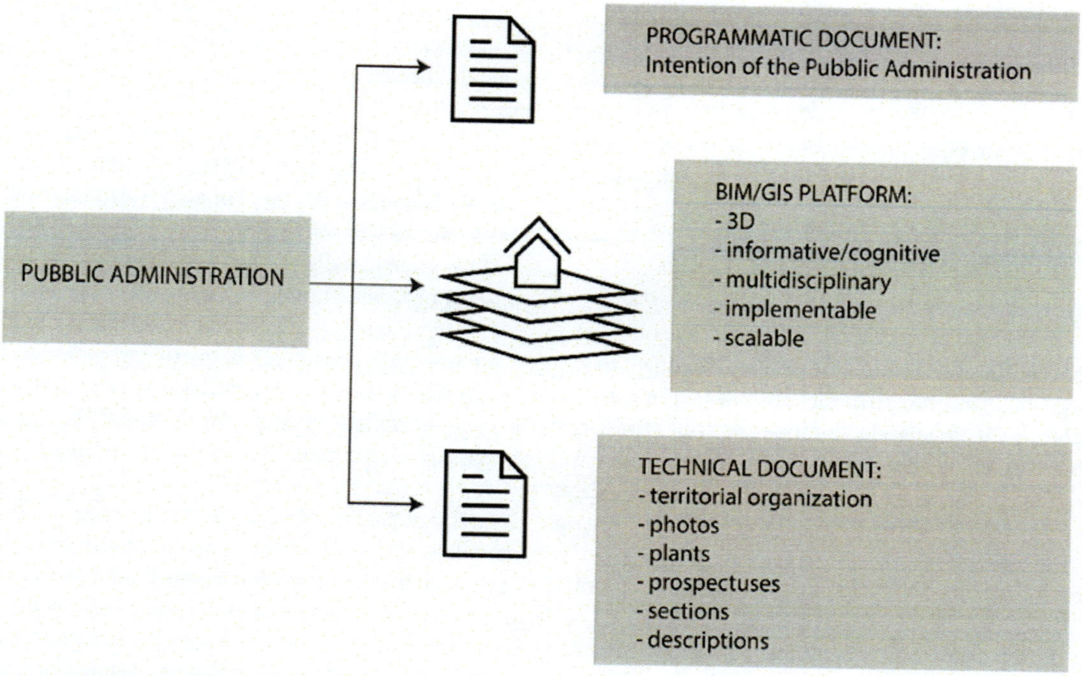

Fig. 4 Digital PMC approach, elaboration by the author

consultation. At this time, on the online platforms of the contracting authority, a series of documents containing the objectives of the operation and a file illustrating the essential information related to the asset subject to the operation are made available to hypothetical investors (Fig. 4).

The ways in which information is made available within the traditional method are therefore through the sharing of a "paper" dossier containing an essential description of the area subject to intervention, through a description of the property with respect to its appearance (photos, floor plans, etc.), its history, the context in which it is inserted, the services available, as well as the urban, landscape and monumental constraints to which it is subjected.

Providing the data in this way, therefore, leads to a disclosure in which they are not amalgamated, but combined with each other, often leading to offering a not sufficiently immediate and clear vision of the asset and its potential or problems. This, therefore, forces the private subject to analyze and remodel the information in this regard, risking compromising participation in the transaction, reducing the number of potential private subjects interested (making the "tender" less competitive) (ii) and lengthening the completion times of the works (i).

3.1 Digitalisation of the Process

The ability of IT to facilitate communication and the development of operations is therefore analysed.

In detail, reference is made to the use of technologies that make it possible to integrate the computation capabilities of geographic information systems (GIS) linked to the scale of the broad territorial context (which makes it possible to better evaluate the information related to the connections and the context in which the work is inserted) and the more precise elements of the elements inserted in it through the computerization of the asset (BIM) with respect to its characteristics.

The ability to offer a greater understanding of the state of assets through the information made ready for use is expected to change the outcome of the tender. In the face of higher commitment by the administration, it is expected that the potential investor will be more aware of all the facets of the transaction and better disposed to risk since, through the already computerized material, he will immediately be able to make fairly precise estimates of what the economic commitment could be as well as simulations of the scenarios that could arise through the tender.

The strategic objective of the project is the development of tools and methods for the proposal of PMC operations from the early planning stages through an integrated BIM-GIS approach, through a request for a project proposal to be included in a defined computational framework, which also makes it possible to compare market responses in a timely manner, allowing the PA to operate according to principles of cost-effectiveness, impartiality, effectiveness and transparency.

The tool offers greater completeness in the management of information relating to a specific asset, using a structured data set. The tool is suitable for operating in complex scenarios and decision-making models with a precise definition for investor risk, able to direct the choices of all those involved in the project and in the management of the work.

3.2 Operational Proposal

Two types of information are therefore identified to be merged into the platform: territorial data, which include all cartographic and large-scale information (in particular, information related to urban plans, infrastructures and the presence of externalities are identified); and the precise data, on a building scale, which includes all the precise studies on the object of interest (Fig. 5).

The essential aim of the research is to define a system:

- Three-dimensional, multidisciplinary and multiscalar, which can offer a coherent and in-depth photograph of the asset
- That guarantees the permanence, consultation and implementation of the data, both by the public administration (owner of the asset) and by the private individual (who has invested in it and manages it to profit from it);
- Where the data is accessible and understandable.

The phases identified are the result of reasoning related to the analysis of what are essential information related to in-depth knowledge of the existing real estate assets (Fig. 6) are:

- Historical and archive analysis and survey phase
- BIM modelling of the building (identification of the typological characteristics of the building and translation of the same into digital families) choice of the appropriate Level of Detail (LoD)
- Territorial analysis of the context of the operation (definition of the constraints and existing plans in the area, analysis of vehicular and pedestrian flows, etc.)
- Implementation of the two systems (definition of a multi-scale and multi-disciplinary platform).

In the request for proposals to be made by the private individual, in view of the amount of timely information made available by the public administration on the asset and the area in which it is inserted, it is required to strictly follow the correct data transmission protocol, in format *.ifc interoperable, paying attention to the characteristics highlighted by the PA (coordinate system, LoD, nomenclature, etc.), so that the comparison between the proposals is easy and consistent and to establish a correct and effective competitive dialogue between the participants.

It is important to underline that in the desire to redevelop an asset, we are faced with the need to know it in depth, in

Fig. 5 Information setting

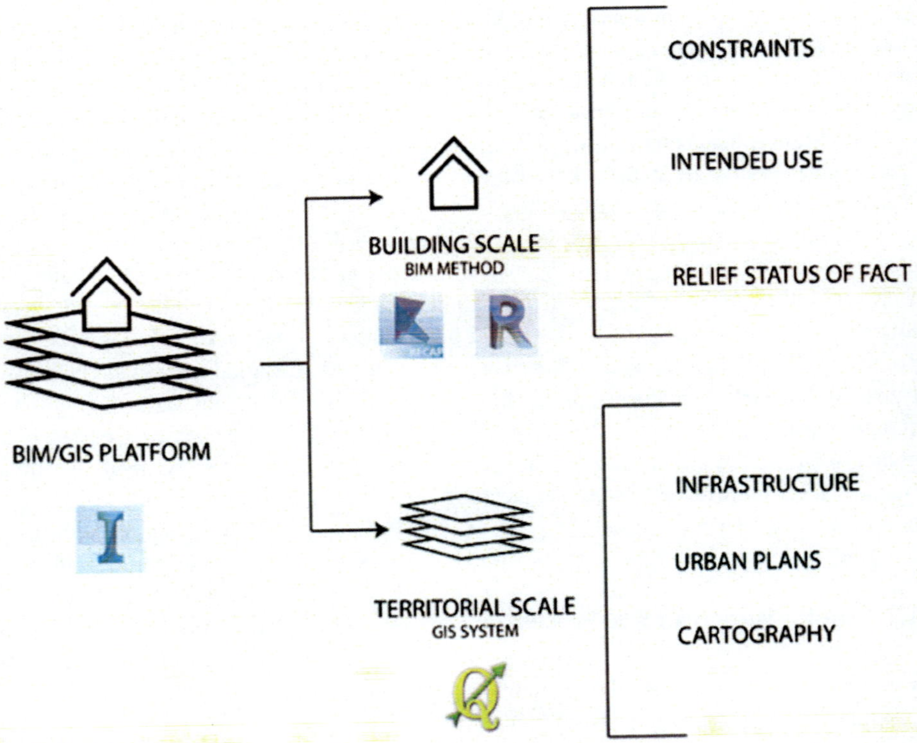

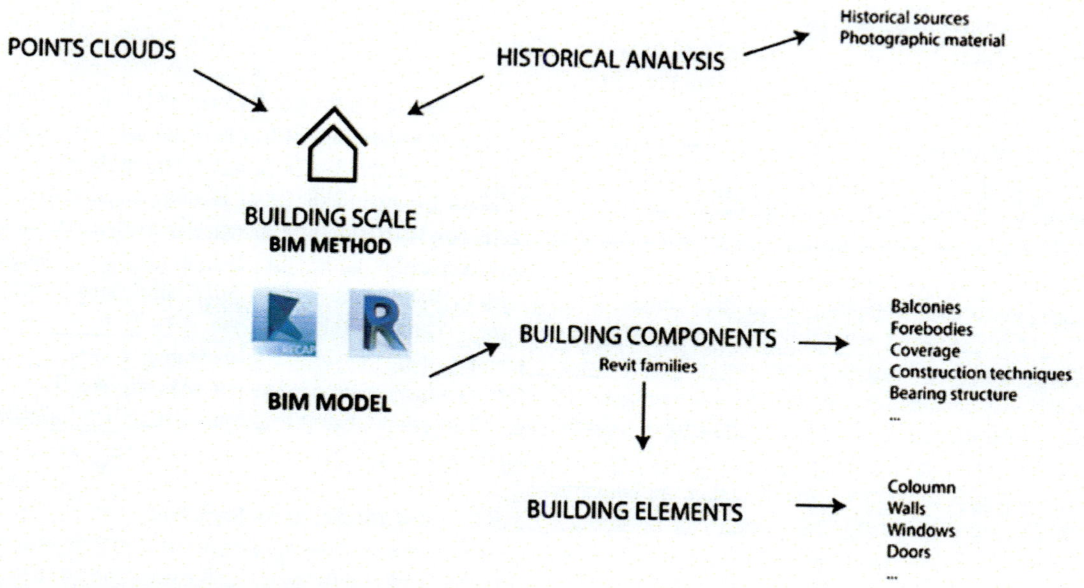

Fig. 6 Operational workflow proposal about building scale

order to be able to make calibrated and sustainable choices. In approaching these concepts, it is therefore essential to make considerations related to the knowledge of the architectural heritage in question.

One of the essential characteristics that we want to note in the definition of the method is that for which the public administration that intends to start an operation of this type, must be completely aware of the architectural characteristics of the asset subject of the operation, leaving the private individual a degree of freedom in the presentation of their own design considerations and proposals, but maintaining the utmost rigour on safeguarding the typological characteristics of the asset.

The main issues to consider in the BIM approach to existing architectural heritage are:

- The uniqueness of the historic buildings, due to the craftsmanship of its construction, the processes of modification and transformation over time and the phenomena of degradation and transformation, which should be kept track in the construction of a dynamic archive of the asset;
- The large amount of information necessary for a complete and exhaustive knowledge of the architectural heritage, requires a mapping of the elements of the asset by defining a database of the components of each building (Bianchini et al., 2016; Cianci et al., 2019; Donato et al., 2017; Giovannini, 2017).

3.3 Case Study. Warehouse 20, Old Port of Trieste

The case study is aimed at an urban survey experience (acquisition, restitution and creation of the spatial model) aimed at the preparation of a multi-scalar and multi-dimensional information model to support a potential preliminary market consultation within the Old Port area in Trieste, in particular on the building called "Warehouse 20".

The objective, as specified above, is not simply to generate a geographic information system of the analysis area, or a 3D model with its own graphic appeal and formal complexity, but to study a dynamic process of acquisition, processing and implementation of spatial and non-spatial information that can be interrogated and functional to planning by the public administration but also to the design of details by private parties.

Initially, the Port of Austrian Trieste at the end of the nineteenth century, it is an area with a regular urban layout typical of the Lagerha users of the ports of northern Europe, with the arrangement of the port structures along three parallel axes. The events that mainly characterize the fate of this space are linked to the fact that it was made a Free Port as a Port of the Austro-Hungarian Empire, subsequently abandoned in favour of the New Port, which better responded to the need for expansion, remained as Porto Franco, essentially abandoned until 2019 When it was desmanialized.

To date, numerous project proposals insist on the area. Of course, this is a space of strong strategic interest, as well as an important opportunity for the relaunch of the city of Trieste on the European scene, and an excellent opportunity for the city, being an unused space of significant size in the immediate vicinity of the city centre.

Within this area there are now numerous examples of building techniques of a historical-identity character of the city.

Therefore, it currently appears to be an area consisting of approximately 617,000 m^2, with 5 piers, 3100 m of quays and 23 large buildings (hangars and warehouses). Since 2019 the area has been handed over by the port authority to the Municipality of Trieste (Caroli, 2002, 2017; Strazza, 2018).

The building types inside are mainly made up of warehouses and hangars (originally 38) which can be divided into:

- One floor above ground (warehouses and deposits)
- Two or three floors above ground with cellar and attic, with balconies between the forepart supported by cast iron columns
- Four floors above ground with cellar, ground floor and four upper floors with balconies, special buildings (hydrodynamic power station, electrical conversion substation and former inns).

3.3.1 3D Survey from Points Clouds

In the traditional PMC, a photographic review of the property is made available to the private subject.

The current technologies of massive data acquisition (laser scanner and digital photogrammetry) allow carrying out data collection of the geometric complexity of the building, even filling the gaps deriving from the lack of information of the archive material or the understanding of the changes operated in later times.

The provision of a point cloud obtained from a survey with a Leica BLK360 laser scanner, available on the market since 2017, is considered as an effective hypothesis, it is a very simple to use and intuitive tool. This has an average accuracy at a range of up to 60 m (Table 1).

Performance studies confirm the 3D point accuracy of approximately 6 mm at a distance of 10 m, as confirmed by the manufacturer. The measurement distance is performed according to the WFD principle (digitalization of waveforms). Scanned with BLK in high density mode, which corresponds to a point's distance of 6 mm at a distance of approximately 10 m. In this mode, a complete scan, including the acquisition of panorama images, takes about 5 min. Through this tool, the data is first stored in the scanner, then transferred to the tablet computer via Wi-Fi. Data acquisition is carried out with the ReCap Pro (Autodesk) program on site (Blaskow et al., 2018; Luhmann et al., 2019) (Fig. 7).

3.3.2 BIM Model Development

A framework is developed that starting from the survey of the object of study, aims at the creation of BIM models useful to offer a coherent and in-depth photograph of the asset, in order to build a starting point for a restoration, maintenance and management project of the asset, defining a coherent and effective tool for the planning of the public administration.

The importance of the theme of modelling within the application of BIM to existing architectures is demonstrated by the presence of numerous studies in this regard, which highlight various problems.

The BIM modelling process mainly focused on the detection of the building characteristics, with the aim of planning conscious interventions, which do not alter the historical morphology of the buildings subject to PPP.

The modelling operation will therefore allow, regardless of the operation that will be carried out, to replicate the object of investigation in its digital twin, allowing to facilitate the study and analysis to better evaluate any future intervention, defining a digital archive of real estate.

Through the detailed basis given by the point cloud, we intend to generate parametric 3D models useful for detecting the typological characters of the buildings under study, with

Table 1 Technical specifications Leica BLK360

Leica BLK360	Specifications
	Scanning unit
Range	0.6–60 m, FoV orizz. 360° | vert.300°
Distance measurement method	Pulse transit time with waveform digitizing
Point of measurement rate	Up to 360.000 pts/sec
3D point accuracy	6 mm @ 10 m/8 mm @ 20 m
Camera system	15 Mpixel 3-camera system, 150Mpx full dome capture, HDR, LED flash Calibrated spherical image, 360° × 300°

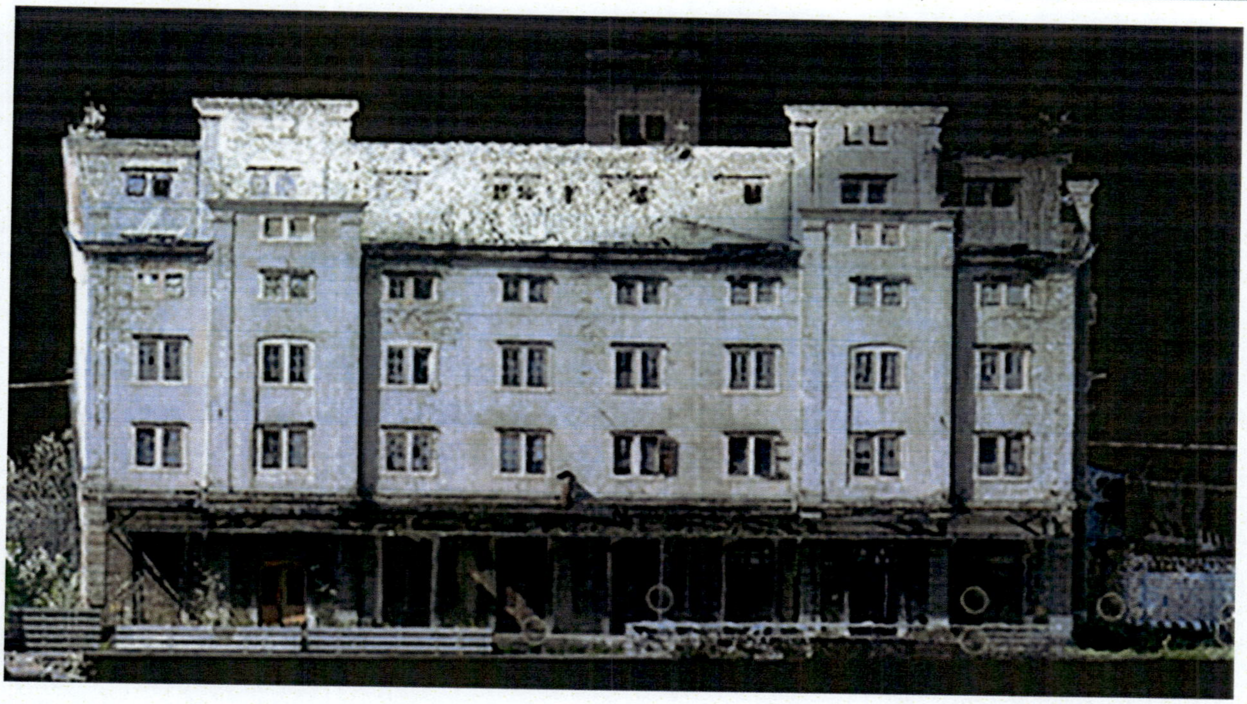

Fig. 7 Warehouse 20's points clouds (Recap PRO)

the aim of defining a document addressing the project that imposes a recovery and enhancement of the same.

The proposed methodology, having as its objective the provision of a basis on which to advance a proposal to the potential investor, identifies the definition of two degrees of detail (LoD) depending on the characteristics identified:

- A simplified model of the entire building, with low geometric accuracy that contains generalized systems with approximate quantities, dimensions, shapes, position and orientation (LoD 200)
- An accurate model with construction requirements and specific elements, which reproduces as much as possible the geometric irregularities of the building's typological characteristics (details of roofs, stairwells, doors, windows, etc.) (LoD 400).

The proposed workflow uses Autodesk Revit software, which guarantees full interoperability with the other software of the autodesk family used in the study. When providing the template, the file is exported in *.ifc interchange format.

In the case proposed, through study and historical analysis, the building was broken down with respect to its technological system typical of the construction of the Old Port of Trieste, dividing it according to its characteristics, and translating them into Revit families.

In detail, the Warehouse 20 has three floors above ground plus an attic and has a rectangular plan and, unlike other warehouses in the port, has a smaller size of the longitudinal fronts. It also features a turret on the roof. The longitudinal fronts present on the ground floor the characteristics of the other buildings, but without the cast iron columns, two slightly projecting bodies (Fig. 8).

To set up the model, the previously mentioned typological characteristics are mapped and translated into the relative Revit families, defining an archive of information on the building. In this way, the model will be easily interrogated and implemented, allowing a quick and coherent evaluation of the project hypotheses, also with respect to the evaluation of a future economic commitment, for the respect of the economic financial equilibrium also in the perspective of the management of the asset (for the calculation expenses) as well as to count the elements.

3.3.3 GIS Analysis

In identifying the data that are considered useful for the success of PPP operations, an in-depth study of the area in which the operation is developed is identified. In detail, it is believed that an analysis of what may be the relationships that insist on the area and the characteristics of the elements within it, to allow an adequate understanding of the possibilities of success of the operation.

GIS systems are tools of high potential for the rational and controlled management of the territory, allowing an immediate understanding of its morphology and the urban elements included in it.

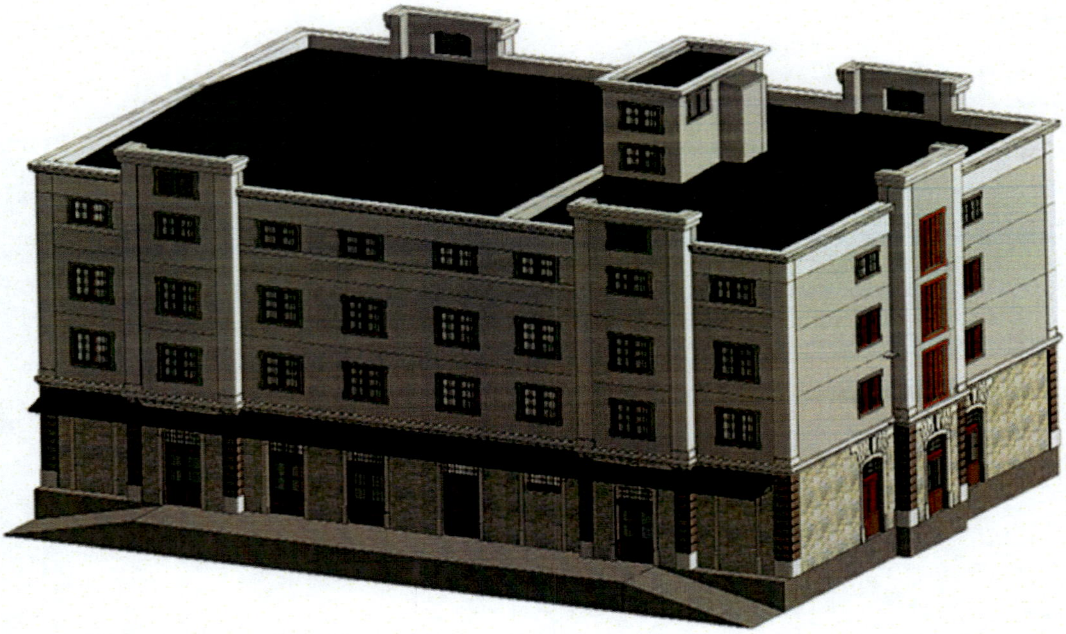

Fig. 8 Warehouse 20's BIM model (Revit)

In the definition of the method, for the definition of the territorial information system, the open source software QGIS was identified, connected to the regional database (Sani et al., 2019).

The choice of this software depends on its ability to support numerous vector and raster files and interact with databases and advanced analysis functions. QGIS is able to display and overlay raster vectors of different formats, using different projection systems at the same time, without any format revision being necessary.

It is therefore possible to create maps and explore spatial data through a graphical interface that simplifies the use of the software. As regards the workflow in description, the analysed and systemized data are provided to the stakeholders through the provision of the standard CityGML data format.

This format allows for the exchange of 3D data, storing different three-dimensional geometries, thus describing the objects in relation to their geometric, topological, semantic and appearance properties. This allows to perform complex data queries in GIS environments, opening the way to spatial analysis of urban models (de Laat et al., 2011).

The analyses are identified by different layers containing the urban plans insisting on the area (urban plan, regional landscape plan and containing the related constraints) flows of the driveways, pedestrian paths, and the precise analysis of the functional destination of the elements of the area, in order to easily analyse the strategic nature of the operations.

The hypothesis of investigating communication technologies capable of operating within web-based information systems is also analysed for future development. The Web-GIS, conceived as extensions towards the network of GIS platforms, allow to increase the number of potential interested parties.

3.4 Implementation BIM/GIS

The integration of BIM models and GIS systems represents a great opportunity to reach a complete level of knowledge of the territory, for the activation of winning operations, as well as the ability to offer an excellent ally to the design of a coherent and potentially digital archive, always updated.

The research in this sector is particularly active, especially with regard to territorial planning, cadastre management and environmental simulations (Basir et al., 2018; Zhu et al., 2018).

With the aim of associating the territorial elements and those on the territorial scale, the Autodesk Infraworks software was identified, which associates the BIM model to the territorial element (Fig. 9). In this way, it is possible to define, within a single window, all the elements of the system, creating a database that holds together all the information entered in it and creating an error-free basis also for the evaluation of proposals arriving from private individuals both from a compositional and performance point of view (Carta et al., 2018; Fai et al., 2011).

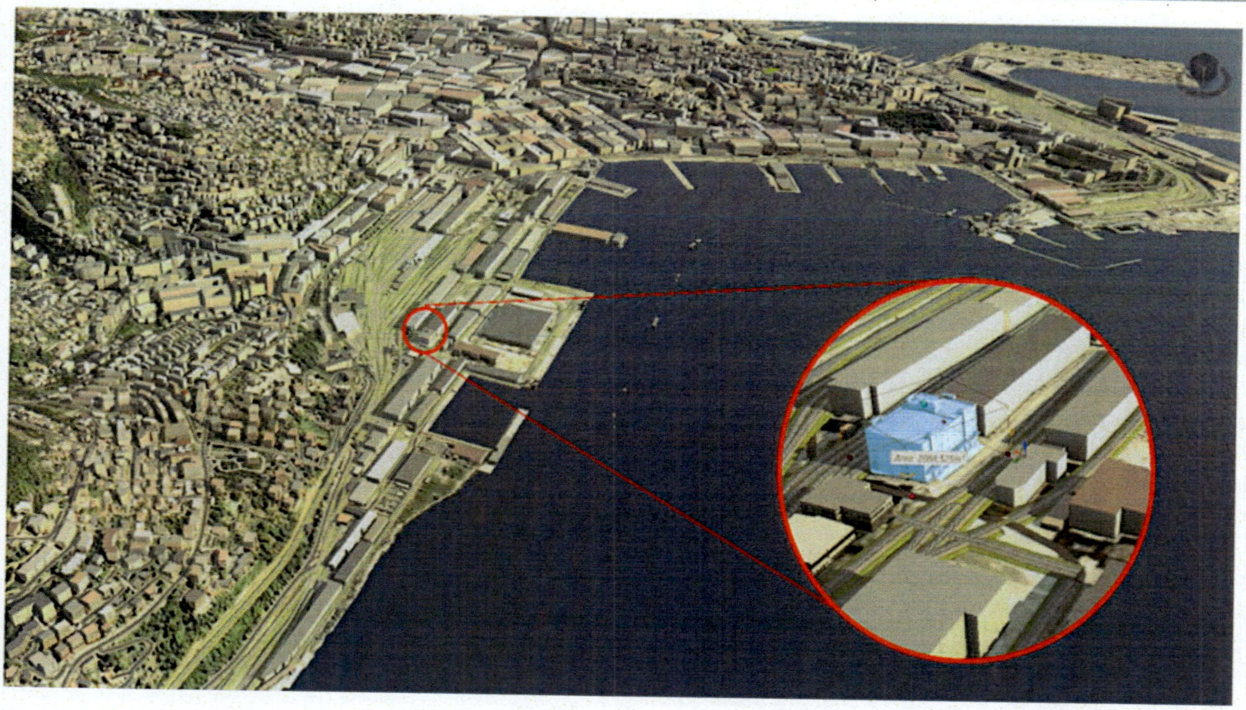

Fig. 9 Warehouse 20 in old port of Trieste's urban model (Infraworks)

In this process, there is currently a need to define a platform that can make the models (BIM and GIS) generated by the public administration viewable (and possibly downloadable). In fact, the illustrated solution can be considered satisfactory if the public administration makes the territorial models generated, complete with the related connected *.ifc models, available to potential interested parties.

BIM and GIS can be understood as complementary solutions in different projects. Specific tools are available to solve different aspects of practical applications, allowing a more complete and interrogable perspective with respect to different scales by the various actors involved.

4 First Results

In the request for the elaboration of the projects by the private individual, it is therefore asked to follow a protocol that is the same for everyone, identifying an information specification, in which all the elements of the file to be acquired are outsourced, with the definition of its own characteristics of the model, such as the coordinate system, the orientation, the interoperable format as well as all the data concerning privacy.

This structured request is also linked to the need to analytically check the response to certain parameters determined within the address document. To do this, Autodesk Nav is works software was identified, capable of verifying the compliance of system models. Through the definition of customized interference rules, it is possible to verify the consistency of the project with the request for the steering document.

At the same time, through the same software, the private individual will be able to effectively determine the quantities of materials and the workflow suggested for the design and quantitative take-off of architectural and structural models, allowing once again to be extremely aware of the commitment that is evaluated to undertake.

This application can also be made available to the Public Administration which will be able to transparently control the operation.

5 Conclusions

As a first methodological test, the data determining the workflow hypothesis for a preliminary market consultation for Warehouse 20 in the Old Port of Trieste were used. It was evaluated how to make available and effective the information useful to a proven who wanted to elaborate a proposal for a possible public–private partnership.

Once the method has been tested through the response of the stakeholders to the procedure, we intend to re-evaluate the data delivery and request protocol, in such a way as to define guidelines that lead to the receipt and delivery of coherent information, for the definition of an archive 3D of the public administration.

The next goal, in the face of a greater commitment by the public administration, is to define an integrated information platform that can contain all the areas and buildings of the Municipality that must be redeveloped with all the useful information, so that it can be private individuals to make proposals for partnership projects which can then be evaluated by the administration.

The experimentation illustrated in this contribution, at a methodological verification stage, has allowed us to arrive at some considerations on the results obtained and on the possible implementations of the research.

The most important result is the awareness of uniquely defining the workflow of the public administration with respect to the modelling and setting of data, in order to work in a targeted manner as necessary to stimulate the market with respect to the assets to be redeveloped.

It is particularly important to think in terms of interoperability for the structuring of the relationships between 3D GIS and BIM, as well as to ensure the complete passage of information between the various stakeholders.

The ability to make the process for retraining analytical through ICT, starting from the provision of information from the beginning of the operation, can also allow for complete transparency by the administration, which in this way will be able to collect and manage a historical and updated digital archive of all phases.

References

ANAC. (2018). *Linee Guida n.9, Monitoraggio delle amministrazioni aggiudicatrici sull'attività dell'operatore economico nei contratti di partenariato pubblico privato.*

ANAC. (2019). *Linee Guida n. 14 recanti "Indicazioni sulle consultazioni preliminari di mercato" Relazione Illustrativa 1.* 1–19.

Bardelloni, S., & Capotorto, D. (2013). *Codice dei contratti pubblici commentato.*

Basir, W. N. F. W. A., Majid, Z., Ujang, U., & Chong, A. (2018). Integration of GIS and BIM techniques in construction project management—A review. *International Archives of the Photogrammetry, Remote Sensing and Spatial Information Sciences—ISPRS Archives, 42*(4/W9), 307–316. 5194/isprs-archives-XLII-4-W9-307-2018

Bianchini, C., Inglese, C., & Ippolito, A. (2016). The role of bim (building information modeling) for representation and managing of built and historic artifacts. *Disegnarecon, 9*(16), 10.1–10.9.

Blaskow, R., Lindstaedt, M., Schneider, D., & Kersten, T. (2018). Untersuchungen zum Genauigkeitspotential des terrestrischen Laserscanners Leica BLK360. *Photogrammetrie, Laserscanning, Optische 3D-Messtechnik–Beiträge Der Oldenburger 3D-Tage 2018, May*, 284–295.

Caroli, A. (2002). *Il Porto di Trieste. Cronaca e Storia delle Costruzioni Portuali.*

Caroli, A. (2017). *Guida storica del Porto Vecchio di Trieste.* Edizioni "Italo Svevo" Trieste.

Carta, E., & Scanu, S. (2018). *Fotogrammetria , GIS e BIM per la gestione del Piano Particolareggiato di un Centro Storico della Sardegna.* 253–260.

Cianci, M. G., & Molinari, M. (2019, February). Information modeling and landscape: intervention methodology for reading complex systems. In *ISPRS—International archives of the photogrammetry, remote sensing and spatial information sciences, XLII-2/W9* (pp. 269–276). https://doi.org/10.5194/isprs-archives-xlii-2-w9-269-2019

Cresme Europa Servizi. (2018). *Il mercato del Partenariato Pubblico Privato in Italia.*

de Laat, R., & van Berlo, L. (2011). *Integration of BIM and GIS: The Development of the CityGML GeoBIM Extension*, 211–225. https://doi.org/10.1007/978-3-642-12670-3_13

Donato, V., Biagini, C., Bertini, G., & Marsugli, F. (2017). Challenges and opportunities for the implementation of h-bim with regards to historical infrastructures: A case study of the ponte giorgini in castiglione della pescaia (grosseto–Italy). *International Archives of the Photogrammetry, Remote Sensing and Spatial Information Sciences—ISPRS Archives, 42*(5W1), 253–260. https://doi.org/10.5194/isprs-Archives-XLII-5-W1-253-2017

EUBIM Taskgroup. (2018). *Manuale per l'introduzione del BIM da parte della domanda pubblica in Europa.*

Fai, S., Graham, K., Duckworth, T., Wood, N., & Attar, R. (2011). Building information modelling and heritage documentation. *XXIII CIPA International Symposium, Prague, Czech Republic.* https://doi.org/10.1136/adc.2010.183327

Garilli, A. (2007). *A. Garilli, La privatizzazione del lavoro nelle Pubbliche Amministrazioni e l'art. 97 Cost.- di alcuni problemi e dei possibili rimedi* (WP C.S.D.L.E. "Massimo D'Antona".IT—49/2007).

Giovannini, E. C. (2017). VRIM workflow: semantic H-BIM objects using parametric geometries. *Progettazione, Design, Proposte per La Ricostruzione.*

Luhmann, T., Chizhova, M., Gorkovchuk, D., Hastedt, H., Chachava, N., & Lekveishvili, N. (2019). Combination of terrestrial laser scanning, uav and close-range photogrammetry for 3D reconstruction of complex churches in Georgia. *ISPRS Annals of the Photogrammetry, Remote Sensing and Spatial Information Sciences, 42*(2/W11), 753–761. https://doi.org/10.5194/isprs-Archives-XLII-2-W11-753-2019

Pasquini, G., Balzarini, P., Cipolletta, D., De Amici, A., Ferraro, M., Lo Burgio, A., Marasco, P., Mazzitti, W., Menduni De Rossi, E., Paparo, T., Pedacchio, C., Proietti, R., Siclari, D. R., Surdi, G., Tranquilli, M., Tretola, L., Forti, C., & Turriciano, M. (2019). *Partenariato Pubblico Privato e Finanza di Progetto- Relazione sull'attività svolta dal DIPE negli anni 2017–2018.*

Perino, G., Sileno, L., & Tresalli, G. (2018). *Partenariato pubblico privato. Presupposti e modelli 2018.*

Ricchi, M. (2016). *L ' Architettura dei Contratti di Concessione e di Partenariato Pubblico Privato nel Nuovo Codice dei Contratti Pubblici d. lgs . 50 / 2016.* 811–828.

Sani, M. J., Musliman, I. A., & Abdul Rahman, A. (2019). Extraction and transformation of Ifc data to citygml format. *ISPRS—International Archives of the Photogrammetry, Remote Sensing and Spatial Information Sciences, XLII-4/W16*(October), 595–601. https://doi.org/10.5194/isprs-archives-xlii-4-w16-595-2019

Siclari, D. (2019). *Il Project Financing nel Codice dei contratti pubblici* (Giappichelli (ed.)).

Strazza, N. (2018). *Riqualificazione sostenibile del sistema urbano fronte-mare di Trieste.* Università degli Studi di Trieste.

Tajani, F., Morano, P., Di Liddo, F., & Locurcio, M. (2018). *Un'interpretazion innovativa dei criteri di valutazione della DCFA nel Partenariato Pubblico-Privato per la valorizzazione del patrimonio immobiliare pubblico.* 53–57. https://doi.org/10.19254/LaborEst.16.09

Unità Tecnica Finanza di Progetto. (2010). *Partenariato Pubblico Privato in Italia. Stato dell'arte, futuro e proposte.*

Vacca, G., Quaquero, E., Pili, D., & Brandolini, M. (2018, October). *Integrating Bim and Gis Data To Support the Management of. XLII*, 1–5.

Zhu, J., Wright, G., Wang, J., & Wang, X. (2018). A critical review of the integration of geographic information system and building information modelling at the data level. *ISPRS International Journal of Geo-Information, 7*(2), 1–16. https://doi.org/10.3390/ijgi7020066

Evaluating the Daylighting and Energy Performance of Container Housing Systems in Coastal Regions: An Iterative Modeling Approach

Md. Obidul Haque, Nabeela Nushaira Rahman, Nusrat Tabassum, and Jayedi Aman

Abstract

The container housing concept is presented as a sustainable housing choice for low-income families in the coastal region in this study. As the goal of reducing the environmental effects of structures becomes clearer, environmentally sustainable construction approaches are becoming increasingly significant. The study used a generative design method that sets off from a building performance optimization process, allowing designers to compare the performance of multiple design possibilities instantaneously and recommend the best options. A computer-assisted Iterative Simulation Modeling (ISM) technique is outlined for calculating energy performance in a Container Housing System (CHS). The framework includes the program analysis, site selection, container housing system generation, performance rating, and subsequent discussion. The findings suggest that the ISM technique has the potential to assess the container housing system's energy performance as well as lower total energy usage. The research concludes with the hope that additional researchers and inventors will use the ISM method to deliver multifunctional, energy-optimized design for adaptation and future resistance to a variety of climate change impacts.

Keywords

Iterative energy modeling • Energy optimization • Useful daylight illuminance • Energy use intensity • Container housing system • Coastal region

Nomenclature

CHS	Container Housing System
ISM	Iterative Simulation Modeling
HBRI	Housing and Building Research Institute
ICESCR	International Covenant on Economic, Social and Cultural Rights
UDI	Useful Daylight Illuminance
EUI	Energy Use Intensity

1 Introduction

Considerably, the most recurrent calamitous spectacles along coastal areas of Bangladesh are cyclones and tidal heaves, which are due to topographical positioning. A typical of 16 of these turbulences occur annually (DDM, 2014) catapulting Bangladesh among the heaviest sufferers of disaster losses. Besides causing extensive health risks, these disasters abolish harvests and impair structures and connections. The hindrances caused by the disasters set back the country in its journey toward development, economic advancements, and self-reliance by occurring almost every year. Cyclones disasters in 1970 and 1991 (0.5 and 0.14 million fatalities respectively) in Bangladesh are considered as some of the deadliest calamities globally (DDM, 2014), and over 4 million people live in regions at threat of cyclones in the country (Minar et al., 2013). An identified reason behind the severity of damages is the deficiency of robust and planned housing. The administration is dedicated to bring about a considerable drop in disaster danger and increased resilience of the vulnerable population in this sector (Shelter Cluster,

Md.O. Haque (✉)
Department of Architecture, Premier University, Chittagong, Bangladesh
e-mail: obidul.arch@puc.ac.bd; ; obidul.line@gmail.com

N. N. Rahman
School of Architecture and Design, BRAC University, Dhaka, Bangladesh

N. Tabassum
The Pennsylvania State University, State College, PA, USA

J. Aman
Department of Architectural Studies, University of Missouri, Columbia, MO, USA

H. Altan et al. (eds.), *Advances in Architecture, Engineering and Technology*,
Advances in Science, Technology & Innovation, https://doi.org/10.1007/978-3-031-11232-4_13

2018), making housing have high-priority in government programs. Housing and Building Research Institute (HBRI), consequently aiming to plummeting the jeopardies connected to natural catastrophes, is resolute to endorse justifiable construction constituents and practices for housing, urban and rural (HBRI, 2018). Addressing these issues, this paper attempts to introduce a system that may be used by academics, designers, architects, and engineers to achieve architectural resilience for poor communities in coastal areas. It is observed from previous studies and casual observation in port areas that Container Housing—housing constructed by shipping containers—is a feasible solution toward affordable housing (Rodrigue et al., 2016). In particularly the hot-humid climate like Bangladesh, it may seem inappropriate to construct this type of housing unit (Kristiansen et al., 2020), specifically if the unit is poorly ventilated since poor communities usually accommodate ventilation in passive design with occasional aid from electric fans (i.e., airconditioning units are not used to ventilate the unit). However, proper insulation, orientation, and architectural techniques could make containers a feasible and sustainable solution for architectural resilience in different contexts (Aman et al., 2021). As a result, this article proposed a recommendation to optimize the system using a computer-aided Iterative Energy Modeling (ISM) approach to quantify the Container Housing System (CHS)'s daylighting and energy performance. Furthermore, it proposes the system resolutions as a means of promoting computer-aided design not only in traditional buildings but also in low-cost informal communities. This aims to fill a research gap identified as a lack of iterative design optimization studies on low-income container housing clusters.

1.1 Aim and Objectives

In order to achieve sustainable development goal 11 of United Nations (2018), i.e., empowered communities, making settlements inclusive, safe, and resilient is mandatory. This research intends to contribute to disaster-resilient architecture through an innovative housing system in Bangladesh's coastal regions and other similar settings.

The following are the research's objectives:

- To promote container housing system (CHS) for mitigating the coastal area's challenges toward environmental sustainability to embolden resilience.
- To establish an iterative energy modeling workflow for best fitting container cluster options per module.
- To propose the optimized design variables for the optimal trade-offs between daylighting and energy performance.

1.2 Scope of the Study

The purpose of this article is to identify the parameters for ISM in the context of the specific chosen site region, which includes climatic research, site surrounding studies, family size studies on the intended users, user needs analysis, existing site condition study, and material exploration study. Land-use patterns, specific interior planning, and building details are not included in the scope of this study.

1.3 Formulation of the Work Process

This research began with identifying the housing needs of disaster-affected and low-income populations in the chosen site in Bangladesh's coastal region of Chittagong and then moved on to climatic, contextual, socio-economical, and demographical analyses via on-site questionnaire surveys. The survey was undertaken to look at the housing patterns, livelihood patterns, space usage patterns, service patterns, vulnerability, requirements, and other pertinent details in the chosen area. Physical perception and meeting locals at a residence, field, fishing location, and market were used to acquire information. However, books, research articles, and the District Fisheries Office provided valuable information on the state of anglers, the site map, and the climate. The results noted below of a related literature review and data analysis revealed concerns with relation to local needs.

- Organization of settlements and future extension-Improper planning and frail infrastructure. No future extension plans were noted.
- Economic Development- The absence of a dedicated income source during the off-season, uneconomical building strategies, and additional costs for repairing two times in a year are the obstacles for economic development.
- Environmental Conditions- Unhygienic environmental approaches and poor lighting.
- Current materials- Mostly vulnerable temporary materials like bamboo, tin, mat, etc. However, it has been observed that there is a trend toward employing shipping containers as housing units, although this trend is not sustainable.
- Resilient Structures- The need for adaptable but sturdy settlements in the face of natural disasters and other threats eliminates the possibility of evacuation.
- Communal Relationship: Need for establishing co-existing community quality.
- Essential endurance requirements: Fresh drinking water, sanitation, electricity, etc.

This aided in deriving from them the required parameters for the ISM process. Following that, the key to this investigation was formulated. The generated results and discussion follows the framework and concludes, indicating future implications.

2 Literature Synthesis

The literature review talks about the significance of disaster resilience in coastal areas, the housing nature of Bangladesh and guidelines of housing in the coastal region, and the properties of the shipping container as a housing unit before moving forward to the research framework.

2.1 Significance of the Coastal Resiliency

On the scale of countries most at risk from disasters, in view of heights of subjection to environmental disasters and diverse points of susceptibility, Bangladesh ranks fifth, wherein about every three years a drastic cyclone or tropical storm hits (World Risk Report 2016). In the Bay of Bengal (Fig. 1), cyclonic movements have converted to be more recurrent in recent times, and near about half of the total demises of tropical cyclones in the world happened in Bangladesh—a country which was hit by these disasters 159 times from 1877 to 2009 (Shelter Cluster, 2018). As it proposes the capability of structures to endure catastrophic natural events, the phrase "coastal resilience" is pertinent in strategy and planning for designing coastal housing. However, in Bangladesh, recent severe tropical cyclones, along the shores specifically, have emphasized the defenselessness of littoral areas and instigated needless damage to natural life and assets on top of huge financial impoverishments.

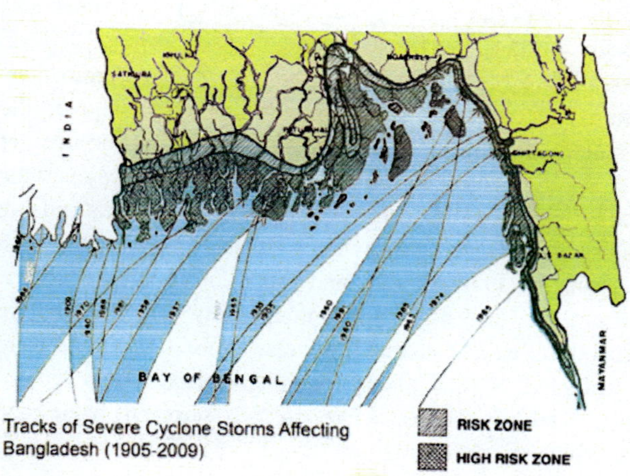

Fig. 1 Coastal region of Bangladesh (after Multipurpose Cyclone Shelter Program, 1993)

The cyclone SIDR in 2007 alone devastated 1.5 million houses, including over 0.5 million structures completely ruined (HBRI, 2018). Global warming is causing a deadly escalation in sea level rise, which affects the rate of recurrence and extremity of cyclones, directing the evaluation of coastal resilience designing and planning to acclimatize to eventual and sudden penalties of these environmental fluctuations (SCR, 2014).

2.2 Traditional Housing Features and Coastal Housing Guidelines in Bangladesh

Bangladesh possesses a rich tradition in the formation and development of its human settlement. The built forms of these settlements have their specific characteristics. Indigenous people of this country have hereditarily been exercising this sort of housing technology for years which reflects ecological and environmental concerns. An open planning pattern is an appropriate response to this warm-humid climate. The basically separated rectangular shape of living zones that elongated along the east–west direction is arranged around a square or rectangular-shaped courtyard. To provide sufficient air circulation, buildings are scattered while being introvert in character, and researches advocate that voluminous shielding protection by foliage and use of a court decreases total heat gain considerably (Hasan, 1985; Olgay, 1963). They are divided by huge open spaces to facilitate circulation and cooling ventilation. The desired direction of the central living section is north–south stretched along the courtyard (Haque et al., 2022a, b). This also allows protection from the sun, which is along the east–west axis (Fatemi & Islam, 2011).

The unattached yet closely-placed dwelling units have gaps of 0.7–1.25 m in between and give the courtyard a defined space. This arrangement construtes the second level of dwelling units typically for one family, and although detached, they are measured as a unit considering the inter-relationships of the functions expressing an integral impression that these detached houses are rooms forming a bigger unit of a home (Ahmed, 2012). A household is comprised of several rooms encircling an "uthan" (courtyard) in rural regions. The yards are the vital socializing places on a family level, on top of being used for numerous domestic goings-on, and typically, bigger families have two courtyards as seen in Fig. 2—the inner as the domain of women and children while the outer one as the mens' (Soud & Haque, 2018).

In Bangladesh, maximum peril considering fatalities and assets reparations is at the coastal region. Site for habitats, located at varying gradations of hazards, spread in an unplanned manner from front of inward bearings to expanses much inland where massive requirement for recuperation

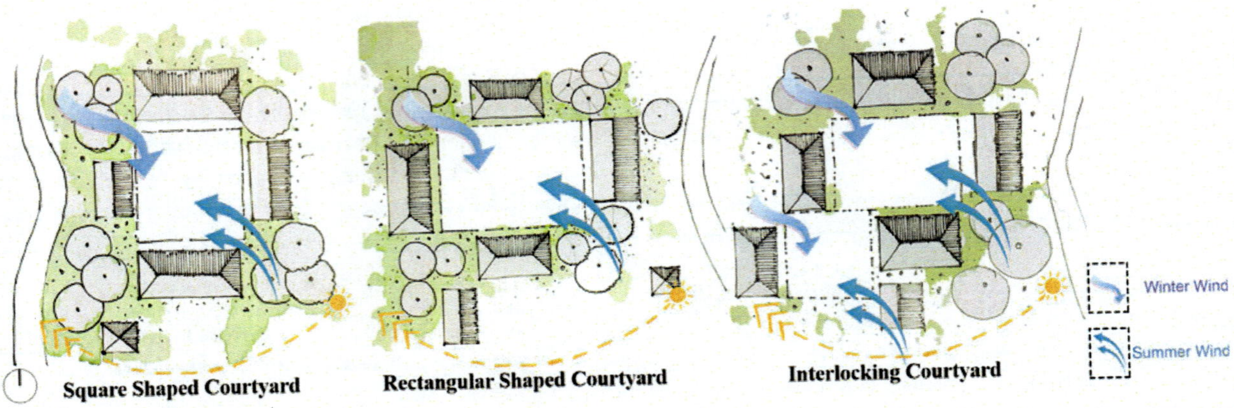

Fig. 2 Different types of the courtyard in Bangladesh (after Rahman, 2017)

and repositioning is a dominant need. The 1966 International Covenant on Economic, Social, and Cultural Rights (ICESCR) defines suitable housing in seven aspects and locations (Shelter Cluster, 2018). They have been used as a foundation to design the ensuing minimum standard for decent housing in Bangladesh, based on earlier Shelter Cluster study in Bangladesh. The following are the standards that were considered for this study.

Research Framework

This section presents a new ISM workflow for energy optimization of CHS design in the coastal areas. This approach was processed through four main phases. The first step was to select the case area, site surroundings, and to pick out the housing energy schedule for the placement of the proposed container cluster system. The second step was the formation of clusters based on the common courtyard types and cluster typologies. Moreover, several design variables were identified to be examined and parameterized for performance simulation. The third step was to integrate the daylight and energy performance simulation analysis, considering the local weather data. The final step was to conduct the automated iterations and to analyze the resulted data and visualizations. The overall workflow of this study is depicted in Fig. 3.

3 Implementation

3.1 Case Area

Kattoli Jelepara (fishermen community), Halishahar, Chattogram was chosen as the case area in the first step. This neighborhood, where the community has lived for over a century, was an ideal location for the investigation. The property is on the outskirts of town, near to Halishar Beach. Another important factor in choosing the location was the low cost of shipping containers. During the physical survey, it was discovered that the unplanned development of the neighborhood had made the built environment unhealthy and hazard-prone (Fig. 4).

The basic housing infrastructure and minimum housing standards are entirely ignored, resulting in an unstable and highly vulnerable structure in terms of government guidelines regarding the housing development on the coastline. Despite these, essential functions of the households like bedrooms, common kitchens, common toilets, living patterns, multifunctional spaces, size, and shape of the houses (i.e., 100–300 ft^2), needs, etc. have been documented which prompted this research to be accounted for energy modeling. Table 1 explains the detailed geographic and climatic information of the case area.

3.2 Parametric Cluster Formation

In this step, the existing site, surroundings, and other predictor attributes were developed in Rhinoceros (McNeel, 2020). The container dimensions were attained from literature and from the available shipping containers of the Chattogram port area. The number of container houses per cluster was considered based on the physical survey and the local regulations for standard living measures in affordable houses. A physical survey showed that an average of 4–6 persons live in one family, and about 200 people live in one cluster at the fishermen's community in Chattogram. Therefore, eight containers per cluster for 192 people were considered. A total of six container cluster system typologies were established to perform automated parametric iterative energy simulation (Fig. 5). The site surrounding information was imported through urban planning and city

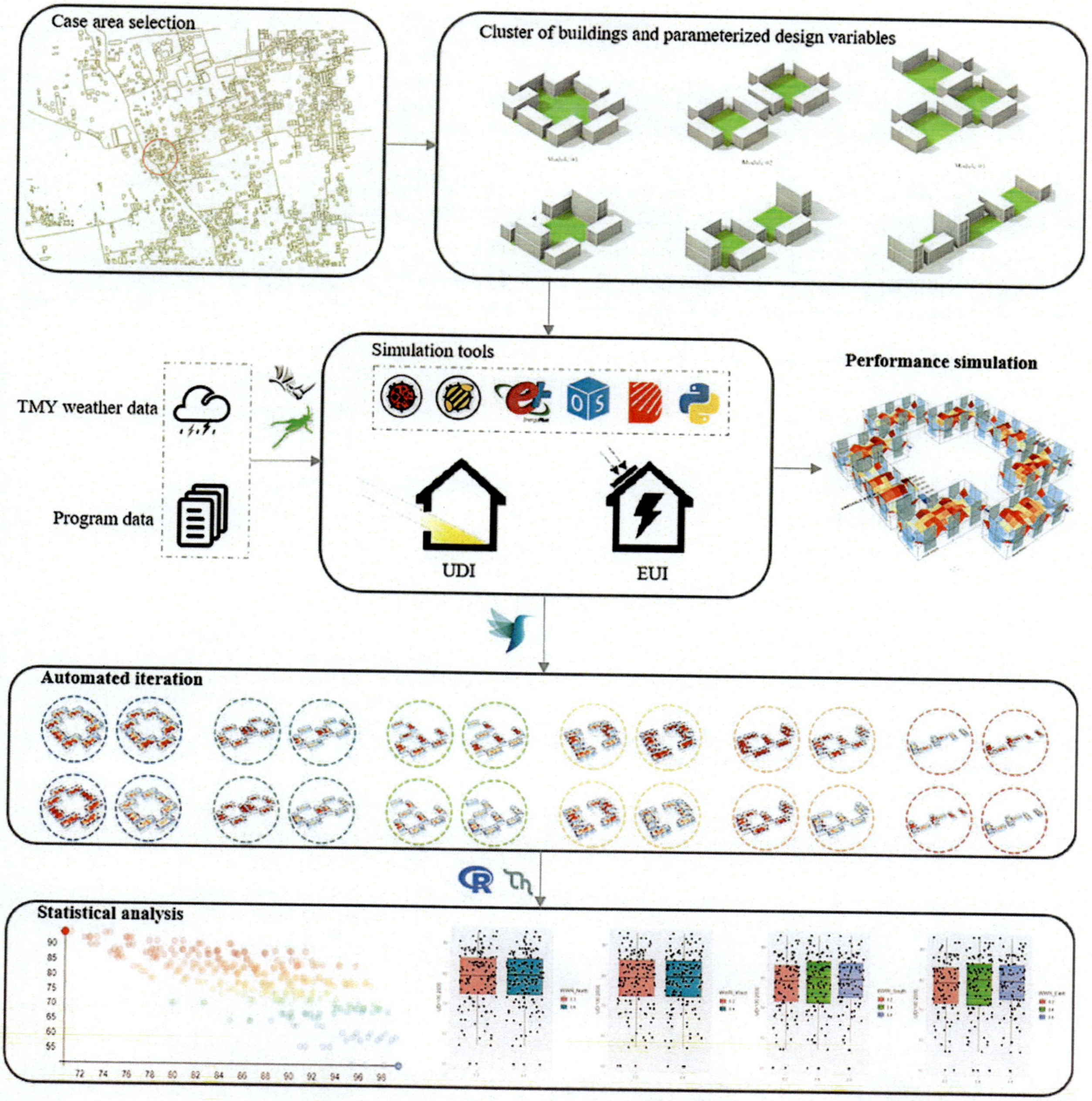

Fig. 3 The research framework of this study

analyzing plugin-Urbano (Dogan et al., 2020) for the visual programming platform Grasshopper (Davidson, 2020). The geographical information was downloaded from the open data source OpenStreetMap (Kloog et al. 2018).

The study focuses on the sustainable and resilient cluster design that would allow the inner spaces of the houses to receive the most beneficial daylight while using the least amount of energy. To achieve this purpose, predictor variables such as the Window-Wall Ratio (WWR) and Shading were selected based on previous studies (Aman, 2017). The WWR is one of the most significant factors to consider when adjusting daylight access, heating, and cooling. This ratio is determined by splitting the total glass area of a structure by the total outside envelope area (ConstruPM, 2017). In this study, the WWR ranged from 0.20 to 0.60 for north and south facades and 0.20 to 0.40 for east and west facades. The shading depth for the iteration process, on the other hand, was set to 0.25 m to 0.50 m. U-values and R-values in the glazing zones ranged from 1.4 to 2.5 w/km^2 and 2.6 to 4.4 km^2/w, respectively. The Useful Daylight

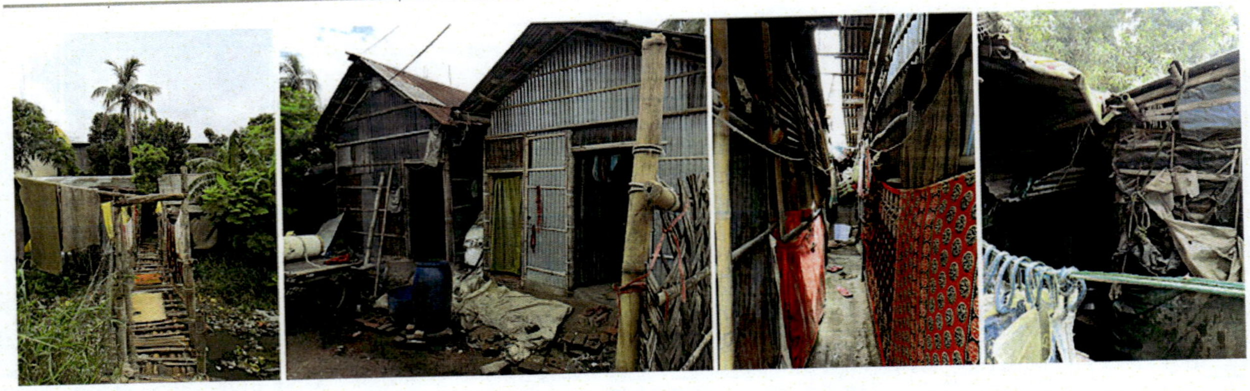

Fig. 4 Current condition of the case area, located in the existing fishermen's community (*Source* Authors)

Table 1 Detailed information of the case area

Sl. No	Information	Case area
01	Location	Kattoli Jelepara, Halishahar, Chattogram
02	Altitude*	29 m
03	Temperature (Min)**	13.9 °C
04	Temperature (Min)**	32.3 °C
05	Relative humidity**	85% (Max), 70% (Min.)
06	Construction materials***	The majority of houses in rural areas are made of local materials such as bamboo, mud, and so on Walls: straw, jute stick, CI sheets, etc Roofs: thatch or CI sheets, with clay tiles, added occasionally

*Source Haque et al. (2021)
***Source* Bangladesh Meteorological Department, Dhaka
***Hasan et al. (2000)

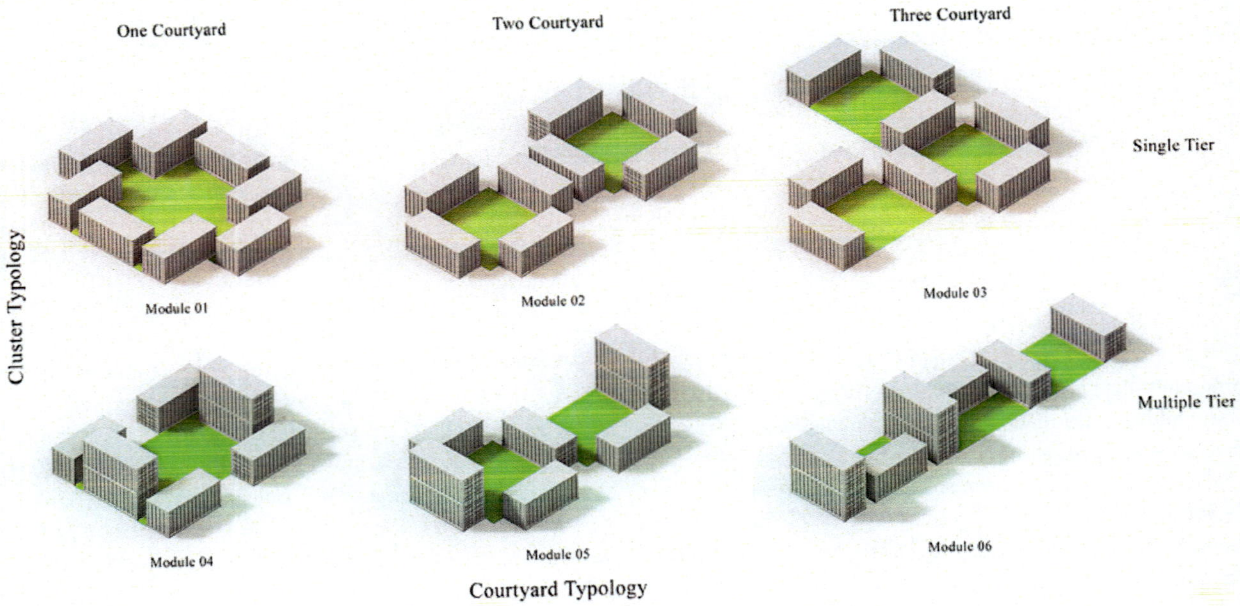

Fig. 5 Container cluster modules

Illuminance (UDI) which predicts the daylighting and Energy Use Intensity (EUI) which defines the energy performance, was chosen as the outcome variables.

3.3 Simulation Performance Analysis

The previous stage assisted in the selection of a suitable matrix for courtyard and container housing typologies, as well as the collection of related attributes for subsequent analysis. In the third step, the built container clusters were sent into the simulation process. The process consists of operating the input for getting the UDI and EUI results and then export the variables for iterations. Figure 6 depicts the Grasshopper script workflow.

The geometries were parametrically connected to the Daysim, a dynamic daylighting simulation plugin (Haque 2019; Bourgeois et al., 2008; Lartigue et al., 2013) using Ladybug and Honeybee (Roudsari et al., 2013). Sensor planes for daylighting were set up at 0.75 m from the floor level. On the other hand, for the energy performance analysis, the modules were allocated to distinct areas. Honeybee has automated construction template for specific design and internal loads for the interior area utilizing a set of home energy schedules gathered in the initial stage. Additionally, the honeybee presents bespoke material properties. The simulation of the Container Cluster System has been depicted in Fig. 7.

The exterior facades of containers, made of steel panels, were a major source of concern. Honeybee's unique material attributes enabled thermal insulation material to be assigned to the inside walls. The inner walls, floors, and ceilings of eight structures were assigned radiance opaque material with reflection, while the glazed zones were allotted glass material. Weather data is another important information that must be fed into the energy simulation process. As weather data input, chattagram data in .epw format was used (EnergyPlus, 2020).

3.4 Automated Iteration Process for Energy Optimization

In the final step of the framework, Colibri (Tomasetti, 2020a) plugin was connected to the script for iterating the inputs and to capture the outputs in a dataset. The plugin has both Iterator and Aggregator batteries to initiate the iteration process. Parametric input variables, such as window-to-wall rations, shading depths, and material properties were used and UDI and EUI were used as the output variables. Table 2 shows the iteration sets and numbers.

Although a high number of iterations may be employed to acquire more precise results, time and resource constraints limited the number of parameters and values examined. The simulation data and visualization outputs were then saved in .csv, .json, and .png formats, which were then analyzed using several open source packages in CORE Studio's statistical software and Thread (Tomasetti, 2020b; The R Foundation, 2020). A total of 288 × 6 = 1728 simulations for six container cluster system typologies were performed in this study. The simulation took roughly two minutes for each reiteration on a DELL Precision XEON PC.

4 Results and Discussion

The design iteration analysis method is demonstrated in this section. It addresses the usefulness and efficiency of this architectural resilience technique for constructing in coastal

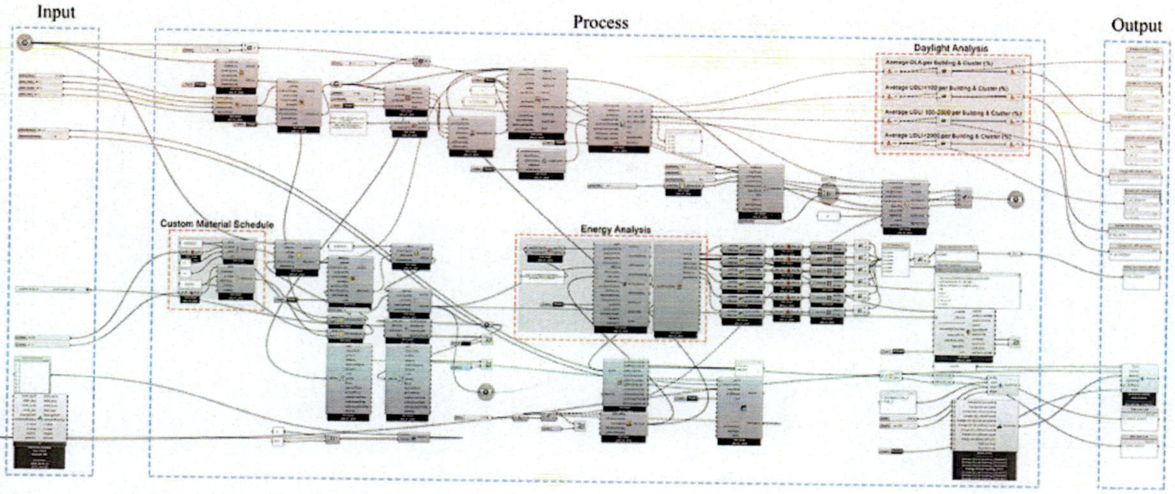

Fig. 6 Grasshopper definition script for iterative simulation modeling

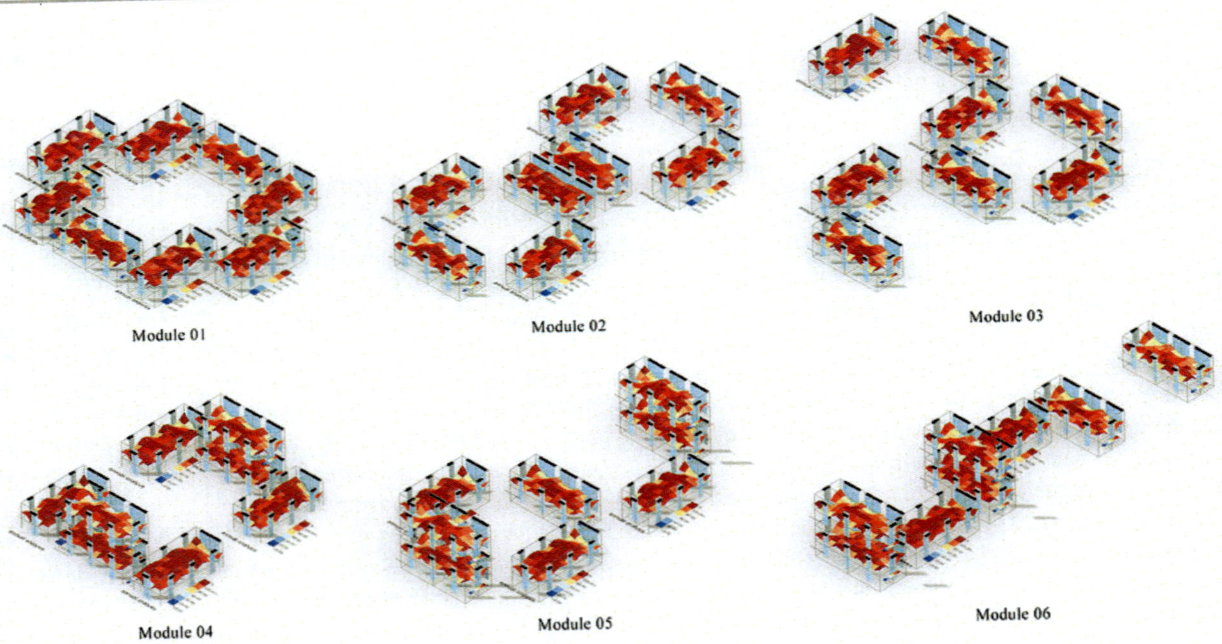

Fig. 7 Simulation performance results

Table 2 Parameter sets for simulation iterations

S. No.	Parameter inputs	Values	Number of iterations
P1	WWR-north	0.20, 0.40, 0.60	3
P2	WWR-west	0.20, 0.40	2
P3	WWR-south	0.20, 0.40, 0.60	3
P4	WWR-east	0.20, 0.40	2
P5	Shading depth	0.25 m, 0.50 m	2
P6	U-value	1.4 w/km^2, 2.5 w/km^2	2
P7	R-value	2.6 km^2/w, 4.4 km^2/w	2
Total iterations per container cluster module			$3 \times 2 \times 3 \times 2 \times 2 \times 2 \times 2 = 288$

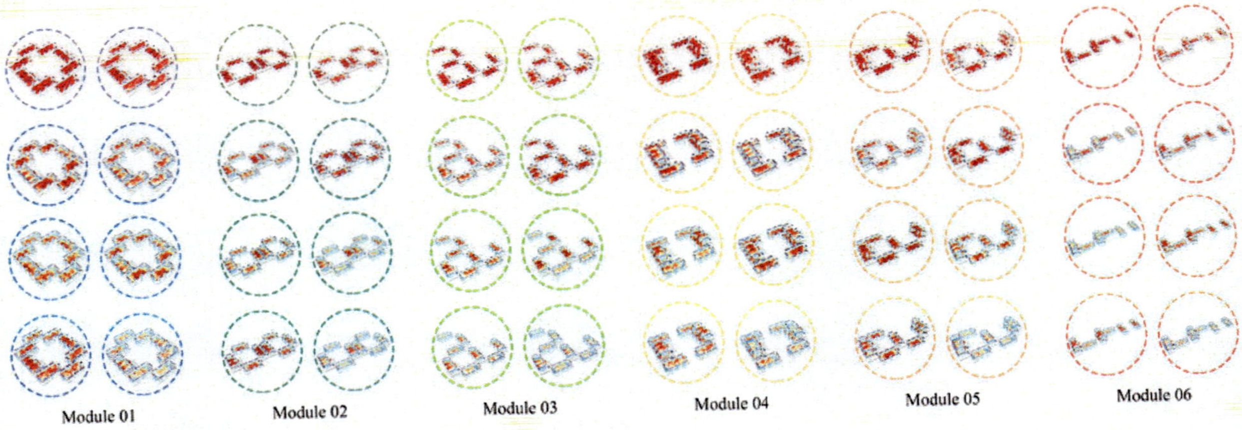

Fig. 8 Iterated energy modeling visualizations for six container cluster modules

areas in the climatic setting of Chattogram or comparable tropical climates. Figure 8 shows a selection of iterated energy modeling representations for six cluster modules created using the Ladybug and Colibri plugins.

The data for module 01 was evaluated for this article in order to project energy optimization trends, determine the optimum container cluster possibilities for each module, and determine the effects of the input factors. Figure 9 showed numerous graphs to demonstrate the iterated 288 design alternatives. The 288 design options were used to create scatterplot and parallel coordinate diagrams. While a comparable coordinate diagram implies a positive and negative relationship between variables through parallel and cross association, scatterplots offer a graphical interpretation of the correlation. There were seven independent variables in this study, making it difficult to interpret using just a scatterplot or parallel diagram. In module 01, interactive charts in Thread were used to determine the best and least acceptable alternatives. A scatterplot was developed to show the association of UDI to EUI for all the iterations. The graphical location of every option is used to demonstrate performance. Because these factors are inversely proportional, the simplest strategy to optimize design possibilities is to identify the dot that consumes the best UDI as well as the minimum EUI. In the plot, the red spot in the higher left-hand corner signified the best possible design option, while the blue dot represented the least viable design option. The right-hand comparison graphic revealed that prediction variables such as window-to-wall ratios, shadings, and material qualities are strong predictors of energy and daylight performance. Finally, the parallel plot revealed comparable results in terms of construction performance.

Although the first objective can be easily obtained from these two figures, getting precise information regarding the influence of the design variables is more difficult. For example, which windows on which sides contributed considerably to the variation in two outputs? Or did shading depth play a major role in determining the optimal design option? Multiple hypotheses were tested using one-way ANOVA statistical testing (Fig. 10). Figure 11 depicts the distribution of UDI 100-2000 values based on north, south, west, and east window-to-wall ratios.

A one-way ANOVA test revealed that the percentage of glazing on the south and east facades had a significant influence on the useful daylighting and energy performance over the year. In contrast, window-to-wall ratios on the west, as well as north sides, revealed no changes. The test also showed that the UDI varied significantly depending on the shade level, with a substantial effect. Homogeneity of Variance was not a worry based on assumptions, although normalcy assumptions could not be made for 0.25 and 0.5 m shading. The results also showed that 0.5 shading outperforms 0.25 m shading.

The research purpose was to propose a computer-aided prototyping technique to determine the optimal container cluster option for each module. Additionally, the technique enabled sorting through the design variables and identifying the alternatives as the most practical for boosting energy efficiency in the complex built environment. For the container housing modules, prototyping offered access to control several parameter values within a minimum and maximum range. Although there are many iterative energy performance studies on conventional structures, there are few research that focuses on daylighting optimization specifically (Natanian & Auer, 2020). However, when examining ISM of daylighting and energy performance in affordable home studies, a gap was discovered (Dara & Hachem-Vermette, 2019). There were no references to container housing system optimization in the vicinity of this research site. This project expects to close that gap through

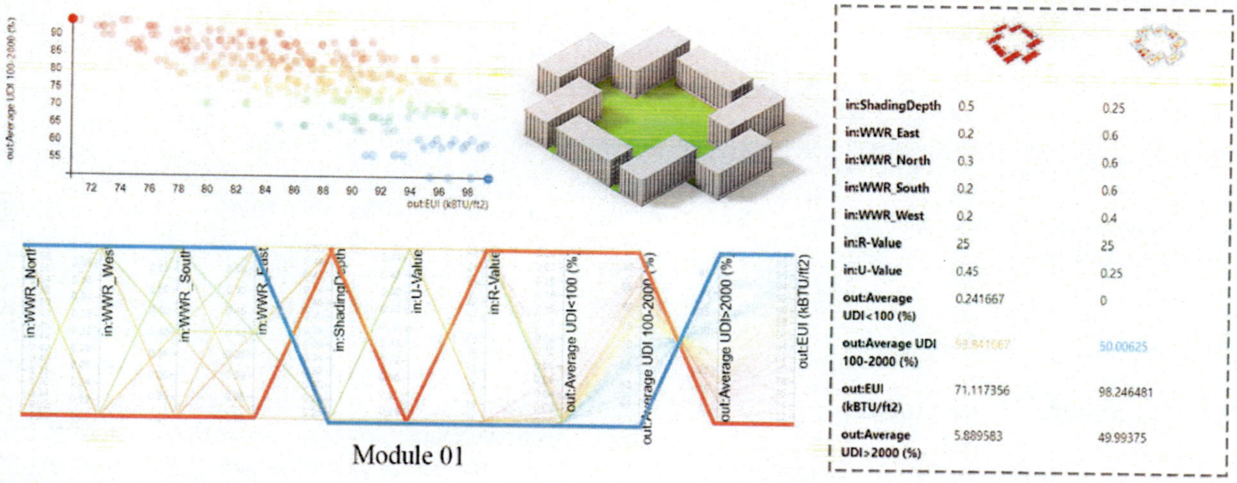

Fig. 9 Statistics for six container cluster modules

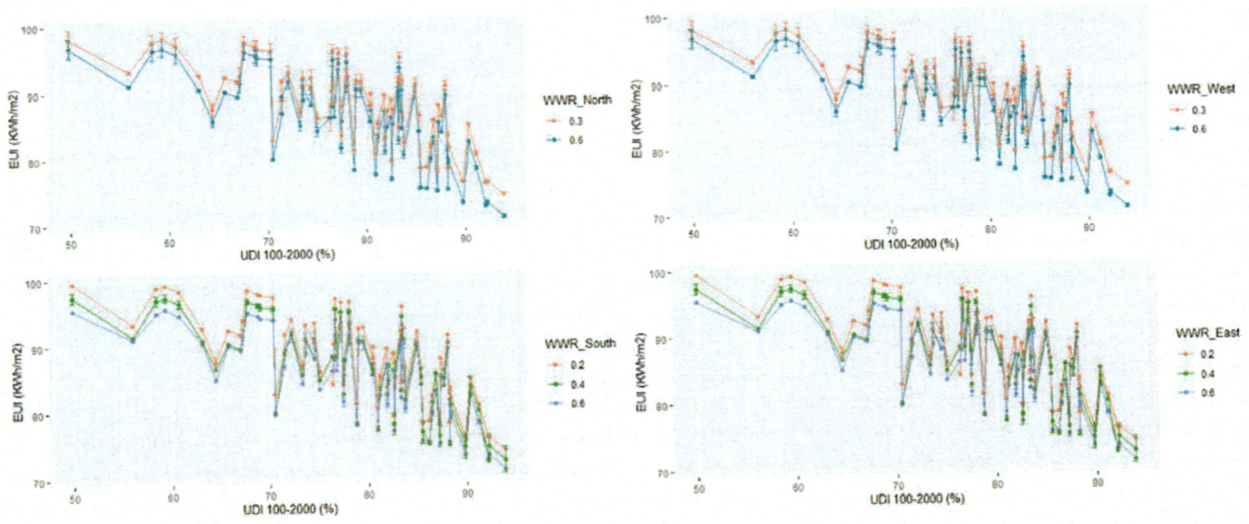

Fig. 10 Lineplots, representing the impact of windows on UDI and EUI

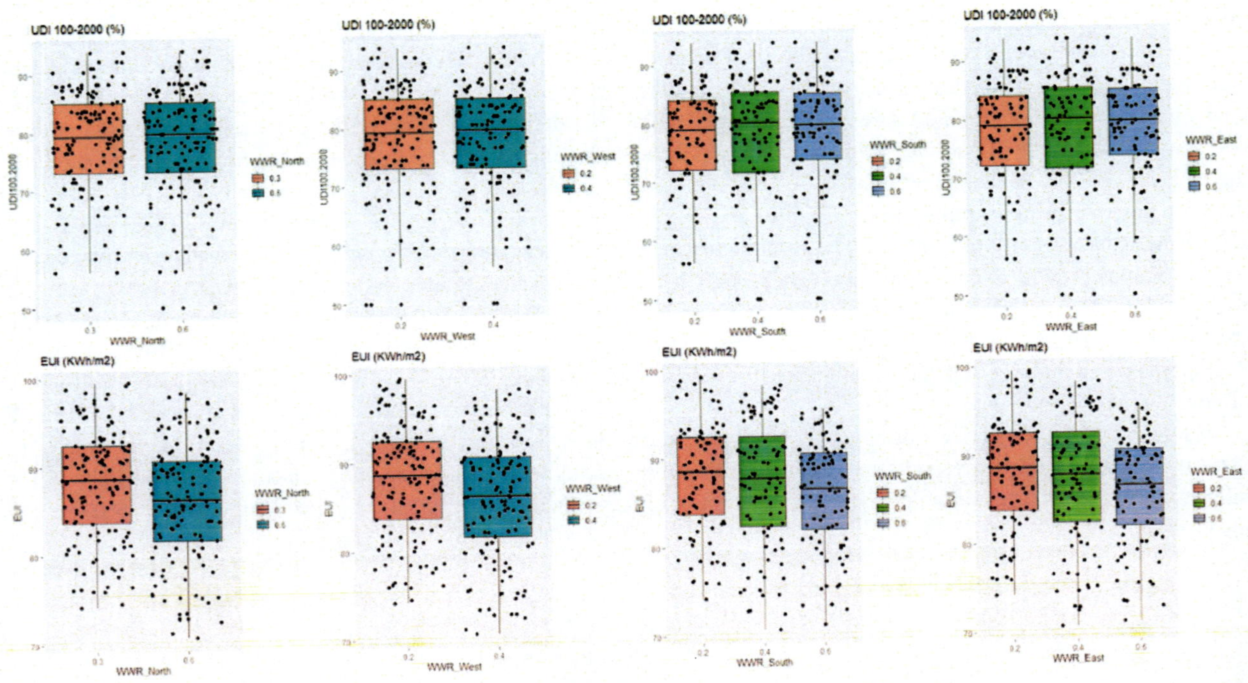

Fig. 11 Boxplots, representing the impact of windows and shading on daylighting

an iterative energy modeling workflow to get the optimal container cluster solutions per module.

5 Conclusion

5.1 Limitations of the Study

The limitations of this paper are chiefly technical and directed by the scope of this study. This study can be made even richer if the following limitations are addressed in the future.

- To simulate 1728 design choices, this prototyping took roughly 60 h. The integration of several simulation engines on a single platform was the key challenge with the analytical tools. A machine learning module such as Artificial Neural Networks, Generative Adversarial Networks, and so on can be added to the physics-based modeling for reducing the time as well as to improve the efficiency of the framework (Duering et al., 2020).

- Sensitivity analysis was necessary to make the design optimization process easier (Fang & Cho, 2019). The link between input and output variables was investigated using a variety of statistical models. Sensitivity analysis, on the other hand, is required to discover critical design variables that influence performance metrics.
- A comparison of all six module combinations would have yielded more optimization options, but because this is the first experiment, just one module was investigated, with intentions for future testing. In the optimization process, different independent variables and values can also be added. For container clusters, for example, sky roof width, length, and placement can be added as well as extra variables within the pre-determined range.

5.2 Future Scope

The research presented in this paper addresses inexpensive housing challenges in Bangladesh's disaster-prone coastal areas by promoting shipping containers as a viable construction unit through ISM to optimize cluster housing modules and provides a framework for achieving so. The researchers conducted a physical assessment in the Chittagong coastal zone to try a middle-out method for adapting the container housing system for architectural resilience, and then they conceptualized a prototype of a cheap and energy-efficient modular housing system. The methodology included architectural design of a single cluster of a house, energy performance simulation for a baseline vs. traditional model, and simulation data analysis to propose the noble framework. A few elements, such as the container's comprehensive technical construction, user response, precise floor designs, etc. were evaluated beyond the scope of this study. Its aim is to promote knowledge and awareness among architects, academics, engineers, and legislators about the growing demands for coastal resilience, design optimization, sustainability, and energy savings while remaining affordable.

Efficient designs tend to have improved daylighting and energy efficiency at the same time. Getting the optimal measurements and solutions is difficult in the traditional design process, though. By analyzing each module in the future, the combination and intra-dynamics can change the results and have the potential to be tweaked as needed in the primary phase of design, using a fast-fail method as a project management practice, saving an enormous amount of resources due to the nature of buildings' permanence (Rahman, 2018). The amount of shading was parameterized but ignored for iterations due to the need to limit the number of iterations in this workflow verification method, and it was maintained as a future study subject.

Along with the automated iteration process, another option that can revolutionize how architects and stakeholders develop future places is co-design using immersive media (Aman et al., 2021). The inhabitants have distinct personal insights and contextual information regarding their spatial choices (Haque et al., 2022a, b). Therefore, immersive technology, such as the iStudio developed by D'souza et al. (2018), can help dwellers and lay people participate in the participatory design process in a variety of ways. End-users might benefit from such tools by having a better spatial awareness of their future home and taking part in the design process. Additionally, utilizing technologies to mediate collocation between architects and end-users can enhance the decision-making process, especially in the early stages of design (Dastmalchi et al., 2021).

This study started by addressing the climatic issues, impact on informal settlements in the coastal areas, and their need to bounce back the natural calamities. It concludes by establishing a novel framework that, with further research, has the potential to become a significant part in the architectural studies regarding climate change, construction sustainability of container modular housing (Haque et al., 2022a, b) and its consequences. It is presumed to provide new perspectives for researchers regarding affordable, resilient, and sustainable housing, particularly in coastal regions facing extreme natural events.

5.3 Practical and Theoretical Implications

Practical implications: The research elaborates an iterative design workflow that can be helpful for the architects and designers to take initial design strategies. Moreover, the stakeholders can get the specifications regarding the spatial quality, energy requirements, and feasibility of the projects.

Theoretical implications: Several research projects in the last decades have achieved significant success in energy and daylighting analysis to attain sustainable architecture. However, the iterative design process is still theoretically developing and lots of sustainable design features are yet to be achieved to play a vital impact factor in the architectural design process.

Acknowledgements The authors are grateful to Dr. Jong Bum Kim and Prof. Fuad H. Mallick, Ph.D. for their contributions to the environmental sustainability section of this study. They also appreciate the help they received from Iftee, Fahim, and Radina during the study. Thanks also go out to the reviewers, whose insightful comments significantly enhanced this paper.

References

Ahmed, I. (2012). The courtyard in rural homesteads of Bangladesh. *Vernacular Architecture, 43*(1), 47–57. https://doi.org/10.1179/0305547712z.0000000005

Aman, J. (2017). *Impact of windows for daylighting on thermal comfort in architecture design studios in Dhaka (thesis).* Bangladesh University of Engineering and Technology (BUET), Dhaka.

Aman, J., Tabassum, N., Hopfenblatt, J., Kim, J. B., & Haque, M. O. (2021). 'Optimizing container housing units for informal settlements: A parametric simulation & visualization workflow for architectural resilience. In *CAADRIA 2021—Hong Kong, 'projections', The proceedings of 26th international conference of the association for computer-aided architectural design research in Asia (CAADRIA).* Hong Kong, Guangdong; Association for Computer Aided Architectural Design Research in Asia (CAADRIA).

Bourgeois, D., Reinhart, C. F., & Ward, G. (2008). Standard daylight coefficient model for dynamic daylighting simulations. *Building Research & Information, 36*(1), 68–82. https://doi.org/10.1080/09613210701446325

ConstruPM. (2017). *Edge green buildings: What's window to wall ratio?* ConstruPM. http://mundobim.com/construpm/edge-green-buildings-whats-window-to-wallratio/#: ~ :text=What's%20Window%20to%20Wall%20Ratio%3F,Let's%20start%20by&text=You%20simply%20divide%20the%20total,window%2Dto%2Dwall%20ratio

D'souza, N., Kress, M., Balakrishnan, B., Hopfenblatt, J., & Dastmalchi, M. R. (2018). iStudio: An interactive form-making environment for architectural studio pedagogy. In *Design Communication Conference*, Cornell University, Ithaca, NY.

Dara, C., & Hachem-Vermette, C. (2019). Evaluation of low-impact modular housing using energy optimization and life cycle analysis. *Energy, Ecology and Environment, 4*(6), 286–299. https://doi.org/10.1007/s40974-019-00135-4

Dastmalchi, M. R., Balakrishnan, B., & Oprean, D. (2021). Exploring the role of transactive memory systems in team decision-making during ideation phase. In *The 2021 international conference on engineering design (ICED),* Gothenburg, Sweden.

Davidson, S., (2020). "Grasshopper". *Grasshopper3d.Com.* https://www.grasshopper3d.com/

DDM. (2014). *Disaster report 2013*. Dhaka: Department of Disaster Management, Ministry of Disaster Management and Relief, Government of the People's Republic of Bangladesh.

Dogan, T., Yang, Y., Samaranayake, S., & Saraf, N. (2020). Urbano: A tool to promote active mobility modeling and amenity analysis in urban design. *Technology|Architecture + Design, 4*(1), 92–105. https://doi.org/10.1080/24751448.2020.1705716

Duering, S., Chronis, A., & Koenig, R. (2020). Symposium on simulation for architecture and urban design, SimAUD 2020. In *2020 Proceedings of the symposium on simulation for architecture and urban design* (pp. 503–509). SimAUD. http://simaud.org/2020/proceedings/109.pdf

EnergyPlus. (2020). "Weather data by location|Energyplus". *Energyplus.Net*. https://cutt.ly/dQxcVNV

Fang, Y., & Cho, S. (2019). Design optimization of building geometry and fenestration for daylighting and energy performance. *Solar Energy, 191*, 7–18. https://doi.org/10.1016/j.solener.2019.08.039

Haque, M. O. (2019). *Biomimicry inspired design for daylighting through roof of multipurpose hall (thesis).* Bangladesh University of Engineering and Technology (BUET), Dhaka.

Haque, M. O., Aman, J., & Mohammad, F. (2022a). Construction sustainability of container-modular-housing in coastal regions towards resilient community. *Built Environment Project and Asset Management, 12*(3), 467–485. https://doi.org/10.1108/BEPAM-01-2021-0011

Haque, M. O., Mohammad, F., & Rahman, R. (2022b). Shipping container housing for architectural resilience in coastal regions: Addressing construction and material sustainability. *ZEMCH 2021 International Conference Proceedings*, ISBN: 978-9948-31-000-6. www.zemch.org

Haque, M. O., Rahman, N. N., & Zaman, T. (2020). Technical intervention on Khona's maxims to design an amphibian house. In *Proceedings of 2nd international conference on smart villages and rural development.* Smart Villages Lab, Faculty of Architecture, Building and Planning, The University of Melbourne. https://smartvillageslab.msd.unimelb.edu.au/#publications

Haque, M. O., Munzarin, B. U. T., Saikia, S. (2021). Implementation of alternative affordable materials in sustainable rural housing design. In *Proceedings of 3rd international conference on smart villages and rural development.* Smart Villages Lab, Faculty of Architecture, Building and Planning, The University of Melbourne.

Hassan, M., Ullah, M. S., & Gomez, C. D. (2000). Rural housing in Bangladesh: An enquiry into housing typology, construction technology and indigenous practices. In *Proceedings of international conference on housing & hazards and the rural community.* Dhaka; Bangladesh University of Engineering and Technology (BUET), Housing and Hazards Group. https://cutt.ly/IQxcZpK

HBRI. (2018). *Standard guideline for housing in disaster prone areas of Bangladesh.* Dhaka: Housing and Building Research Institute, Ministry of Housing and Public Works, Government of the People's Republic of Bangladesh.

Kloog, I., Kaufman, L., & de Hoogh, K. (2018). Using open street map data in environmental exposure assessment studies: Eastern Massachusetts, BERN region, and South Israel as a case study. *International Journal of Environmental Research and Public Health, 15*(11), 2443. https://doi.org/10.3390/ijerph15112443

Kristiansen, A. B., Satola, D., Lee, K., Zhao, B., Ma, T., Wang, R. Z., Gustavsen, A., & Novakovic, V. (2020). Feasibility study of an off-grid container unit for industrial construction. *Sustainable Cities and Society, 61*, 102335. https://doi.org/10.1016/j.scs.2020.102335

Lartigue, B., Lasternas, B., & Loftness, V. (2013). Multi-objective optimization of building envelope for energy consumption and daylight. *Indoor and Built Environment, 23*(1), 70–80. https://doi.org/10.1177/1420326x13480224

McNeel, R. (2020). *McNeel Asia: Robert McNeel & Associates*. McNeel North America. https://www.mcneel.com/

Minar, M. H., Hossain, M. B., & Shamsuddin, M. D. (2013). Climate change and coastal zone of Bangladesh: Vulnerability, resilience and adaptability. *Middle-East Journal of Science and Research, 13* (1), 114–120. https://cutt.ly/QQxcGpT

Natanian, J., & Auer, T. (2020). Beyond nearly zero ENERGY urban design: A HOLISTIC microclimatic energy and environmental quality evaluation workflow. *Sustainable Cities and Society, 56*, 102094. https://doi.org/10.1016/j.scs.2020.102094

Rahman, R. (2017). Re-defining courtyard to re-vitalize urban community. In *Proceedings of UIA 2017 Seoul world architects congress* (p. 231). Seoul, UIA. https://homedocbox.com/Landscaping/103037089-Re-defining-courtyard-to-re-vitalize-urban-community.html

Rodrigue, J. P., Comtois, C., & Slack, B. (2016). *The geography of transport systems*. Routledge.

Roudsari, M. S., Pak, M., & Smith, A. (2013). Ladybug: A parametric environmental plugin for grasshopper to help designers create an environmentally-conscious design. In *Proceedings of the 13th international IBPSA conference held in Lyon, France Aug* (pp. 3128–3135). https://www.ibpsa.org/proceedings/bs2013/p_2499.pdf

The R Foundation. (2020) R: The R project for statistical computing. *R-Project.Org*. https://www.r-project.org/

The United Nations. (2018). *Measuring progress towards the sustainable development goals—sdg tracker*. Our World in Data. https://sdg-tracker.org/

Tomasetti, T. (2020a). CORE studio. *Core.Thorntontomasetti.Com*. http://core.thorntontomasetti.com/

Tomasetti, T. (2020b). Thread. *Thread.Thorntontomasetti.Com*. https://thread.thorntontomasetti.com/welcome

World Risk Report. 2016. Bündnis Entwicklung Hilft & United Nations University – EHS, ISBN 978-3-946785-02-6. www.sciencedaily.com/releases/2016/08/160825084651.htm

The Application of Digital Technologies into Utilizing Urban Voids

Ebtesam Mohamed Ahmed Elgizawy and Ali Khaled Ahmed Elewa

Abstract

As time passes and cities gradually expand, an urban void resulting from land-use transformation has been with the inadequacy of the current urban situation. However, by generating appropriate urban voids, urban quality of life can be improved from social, economic, and environmental aspects. Thus, this paper focuses on the role of digital technologies in utilizing abandoned urban voids having the potential for applying Landscape Urbanism principles. This paper uses an exploratory methodology that involves both a literature review and a case study, analyzing urban void exploitation projects and their practical approaches. This study largely depends on selectively collected data, while it contributes to a better understanding of digital technologies to utilize urban voids. Furthermore, the developed parameters can guide urban planners responsible for the creation of urban voids and their re-imagining process. Finally, this paper aims to improve the quality of urban life in cities and to avoid past weaknesses and deficiencies in the field.

Keywords

Urban voids • Virtual reality • Augmented reality • Collaborative virtual environments • Media facades • Interactive public displays

1 Introduction

Recently, urban voids have gained great attention in urban design due to their impact. Urban voids can be identified as the spaces among public and private realms, either on the urban macro-scale or on the architectural micro-scale. It means that urban voids are areas that are neither city nor nature (Carmona, 2010). Accordingly, urban voids are an opportunity to apply landscape urbanism (Klever, 2017). Architectural scale and urban scale are the two scales that are typically used to categorize urban voids. Life activities occur in an area bounded by structures to form a built environment (Gehl, 2011). In the same context, quality of life could be regarded as a tool for measuring and guaranteeing the optimal exploitation for existing urban voids, as the quality of life, as it focuses on social relationships, surroundings, housing, community services, income, health, nutrition, resource management, recreation, unemployment, and family harmony. Shortly, it concerns both social and economic constraints (Lyles-Chockley, 2008).

Importantly, the process of transforming urban voids into public spaces has a great impact on the quality of life in cities. The real challenge is not only in the transformation process but in keeping the public space alive. Therefore, it is essential to take advantage of the growing digital technologies to maintain the vitality of the public space. Correspondingly, as technological advancement accelerate, several advanced information technology approaches arise. Moreover, there is a gap in coping with the increasing diversity and its applications in urban design, especially in the Middle East and North Africa (MENA) region. Additionally, urban void as terminology appeared in several types of research before. However, most of the researches discussed it either in terms of socio-economic impacts or the challenges that obstruct the exploitation process of urban voids. That is to say, the technological aspect of exploiting urban voids still needs to be redefined. In this context, the study aims to identify a range of applications in which digital technologies can support the generation of valuable urban voids.

This paper consists of five components. Following the methodology section, this paper describes urban voids and their origin, taking into account their classification. Secondly, the impact of the rapid development of digital

E. M. A. Elgizawy · A. K. A. Elewa (✉)
Faculty of Fine Arts, Helwan University, Cairo, Egypt
e-mail: ali.khaled.elewa@gmail.com

H. Altan et al. (eds.), *Advances in Architecture, Engineering and Technology*,
Advances in Science, Technology & Innovation, https://doi.org/10.1007/978-3-031-11232-4_14

technologies is investigated. Besides, various techniques for applying digital technologies in exploiting urban design projects are discussed. The third describes a case study addressing the role of digital technologies on urban voids and their effect on the user experience. A potential urban void (Ghamra's Urban Void, Cairo, Egypt) is then selected to apply digital technologies to enhance both the visual and contextual experiences of the user to exploit the urban void using a proper contemporary approach. Finally, the research findings are focusing on the role of digital technologies in exploring urban voids implementation.

1.1 Research Methodology

This exploratory research identifies how digital technologies can support the utilization of urban voids. Since the research is designed to provide a better understanding of the role of digital technologies and their applications in urban design, it selectively conducts a literature review on five digital technologies: virtual reality (VR), augmented reality (AR), collaborative virtual environment (CVE), digital media facades, and interactive public displays.

A case study approach is then conducted to learn lessons from similar cases. This paper investigates three cases: Fresh Kills landfill at New York City, the Millennium Park in Chicago City in the United States, and the AR journey at Miami City, United States. In this exploratory research, open-ended questions are more suitable because using digital technologies for exploiting urban voids can be achieved by several means.

2 Urban Voids Classification

Urban voids can be classified into seven types; landfills, Brownfield lands, train yards, parking lots, passenger terminals, abandoned railways, and slaughterhouses. See Fig. 14.1. The figure is developed from several literature reviews. For example, Lethugile (2011) claims that both landfills and brownfields are lost spaces. Additionally, Rose (2008) suggests that unused infrastructural roads and facilities are considered wasted treasures. In this context, the urban void type should affect the method of its utilization. Linear infrastructural urban voids could be a potential virtual journey, using a proper intervention.

Generally, urban voids harm their surrounding communities (Yuan & Norford, 2014). On this occasion, those urban voids need reusing to have a positive impact on their communities. Accordingly, urban voids are defined here as useless, unused, and underutilized urban spaces (Lee, 2015). However, the unused metro railway of Misr Al-Jadida could be a potential for applying Landscape Urbanism concepts to exploit the existing urban void. Al-Azhar Park is a landscape urbanism project that was a landfill full of poisonous and salty materials since the late Mameluke Era in old Cairo. It was transformed into an urban park (Salama, 2011). In other words, converting this large-scale urban void into a useful park is a method for improving the quality of urban life in Cairo city, which in turn affects environmental, socio-economic, and cultural aspects of life.

To put it differently, landfills and Brownfield lands are usually located outside the border of the cities. However, these locations were relocated inside the city itself as a result of the city's gradual expansions. Thus, the new situation is inappropriate for the quality of life in cities (Gul, 2016). Urban voids can also be classified into three categories according to the extent of contamination, which can be either contaminated, uncontaminated, or potentially contaminated urban voids (Bullock, & Gregory, 2009). However, this classification may matter in the case of Brownfields and landfills in particular.

3 Digital Technologies Applied in Urban Design

First of all, the rapid development of digital technologies affects all aspects of life (Siemens, 2014). Accordingly, urban design is influenced by such technologies. This section discusses the impact of digital technologies on urban design. In this section, a set of digital technologies is chosen to discuss its validation to be applied in urban design in general, and exploiting urban voids in particular, these technologies are VR, AR, CVE, digital interactive media facades, and interactive public displays.

3.1 Virtual Reality (VR)

VR is a computer-generated simulation as a replacement of one's environment, where a headset is worn to isolate the user from the real-world view and replaces it with a virtual environment (Carmigniani, & Furht, 2011). In other words, VR can be considered as mental teleportation via fully immersive with no real-world stimuli (Partala, 2011). VR is a digital approximation of our reality, but not the reality itself, where the user feels that he is in another place (Bailenson et al., 2008). Furthermore, VR technology started to take place in 1962, when the filmmaker Mort Heling created the Sensorama, which is an arcade-style cabinet. Its features include a 3D display with a vibrating seat and stereo sound. However, this technology did not gain the expected popularity (Hirota et al., 2011).

Fig. 14.1 Urban voids classification

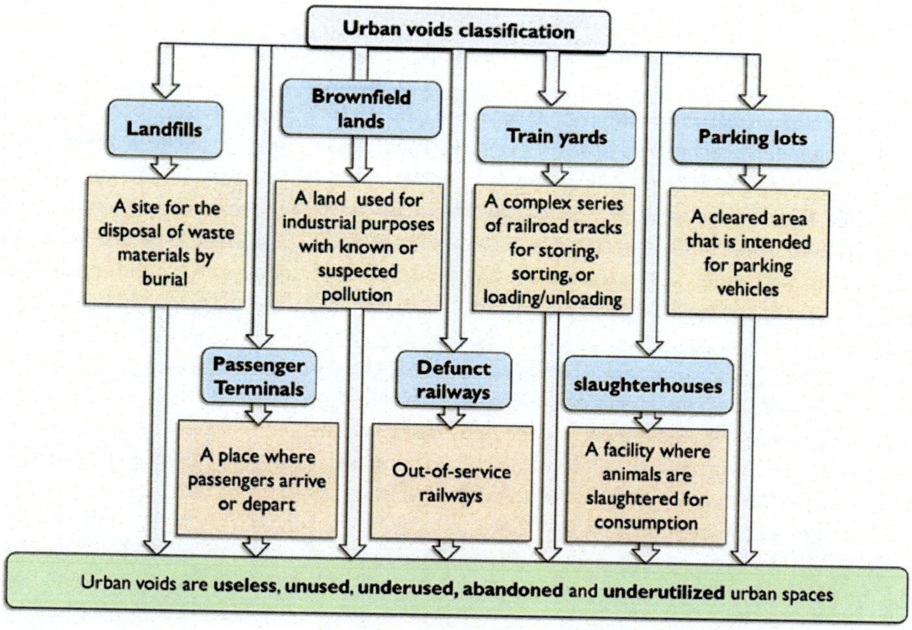

However, the practical beginning of VR was in 1966 by the scientist Ivan Sutherland, who invented the first VR headset called The Sword of Damocles. However, the scientist Jaron Lanier was the first to coin the term itself in 1987 (Carmigniani et al., 2011). To sum up, VR is a process where a user can be embedded within an imaginary context, either fully or partially. It allows the user to feel the context through the specific device (Black et al., 2012). In conclusion, VR helps the designer to express his ideas to the users in an interactive way.

3.2 Augmented Reality (AR)

AR is also known as mixed reality (MR). Moreover, its purpose is neither to cut out the real-world nor transport it to another one, but rather to be an enhancement of our real world by introducing a set of virtual components in it (Wang & Dunston, 2011). Moreover, AR was coined by the researcher Thomas Corder in 1990. The word augment is derived from the Latin word Guerra which means increase or add. Accordingly, AR means adding virtual objects to our real world (Haugstvedt & Krogstie, 2011).

AR is a digital addition to reality. Both VR and AR affect the perceptions and processes of the human mind. In conclusion, AR helps in creating an enjoyable experience for urban space users through mixing existing reality with additional digital and virtual inputs. To point out, the street museum in London is an example of applying AR in urban design (Styliani et al., 2009).

3.3 Collaborative Virtual Environment

Collaborative virtual environment (CVE) is a technology in which multiple participants or users can virtually share and interact in the same virtual space. These participants are not obliged to maintain a certain distance between them (Naughton & Redfern, 2002). Also, collaborative virtual environments are most popular and common in digital online multiplayer interactive games, where each player may have a specific avatar and identity. CVE applications extend to architecture, urban design, and other disciplines (Erickson et al., 2011). Another key point, CVE started to appear in the late eighties of the past century, as an attempt to share one virtual space that allows remote collaboration and interaction via connecting multiple computers (Geers, 2011).

Many CVE engineering software has also been introduced, including CollabCAD, CoCAD, and CyberCAD. The software integrates collaborative environments to enhance urban and architectural design processes (Wang & Tsai, 2011). It is important to realize that the main role of collaborative virtual environments is to provide a three-dimensional virtual multi-user space for the planning process. This system allows teleconferencing (audio, video, and shared boards), cooperation between users in the form of avatars (Churchill et al., 2012).

3.4 Digital Media Façades

Media Facades are associated with the so-called media architecture, which indicates any building that has the potential to include a media façade, regardless of the building type, which could be an office building, shopping mall, cultural or industrial building (Stalder, 2011). In this context, media facades should be an integral part of the building that depends on an information and communication technology platform (Clements-Croome & Croome, 2004).

Media facades can be categorized into four main types according to how they integrate the urban environment: sender-recipient mode, interactive mode, environment mode and organicistic mode (Dalsgaard & Halskov, 2010). On one hand, the sender-recipient mode is a mode in which the media façade contains certain predefined content. On the other hand, the interactive mode allows the users to communicate directly to the media façade via smart mobile phones, where users can play multiplayer games, or transfer content, as in the case of public museums and exhibitions, or get involved in a voting process (Boring et al., 2011).

In this context, the environment mode is a mode in which the media façade is affected and has sensors to react to external effects including traffic density, weather, and temperature changes, light exposure, or even daytime and traffic density. The Tower of Winds, implemented in 1986 in Yokohama City in Japan, is an example of the environment mode media façade where wind intensity affects the light pattern used in the media façade (Dalsgaard & Halskov, 2010) whereas the organistic mode is the mode that depends on an intelligent skin that can identify various parameters from the built environment around, then using its computing power and memory. It can interact with the surrounding media facades on adjacent buildings to create what is known as a digital urban reality (Boring et al., 2011).

3.5 Interactive Public Displays

According to (Müller et al., 2010), interactive public displays can be classified into four categories according to their mental models, depending on the context and environment: poster, windows, mirror, and overlay. There is another classification of interactive public displays that are split into two categories: passive and active engagements, in which the first is when people interact with digitalized items, and the second is where people interact with both digitalized objects and each other (Ylipulli et al., 2014). One of the examples is digital interactive signage paired with smartphones (Müller et al., 2010).

Table 14.1 summarizes the digital technologies recommended for exploiting urban voids. The table introduces information technology features and their possible exploitation in urban design in general and urban voids in particular. In the same context, the information technologies are categorized into five main sections: VR, AR, CVE, digital media facades, and interactive public displays.

4 Case Study

This section explores some cases that witnessed the exploitation of the urban void affected by digital technologies to create interactive public urban spaces. Digital technologies play a role in the exploitation of urban voids, using the techniques mentioned above. In addition, many of these ideas are expected to be applied on site. For example, West Covina's landfill in Los Angeles in the United States, where a concept is adopted to convert this landfill into a VR park. VR roller coaster and Arena-style CVE shooting range are also expected to be done.

4.1 Fresh Kills Landfill, New York City, USA

Fresh Kills landfill was the largest landfill in the world. It was situated on the western edge of Staten Island in New York City in the United States. The project opened in 1948 covered about nine million-meter squares. In 2001, Fresh Kills landfill was closed and its exploitation process started in 2008 due to the large value of its land, as it contains large amounts of wildlife and its harmful effect on the built environment of New York City. Therefore, the landfill was to be transformed into a public recreational park (Molnar, 2015). Figure 14.2 illustrates the smartphone application developed to view AR at Freshkills Park.

This AR application compatible with smartphones allows Fresh Kills park visitors to imagine the expected development for the landfill after completing all its transformation phases into a public recreational park. It allows observing future modifications and understanding the features and the characteristics of the site (Molnar, 2015). This application does not only provide AR but also provides georeferencing for various site features. It also allows exploring construction stages. AR interact with the site using georeferencing, articles, photos, panoramic views, and videos (Pham, 2011). Furthermore, regarding the relative vastness of the park, global positioning system (GPS) is used to determine user location and suggests routes to any location at the park. The purpose is to facilitate wayfinding, either via bike or on foot. This project is supposed to be finished by 2035 (Molnar, 2015). Figure 14.2 shows the navigation capabilities of the application.

4.2 The Millennium Park, Chicago, USA

The land occupied by Millennium Park in Chicago was a transportation hub consisting of Illinois central railway yards and a large parking lot. The new land-use applied to this land was a public recreational space acting as a platform for public art. The roof of the existing building was transformed

Table 14.1 Digital technologies recommended for exploiting urban voids

Digital media type	Role in exploiting urban voids
Virtual reality (VR)	• Helping the urban designers to create three-dimensional walk-through models to simulate urban spaces • Allowing interaction via the internet, either immersive or non-immersive (Black, et al., 2012)
Augmented reality (AR)	• Helping in creating an enjoyable experience for urban space users through mixing existing reality with digital inputs • Example: street museum in London (Styliani et al., 2009)
Collaborative virtual environment (CVE)	• Providing a three-dimensional simulated multi-user planning space teleconferencing and avatar-based collaboration are all possible with this system (Schroeder, 2010)
Digital media facades	• Increasing the interaction between users and context, where users can interact with digitals facades via smartphones by either sending messages or even play games (Clements-Croome & Croome, 2004) • Examples: "Haus des Lehrers" in Berlin, where users can control façade lights; "Aarhus by light" in Denmark, where users' movement in urban spaces are captured and simulated as a silhouette via large media façade (Dalsgaard & Halskov, 2010)
Interactive public displays	• Interactive public displays are divided into two categories; passive and active engagements, the first where people interact with digitalized objects, whereas the second is where people interact with each other and with digitalized objects (Ylipulli et al., 2014) • Example: digital interactive signage, which can be paired with smartphones (Müller et al., 2010)

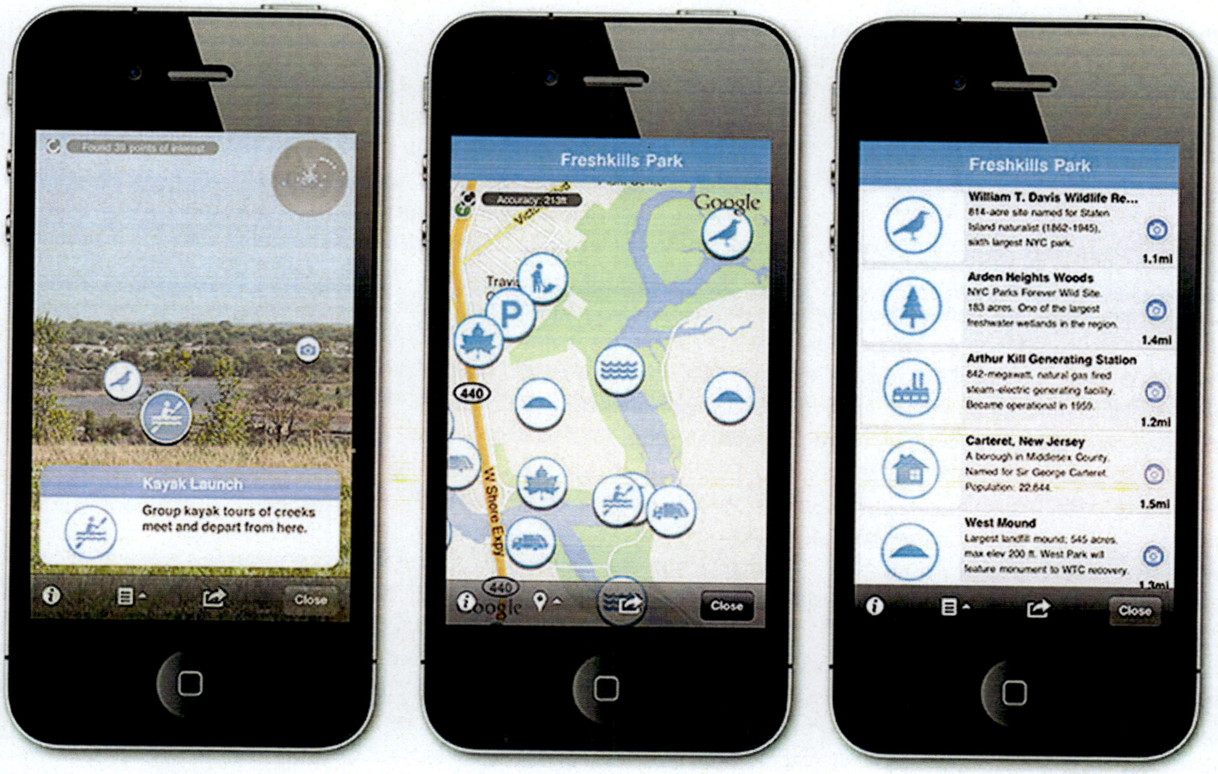

Fig. 14.2 The visitors are allowed to interact with the expected exploitation features of Fresh Kills landfill. *Source* https://www.dexigner.com/news/23925

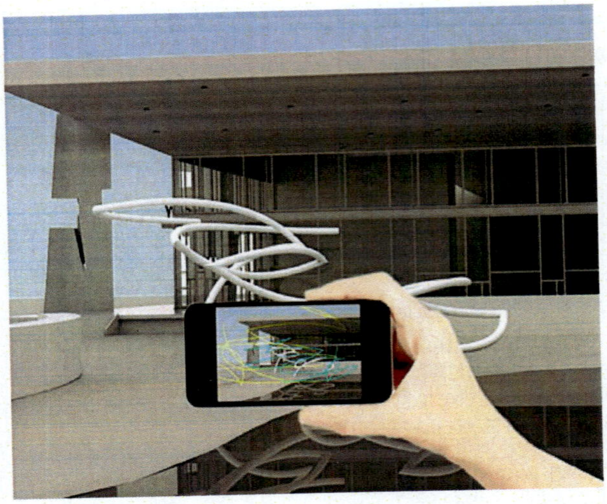

Fig. 14.3 The visitors are allowed to watch an indication for the wind passing the arc sculpture at Aventura, Miami. *Source* https://www.zoesoroko.wordpress.com/2017/03/31/lapse-augmented-reality-art-app

into one of the largest green roofs in the world (Francis & Lorimer, 2011).

Millennium Park serves several interactive public displays, one of them is the Crown Fountain, which is an interactive public display that consists of two units, each one consists of a glass brick transparent tower that has an embedded LED lights to display digital videos on it. The nozzle pours water into a reflecting pool, which is coated with black granite tiles (Mathew, 2014).

4.3 Augmented Reality Journey, Miami, USA

An AR journey along Miami's urban voids was invented by Ivan Toth Depena. This application was called the lapse. It is spatially defined by global positioning system (GPS) and connected to mobile smartphones using a specific application (Papagiannakis et al., 2008). The AR application starts from the Aventura district, where an urban void exists in a site hosting a destroyed public library due to a hurricane (Furht, 2011). An art sculpture was designed using the smartphone AR. Visitors are allowed to watch a virtual representation of the wind currents passing around the arc sculpture in Fig. 14.3.

The second interactive sculpture is activated with a cell phone, and when applying augmented reality to see virtual sculptures, black squares indicators are triggered. Utilizing Miami's Metro mover, a public elevated train system activates the third interactive component, termed the noises, which is a soundtrack tied to the surroundings (Papagiannakis et al., 2008).

5 The Application of Digital Technologies to Ghamra's Urban Void, Cairo, Egypt

Ghamra is a residential district in Cairo, in which exists an urban void represented in the underutilized gap resulting from the existence of metro, trainyard, and warehouse buildings. The district itself is located nearly in the central region of Cairo City, where its northern side is bordered by the 6th of October Bridge. Moreover, the district is considered an important transportation node due to its location near Nasr city and Misr Al-Jadida toward the east. In addition to its adjacency to Ramsis district towards the west.

Ghamra's urban void bisects Ghamra district into two semi-isolated parts, in terms of urban connectivity for pedestrian circulation. As a result, this urban void has a spatial impact on its surrounding. In addition, it can be considered as a wasted vacant space that needs utilizing to have a better impact on its surroundings. Furthermore, Ghamra's urban void bisects Ghamra district into two semi-isolated parts, leading to the lack of urban connectivity for pedestrian circulation. As a result, this urban void has a spatial impact on its surroundings, besides, it can be considered as a wasted vacant space that needs to be utilized to have a better impact on its surroundings as shown in Fig. 14.4.

In the light of the previously discussed case studies and ideas, this node and linear axis could be converted from an underutilized urban void into an urban recreational park to improve the quality of life in Ghamra regarding social, economic, and cultural aspects of life. Collectively, Table 14.2 introduces a list of suggested information and communication featured to be used in the exploitation process of Ghamra, in addition to the idea or objective of using that technology, including creating an AR mobile application for that purpose. Moreover, applied cases are mentioned to ensure the probability of applying those cases to Ghamra's urban void.

6 Discussion

This section discusses the outputs of this research, where the implementation of digital technologies is reviewed depending on the past sections. The selected digital technologies for this research are VR, AR, CVE, digital media facades, and interactive public displays. First of all, VR can interfere in exploiting urban voids in several approaches. For instance, VR coasters add VR experience to roller coasters, especially at theme parks, depending on positional tracking sensors. The purpose here is to enhance user experience and entertainment. One of the examples: Shaman roller coaster at

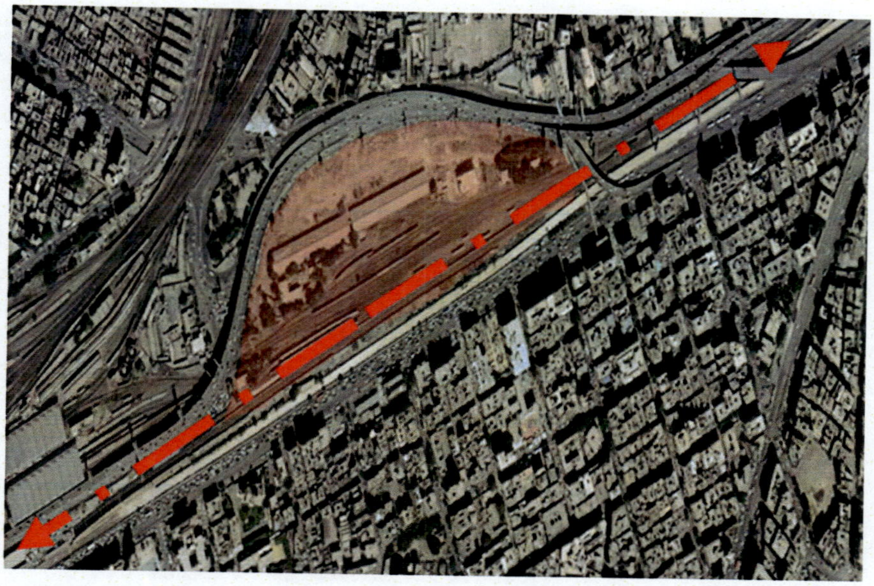

Fig. 14.5 Ghamra's Urban Void, Cairo, Egypt. *Source* https://www.google.com.eg/maps/@30.0651673,31.255182,16z

Table 14.2 Ghamra's urban void suggested exploitation via digital technologies

Digital media type	Application	Direction	Example
Virtual reality (VR)	VR coasters	Applying VR for train passengers to increase their cultural knowledge via creating virtual walk-throughs with historical and infographic background	Six flags magic mountain amusement park, near California, USA
Augmented reality (AR)	Street museum	Developing an AR application that virtually restores vintage early nineties residential buildings in black and white mode	Street museum in London, UK
	AR as a design phasing tool	The urban park phasing could be virtually visualized via an AR application, to allow the user to predict the future designs for the park, as well as the metro station and trainyard	Freshkills Park in New York, USA
	The sound-scape	Creating an AR mobile application that interacts with the surrounding architectural context along with the train passengers" journey	Lapse in Miami, USA
	AR sculptures and artworks	Adding wall scannable triggers to be detected via mobile smartphone's specific application to generate AR sculptures and artworks at the metro waiting areas	
Digital media facades	Media facades as a public participation tool	Using smartphone applications, media façades could be a tool for creating surveys and voting, where answers and responses could appear instantly in an interactive way	Media façade festival in Berlin, Germany in 2008
Interactive public displays	Interactive parking detector	Providing parking spaces at the urban void node. This parking could be a potential for applying interactive applications for detecting available parking spaces, todecrease the traffic congestion in Ghamra	A application developed by San Francisco Municipal transportation agency in USA

Gardaland, Verona, Italy, where VR technology creates a unique experience to this roller coaster since 2017. (Wei et al., 2019). Besides, VR simulators can be used as a gaming tool, to make the user feel an enjoyable informative experience of being fully immersed in a multi-sensory story or experiment. One of the examples is Derren Brown's VR Experience at Thorpe Park in the United Kingdom. VR Simulators are used as a public participation tool. This technique was applied in Lithuania as a tool to engage users to interact with the urban voids redesign proposals. Also, VR is used as a design tool, because it is used to pre-visualize the urban void before utilizing it, in addition to the possibility of dividing the design into phases (Wu et al., 2010). Similarly, AR can be used as a smartphone-based AR. However, this type depends on adding virtual objects to the existing urban void reality. Besides, AR can be applied by adding virtual objects via periscopes (Styliani et al., 2009).

CVE can be used in exploiting urban voids by using arena-style CVE shooting. This process depends on multi-player gaming, where MR is used. On the other hand, a collaborative virtual environment can be used as a design tool. This process depends on utilizing specific software for intruding design stakeholders as avatars in the same virtual

Table 14.3 Classification of digital technologies recommended for exploiting urban voids

Digital media type	Application	Its implementation in exploiting urban voids
Virtual reality (VR)	VR coasters	• Adding VR to roller coasters, especially at theme parks, depending on positional tracking sensors • Enhancing user experience and entertainment • Example: Shaman roller coaster at Gardaland, Verona, Italy, where the VR technology was added to this roller coaster in 2017 (Wei et al., 2019)
	VR simulators as a gaming tool	• VR Simulators can be used to make the user feel an enjoyable informative experience of being fully immersed in a multi-sensory story or experiment (Wei et al., 2019) • Example: Derren Brown's VR experience at thorpe park in the United Kingdom
	VR simulators as a public participation tool	• VR was used in Vilnius, Lithuania as a tool to engage users to interact and be updated with the urban voids redesign proposals (Wu et al., 2010)
	VR as a design tool	• Pre-visualizing the urban void before utilizing, in addition to the possibility of dividing the design into phases (Wu et al., 2010)
Augmented reality (AR)	Smartphone based AR	• Adding virtual objects to existing urban void reality (Styliani et al., 2009)
	AR periscopes	• Adding virtual objects via periscopes (Styliani et al., 2009)
Collaborative virtual environment (CVE)	Arena-style CVE shooting	• Depending on multiplayer gaming, which employs MR (Pearce, 2011)
	CVE as a design tool	• Depending on utilizing specific software for intruding design stakeholders as avatars in the same virtual space to perform the design process for urban voids and decision making • Example: a study on three groups of architects to analyze their collaboration in design process using CVE (Maher et al., 2006)
Digital media facades	Sender/recipient mode digital media facades	• Example: the Bayer headquarters building at Leverkusen, Germany: T-mobile headquarters at Bonn, Germany (Dalsgaard & Halskov, 2010)
	Interactivesssss mode digital media facades	• Example: the media façade at Paris national library (Dalsgaard & Halskov, 2010)
	Environment mode digital media facades	• Example: the tower of winds in Yokohama (Dalsgaard & Halskov, 2010)
	Organicistic mode digital media facades	• Example: the weather tower, Brussels in Belgium (Dalsgaard & Halskov, 2010)
Interactive public displays	Poster interactive public displays	• An electronic interactive paper of variable content. That is to say, digital interactive signage, which can be paired with smartphones • Example: the city wall in Helsinki, Finland (Müller et al., 2010)
	Window interactive public displays	• Creating a virtual connection or port to a virtual remote location (Müller et al., 2010)
	Mirror interactive public displays	• They look like a mirror that reflects the users and surroundings, in addition to augments like making a ribbon effect when touching the screen (Müller et al., 2010)
	Overlay interactive public displays	• Characterized by creating overlays, which integrates the surrounding environment (Müller et al., 2010)

space to perform the design process for urban voids and decision-making (Maher et al., 2006).

Digital media facades have many applied cases. To put it differently, the Bayer headquarters building at Leverkusen and the T-Mobile headquarters at Bonn, Germany, are both two examples of a sender/recipient mode media façade. Furthermore, the media façade at Paris national library is an example of interactive-mode media façades. Also, the tower of winds in Yokohama is an example of interactisssssve-mode media façades. Besides, the weather tower, Brussels in Belgium is an example of organicistic mode media façades (Dalsgaard & Halskov, 2010).

Public displays can be used as electronic interactive paper of variable content. That is to say, digital interactive signage can be paired with smartphones. For example, the city wall in Helsinki, Finland. This type is considered as a poster interactive public display (Müller et al., 2010).

In addition, interactive public displays can be used to create a virtual connection or port to a virtual remote location. This type is called a window interactive public display. In the same context, mirror interactive public display is also a notable type of interactive public display, where they look like a mirror that reflects the users and surroundings. Besides, making a ribbon effect when touching the screen is an additional possible feature. Moreover, overlay interactive public displays are characterized by creating integrating overlays with the surrounding environment (Müller et al., 2010). To sum up, a classification of digital technologies recommended for exploiting urban voids is illustrated in Table 14.3.

7 Conclusion

Urban voids have been classified into two main categories, contaminated and infrastructural urban voids. Contaminated urban voids include landfills and brownfield lands. The infrastructural urban voids include train yards, parking lots, passenger terminals, abandoned railways, and slaughterhouses. The case study in this paper has highlighted some implementations of digital technologies in transformed urban voids. It was concluded that their usage became indispensable for urban voids operators to maintain their vitality.

Consequently, the reviewed cases had an impact on their context, including social, economic, environmental, and spatial aspects. First, the application of digital technologies into utilizing urban voids affects social life within its context, where people are allowed to gather and interact together using unconventional approaches. Second, the utilization of those techniques helps the created public space to be remunerative. People will be attracted to pay and try those playful attractions. The revenue can be used for paying the running costs.

From the environmental point of view, those techniques could be used to raise awareness of environmental issues. Moreover, virtual attractions could gradually replace several printed and wasted raw materials. Furthermore, the spatial user experience for the utilized urban voids would be full of diversity, as it provides a unique virtual dimension for the existing reality that enriches the spatial experience of the users. To sum up, using digital technologies in the exploited urban voids maintain their livability, by adding playful experiences for the users. This will create an added value to the public space.

References

Bailenson, J. N., Yee, N., Blascovich, J., Beall, A. C., Lundblad, N., & Jin, M. (2008). The use of immersive virtual reality in the learning sciences: Digital transformations of teachers, students, and social context. *The Journal of the Learning Sciences, 17*(1), 102–141.9.

Black, J. B., Segal, A., Vitale, J., & Fadjo, C. L. (2012). Embodied cognition and learning environment design. *Theoretical Foundations of Learning Environments, 2*, 198–223.

Boring, S., Gehring, S., Wiethoff, A., Blöckner, A. M., Schöning, J., & Butz, A. (2011). Multi-user interaction on media facades through live video on mobile devices. In *Proceedings of the SIGCHI conference on human factors in computing systems* (pp. 2721–2724). ACM.

Bullock, P., & Gregory, P. (2009). *Soils in the urban environment*. John Wiley & Sons.

Carmigniani, J., & Furht, B. (2011). Augmented reality: an overview. In *Handbook of augmented reality* (pp. 3–46). Springer.

Carmigniani, J., Furht, B., Anisetti, M., Ceravolo, P., Damiani, E., & Ivkovic, M. (2011). Augmented reality technologies, systems and applications. *Multimedia Tools and Applications, 51*(1), 341–377.

Carmona, M. (2010). Contemporary public space: Critique and classification, part one: Critique. *Journal of Urban Design, 15*(1), 123–148.

Churchill, E. F., Snowdon, D. N., & Munro, A. J. (2012). *Collaborative virtual environments: Digital places and spaces for interaction*. Springer Science & Business Media.

Clements-Croome, D., & Croome, D. J. (2004). *Intelligent buildings: Design, management and operation*. Thomas Telford.

Dalsgaard, P., & Halskov, K. (2010). Designing urban media façades: cases and challenges. In *Proceedings of the SIGCHI conference on human factors in computing systems* (pp. 2277–2286). ACM.

Erickson, T., Shami, N. S., Kellogg, W. A., & Levine, D. W. (2011, May). Synchronous interaction among hundreds: an evaluation of a conference in an avatar-based virtual environment. In *Proceedings of the SIGCHI conference on human factors in computing systems* (pp. 503–512). ACM.

Francis, R. A., & Lorimer, J. (2011). Urban reconciliation ecology: The potential of living roofs and walls. *Journal of Environmental Management, 92*(6), 1429–1437.

Furht, B. (2011). *Handbook of augmented reality*. Springer Science & Business Media.

Geers, K. (2011). *Strategic cyber security*. Kenneth Geers.

Gehl, J. (2011). *Life between buildings: Using public space*. Island Press.

Haugstvedt, A. C., & Krogstie, J. (2011). *The historical tour guide augmented reality*. Faculty of Information Technology, Mathematics and Electrical Engineering, NTNU.

Hirota, K., Ebisawa, S., Amemiya, T., & Ikei, Y. (2011, July). A system for creating the content for a multi-sensory theater. In *International Conference on Virtual and Mixed Reality* (pp. 151–157). Springer, Berlin, Heidelberg.

Klever, P., Klotz, S., Pappert, L., & Schulze, J. (2017). *Spatial commons*. Universitätsverlag der TU Berlin.

Lee, S. J. (2015). Urban voids: As a chance for sustainable urban design. In *Proceedings of the 8th conference on international forum urban*.

Lethugile, G. (2011). When a city embraces its paradox: the exploration of incremental waste mining of a decommissioned landfill site and its gradual transformation into a productive public space (Master's thesis, University of Cape Town).

Lyles-Chockley, A. (2008). Building livable places: The importance of landscape in urban land use, planning, and development. *Buffalo Environmental Law Journal, 16*, 95.

Maher, M. L., Bilda, Z., & Gül, L. F. (2006). Impact of collaborative virtual environments on design behaviour. In *Design computing and cognition'06* (pp. 305–321). Springer.

Mathew, A. (2014). *Interactive placemaking: Creativity and user experience at urban installations* (Doctoral dissertation, The Open University).

Molnar, D. (2015). *Anatomy of a park: Essentials of recreation area planning and design*. Waveland Press.

Müller, J., Alt, F., Michelis, D., & Schmidt, A. (2010). Requirements and design space for interactive public displays. In *Proceedings of the 18th ACM international conference on Multimedia* (pp. 1285–1294). ACM.

Naughton, N., & Redfern, S. (2002). Collaborative virtual environments to support communication and community in internet-based distance education.

Papagiannakis, G., Singh, G., & Magnenat-Thalmann, N. (2008). A survey of mobile and wireless technologies for augmented reality systems. *Computer Animation and Virtual Worlds, 19*(1), 3–22.

Partala, T. (2011). Psychological needs and virtual worlds: Case second life. *International Journal of Human-Computer Studies, 69*(12), 787–800.

Pearce, C. (2011). *Communities of play: Emergent cultures in multiplayer games and virtual worlds*. MIT Press.

Pham, T. (2011). Resistance through transformation: a transitional approach for sustainable peri-urban environments.

Rose, C. M. (2008). Big roads, big rights: Varieties of Public infrastructure and their impact on environmental resources. *Arizona Law Review, 50*, 409.

Salama, A. (2011). Media coverage and users' reactions: Al Azhar Park in the midst of criticism and post occupancy evaluation. *QNRS Repository, 2011*(1), 16.

Schroeder, R. (2010). *Being there together: Social interaction in shared virtual environments*. Oxford University Press.

Siemens, G. (2014). Connectivism: A learning theory for the digital age.

Stalder, U. (2011). Digital out-of-home media: means and effects of digital media in public space. In *Pervasive advertising* (pp. 31–56). Springer.

Styliani, S., Fotis, L., Kostas, K., & Petros, P. (2009). Virtual museums, a survey and some issues for consideration. *Journal of Cultural Heritage, 10*(4), 520–528.

Wang, X., & Tsai, J. J. H. (2011). Collaborative design in virtual environments (Vol. 48). Springer Science & Business Media.

Wang, X., & Dunston, P. S. (2011). Comparative effectiveness of mixed reality-based virtual environments in collaborative design. *IEEE Transactions on Systems, Man and Cybernetics, Part C: Applications and Reviews, 41*(3), 284–296.

Wei, W., Qi, R., & Zhang, L. (2019). Effects of virtual reality on theme park visitors' experience and behaviors: A presence perspective. *Tourism Management, 71*, 282–293.

Wu, H., He, Z., & Gong, J. (2010). A virtual globe-based 3D visualization and interactive framework for public participation in urban planning processes. *Computers, Environment and Urban Systems, 34*(4), 291–298.

Ylipulli, J., Suopajärvi, T., Ojala, T., Kostakos, V., & Kukka, H. (2014). Municipal WiFi and interactive displays: Appropriation of new technologies in public urban spaces. *Technological Forecasting and Social Change, 89*, 145–160.

Yuan, C., Ng, E., & Norford, L. K. (2014). Improving air quality in high-density cities by understanding the relationship between air pollutant dispersion and urban morphologies. *Building and Environment, 71*, 245–258.

Systems Integration in Architecture, Engineering and Construction

Energy Efficiency Analysis in the Residential Sector of the Urban Area of Canton Saraguro, Ecuador

Cesar Morocho, Gerardo Campoverde, and Javier González

Abstract

Energy consumption at a national level shows an important growth, the demographic increase, the sum of the population density, the comfort level and lifestyle attributed by the urban planning conditions of each city, insinuating that day by day we are consumers of resources energy sources, such as electricity, liquefied petroleum gas, and water, which are currently essential to carry out daily activities and the firewood that is still present in the cooking processes in some homes in our environment. In this virtue, this work presents the descriptive-inferential statistical analysis of the energy resources of a certain locality, such as the urban area of the Saraguro canton of the Loja province, Ecuador. This analysis is done to determine the consumption of the four types of energy: electric, LPG, water, and firewood since these are the main energy resources used in this parish. The study is based on door-to-door surveys, they were tabulated, graphed, and interpreted, showing the reality of consumption in the residential sector, thus identifying the greatest energy resource used in homes. Several recommendations based on energy efficiency guidelines are presented that help reduces the consumption of the energy resources in the parish, managing to propose technical–economic solutions for energy saving and efficiency, and as such contributing to the conservation of the environment.

Keywords

Consumption • Energy efficiency • Energy resources • Household

1 Introduction

In society, energy occupies a strategic sector due to the dependency people have on it. Energy is needed for the progress and development of cities and urban areas in sectors like transport, industry, agriculture, and mining, among others that are fundamental for the economic and scientific advance (Poveda, 2007). Enjoying a good life quality based on comfort requires elevated energy consumption, which aggravates air, ground, climate, and water contamination (Recalde et al., 2018). This impact should be tried to be minimized as much as possible (Arízaga, 2015).

In the last 25 years, globally, in the residential sector, the energy consumption has been duplicated, due to: world population's increment of socio-economic needs (Ding et al., 2017); the growth of the population; and the daily use of basic services. All of these factors have provoked the increase in energy demand at a residential level (Arízaga, 2015).

Based on the experience of developed countries, it has been demonstrated that rational use (Lauretis et al., 2017) and energy saving are the base of plans and strategies that can be done in developing countries (Arcos López, 2016; Jara & Isaza, 2014), facing a scenario of limited energy resources and increasing precautions regarding protection and conservation of the environment (Poveda, 2007). The balanced use of energy implies lower costs and a lower negative impact on the world, without altering the human wellbeing (Riofrío et al., 2014).

It is precise to know the demand and its final use to focus and center energy efficiency and consumption plans on it (Arízaga, 2015). In 2015, the residential sector had a

C. Morocho (✉) · G. Campoverde · J. González
Universidad Católica de Cuenca, Cuenca, Ecuador
e-mail: cmorochop@gmail.com

G. Campoverde
e-mail: gcampoverdej@ucacue.edu.ec

J. González
e-mail: tjgonzalezr@ucacue.edu.ec

H. Altan et al. (eds.), *Advances in Architecture, Engineering and Technology*,
Advances in Science, Technology & Innovation, https://doi.org/10.1007/978-3-031-11232-4_15

participation of 14.66% of the final energy consumption, which was distributed in: 1.53 MBOE of firewood; 4.29 MBOE of electricity; and 6.29 MBOE of liquefied petroleum gas (LPG) (Araujo & Robalino-López, 2019). Also, in 2013, the final uses were: refrigeration with 51%, lighting with 18%, cooking and heating with 12%, air conditioning with 7%, and other uses with 12% (Arízaga, 2015).

Electric energy has become a part of our daily basis, such is so that from 1971 to the day, the consumption has triplicated and this is strongly related with society's actual needs. In Ecuador, the electric energy consumption is dispersed in lighting and powering of household appliances (Baquero & Quesada, 2016), such as electric showers and induction cooking (Arízaga, 2015). Each city of Ecuador has its electric utility that manages consumption, distribution, and transmission (Arízaga, 2015).

Until today, in Ecuador the government subsidies the liquefied petroleum gas (LPG); and it is used in the residential sector for cooking labor, and water heating uses (Samaniego et al., 2016), and local companies are in charge of the distribution. The local company PETROECUADOR EP, in a report, showed that the LPG demand incremented from 9.743 tons/year into 1,016,757 in 2012, thus exceeding consumption of other fossil fuels (Abad Ordóñez & Aguaiza Viñanzaca, 2019).

The water consumption as an essential service is essential for the well-living of the human being. In this manner, according to the ordinary activities by the society of Cuenca, it's determined they consume approximately 220 L per day of water (María Teresa Baquero, 2013). Water consumption activities are: personal hygiene, home cleaning, and feeding. Each city of Ecuador has its local utility in charge of the resource distribution (Tipán Jinde, 2017).

Some households, however, consume firewood as an energy resource, due to the socio-economic conditions of the population; accessibility to the supply, and logging. In Cotacachi canton, it's estimated that around 2.2 kg/person/day is consumed in households (Narváez & Cabrera, 2007). In this canton, the resource is utilized in food cooking and water heating (Arízaga, 2015).

Since Saraguro is a parish that has a constant population growth, and this implies a residential development and new technologies, the energetic consumption of the households is every time bigger. This is why it is precise to do an energy efficiency analysis of the sector, allowing the establishment of guidelines looking for energetic and economic savings, that will allow the reduction of carbon emissions helping the environment and house owners with lower bills.

This paper will provide socio-economic solutions that can help the residential development of the parish in study. It also aims to delete some bad habits when using the energy resources such as wasting too much water or electricity in the daily activities. The guidelines that will be proposed will help the household owners manage and use better the resources, having some savings in their monthly bills.

Case study Background.

Saraguro's households have had a significant population and urbanistic growth. In this context, it's the third canton with more population of Loja province. This implies a more excessive consumption of vital supplies, thus superior demand of energy resources (Saraguro, 2016). In this paper, a household's energy efficiency analysis is done where the reality of the urban sector regarding energy consumption of water, firewood, electric and LPG energy and its impact on a social, economic, and environmental level is determined. These four types of energy are the most used in the parish for the daily activities (Gould et al., 2020).

Saraguro canton belongs to Loja province and has a territorial extension higher than 1000 km^2. This canton has a single urban parish named Saraguro and 10 rural parishes: Manú, Sumaypamba, Lluzhapa, El Paraíso de Celén, San Sebastián de Yuluc, Selva Alegre, San Pablo de Tenta, Cumbe, Urdaneta and el Tablón as seen in Fig. 15.1. It has a population of 30,183 inhabitants. It is one of the cantons of Loja with the greatest population growth of the latest years (Mendieta, 2012).

2 Research Methodology

In the present paper, the observational type of research was employed, since a recompilation of data of the different households of the urban area of Saraguro was done, without the manipulation of any variable. It is equally significant to

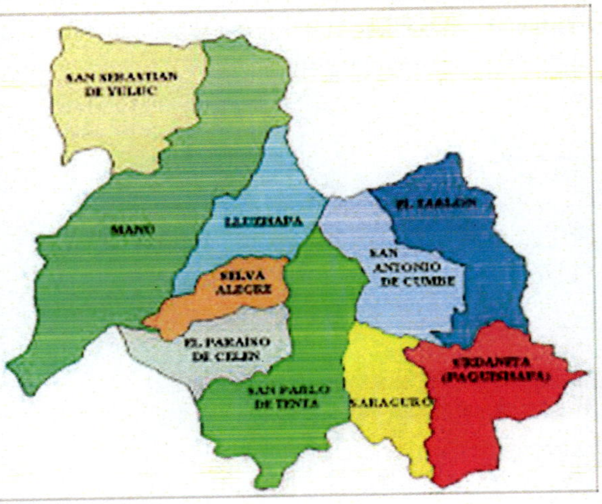

Fig. 15.1 Saraguro canton map

mention that the subtype is a case study, specifically Saraguro parish.

Likewise, the case study of different energetic resource consumptions, such as electricity, LGP, water, and firewood, and its monthly billing was done.

2.1 Sample Size Determination

Saraguro parish has a total of 1035 households (Saraguro, 2016), to determine the sample size a random inferential statistical study must be done, where a confidence level Z of 95%, an error of 5%, and a binomial distribution of 0.5 were applied, and for the calculation refer to (1) (Moran Moran & Ramírez Añazco, 2015). The households in the urban area where the research was done were selected randomly.

$$n = \frac{NZ^2pq}{(N-1)E^2 + Z^2pq} \tag{1}$$

where:

- N Total number of households
- E Error
- Z Confidence level
- p Probability of success
- q Probability of failure

Having a total of 280 samples, door-to-door surveys were applied. The quantifiable data was obtained by close-ended questions and thus verifying the population in each household.

In the case of the electricity and water costs, the monthly consumption bills of each household were asked for, determining that way, the kWh of electricity and m^3 of water consumed. For the calculation of LPG consumed, it was determined according to the number of cylinders used monthly and for firewood according to the amount in kg used per month.

The qualitative variables were obtained with questions about their energy resource use habits. A crucial factor that was used to evaluate the electric consumption was the number of domestic appliances in each household, and if these devices had an energy efficiency tag.

The sample was asked also about the type of illumination used in their houses, whether they implemented electric showers, about the presence of water heaters, and also, the use of induction or gas cooktops.

Finally, with the results of consumption of every type of energy resource, the household's most used resource was determined, likewise, which generates a higher bill and finally propose a guide based on good use of energy, which serves to obtain economic and environmental savings.

The energy calculation method depends on the type of energy being analyzed; this is because each one has a different way to be measured. First, the electric energy will be calculated where the time and power of the household is needed; secondly, the LPG consumption will be estimated, next the water resource, and lastly the amount of firewood that is monthly consumed by the house owners.

2.2 Electric Energy Calculation

In Saraguro canton, the utility company Empresa Eléctrica Regional del Sur EERSSA is in charge of billing and distribution of electric energy (Saraguro, 2016). The utilities implement digital meters that ease the monthly measurements (Arízaga, 2015).

For the calculation of electric energy, refer to (2)

$$E = P_{[W]} * t_{[h]} \tag{2}$$

2.3 LPG Consumption Calculation of a Standard Household of Saraguro Parish

The utility company Loja Gas handles the distribution of LPG in Saraguro canton. This company has different sub-distributors spread throughout the canton (Saraguro, 2016). The LPG tank weighs 15 kg, and for the calculus of its consumption the following concept is used (Samaniego et al., 2016):

Calorific value: The calorific value of a fuel represents the amount of heat that is obtained after a complete oxidation process, at atmospheric pressure and 25 °C temperature of the components of the unit mass of the fuel kWh/kg (Salazar Masson, 2010).

Considering the calorific value of electricity and LGP, it is determined that 1 kg of LPG is equivalent to 13.66 kWh (Arízaga, 2015). Since the tank of gas in Ecuador weighs 15 kg, its calorific value is 205 kWh (Arízaga, 2015).

2.4 Water Consumption Measurement

The municipality of Saraguro canton has its drinking water utility called EMAPASA, which is in charge of the control, supply, and billing of the resource to the city (Saraguro,

2016). It is important to consider that, to control water consumption, digital volumetric meters are used (Tipán Jinde, 2017; Villalba et al., 2014).

Volumetric meter: Its measurement principle is the positive displacement by flat or conical disk, placed in a chamber (Tipán Jinde, 2017). The volume of water that goes through the meter is calculated based on the number of cycles. Each cycle takes a certain amount of water volume, allowing an accurate measurement in a wide range of flow rates (Tipán Jinde, 2017).

2.5 Firewood Consumption Calculation of a Standard Household of Saraguro Parish

The consumption of firewood in Saraguro is performed informally, i.e., in the urban area, it is used in cooking food, after felling (Saraguro, 2016).

15 kg sack: It corresponds to a certain amount of dry (or green) wood chips placed systematically in a nylon bag until a weight of 15 kg is completed (Área).

On average, 17.85 MJ of energy is produced per kilogram of firewood, knowing that in general the consumption is measured in kWh and that 1 MJ is equivalent to 0.2778 kWh, it is determined that per kg of firewood, 4.985 kWh of energy is produced (Lemes Martín et al., 2016; Ministerio de Energía, 2015).

3 Analysis of Results

3.1 Number of Inhabitants Per Household Saraguro, Ecuador

As seen in Fig. 15.2, in the Saraguro parish 70% of households have between 4 and 6 inhabitants, 29% have between 1 and 3, and 1% have between 6 and 9 persons; obtaining an average of 4.16 persons per household.

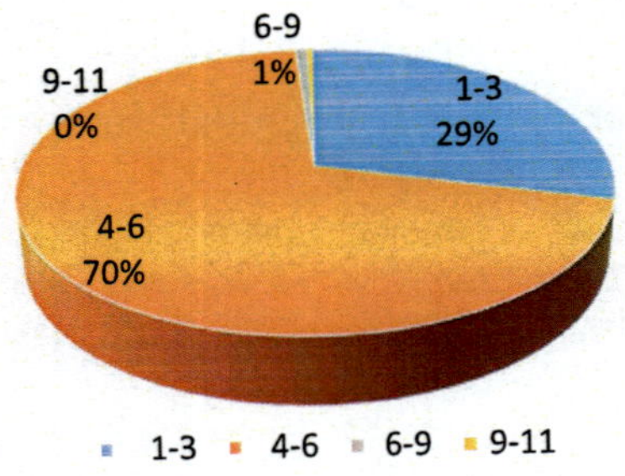

Fig. 15.2 Number of inhabitants per household

3.2 Home Appliances Used in Households

In Fig. 15.3, it is shown that the home appliances most used in households are the refrigerator, television, computer, and washing machine. We can also appreciate that there is not any use of inductor cooktops or dryers. Likewise, it can be seen a significant use of electric showers, since 85 households have in their houses.

3.3 Energy Efficiency Tag in Home Appliances

From all the surveys taken, 80% of the households ignore energy efficiency tags and their content, the 20% remaining are distributed between the knowledge of tags in refrigerators, washing machines, and microwaves, as seen in Fig. 15.4.

3.4 Type of Lighting in Households

As seen in Fig. 15.5, every house in the parish use saving light bulbs.

3.5 Electric Energy Consumption According to Monthly Bill

In Fig. 15.6 can be seen that 62% of households consume between 100 and 200 kWh, meanwhile, 16% of surveys use 200–300 kWh, and 15% consume between 0 and 100 kWh, and only 7% from 300 to 400 kWh per month. From these results, the average consumption per month of the households is 165.35 kWh.

3.6 Use of Gas Water Heater

From the surveys, 70% of inhabitants use gas water heaters, compared to the remaining 30% who lack this type of equipment (Fig. 15.7).

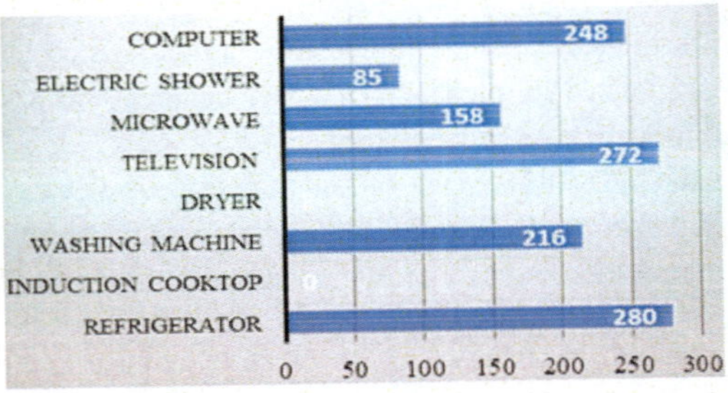

Fig. 15.3 Home appliances used in households

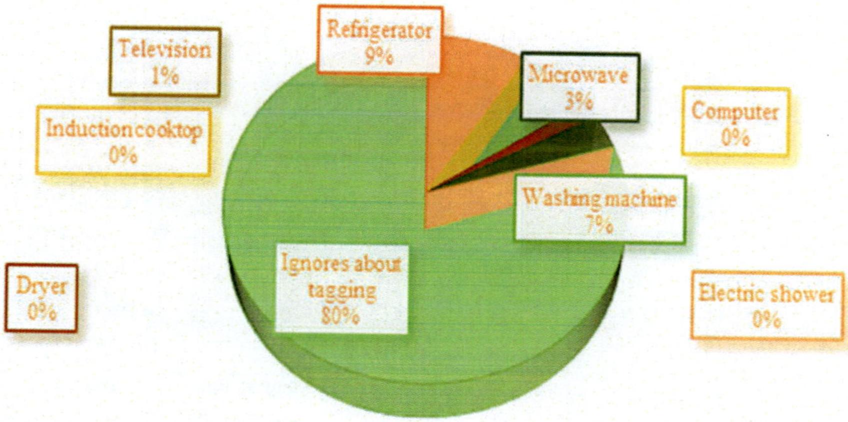

Fig. 15.4 Home appliances energy efficiency tag

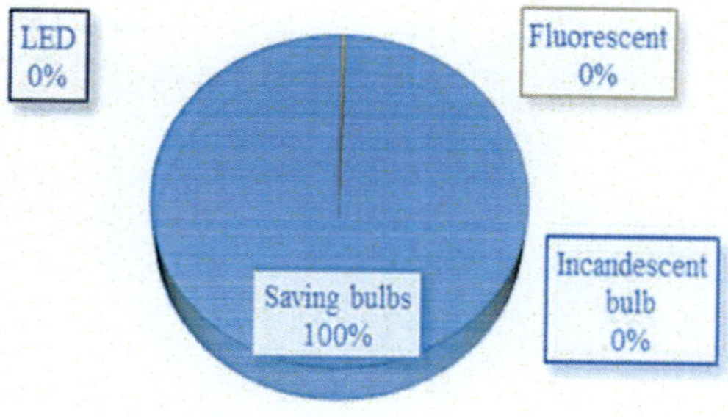

Fig. 15.5 Type of lighting in households

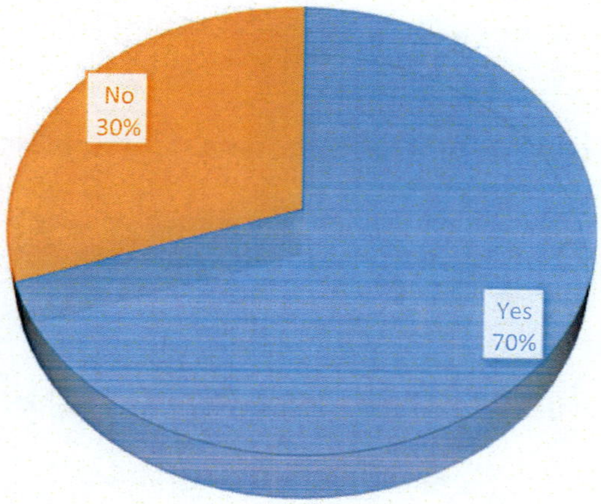

Fig. 15.7 Presence of a gas water heater in the household

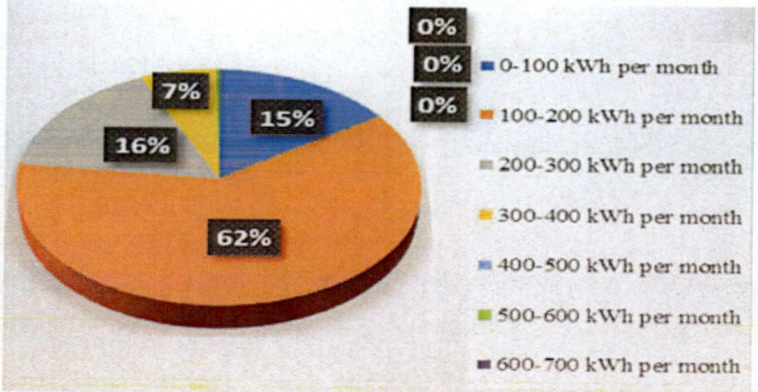

Fig. 15.6 Electric energy consumption

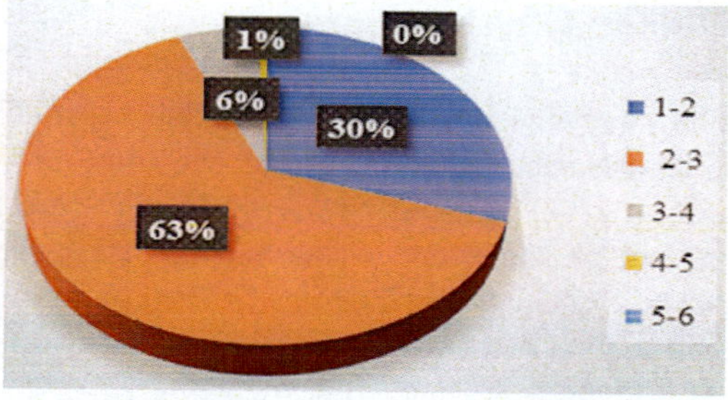

Fig. 15.8 Gas cylinders consumption

3.7 Use of Gas Cooktops for Cooking and Water Heating

The totality of the households surveyed in this research use gas cooktops, they do not employ any other kind of technology.

3.8 Monthly Consumption of LPG Cylinders

Figure 15.8 represents the consumption of this resource in the Saraguro parish. 63% of the households consume from 2 to 3 cylinders per month; meanwhile, 30% consume between 1 and 2, likewise, 6% use 3–4, and only 1% of the households consume between 4 and 5 cylinders per month. From this result, it can be determined that an average of 2.2 cylinders are consumed per month in Saraguro's households.

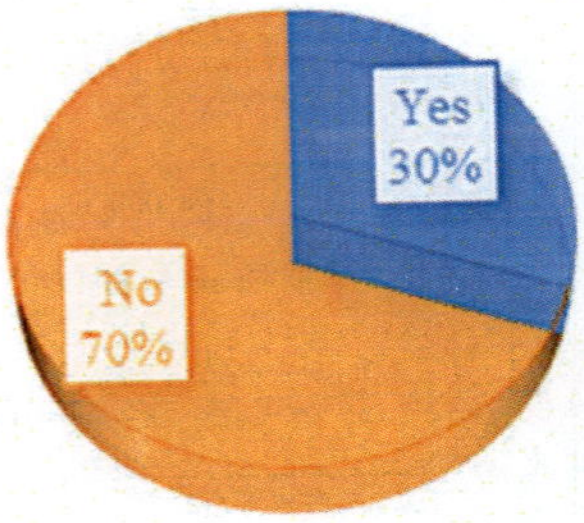

Fig. 15.9 Use of potable water for crop irrigation

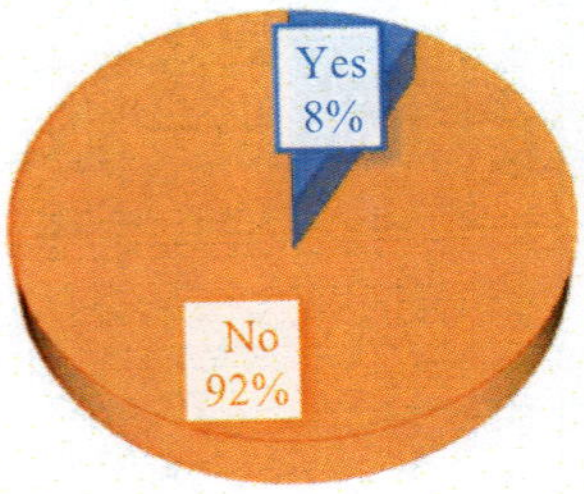

Fig. 15.10 Drinking water for car washing

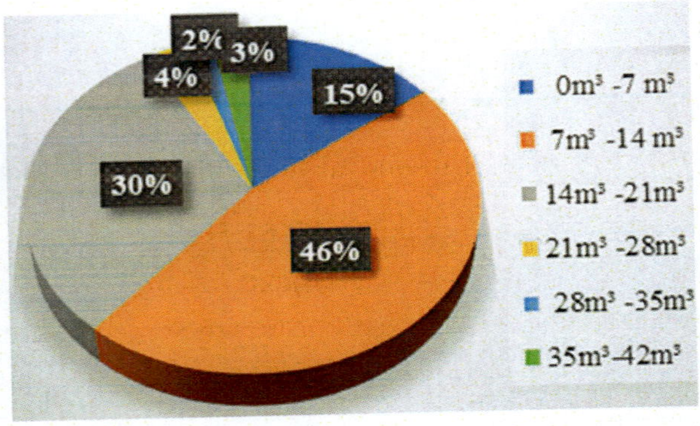

Fig. 15.11 Monthly water consumption

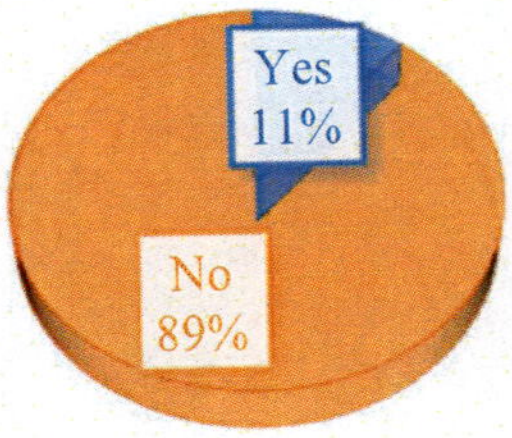

Fig. 15.12 Firewood consumption for cooking and water heating

3.9 Use of Potable Water for Crop Irrigation

30% of households in Saraguro use the drinking water system for irrigation of their crops, meanwhile, 70% use other methods, as seen in Fig. 15.9.

3.10 Use of Drinking Water for Car Washing

In Fig. 15.10 can be observed that there is a total of 92% of the sample that do not use potable water for car washing, compared to the remaining 8% that do use it.

3.11 Monthly Water Consumption

After the surveys were analyzed, it was found that 46% of the households consume between 7 m^3 and 14 m^3 of water; 30% use 14 m^3–21 m^3; 15% between 0 and 7 m^3; likewise, 4%, 3%, and 2% correspond to consumption in the ranges of 21 m^3–28 m^3; 35 m^3–42 m^3 and 28 m^3–35 m^3, respectively. It was also found that an average of 13.25 m^3 of water was consumed by the households (Fig. 15.11).

3.12 Use of Firewood for Cooking and Water Heating

According to Fig. 15.12, 89% of the household owners do not consume firewood for cooking or water heating, compared to the 11% who do use this resource.

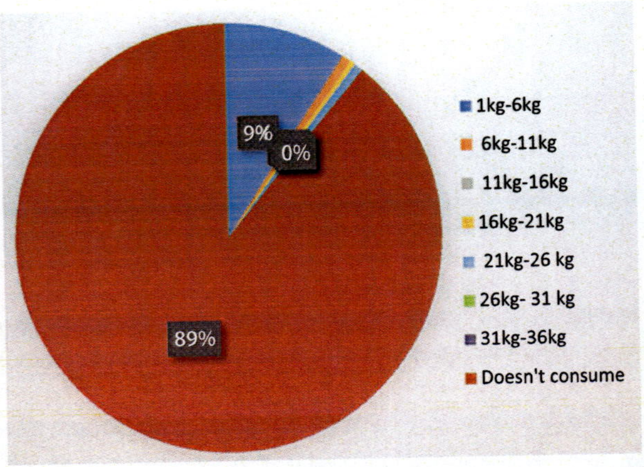

Fig. 15.13 Monthly firewood consumption

3.13 Monthly Firewood Consumption

In Fig. 15.13, it can be seen that 84% of the households who consume firewood, use between 1 and 6 kg, 7% between 6 and 11 kg, 6% between 21 and 26 kg, and only 3% consume between 16 and 21 kg. This way, an average consumption of 5.6 kg is determined.

3.14 Determination of the Resource with Greatest Energy Consumption

- Saraguro parish households' average consumption of electric energy is 165.35 kWh.
- Having as a reference that one LPG cylinder of 15 kg is equivalent to 205 kWh (Arízaga, 2015), and that the average monthly cylinders consumed by households are 2.2:
- Saraguro parish households' average consumption of LGP in kWh is 451 kWh.

Knowing that 1 kg of firewood is equivalent to 4.985 kWh (Lemes Martín et al., 2016) of energy and that the average consumption of the households in Saraguro is 5.6 kg we have that:

- Saraguro parish households' average consumption of firewood in kWh is 27.91 kWh.

From this analysis, it was determined that the resource with the monthly greatest consumption in the Saraguro parish is the LPG with 451 kWh, with 70% of the total energy consumption.

Since the water consumption is in volume units, and cannot be compared to the rest of the energy resource consumption units, it is determined at 13.25 m^3 per month.

3.15 Energetic Resources Economic Expenditure

The economic expenditure of essential services has subsidies on the electric, LPG, and water rates, knowing that only residential level consumptions are considered in this research.

The official household electric rate for Ecuador is of \$0.09 (Arízaga, 2015), the value of an LPG tank can be as high as \$2.50 (Salazar Masson, 2010), and for each m3 consumed in Saraguro, the household owner has to pay \$0.335 (Parra, 2016), as exposed in Table 15.1.

Due to that only 11% of the sample use firewood, and since it is a resource available in nature itself, it is analyzed separately from the other resources. The estimated price per kilogram of firewood is \$2.50 (León Armas & Quezada Muñoz, 2016), seen in Table 15.2.

Table 15.1 Monthly cost of basic services

Energetic resource	Average consumption	Resource cost [\$]	Total cost per month [\$]
Electricity	165.35 kWh	0.09	14.88
LPG	2.2 cylinders	2.5	5.5
Water	13.25 m^3	0.335	4.44

Table 15.2 Monthly consumption of firewood

Energetic resource	Average consumption	Resource cost [\$]	Total cost per month [\$]
Firewood	5.6 kg	2.50	14

Thus, the resource that has the greatest household energy consumption cost in Saraguro is electricity with a referential monthly value of \$14.88.

4 Energy Efficiency

Energy efficiency is the set of actions that allow the optimization of the relationship between the amount of energy consumed and the final uses. Have as the highest priority is the good use of energy, careful with the environment and the billing savings, it is recommended to establish mitigation measures plans, aimed at achieving greater efficiency in the use of energy assets (Rosenfeld et al., 2007). The measures recommended are:

- Change the saving bulbs to LED technology, since these consume less and have higher lighting levels.
- Replace conventional home appliances for those that have energy efficiency tagging, which would have a lower energy consumption. The tag contains information on the energy consumption of the home appliance indicating its position in a scale from A to G, where A is the most efficient and G the worst (Baquero, 2013).
- Exploitation of natural light from the sun.
- When home appliances aren't being used, they should be disconnected until needed. Likewise, fiery food shouldn't be inserted into the refrigerator, and the washing mashing shouldn't be used at its maximum capacity (Baquero, 2013).
- Use a solar water heater instead of gas heaters.
- Reuse rainwater in irrigation systems of crops.
- Control water leaking in the pipe system.
- Take short showers and close the faucets when they're not being used.
- Promote energy and resource use rational habits, which should be sustainable.

With the application of these measures, the is a 40% savings in electric consumption, 30% for water consumed, and 60% of LPG; and eliminating the consumption of firewood as seen in Table 15.3.

With these parameters applied, the household should save around 3834 kWh of energy per year, and 47.76 m^3/year.

Table 15.3 Monthly cost of basic services after application of measures

Energetic resource	Average consumption	Resource cost [$]	Total cost per month [$]
Electricity	99.21 kWh	0.09	8.92
LPG	1.1 cylinders	2.5	2.75
Water	9.27 m^3	0.335	3.10
Electricity	99.21 kWh	0.09	8.92

Economically, the household owner would save $120.60 each year, likewise savings of $168 in households that use firewood. Following an emission factor of 0.57 $kgCO_2/kWh$ and 0.5 $kgCO_2/m^3$ (Alvarez & Heras, 2008), it would evade a total of 2209.6 kg of CO_2.

5 Conclusions and Recommendations

Saraguro is an urban parish that has an elevated consumption of energy resources, where households have extensive tracts of land. Most of these terrains are used for sowing, and for irrigation of these, owners of the houses use potable water, as well as during car washing, therefore reflecting the bad habits of water consumption.

According to the calculated sample of 280 households, it was determined an electric consumption average in Saraguro's homes of 165.35 kWh per month, which belongs to home appliances like refrigerators, washing machines, televisions, and lighting. The light bulbs of the homes in their totality are saving bulbs.

The government subsidies to LPG imply directly in the use of gas cooktops, instead of another technology. The 100% of households use this type of cooking, likewise happens with water heaters, 70% of them consume gas. But, without the subsidies, this resource would be overly expensive.

Electricity is the most expensive type of energy consumed, meanwhile LPG is the most consumed type of energy by households with 451 kWh per month. The firewood consumption is low, nearly 11% of the population uses this resource, which in a certain way is better for the environment, preventing the chop of trees and reducing the environmental impact.

Implementing a correct energy efficiency guide leads to energy and economic savings, as well as contributing to the conservation of the environment, this is so that around 3800 kWh of energy per year is saved, meaning that each household owner saves $120.60 yearly.

References

Abad Ordóñez, L. E., & Aguaiza Viñanzaca, D. S. (2019). *Análisis de escenarios de eficiencia energética en el sector residencial del cantón Cuenca utilizando el modelo LEAP.*

Alvarez, N. L., & Heras, D. (2008). *Metodología para el Cálculo de la huella ecológica en universidades.* Paper presented at the Congresso Nacional del Medio ambiente, Cumbre del Desarrollo Sostenible, Congresso Nacional del Medio ambiente, Cumbre del Desarrollo Sostenible, Madrid.

Araujo, G., & Robalino-López, A. (2019). Aportes desde el enfoque analítico: Consumo energético del sector residencial del Ecuador.

Arcos López, E. R. (2016). *Estrategias de eficiencia energética en usuarios residenciales.*

Arízaga, E. S. U. (2015). Eficiencia del consumo electrico en el sector residencial urbano de Cuenca. *Ingeniero Tesis Previa a la Obtención del Título de Magister).* Universidad de Cuenca

Baquero, M. T. (2013). Ahorro de agua y reutilización en la edificación en la ciudad de Cuenca, Ecuador. *Estoa. Revista de la Facultad de arquitectura y Urbanismo de la Universidad de Cuenca, 2*(3), 71–81.

Baquero, M. T., & Quesada, F. (2016). Eficiencia energética en el sector residencial de la Ciudad de Cuenca, Ecuador. *Maskana, 7*(2), 147–165.

Ding, Q., Cai, W., Wang, C., & Sanwal, M. (2017). The relationships between household consumption activities and energy consumption in china—An input-output analysis from the lifestyle perspective. *Applied Energy*, 520–532.

Gould, C., Schlesinger, S., Molina, E., Bejarano, L., Valarezo, A., & Jack, D. (2020). Household fuel mixes in peri-urban and rural Ecuador: Explaining the context of LPG, patterns of continued firewood use, and the challenges of induction cooking. *Energy Policy*.

Jara, N. G., & Isaza, C. (2014). Programas de Eficiencia Energética y Etiquetado en el Ecuador-Revisión del Estado Actual. *Memorias Ponencias, 2014*, 1–12.

Lauretis, S. D., Ghersi, F., & Cayla, J.-M. (2017). Energy consumption and activity patterns: An analysis extended to total time and energy use for French households. *Applied Energy*, 634–648.

Lemes Martín, O. S., Machín Reid, J. D., & González Díaz, J. J. (2016). *La extracción de leña en los montes públicos de Canarias entre 1877 y 1900.*

León Armas, L. A., & Quezada Muñoz, D. M. (2016). *Plan de negocios para la comercialización de un combustible ecológico alternativo a base de cascarilla de arroz, en la Ciudad de Guayaquil.* Universidad Estatal de Guayaquil Facultad de Ciencias Administrativas.

Mendieta, M. R. (2012). *Dinámica poblacional comparativa de la provincia de Loja.* Instituto de Investigaciones Económicas.

Ministerio de Energía (2015). *Medición del consumo nacional de leña y otros combustibles sólidos derivados de la madera.*

Moran Moran, M. J., & Ramírez Añazco, B. J. (2015). *Modelamiento de eficiencia energética en el consumo de energía eléctrica en Guayaquil a lo largo del tiempo, zonas y grupos.* Espol.

Narváez, J. L. P., & Cabrera, R. S. R. (2007). *Consumo de leña en el área rural del Cantón Cotacachi y propuesta de plantaciones energéticas.*

Parra, P. F. V., & Martha, E. E. A. (2016). ESTUDIO SOBRE EL NIVEL DE SATISFACCIÓN DE LA CIUDADANÍA DEL SERVICIO DE AGUA POTABLE POR EL GOBIERNO AUTÓNOMO DESCENTRALIZADO CANTONAL DE

SARAGURO EN EL AÑO 2012. Universidad Nacional de Loja, Loja.
Poveda, M. (2007). *Eficiencia energética: recurso no aprovechado.* OLADE.
Recalde, M., Zabaloy, F., & Guzowski, C. (2018). El Rol de la Eficiencia Energética en el Sector Residencial para la Transición Energética en la Región Latinoamericana. *Trayectorias Año, 20,* 77–102.
Riofrío, A., Vaca, D., Orozco, D. C. M., & Martínez, J. (2014). *Análisis del consumo energético en procesos de cocción eficiente para el sector residencial.* Paper presented at the Memorias del Congreso latinoamericano de ingeniería mecánica Colim.
Salazar Masson, J. F. (2010). *Estudio técnico-comparativo para la introducción de cocinas eléctricas de inducción magnética en el Ecuador.* QUITO/EPN/2010.
Samaniego, C., Alvarez, O., & Maldonado, J. (2016). Emisiones provocadas por combustión de GLP a partir de calefones en la ciudad de Loja y su posible relación con enfermedades respiratorias agudas (ERA's). *Revista CEDAMAZ 2016, 6*(1), 60–67.
Saraguro, G. (2016). *Plan De Desarrollo Y Ordenamiento Territorial Del Cantón Saraguro Periodo 2014–2019.*
Tipán Jinde, J. C. (2017). *Estudio del consumo de agua potable en sectores residenciales de la zona centro de la ciudad de Ambato y su incidencia en la curva de consumo diario.* Universidad Técnica de Ambato. Facultad de Ingeniería Civil y Mecánica ...
Villalba, D. O., Proaño, J. L., Riera, O. J., & Amores, G. L. (2014) SISTEMA DE MEDICIÒN INTELIGENTE PARA LA GESTIÓN DE LA ENERGÍA ELÉCTRICA Y MEDICIÓN DE AGUA POTABLE EN LOS HOGARES.

Forecast of Short-Term Energy Production in Ecuador 50 MW Huascachaca Mines Wind Farm

Anibal Espinoza and Pablo Arias-Reyes

Abstract

In Ecuador, the last 40 years, the energy matrix has been based exclusively on the production of electrical energy from fossil fuels and hydroelectric resources. The development of the electrical matrix is based on the accessibility to the different energy sources when there is an adequate level of availability; therefore, wind generation allows increasing the level of security in the supply of electrical energy. To determine the generation of Huascachaca mines wind farm of 50 MW, computational tools such as the Windographer for the modeling of the wind flow, energy production and also the IBM SPSS statistics software for the short-term prediction through the classic auto-regressive statistical models, like Box and Jenkins, will be used. Elements such as the orography and topography of the terrain such as the wind at the height of the hub of the wind turbines are necessary to carry out the complete simulation of the wind farm. The main objective is determining the energy production of wind farm, using the computational tool, characterizing the wind variable and applying a short-term methodology on the timescale, to support the use of the wind resource in the country.

Keywords

Electrical energy • Prediction • Power • Wind energy • Wind speed

1 Introduction

1.1 Location

The Huascachaca mines wind farm is located 84 km southeast in the Giron-Pasaje road, in Loja Province, Saraguro canton, San Sebastian de Yuluc parish. In Table 16.1, the wind data of four weather stations is shown **(ELECAUSTRO)**.

The data was taken every 10 min between the years 2008 and 2020, and this same information was inserted in the computational tool Windographer v4.2.18 (ISAÍAS, 2016; Mur-Amada, 2001; Villarubia, 2012; Windographer Software, 2020).

1.2 Wind Power

For the wind power evaluation, the average wind speed was taken with a frequency of 10 s. An average of these samples was done every 10 min and every hour, for the lapse of 5 years (Burton et al., 2011; Cuco Pardillos, 2017; ISAÍAS, 2016; Mur-Amada, 2001; Villarubia, 2012). The data was obtained from a town that meets speed measurements and wind directions standards and also incorporates data-logger instrumentation.

Refer to (1) for the calculation of the wind power available in the transversal area and perpendicular to the wind flux movement as seen in Fig. 16.1, with a velocity $\overline{v}$ [m/s] with an air density ρ [kg/m^3].

$$P_{\text{wind}} = \frac{1}{2}\rho\overline{v}^3 \text{ [W/m}^2\text{]} \quad (1)$$

where the air density is equivalent to 1.225 kg/m^3. The fraction of the total power, captured by the aerogenerator is the usable wind power. This usable power is determined by an important factor named power coefficient C_p expressed in

A. Espinoza (✉)
Master's Program in Renewable Energy, Catholic University of Cuenca, Cuenca, Egypt
e-mail: anibal.espinoza@est.ucacue.edu.ec

P. Arias-Reyes
Research, Innovation, and Technology Transfers Centre, Catholic University of Cuenca, Cuenca, Egypt
e-mail: pariasr@ucacue.edu.ec

H. Altan et al. (eds.), *Advances in Architecture, Engineering and Technology*,
Advances in Science, Technology & Innovation, https://doi.org/10.1007/978-3-031-11232-4_16

Table 16.1 Location of weather stations

Station	Location	Height [m]	Coordinates *x*–*y*	Altitude [m]
Torre_1	Uchucay	40	681,381 9,629,250	1106
Torre_2	Uchucay	50	681,203 9,628,043	1147
Torre_3	Yuluc	50	679,447 9,628,666	1137
Torre_4	Yuluc	80	679,446 9,628,522	1181

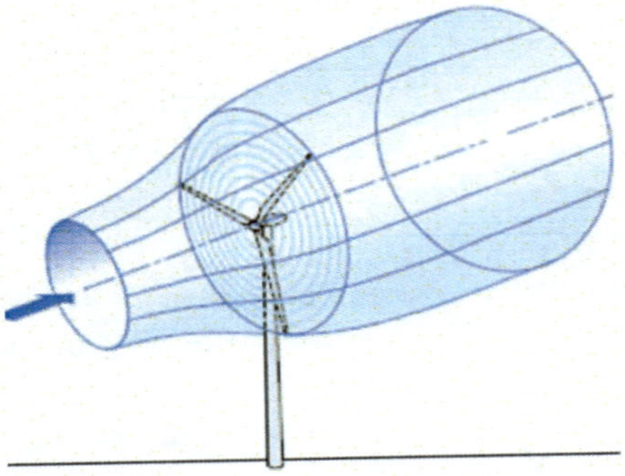

Fig. 16.1 Wind power extraction (Boveri & ABB, 2012)

percentage. A is the area exposed to the wind expressed in m^2 (Burton et al., 2011; Cuco Pardillos, 2017; ISAÍAS, 2016). Refer to (2) for the calculation of the usable wind power.

$$P = \frac{1}{2}\rho A\overline{v}^3 C_p \text{ [W]} \quad (2)$$

The coefficient factor is used to characterize the efficiency of the wind rotor. Since the whole kinetic wind energy cannot be transformed to mechanic energy, the coefficient has a limit of 0.5 (Burton et al., 2011; Cuco Pardillos, 2017; Mur-Amada, 2001).

2 Materials and Methods

The steps followed for the characterization of the wind are observed in Fig. 16.2.

2.1 Topography

The terrain has a thoroughness of 0.005 since it is bare and smooth (Cuco Pardillos, 2017). In Fig. 16.3, the weather stations can be seen as located in the study zone.

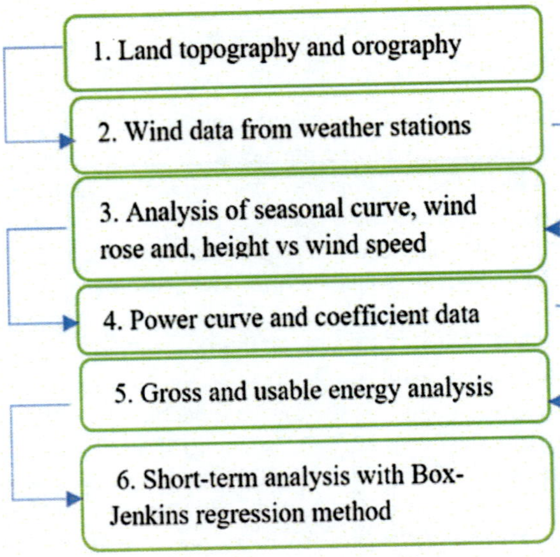

Fig. 16.2 Flow diagram for wind characterization

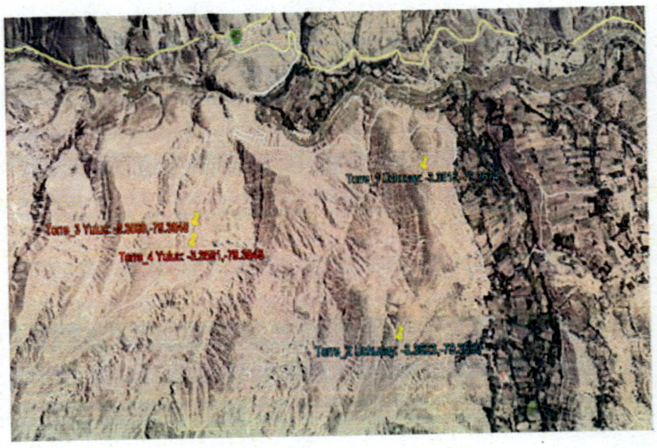

Fig. 16.3 Location of weather stations

2.2 Wind Data

The weather stations are equipped with autonomous power feeding, designed for 12 V batteries, with a charge controller. The data is sampled every second, averaged and marked in an interval of 10 min. This data is inserted in Windographer for its tabulation, obtaining the following information:

2.2.1 Uchucay Station

Figures 16.4, 16.5, 16.6 and 16.7 represent the annual mean wind speeds, mean diurnal wind profile, the wind frequency rose and the vertical wind shear profile, respectively, obtained in Windographer for Uchucay station.

2.2.2 Yuluc Station

Same as for Uchucay, for the Yuluc station, Figs. 16.8, 16.9, 16.10 and 16.11 describe the different graphics obtained by the software Windographer.

2.3 Seasonal Curve, Wind Frequency Rose and Speed with Height

From Figs. 16.4, 16.5, 16.6, 16.7, 16.8, 16.9, 16.10 and 16.11, it is clear to see that for the Yuluc towers 3 and 4, the prominent wind direction is from the northwest, and that most resource exist from 9 am to 6 pm, with a peak at 1 pm. Likewise, the prominent monthly mean wind speeds are from July to October, with a difference that for tower 2, the influent direction is west–northwest (Cuco Pardillos, 2017). In Figs. 16.12 and 16.13, the comparison of the four stations can be observed in a single graphic and can be seen that the prominent wind direction is from the west and that the wind curve is seasonal for most of the year and also its high predictability.

The results of the wind resource can be observed in Table 16.2.

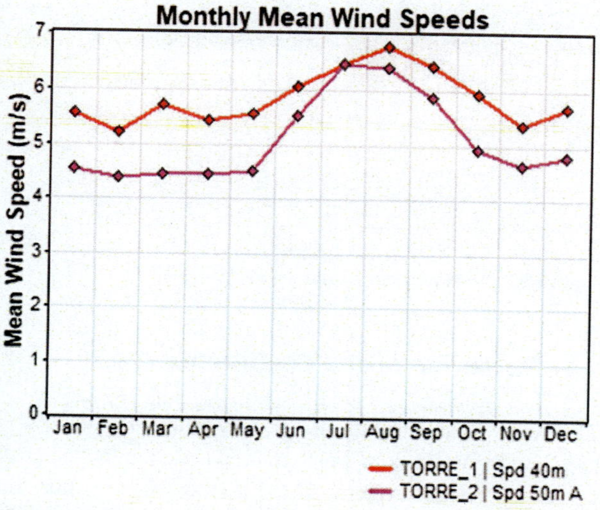

Fig. 16.4 Annual mean wind speed Torre_1, Torre_2

As seen in the table, tower 2 has a wind speed of 5.08 m/s, with a northwestern direction. Also, the wind frequency distribution confirms the occurrence of the wind

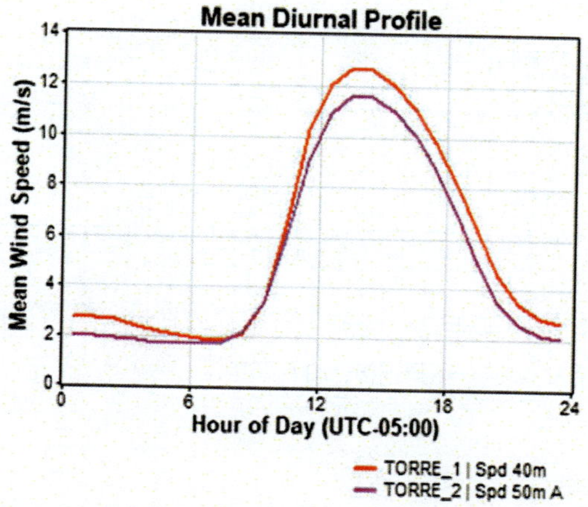

Fig. 16.5 Mean diurnal wind profile Torre_1, Torre_2

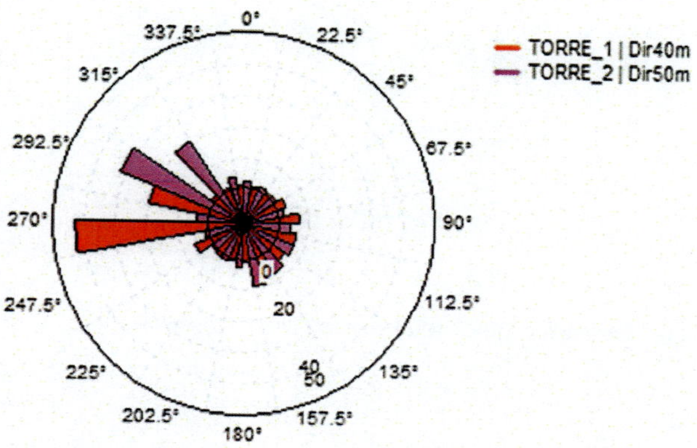

Fig. 16.6 Wind frequency rose Torre_1, Torre_2

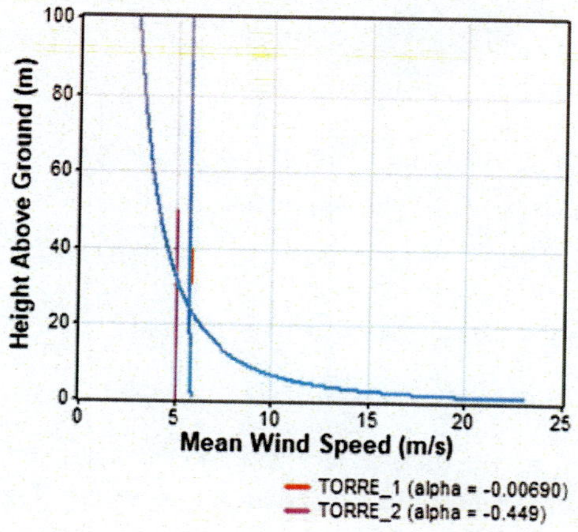

Fig. 16.7 Vertical wind shear profile Torre_1, Torre_2

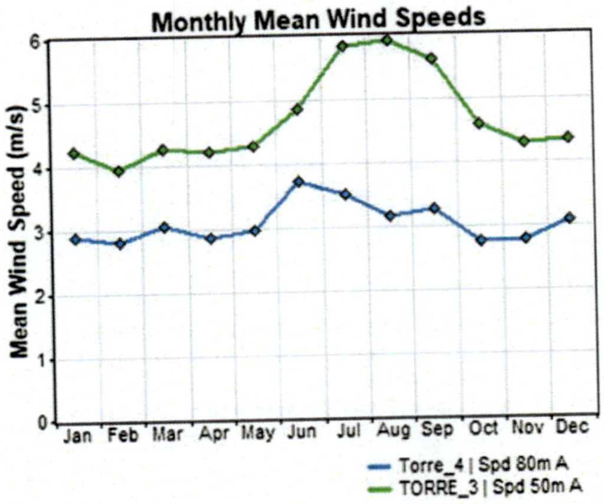

Fig. 16.8 Annual mean wind speed Torre_3, Torre_4

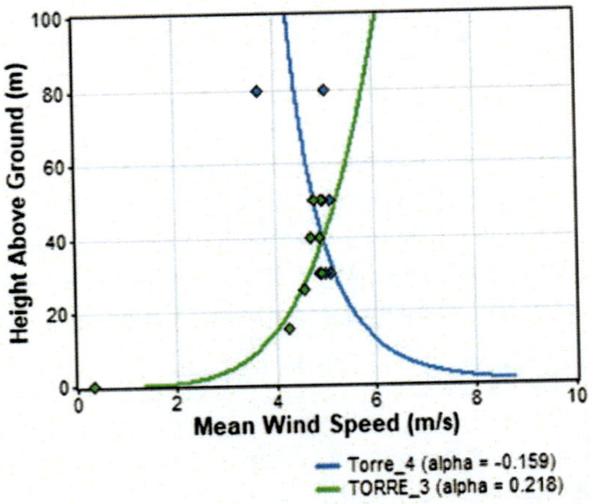

Fig. 16.11 Vertical wind shear profile Torre_3, Torre_4

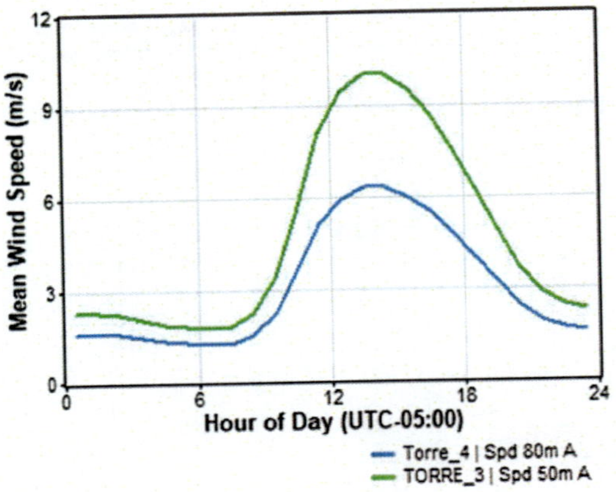

Fig. 16.9 Mean diurnal wind profile Torre_3, Torre_4

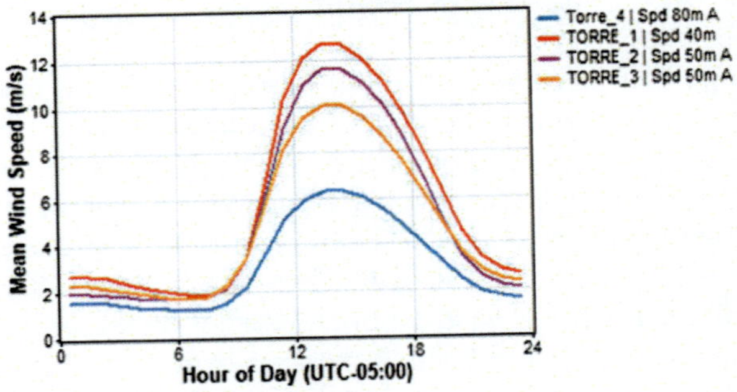

Fig. 16.12 Mean diurnal profile

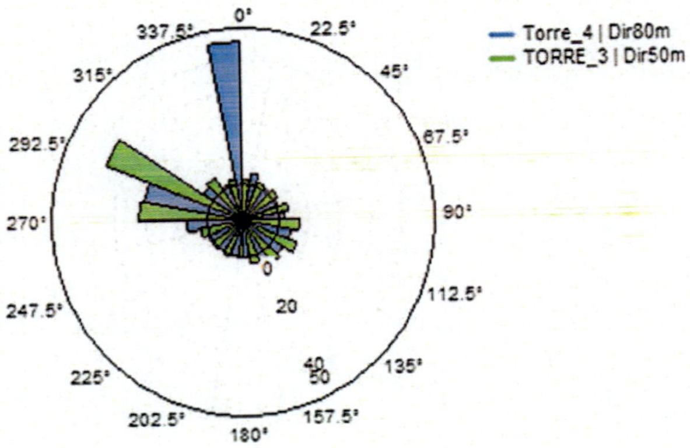

Fig. 16.10 Wind frequency rose Torre_3, Torre_4

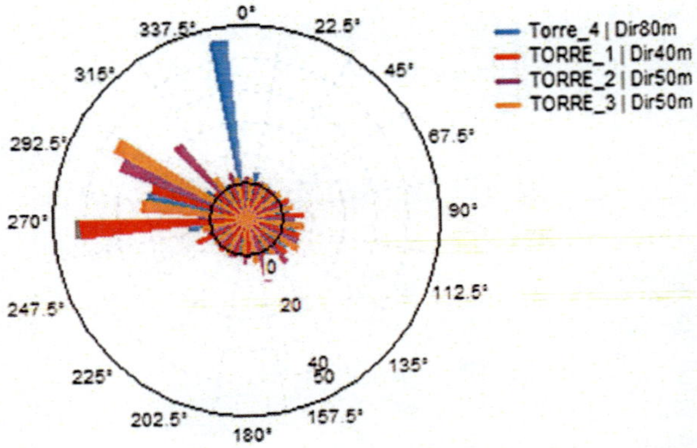

Fig. 16.13 Wind frequency rose

speed in a particular month and for the maximum if the month of the most representative year as seen in Fig. 16.14 (Marcelo & Dioses, 2014).

2.4 Power Curve and Coefficients

The wind measurements at the aerogenerator rotor hub's height in the intervals of time are used to determine the gross power output (Mosquera, 2015; Ramos, 2019). For this calculation, the turbine's power curve, the wind speed at the

Table 16.2 Wind resource results

Item	Tower 1	Tower 2	Tower 3	Tower 4
Height [m]	40	50	50	80
Wind speed [m/s]	5.86	5.08	4.71	3.07
Wind direction	West	West–Northwest	West	West–Northwest

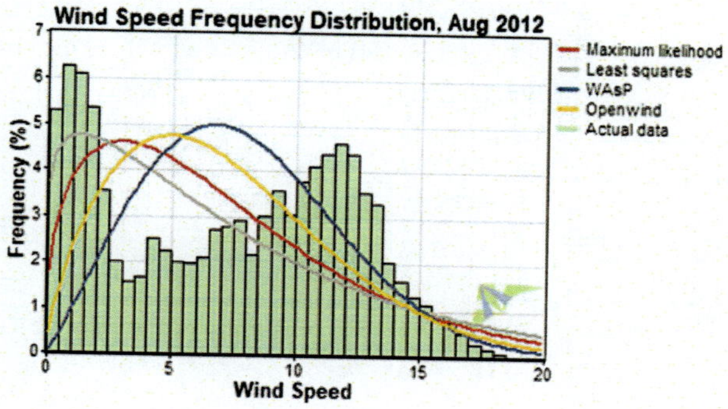

Fig. 16.14 Weibull distribution for tower 2

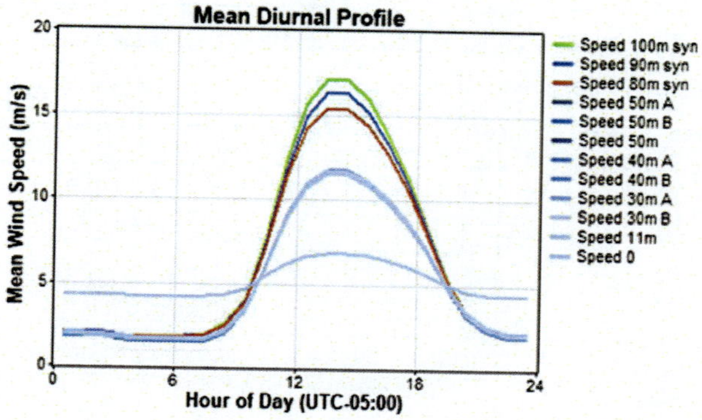

Fig. 16.16 Extrapolated mean diurnal wind speed

hub's height, and the air density are used (Cuco Pardillos, 2017; Villarubia, 2012).

To calculate the power of the aerogenerators, the data obtained from the tower 2 is used. For this purpose, an extrapolation was made at 80,90 and 100 meters for the simulation of the rotor hub of the wind turbine, (Barragán, 2012; Marcelo & Dioses, 2014). Once the extrapolation is done, the mean speed values of this tower will be used to project for the else of the weather stations. The extrapolation data from tower 2 can be seen in Figs. 16.15 and 16.16.

The different results obtained at the rotor hub's height can be observed in Table 16.3. The most important is that the annual mean speed at 80 m is 6.147, 6.39 m/s at 90 and

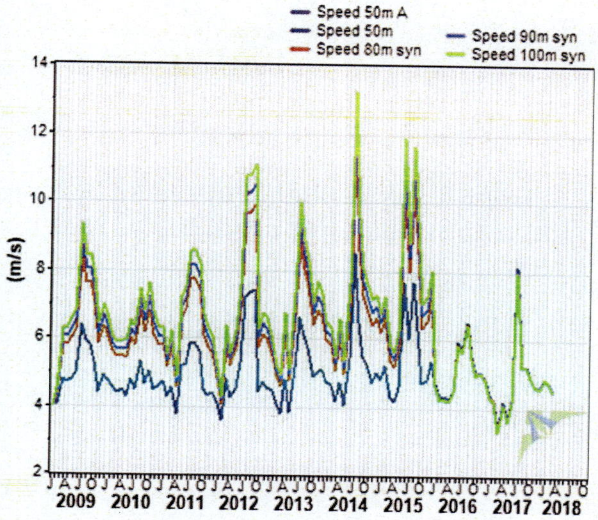

Fig. 16.15 Extrapolated annual mean wind speed

6.623 m/s at 100 m of height (Boveri & ABB, 2012; Cuco Pardillos, 2017). Thus, the aerogenerator power considers the extrapolation of the three possible aerogenerator heights. Table 16.4 shows the wind types in standard conditions (Moorthy et al. 2017). After the extrapolation, the existence of low wind speeds is confirmed, and this has to be considered for the aerogenerator selection. Also, there are different turbulence conditions to take into account, in this specific case, of A:18% and B:16% to take advantage of the wind resource.

After comparing with three aerogenerators at 100 m high, it is found that the 3 MW device has the best power curve, and thus, providing the best response to the wind speeds. Figure 16.17 shows that, with an IEC III wind type, the aerogenerator starts with wind speeds lower than 5 m/s and reaches its nominal power at 10 m/s (Cuco Pardillos, 2017). In Fig. 16.18, the gross power output curve in kW of the 3 MW aerogenerator can be observed.

In Table 16.5, the results of the comparison of the most optioned wind turbines are indicated. These results are the net power at 90 and 100 m high (Cuco Pardillos, 2017; Marcelo & Dioses, 2014).

2.5 Gross and Usable Energy Results Analysis

The results of the gross power of each year of tower 2, with the three types of wind turbines, can be seen in Table 16.6. These are calculated with the rotor at 100 m high (Carta González, 2019; Párraga Palacios et al., 2019). The year 2015–2016 is the best period, meanwhile 2010 the worst.

Table 16.3 Extrapolation results

	Year	Mean (m/s)	Median (m/s)	Min. (m/s)	Max. (m/s)	Std. Dev. (m/s)	Weibull k	Weibull A (m/s)
Height 80 m								
1	2009	6.269	2.772	0.341	33.896	6.754	1.008	6.29
2	2010	5.978	2.465	0.478	33.985	6.638	0.978	5.92
3	2011	6.283	2.553	0.477	33.38	6.751	1.019	6.332
4	2012	6.676	3.226	0.478	33.522	6.85	1.063	6.837
5	2013	6.45	2.631	0.478	34.664	6.994	1.007	6.469
6	2014	6.907	3.277	0.478	34.682	7.169	1.053	7.049
7	2015	7.254	3.69	0.474	35.424	7.324	1.085	7.484
8	2016	5.045	3.103	0.298	20.467	4.852	1.15	5.3
9	2017	4.736	2.607	0.298	19.581	4.672	1.123	4.941
10	2018	4.799	3.037	0.303	18.531	4.571	1.147	5.04
All data		6.123	2.867	0.298	35.424	6.495	1.023	6.179
Mean of monthly means		**6.147**						
Height 90 m								
1	2009	6.562	2.703	0.339	37.728	7.299	0.978	6.497
2	2010	6.257	2.394	0.464	37.857	7.178	0.949	6.11
3	2011	6.588	2.479	0.463	37.15	7.289	0.991	6.564
4	2012	7.006	3.146	0.464	37.322	7.407	1.033	7.099
5	2013	6.774	2.554	0.464	38.673	7.574	0.979	6.71
6	2014	7.267	3.199	0.464	38.668	7.769	1.024	7.336
7	2015	7.642	3.625	0.46	39.531	7.943	1.056	7.808
8	2016	5.03	3.075	0.29	20.438	4.854	1.146	5.28
9	2017	4.723	2.581	0.29	19.558	4.676	1.12	4.924
10	2018	4.796	3.025	0.294	18.521	4.576	1.146	5.035
All data		6.363	2.796	0.29	39.531	6.97	0.99	6.335
Mean of monthly means		**6.390**						
Height 100 m								
1	2009	6.842	2.642	0.336	41.521	7.827	0.952	6.693
2	2010	6.524	2.331	0.452	41.693	7.698	0.924	6.289
3	2011	6.88	2.414	0.451	40.881	7.806	0.968	6.783
4	2012	7.323	3.076	0.452	41.085	7.944	1.008	7.346
5	2013	7.084	2.488	0.452	42.652	8.133	0.955	6.939
6	2014	7.611	3.131	0.452	42.621	8.347	0.999	7.607
7	2015	8.015	3.569	0.448	43.607	8.541	1.031	8.114
8	2016	5.016	3.053	0.282	20.411	4.855	1.144	5.262
9	2017	4.712	2.556	0.282	19.537	4.678	1.117	4.908
10	2018	4.794	3.015	0.286	18.513	4.58	1.145	5.031
All data		6.593	2.735	0.282	43.607	7.433	0.962	6.481
Mean of monthly means		**6.623**						

Table 16.4 Wind types in standard conditions

Wind classes of wind turbines	IEC I high wind	IEC II medium wind	IEC III low wind
Average annual speed (m/s)	10	8.5	7.5
Extreme gust speed every 50 years (m/s)	70	59.5	52.5
Turbulence classes (%)	*A*:18 *B*:16	*A*:18 *B*:16	*A*:18 *B*: 16

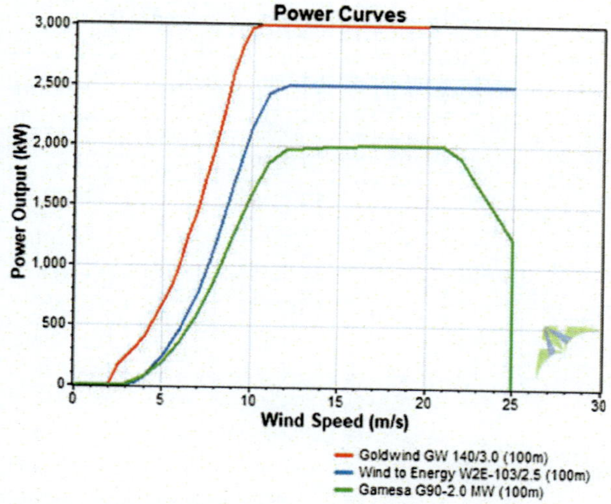

Fig. 16.17 Wind turbines power curves

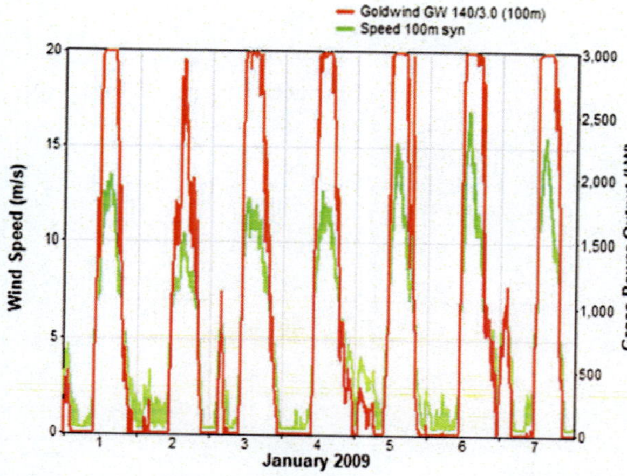

Fig. 16.18 Gross power in kW

The results for the turbines mentioned previously at a height of 90 m are presented in Table 16.7, where the gross power is even better than those at 100 m, especially with the first aerogenerator. This demonstrates that the optimal height for the turbine is 90 m.

The gross power curve can be seen in Fig. 16.19, where the first device receiving a wind speed of 20 m/s, at 90 m of height and a power of 3 MW has an increase of plant factor from 25.73 to 26.81%.

2.6 Short-Term Analysis with Box–Jenkins Regression Method

Box–Jenkins methodology only takes into account the pattern of the time series in the past, ignoring any information that could be included in a regression model (De Arce & Mahía, 2003; Pineda et al., 2017). For this, he uses the most recent observation as the initial value and then analyzes recent forecast errors to select the most appropriate setting.

The advantages of this methodology are as follows (De Arce & Mahía, 2003; Shumway & Stoffer, 2017):

- It uses a lot of information from the time series, and for this, it requires a small number of parameters
- It allows greater flexibility of the method.
- It uses a stational series of time.
- It uses moving averages, self-regulating and mixed models.

The steps in IBM SPSS v.21 statistical software that will be used to perform the self-regression model (Stadistics) are described in Fig. 16.20.

The wind speed values entered in the software are from tower 2, which are synthesized, i.e., with extrapolation to 90 m which is the height of the wind turbine rotor bushing.

2.6.1 Wind Speed

The data is taken from January 2009 to December 2018. The average monthly wind speed values as the dependent variable are exported from Windographer for tower 2 (Verde et al., 2018).

2.6.2 SPSS Statistical Method Tool

The first procedure is to identify an ARIMA model (p, d, q), realizing a sequence model with the wind data as seen in Fig. 16.21.

It is determined to be a stationary time series with the seasonal decomposition process; also, with autocorrelation processes, the interdependence model will be with a

Table 16.5 Net power results and annual energy production

Wind turbine	Valid	Hub height	Percentage of time at		Simple mean
	Time steps	Wind speed (m/s)	Zero power	Rated power	Net power (kW)
Goldwind GW 140/3.0 (90 m)	492,710	6.36	49.38	19.63	803.4
Wind to energy W2E-103/2.5 (90 m)	492,710	6.36	52.86	18.28	654.1
Gamesa G90-2.0 MW (90 m)	492,710	6.36	45.77	4.79	487.9
Goldwind GW 140/3.0 (100 m)	492,710	6.59	51.25	18.87	771.5
Wind to energy W2E-103/2.5 (100 m)	492,710	6.59	53.91	18.39	643.6
Gamesa G90-2.0 MW (100 m)	492,710	6.59	47.21	4.82	475.9

Wind turbine			Mean of monthly means		
	Net AEP (kWh/year)	NCF (%)	Net power (kW)	Net AEP (kWh/year)	NCF (%)
Goldwind GW 140/3.0 (90 m)	7,037,761	26.78	804.3	7,045,612	26.81
Wind to energy W2E-103/2.5 (90 m)	5,729,988	26.16	657.4	5,758,399	26.29
Gamesa G90-2.0 MW (90 m)	4,273,832	24.39	489.9	4,291,165	24.49
Goldwind GW 140/3.0 (100 m)	6,758,269	25.72	772	6,763,101	25.73
Wind to energy W2E-103/2.5 (100 m)	5,637,890	25.74	646.6	5,664,164	25.86
Gamesa G90-2.0 MW (100 m)	4,168,686	23.79	477.6	4,183,978	23.88

Table 16.6 Gross power at a height of 100 m

Year	Gross mean power output (kW)	Gross mean power output (kW)	Gross mean power output (kW)
	Goldwind GW 140/3.0 (100 m)	Wind to energy W2E-103/2.5 (100 m)	Gamesa G90-2.0 MW (100 m)
2009	869.7	751.4	553.4
2010	816.8	709.6	520.9
2011	898.1	796.3	590.1
2012	958.8	842.2	621.5
2013	861	771.4	566.4
2014	933.5	840.5	615
2015	947.7	885.3	645.5
2016	1,011.20	703.1	530.2
2017	955.3	652.5	489.8
2018	932.2	629.5	476.1
Overall	918.5	766.2	566.6

maximum of 36 delays, included in the SPSS, to obtain the estimated autocorrelation (ACF) and partial autocorrelation (ACFP) functions, which validate the seasonal periods (Mauricio, 2007; Torres et al., 2005), as observed in Figs. 16.22 and 16.23.

It can be determined that, for seasonality, the ACF and ACFP of the delays are nonzero and therefore configured as a fan that completes its cycle by rotating on the axis of the abscissa.

3 Results

The evaluation of the wind resource has, as a result, the prominent wind direction which comes from west–north-west, with a speed of 5.08 m/s at a height of 50 m. After the extrapolation to 90 m, it has an annual mean wind speed of 6.39 m/s. The gross power at 90 m high in the tower 2 is 1018 kW in 2015, being the most representative year and a annual mean energy of 7,045,612 kWh per year.

Table 16.7 Gross power at a height of 90 m

Year	Gross mean power output (kW) Goldwind GW 140/3.0 (90 m)	Gross mean power output (kW) Wind to energy W2E-103/2.5 (90 m)	Gross mean power output (kW) Gamesa G90-2.0 MW (90 m)
2009	913.9	767.4	569.5
2010	856.6	723.6	536.4
2011	947.2	805.3	605.4
2012	1006.80	855.1	638.1
2013	910.6	792	585.6
2014	985.7	862.4	637.8
2015	1018.00	906.9	672.9
2016	1012.70	704.1	531
2017	956.3	653	490.3
2018	931.7	628.9	475.6
Overall	956.5	778.7	580.8

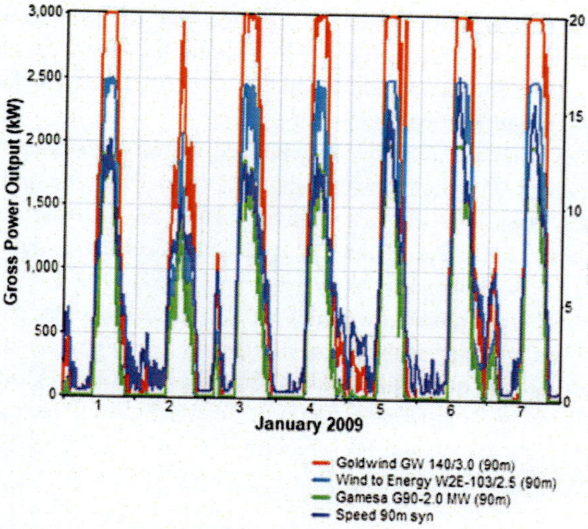

Fig. 16.19 Gross power at 90 m

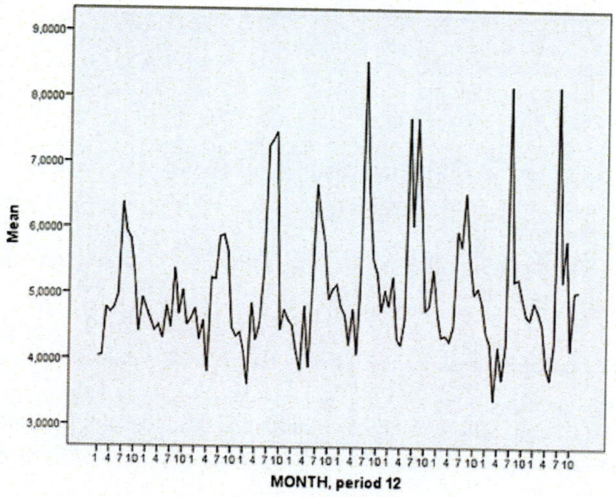

Fig. 16.21 Wind curve

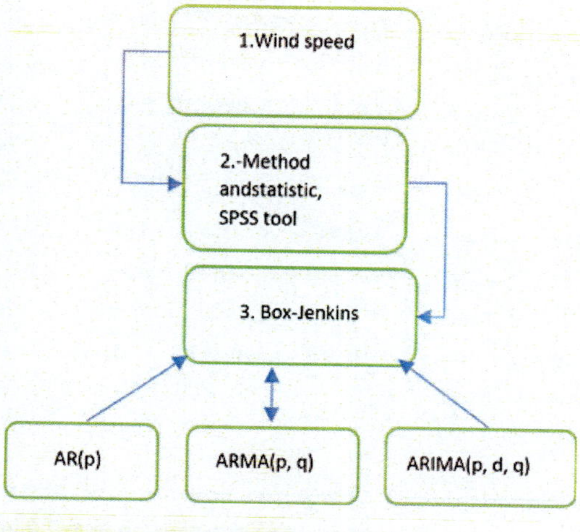

Fig. 16.20 Box–Jenkins methodology diagram

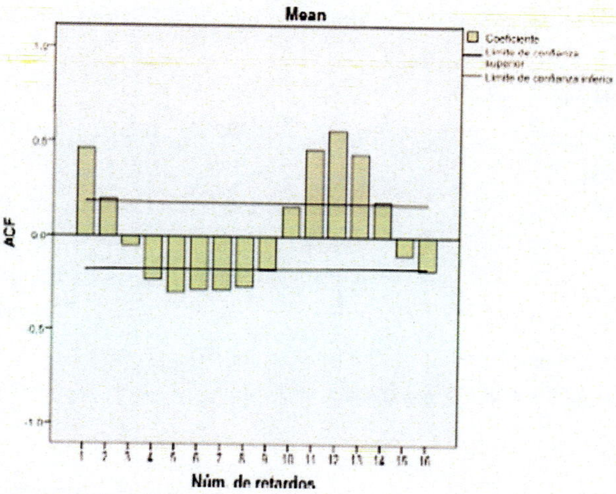

Fig. 16.22 Autocorrelations ACF

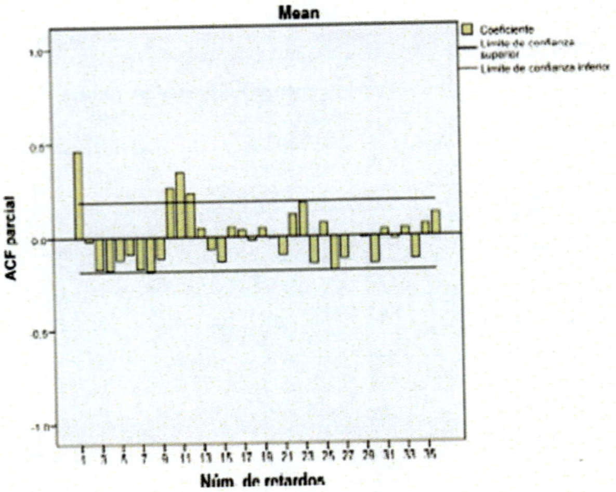

Fig. 16.23 Autocorrelations ACFP

Table 16.8 Model description

	Model type
Model ID mean model_1	ARIMA (1,0,0) (0,0,1)

The best ARIMA model (*p*, *q*, *q*) to predict power generation from wind speed with Torre_2 data is met according to Table 16.8.

Determining the model adjustment is described in Table 16.9.

The prediction of the values from January 2019, having as predecessor data those of the Torre_2, was calculated for 10 months and shown in Table 16.10. For these values, the wind seasonality variable is maintained and does not become an uncertainty for the energy (González et al., 2018; Rangel et al., 2017).

In this time interval, it can be indicated that the ARIMA model (1,0,0) satisfies what is expected with the prediction of the wind variable, and Fig. 16.24 shows how the forecast takes the approximate shape of the curve obtained by the method used (Nanda, 1988).

The wind variable can be predicted within a confidence interval of 95% LCI (lower) and LCS (top) where the predicted wind has the approximate shape of the original average wind value of the Torre_2 synthesized or extrapolated to 90 m, which is shown in Fig. 16.25.

However, a prediction model with the necessary iterations achieves reliable useable values in the future, as long as the wind variable meets the seasonal condition. An example of prediction with the same data up to Dec 2025 can be seen in Fig. 16.26; this would be achieved with an ARIMA model (5,0,0) (De Arce & Mahía, 2003; Sánchez Rosas, 2018).

Table 16.9 ARIMA model adjustment

Estadístico de ajuste	Media	Mínimo	Máximo	Percentil						
				5	10	25	50	75	90	95
R-cuadrado estacionaria	0.451	0.451	0.451	0.451	0.451	0.451	0.451	0.451	0.451	0.451
R-cuadrado	0.437	0.437	0.437	0.437	0.437	0.437	0.437	0.437	0.437	0.437
RMSE	0.747	0.747	0.747	0.747	0.747	0.747	0.747	0.747	0.747	0.747
MAPE	9.900	9.900	9.900	9.900	9.900	9.900	9.900	9.900	9.900	9.900
Max APE	43.644	43.644	43.644	43.644	43.644	43.644	43.644	43.644	43.644	43.644
MAE	0.518	0.518	0.518	0.518	0.518	0.518	0.518	0.518	0.518	0.518
Max AE	2.797	2.797	2.797	2.797	2.797	2.797	2.797	2.797	2.797	2.797
BIC normalizado	− 0.463	− 0.463	− 0.463	− 0.463	− 0.463	− 0.463	− 0.463	− 0.463	− 0.463	− 0.463

where $p = 1$, $d = 0$, $q = 0$ and $R^2 = 0.451$

Table 16.10 January 2019–October 2019 forecast

Modelo		Ene 2019	Feb 2019	Mar 2019	Abr 2019	May 2019	Jun 2019	Jul 2019	Ago 2019	Sep 2019	Oct 2019
Previsión											
Mean-Modelo_1	Previsión	5.2130	4.9337	5.6128	4.5344	4.5645	4.7176	5.9757	5.2219	5.4675	4.5397
	LCS	6.6911	6.3951	7.2810	5.8825	5.9216	6.1203	7.7523	6.7745	7.0930	5.8894
	LCI	3.9931	3.7370	4.2474	3.4311	3.4538	3.5697	4.5216	3.9513	4.1371	3.4351

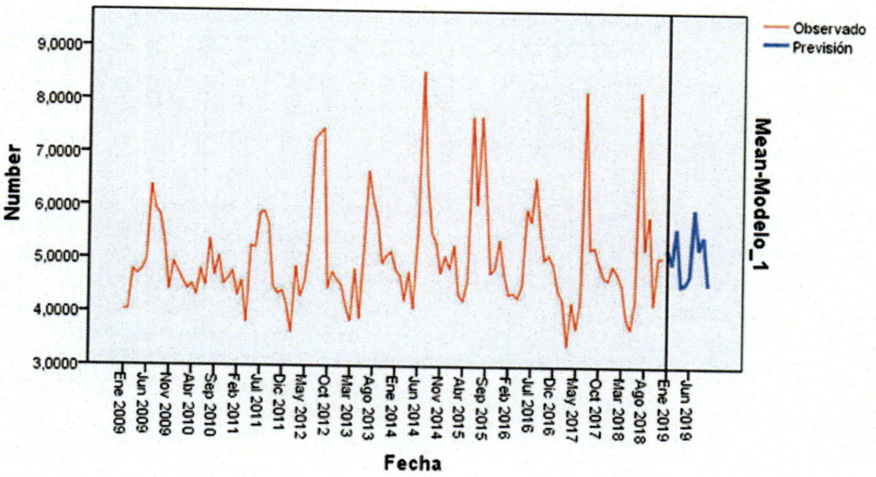

Fig. 16.24 2019 forecast

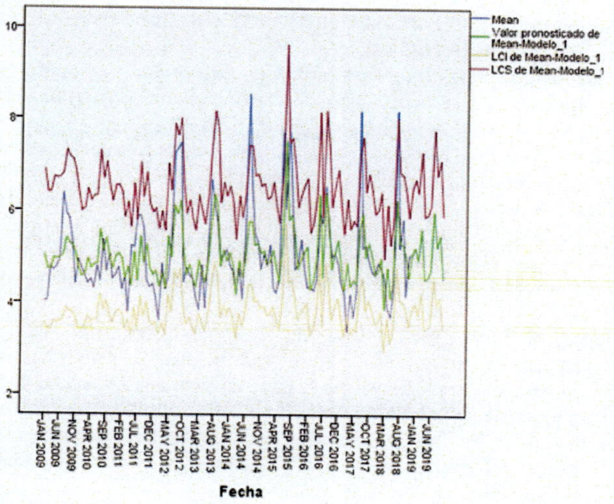

Fig. 16.25 Predicted wind value

4 Conclusions

The characterization of the wind resource is necessary for energy production in a wind farm. Extrapolation at different heights such as those analyzed shows that greater power can be achieved for wind turbines with the same wind conditions.

The gross power calculated at the rotor bushing height at 90 m is available for 3 MW power wind turbines. The most representative years are 2012, 2015 and 2016 which must be taken into account for the energy production of the wind farm. The Windographer tool is a software that estimates the average annual net energy production. For this research, this value is 134.06 GWh and the capacity factor is 26.78%.

Box–Jenkins statistical methodology demonstrates that with a series of seasonal time, it is feasible to predict future

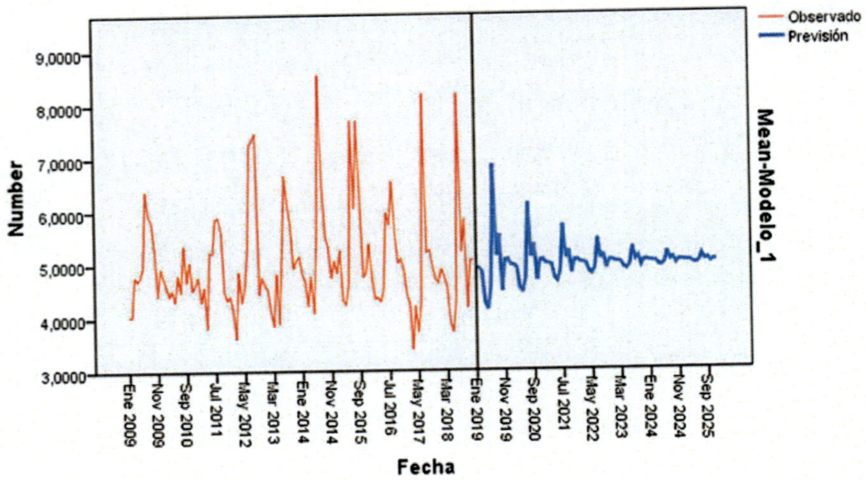

Fig. 16.26 Forecast–December 2025

wind values with ARIMA self-regressions. Taking into account the exempt dependence between the data, it is understood that each observation at any given time is modeled based on the previous values.

The methodology of short-term prediction is considered important in the design of a wind farm and should be taken into account for energy production planning. IBM's SPSS computational tool proves to be accurate for the self-regression calculation by obtaining for ARIMA short-term model prediction (1,0,0).

In summary it can be concluded that, for the forecast of energy production, wind characterization, power curve and coefficients, statistical methodology SPSS tool, and Box–Jenkins methodology should be considered.

Acknowledgements A special thank is dedicated to the company Electro Generadora del Austro ELECAUSTRO S.A for their contribution to this article.

References

Barragán, M. I. A. (2012). ANÁLISIS, ESPECIFICACIÓN Y DESARROLLO DE PROCEDIMIENTOS DE OPERACIÓN PARA LA GESTIÓN DE LA ENERGÍA EÓLICA EN EL ECUADOR.

Boveri, A. B., & ABB, S. (2012). *Cuaderno de aplicaciones técnicas n. 12 Plantas eólicas*. Barcelona, España.

Burton, T., Jenkins, N., Sharpe, D., & Bossanyi, E. (2011). *Wind energy handbook*. John Wiley & Sons.

Carta González, J. A., Calero Pérez, R., Colmenar Santos, A., Castro Gil, M. A., & Collado Fernández, E. (2019). Centrales de energías renovables generación electrica con energias renovables.

Cuco Pardillos, S. (2017). Manual de energía eólica. Desarrollo de proyectos e instalaciones. *Colección Manual de referencia*.

De Arce, R., & Mahía, R. (2003). Modelos Arima. *Programa CITUS: Técnicas de Variables Financieras*.

ELECAUSTRO. ELECTRO GENERADORA DEL AUSTRO. https://www.elecaustro.gob.ec

González, G. L. G., Martínez, A. R., Figueroa, G. C., & Lopez, U. L. J. P. E. (2018). PRONÓSTICO DE GENERACIÓN ELÉCTRICA DE CORTO PLAZO DE UN PARQUE EÓLICO, UTILIZANDO TÉCNICAS NUMÉRICAS DE PREDICCIÓN DEL CLIMA (Short-term electricity generation prognostic of a wind farm, using weather prediction numerical techniques), *40*(130).

ISAÍAS, G. T. O. (2016). ANÁLISIS,DISEÑO,CONSTRUCCIÓN Y OPTIMIZACIÓN DE AEROGENERADOR PARA MICRO GENERACIÓN ELÉCTRICA.

Marcelo, D., & Dioses, I. (2014). Análisis del potencial eólico para un emplazamiento seleccionado en la región Piura. *Revista de la Facultad de Ciencias Químicas*, 19.

Mauricio, J. A. (2007). *Análisis de series temporales*. Universidad Complutence de Madrid.

Moorthy, C. B., Balasubramanian, C., & Deshmukh, M. (2017). Wind turbine output estimation using Windographer Software.

Mosquera, J. L. E. P. M. (2015). ENERGIASRENOVABLESen ECUADORlibro2015 (1).pdf.

Mur-Amada, J. (2001). *Master europeo en energías renovables y eficiencia energética: Curso de energía eólica*. Universidad de Zaragoza.

Nanda, S. J. V. (1988). Forecasting: Does the Box-Jenkins method work better than regression? *Vikalpa, 13*(1), 53–62.

Párraga Palacios, Á. G., Intriago Rodríguez, S. A., Velasco Fuentes, E. D., Cedeño Quinto, V. M., Cedeño Quinto, V. M., Murillo Párraga, N. L., & Zambrano Gavilanes, F. E. (2019). Producción de energía eólica en Ecuador. *Ciencia Digital, 3*(3), 22–32. https://doi.org/10.33262/cienciadigital.v3i3.610

Pineda, S. E. P., Aguilar, J. A. H., & Arroyo-Figueroa, G. (2017). Aplicación de modelos auto regresivos para la predicción de generación de energía eléctrica a partir de datos eólicos. *Research in Computing Science, 139*, 59–70.

Ramos, J. G. E. (2019). Estimación del viento cálculo del potencial eólico del distrito de Salaverry mediante el uso de software computacionales y la estadística.

Rangel, H. R., Carrillo, N. A. G., Flores, J. J., Rosales, L. A. M., & Montelongo, G. M. (2017). Pronóstico a corto plazo de velocidad del viento a partir de datos incompletos. *Research in Computing Science,139*, 81–95.

Sánchez Rosas, Y. S. (2018). *Predicción probabilística de corto plazo en generación eólica*. Universidad Nacional de Colombia-Sede Bogotá.

Shumway, R. H., & Stoffer, D. S. (2017). *Time series analysis and its applications: With R examples*. Springer.

Statistics Software. SPSS statistics.

Torres, J. L., Garcia, A., De Blas, M., & De Francisco, A. (2005). Forecast of hourly average wind speed with ARMA models in Navarre (Spain). *Solar Energy, 79*(1), 65–77.

Verde, A. V., Rodríguez, R. C. C., & Rodríguez, A. R. (2018). Evaluación del pronóstico de viento del modelo Weather Research Forecast (WRF) en torres de prospección eólica. *Revista Cubana de Meteorología*, *21*(2), 16–28.

Villarubia, M. J. B. (2012). Ingeniería de la energía eólica.

Windographer Software (2020). Windographer Software. https://www.windographer.com/

Sustainable Architecture from Proper Recycling: Renewable Energy Integration and Housing Automation

Diego X. Morales, Julio Berzosa, and Santiago Moscoso

Abstract

The increase in energy consumption from non-renewable resources causes large amounts of polluting gases, which ones are emitted into the environment, in addition, it is known that low-income people have serious complications to access their own home and basic services that guarantee a decent lifestyle, which is why this work presents a proposal that seeks to change certain paradigms with the introduction of alternative energies, with the implementation of self-sustaining homes that can supply their consumption by significantly reducing the costs of buying electricity. On the other hand, the cost generated during construction can be reduced by using recycled plastic for the manufacture of masonry blocks, also the first results of the automation carried out inside the house are presented, whose function is to regulate the temperature.

Keywords

Sustainable architecture • Solar panel • Telegram • Automation • Recycled plastic

1 Introduction

Energy from fossil resources needs to be replaced with renewable and inexhaustible resources to meet future energy demand. Ecuador has places that are privileged in terms of a resource for electricity generation (Neira González & Velecela Zhindon, 2014). Therefore, the efficient use of energy is one of the most important technological options to face the future, which is not a new theory, since in the early 1970s, the vast majority of industrialized countries adopted energy rationalization policies, to deal with severe increases in oil prices (Ulloa Arizaga, 2015).

Similarly, another great problem in society is related to the difficulty of accessing decent and affordable housing. Hence, the design of this house is oriented to confer a high degree of self-sustainability, capable of meeting the basic needs of the inhabitants autonomously and uninterruptedly, which implies that it contains the necessary resources to generate its energy and the subsequent disposal of waste generated by the fact of inhabiting it (Hodosi & Losada, 2016).

2 Case Study

The house under study is located in the postgraduate headquarters of the Catholic University of Cuenca located in La Estancia Luis Cordero in the city of Cuenca. See Table 1.

Figure 1 depicts the location of the home and a quick view of the campus.

2.1 Materials and Methods

A very efficient technique in obtaining data is observation. "It is a technique that allows obtaining information by registering the characteristics or behaviors of a group of individuals or elements without establishing a communication process and therefore without the need for collaboration by the analyzed collective" (Giler Chango, 2018). As well as the approach and combination of various technologies apply the automation of housing to contribute to improving the housing efficiency parameters. A bibliographic review will

D. X. Morales (✉)
CIITT—Laboratorio de Economia Circular—Ingeniería Eléctrica, Universidad Católica de Cuenca, Cuenca, Ecuador
e-mail: dmoralesj@ucacue.edu.ec

J. Berzosa
Maestria en Energias Renovables, Universidad Católica de Cuenca, Cuenca, Ecuador

S. Moscoso
Ingeniería Eléctrica, Universidad Católica de Cuenca, Cuenca, Ecuador

H. Altan et al. (eds.), *Advances in Architecture, Engineering and Technology*,
Advances in Science, Technology & Innovation, https://doi.org/10.1007/978-3-031-11232-4_17

Table 1 Location of the house

Coordinates	−2.872	−78.918
Elevation	2508 msnm	

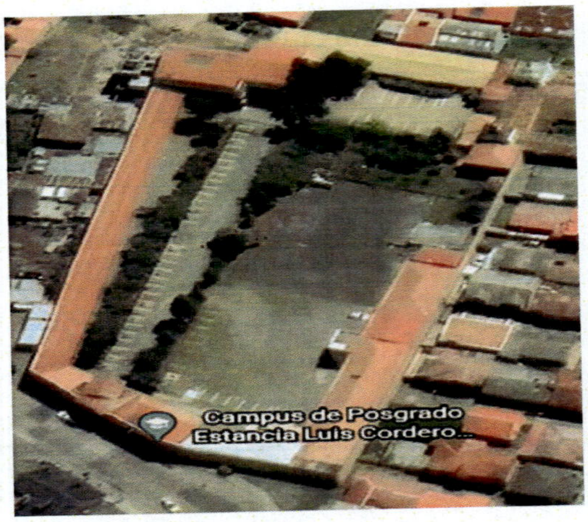

Fig. 1 Geographical location of the home

also be carried out, which will serve to make an adequate selection of criteria that will serve in the sizing and choice of solar panels. Subsequently, the information will be obtained from the solar radiation of the area and the estimated consumption (kWh) of the house, and information collected from the official pages of the Empresa Eléctrica Regional Centrosur S.A. See Fig. 2.

For a home to be self-sustaining, it must be designed and built considering respecting the standards of comfort and quality, it must be practical and economically viable, in addition to having characteristics that respect the environment. To this end, solar technology will be used, considering that it is consolidated worldwide and in Ecuador the solar resource is abundant. According to Sánchez Quiroga (fig), a solar panel is "an electronic device that allows transforming the luminous energy into electrical energy, through the photoelectric effect. The union of these cells gives rise to a photovoltaic panel". The origin of solar radiation can be defined as "it originates from the fusion of hydrogen in the core of the sun, and this causes radiation to spread in all directions without the need for a physical medium for its propagation" (Romero Crespo & Flores Peralta, 2017). The unit is the W/m^2.

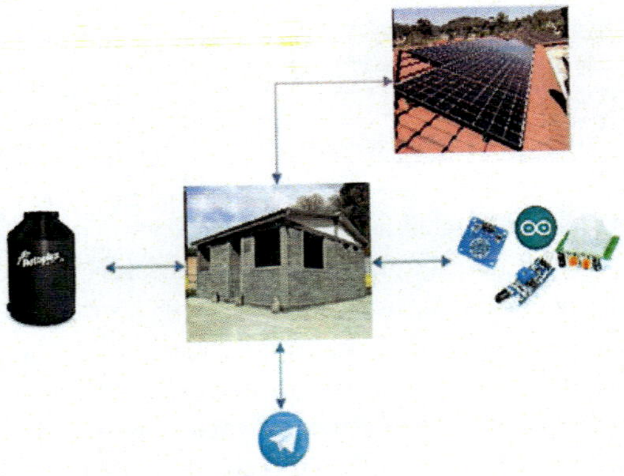

Fig. 2 Proposal for the integration of renewable energy and automation

2.2 Obtaining Solar Radiation Information

Ecuador is in a privileged geographical position so most of the year there is the presence of solar radiation, the solar rays affect perpendicularly on the surface, allowing the use of the resource for power generation. With the help of PHOTOVOLTAIC GEOGRAPHICAL INFORMATION SYSTEM (PVGIS) software, you can obtain radiation and temperature information in the area of influence. This data obtained will be useful in the calculation of the panels that are required to meet the demand of the home. It is necessary to take into account the historical temperature values of the city of Cuenca, and Fig. 3 shows the historical temperatures.

2.3 Panel Calculation

To calculate the solar system, the technical and constructive parameters of the solar modules must be taken into account, see Table 2, in the same way, the climatic conditions of the place where the house is located should be considered.

Table 3 gives the estimated energy consumption in housing built from recycled plastic.

Taking into account the information in the table above, it is determined that the number of panels required is 2. Table 4, you can see the calculations made.

3 Home Automation

For the implementation of the home automation system (see Fig. 4), it is necessary to consider the following criteria.

- Design elaboration: For the elaboration of the design, which is considered the necessary components as well as a strategic location in the house. For this study, the sensors were located in windows, roofs, and the main door.
- Mounting sensors and actuators: The components are placed taking into account the design. "Devices used by the centralized control system, to modify the status of certain equipment or installations In some cases, the

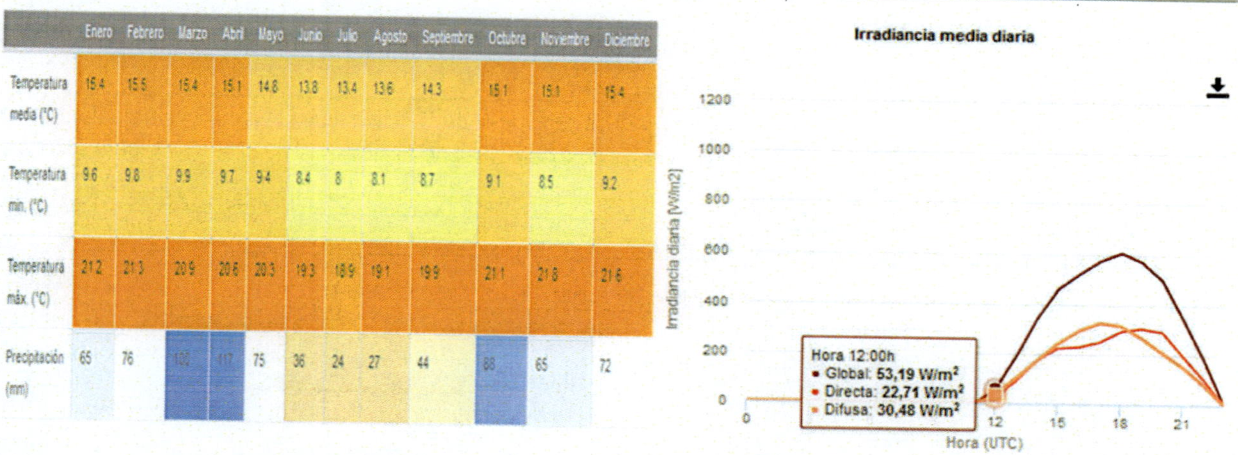

Fig. 3 Table of historical temperatures and radiation values for July in the city of Cuenca (es.climate-data.org)

Table 2 Panel and inverter technical data

Solar Best SE-P230-60	230	W
Voc	36.6	V
Isc	8.42	A
V_{MPP}	29.5	V
I_{MPP}	8.7	A
β%	−0.28	%
δ%	−0.37	%
γ%	0.04	%
Inverter sungrow	300	W
Voc	40	V
Vmin	30	V
Varranque	30	V
V_{MPP}	36–50	V
Imax	7.5	A

Table 3 Calculation of housing consumption

Household appliances consumption	Power (W)	Quantity	Hours of use	Energy (Wh/day)	Energy (kWh/year)
Fridge	60	1	12	720	
Laptop	65	1	4	260	
PC	50	1	5	250	
Lighting	72	1	4	288	
Fans	40	2	3	240	
Total	**287**			**1758**	**641.67**

sensor and actuator are integrated into the same device" (Rodriguez & Fernández, 2012).

- Design: The programming that activates the sensors. In this step, we will select the necessary commands and programs that will serve for the activation of the sensors and components of the circuit, for we will use the Telegram application that will fulfill the messaging function; together with Arduino which is the program selected to execute the different commands.
- Technical test: Multiple tests were conducted to verify the proper functioning of the components in their different applications.
- Low environmental impact system (LID) is a technique that bases its operation on rainwater collection through

Table 4 Panel number

Month	July	February
Temperature	8	21.3
Hour	7:00	13:00
Irradiance (W/m^2)	53.19	682.34
T Panel °C	9.46	40.06
V_{OC}	38.08	36.60
N_{MAX}	1	1
V_{OC} max	38.08	36.60
V_{MPP}	30.69	28.34
N Panels	1	2
V_{MPP} max/min	30.69	56.69
I_{MPP}	8.66	8.74
P_{MAX} Panels (W)	460	

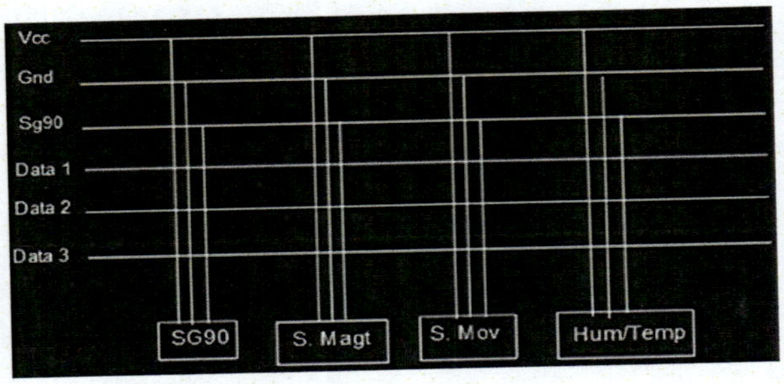

Fig. 4 Automation diagram

calculation processes, roof extension, and the storm level of the area. The collected water is stored in a reserve tank and through a pumping system is distributed to the sanitary batteries and gardening irrigations, optimizing water consumption in the medium and long term.

4 Results and Discussion

The following graphs show the behavior recorded by the temperature and humidity sensors inside the house (see Fig. 5), which will allow us to operate the ventilation and heating mechanisms, allowing us to balance the thermal conditions contributing to a stable environment in temperature and humidity.

As you can see there is a stable temperature behavior on each day of the week. At noon, you can see temperature spikes (about 32 °C) and at night, a drop below 16 °C. It is important to note that the fan installed inside the house is configured to turn on by exceeding 25° to maintain acceptable levels inside the house.

In Fig. 6, you can see how the house is controlled from the Telegram app, in the same way, the daily report generated, the temperature at any time of consultation, and movement alerts.

4.1 Energy Supplied to the House

The house has two solar panels connected in series with an active power of 230 W every one, the energy they deliver to the house is in the order of 839.5 kWh/year, and the consumption of the house is 641.67 kWh/year. That is, there will be a surplus of energy that can be injected into the distribution network to improve economic income. See Table 5.

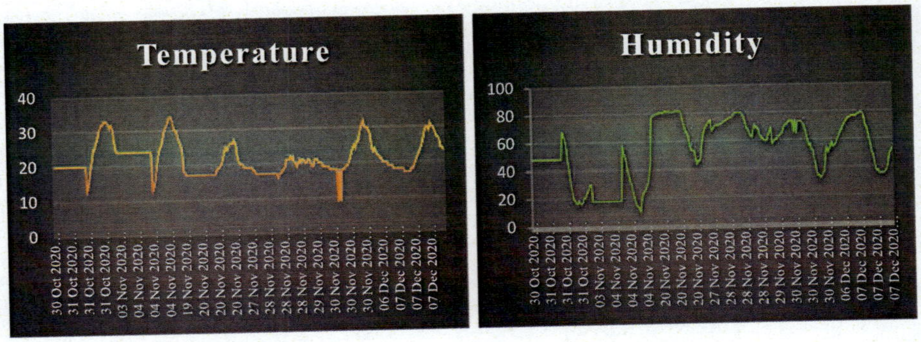

Fig. 5 Temperature and humidity values inside the house

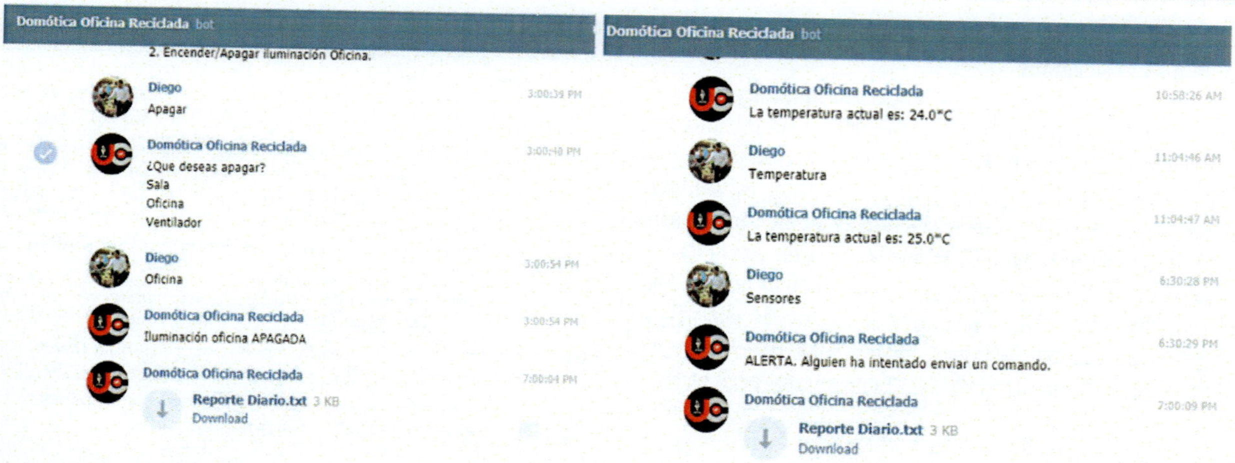

Fig. 6 House sensor control program

Table 5 Data on energy consumption and supply in the home

Home consumption	641.67	kWh/year
Power supplied by the panels	839.5	kWh/year

4.2 Finished House with Panels, Inverter, and Automation

Figure 7 shows the location of the panels (left corner), as well as the arrangement of the automation system inside the house, which consists of a lighting system, humidity and temperature control, window drive mechanisms, and motion sensors.

5 Conclusions

The implementation of self-sustaining housing initially should be focused on low-income people who cannot access their housing and rural sectors where the conventional electricity grid does not have access so that service costs become high. For this reason, an economical, easily accessible, and technologically equipped housing alternative is proposed to minimize construction costs and reduces the negative impact on the environment with the emission of polluting gases.

Today the increase in electricity consumption in households has led to the relentless search for alternatives that suggest savings it, with the advancement of technology in different electrical, electronic devices, and the reduction in cost makes possible the implementation of various automata systems that guarantee a high rate of safety and control of buildings through applications that allow real-time visualization from anywhere in the Internet access is available.

Based on the climatic conditions of the city of Cuenca, it is determined that it is possible to develop sustainable housing in terms of energy, by combining various energy sources, as well as the use of raw material that usually ends up polluting rivers, broken or sent to landfill. The housing proposal has a positive impact, as it raises the level of comfort and safety allowing people to access and monitor

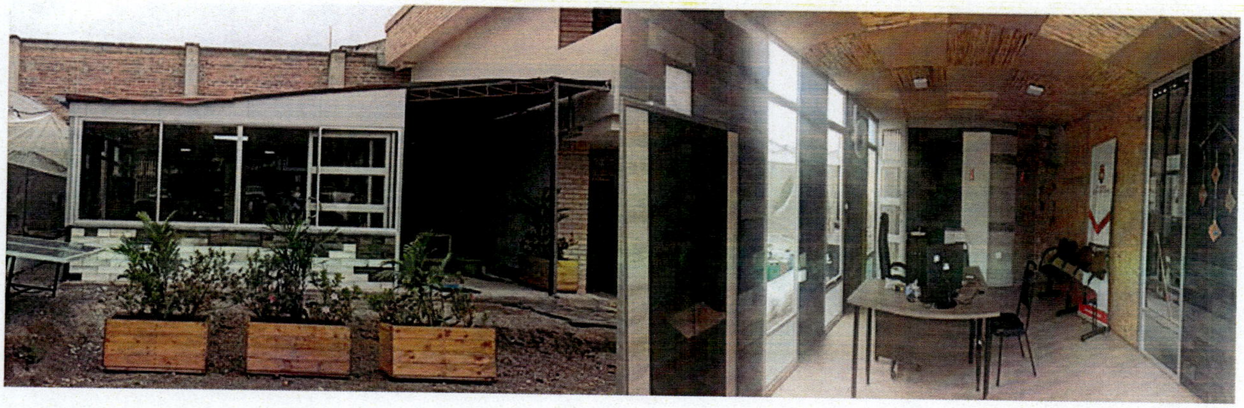

Fig. 7 Main facade and inside view of the house

the different parameters of the house from an application known as a Telegram.

From the first data of temperature and humidity, it can be concluded that the house is comfortable to be inhabited, in the same way, it offers advantages in terms of the cost of construction since savings of up to 30% can be achieved concerning conventional construction systems. Emissions to the interior of the home will be measured in the next phase of the project, and sismoresistence conditions will be checked.

It is necessary to collect data for a longer period and integrate the energy delivered by the solar panels into the monitoring system. Similarly, evaluate energy performance.

References

Giler Chango, J. L. (2018). Sistema automatizado para el control domótico de un a vivienda. *Director, 15*(29), 7577–7588. Retrieved from https://www.uam.es/gruposinv/meva/publicacionesjesus/capitulos_espanyol_jesus/2005_motivacionparaelaprendizajePerspectivaalumnos.pdf; https://www.researchgate.net/profile/Juan_Aparicio7/publication/253571379_Los_estudios_sobre_el_cambio_conceptual_

Hodosi, S., & Losada, D. (2016). *Diseño De Una Vivienda Tipo Autosustentable Con Implementacion De Diversas Tecnologias.*

Mettam, G. R., & Adams, L. B. (1999). How to prepare an electronic version of your article. In B. S. Jones & R. Z. Smith (Eds.), *Introduction to the electronic age* (pp. 281–304). E-Publishing Inc.

Neira González, R. H., & Velecela Zhindon, M. V. (2014). *Estudo de factibilidad de generación eléctrica mediante energía eólica y energía solr fotovoltaica para el sector de Garauzhí de la parroquia Quingeo perteneciente a la ciudad de Cuenca.* 113. Retrieved from http://dspace.ups.edu.ec/bitstream/123456789/6756/1/UPS-CT003484.pdf

Rodriguez, A., & Fernández, M. (2012). *La casa inteligente*. Universidad Carlos Lii de Madril, 8.

Romero Crespo, J. F., & Flores Peralta, J. I. (2017). *Estudio e implementación de un sistema de micro generación solar fotovoltaico para autoconsumo* (caso de estudio considerando la regulación Nro. ARCONEL- 003/18).

Sánchez Quiroga, D. (2012). *Sistema de energía solar fotovoltaica aislado para vivienda unifamiliar aislada*, p. 103. Retrieved from http://hdl.handle.net/10016/16839

Strunk, W., Jr., & White, E. B. (1979). *The elements of style* (3rd ed.). Macmillan.

Ulloa Arizaga, E. S. (2015). *Eficiencia del consumo eléctrico en el sector residencial uirbano de Cuenca.*

Van der Geer, J., Hanraads, J. A. J., & Lupton, R. A. (2000). The art of writing a scientific article. *Journal of Science Communication, 163*, 51–59.

Research in Policy and Management of Urban Underground Space Utilization

Lijun Sun and Jiawei Leng

Abstract

As a kind of space resource, underground space plays an increasingly meaningful role in modern urban construction. The policy and management system of urban underground space utilization is an essential guarantee for underground space construction. This paper aims to sort out the policies and management of international underground space and put forward relevant suggestions for China's underground space management. China has made some achievements in the utilization and practice of underground space. However, there is still a gap between the utilization of underground space in China and the developed countries in terms of law and management. Canada, Japan, the United States, and Finland are selected in this paper as the research scope and specific cases in the four regions are taken as the research object. Combined with specific cases, the characteristics of the four regions in underground space utilization are analyzed from three aspects: planning, regulations and policies, management system. The problems in policies and governance of urban underground space utilization in China are compared and summarized. Finally, suggestions are made for the policy formulation and management system of urban underground space in China: 1. Adhere to the concept of planning first, and integrate underground space into the overall urban planning. 2. Formulate special legislation and auxiliary legislation to clarify the ownership of underground space. 3. Establish an authoritative and efficient management system and management organization.

Keywords

Underground space • Planning • Regulations and policies • Management system

1 Introduction

As a valuable space resource, underground space has received more and more attention in the process of urban development. Urban development, population growth, and further concentration of commerce and industry all require more space to accommodate these activities. Therefore, seeking more space in metropolitan areas is a global phenomenon (Admiraal, 2006). Underground space is more and more popular for commercial activities, transportation systems and storage purposes (Beroggi, 2000) because of its convenient reservoir, additional urban space, natural protection, restraint, and opacity (Godard & Sterling, 1995).

In China, the exploitation and usage of underground space have developed rapidly in the past 20 years (Cui et al., 2013a; Delmastro et al., 2016a). Both the scale and the number are growing rapidly. For instance, from 2007 to 2014, the Beijing subway has expanded by more than 385 km, underground tunnel and subway construction has been increasing speedily, and underground infrastructure construction has been improving. China is at the top of all the statistical categories of underground construction (Kaliampakos et al., 2016).

However, the development and utilization of urban underground space are quite distinct from urban ground space because of its closed space and difficult reconstruction (Xie et al., 2020). Many underground spaces have problems due to a lack of appropriate planning strategies, overlapping layouts and management, resulting in the waste of underground non-renewable resources (Delmastro et al., 2016a). In the meantime, the airtight nature of underground space

L. Sun · J. Leng (✉)
School of Architecture, Southeast University, No.2 Sipailou, Nanjing, 210096, PR China
e-mail: jiaweijw_leng@seu.edu.cn

H. Altan et al. (eds.), *Advances in Architecture, Engineering and Technology*,
Advances in Science, Technology & Innovation, https://doi.org/10.1007/978-3-031-11232-4_18

makes the underground environmental problems related to health urgently in need of improvement (Wen et al., 2020).

Therefore, in order to avoid these problems, improved management systems, planning regulations, and policy applications are needed (Hunt et al., 2016, Xie et al., 2020). In fact, urban underground space has been regarded as a policy issue in many countries (Hunt et al., 2016). This is because the policy directly determines the development direction, use mode, and operation mode of underground space. Therefore, on the basis of the analysis of international mature underground space utilization planning, legal policy, and management experience, this paper puts forward suggestions for the management system of underground space in China.

The methods adopted in this paper are literature reading, case comparison, consultation, and interview with relevant administrative departments. The policy and management of urban underground space utilization in Canada, Japan, the United States, and Finland were studied by literature reading and case comparison, and policy and management in Nanjing (China) were consulted and interviewed by relevant administrative departments.

2 International Experience

2.1 Canada

In order to cope with the harsh climate, Toronto and Montreal in Canada have developed an underground pedestrian system (Cui et al., 2013b; Delmastro et al., 2016b). It provides convenient and fast traffic and living space for the city to cover the wind and rain, block the cold in winter.

Toronto's underground space can be considered as an independent city. The underground is about six blocks wide and ten blocks long. It connects more than fifty office buildings and buildings, six major hotels, two major department stores and more than twenty underground parking lots. It is a complete pedestrian network (Bélanger, 2007).

Montreal has longer underground spaces and more branches, with a total of 20 miles (33 km) of corridors. Montreal has a well-developed underground network, with shops on both sides of the trail. It is spread across the lower floors of the city, and it connects hotels, offices, apartments, train stations and subway stations (Labbé, 2016).

- Planning

In fact, Toronto's underground development didn't start with urban planning (Bélanger, 2007). Similarly, Montreal's so-called indoor city was not initially based on a pre-determined overall development plan. Its expansion was the result of a series of projects implemented by Montreal since 1962 (Jacques, 2007). Montreal has carried out five underground space planning. Three of them were established by the private sector and two by the government. Since there was no officially approved urban master plan in the early days, the continuous network of indoor cities is essentially due to the continuous participation of the government in the development process (Boivin, 1991; Jacques, 2007).

For this reason, the underground space form has been able to continue the earliest planning pattern. It was not until 1992 that Montreal had its first official master plan. Later, the development of Montreal's underground network was linked to the city's master plan. In the new version of the master plan in 2002, the underground pedestrian network was incorporated into the city's overall strategy.

In addition to the master plan for land use, Canada's planning system also includes the implementation of planning intent in urban development through zoning regulations, building permits and other methods (Jacques, 2007).

The government's statutory plan proposes zoning measures to incorporate the underground pedestrian network into the city's development strategy, as part of the overall urban planning for research, and the landowners' federation established by the owners spontaneously is responsible for communicating with the government (Gu & Yu, 2014). Zoning regulations are a kind of legal document that accurately describes the purpose of each piece of land, and puts forward standards for the size of the land, building height, density, retreat distance, parking, and signs (Jacques, 2007).

- Regulations and Policies

Land ownership in Canada includes the space above the ground, the ground itself and underground space, which is quite different from China's land policy. In Canada, the land is actually privately owned, but the public space between the parcels is controlled by the government. And in the actual use of the land, it should be restricted by the zoning and public interests.

- Management System

Due to the land ownership, the Canadian government's management of underground space is primarily reflected in the connection between the various plots. The government coordinates the interests of all parties through policies to complete the management of underground public passages.

Montreal has a unique "underground city" development model, with the cooperation of the private and public sectors (Li et al., 2016a). The important role that the government plays in the construction of underground space is coordination, through more flexible management methods, also known as development agreements. For example, the government can increase the height of buildings to reward developers in order to increase development profitability (Jacques, 2007). In another example, in the process of developing the tunnel, the developer is responsible for the daily maintenance of the underground passage he constructed. The municipal government is mainly responsible for coordinating and adjusting laws and regulations to meet various requirements in the development of ground and underground facilities. In this way, the safety of the public and unobstructed use are ensured. (Bazinet, 2004).

Although the government has not invested in the construction of underground space, this method of coordinating all parties has greatly increased the enthusiasm for urban underground space development. In addition, Montreal has a set of reasonable management mechanisms, such as the landowners' federation. The land owners' federation is responsible for communication and dialogue with the government, independent construction and development, and regulated by the government's policies (Gu & Yu, 2014).

3 Japan

Because of its small land area and dense population, Japan pays excellent attention to underground space development in urban space utilization. Although Japan's underground space started later than Europe and the United States, it has led the underground space to the world's advanced level with its relatively complete legal system, flexible policies, and reasonable management system.

- Planning

Japan's underground space planning includes four parts: the Master plan of underground utilization, the Guide plan of underground utilization, the underground transportation network planning, and the underground street planning. (Yuan et al., 2019) Among them, the Master-plan of underground utilization is a control plan at the overall level, including the planning of urban level facilities such as metro stations and other transportation hubs. While the Guide plan of underground utilization, underground transportation network planning, and the underground street planning are district-level planning, focusing on different aspects (Yuan et al., 2014). In 1991, the Japanese government formulated the "Guidelines for the Compilation of Basic Plans for Underground Public Utilization", the main contents of which include: underground space is a substantial part of urban space, and above ground and underground space planning is equally important (Gu & Yu, 2014). In 1992, the Tokyo Metropolitan Planning Bureau promulgated the "Underground Space Planning of Tokyo Metropolitan Area", which made a further unified plan for the utilization of underground space in Tokyo.

- Regulations and Policies

Due to the pressure of land use and opportunities for urban development, Japan has taken the utilization of underground space as a policy issue of high concern (Sterling et al., 2012). In 2000, the Japanese Diet promulgated the "Deep Underground Utilization Law" (Final revision in 2003). The law guarantees the public attribute of deep underground space for urban services.

The "Deep Underground Utilization Law" promoted legislative changes for more valid underground development. The law allows the utilization of "deep" underground space for public infrastructure (Sterling et al., 2012).

The law applies to the three cities: Tokyo, Osaka, and Nagoya. It defines "deep underground" as (1) the depth at which a basement is not normally constructed (40 m below the ground, the depth of private property is usually within 30–40 m underground) or (2) The space which is 10 m or more deep from the surface of the so-called holding layer. This has resulted in allowing the public to use the deep underground space without having to purchase ground lots in accordance with the government's "basic policy on the use of deep underground". In other words, the connotation of the law is that traditional private land ownership does not interfere with public use in deep underground. The utilization of underground space in shallow and deep layers can be seen in Fig. 18.1 (Chew, 2017; Kishii, 2016).

- Management System

When underground space is used as a section of an urban area, it is a challenge for all administrative departments to promote the integrated management of various administrative districts or regions (Nishida & Uchiyama, 1993). In order to further advance the utilization of underground space, there are still a lot of problems that need to be resolved, not only in terms of construction, but also in cooperation with relevant laws and administrative management. In 1995, the "Special Committee for the Utilization of Deep Underground Space" was established in the Prime Minister's Office to quickly deal with legal and administrative issues related to the promotion of social infrastructure in major cities (Working Group No. 4, 2000).

The "Special Committee for the Utilization of Deep Underground Space" appointed the Minister of land and communications as the president of the agreement, and the heads of the main functional departments of 13 provinces and departments in the cabinet of the government are members of Parliament. To be responsible for the

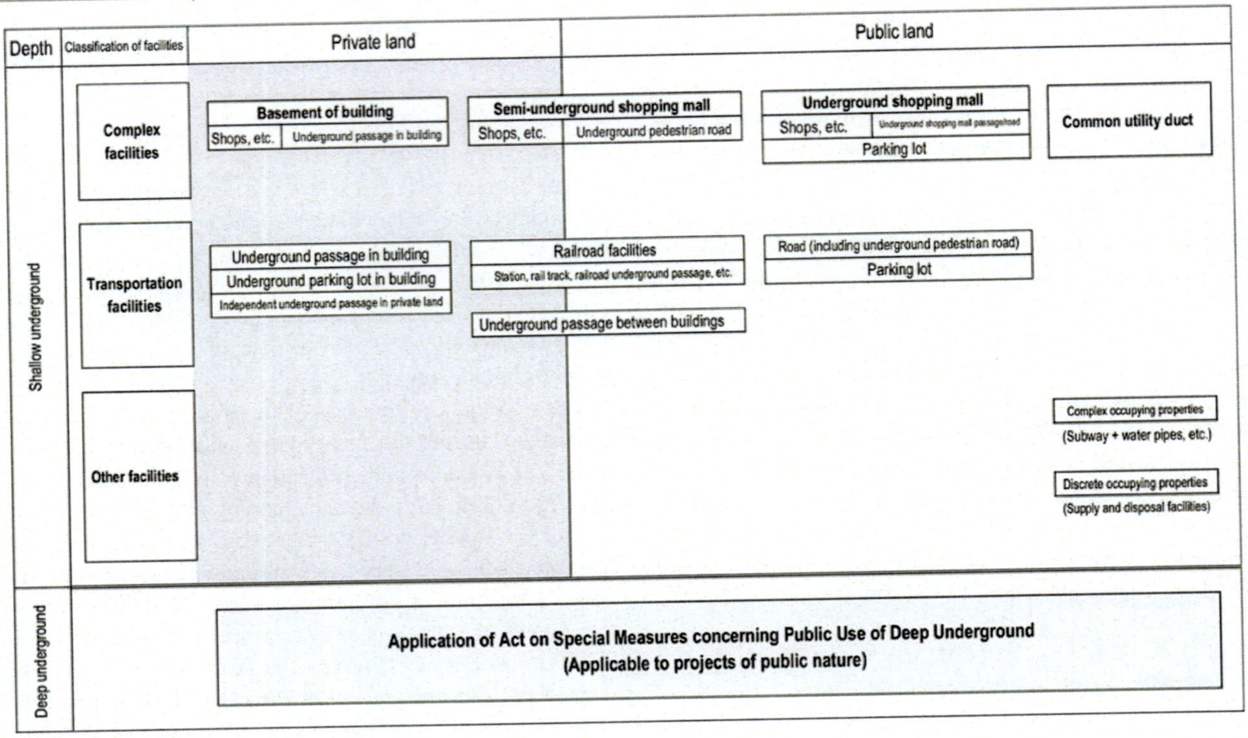

Fig. 18.1 Utilization of underground space in Japan at different depths. *Source* Kishii (2016)

deliberation and decision-making of major national guidelines, policies, and regulations. The Ministry of land and communications is responsible for the organization, coordination and management of the exploitation and usage of underground space throughout the country. Local governments and prefectures are directly responsible for the organization, coordination and leadership of matters related to the exploitation and usage of underground space within their jurisdiction, and set up special functional agencies in the urban planning and construction departments of local governments.

At the same time, a local level "Special Committee for the Utilization of Deep Underground Space" has been set up to be responsible for the deliberation, coordination and supervision of the development and utilization of large-scale underground space within its jurisdiction (Zhang et al., 2011).

In addition, the research on technology is an important means of strengthening management. Underground information database is an important platform for underground space management. A database of map related information for facility, stratigraphic, and soil information is being developed as a section of the development—CALS continuous acquisition and life-cycle support (CALS) system. The study of underground technology is jointly administered by the Ministry of construction and the Ministry of international trade and industry, in cooperation with the private sector, the Japan Tunnel Association, the Japan Society of civil engineering and other organizations (Takasaki et al., 2000).

4 The United States

In terms of climate, metro construction, land use, and economic situation, the reasons for developing underground space in the United States and Japan are different (Cui et al., 2013a). In cities like Chicago, underground space is mainly used to alleviate the congestion of people and vehicles in the high-density central business district.

The Chicago Metropolitan Agency for Planning (CMAP) made a population projection for the go-to 2040 comprehensive regional plan. On the basis of their estimates, the population increase will achieve 380,109 by 2040.The increase requires more space for habitancy, business, entertainment, and other requirements (Zhang, 2012).Therefore, it is particularly important to develop underground space in cities like Chicago to relieve the pressure of traffic, office and parking.

- Planning

There is no separate underground space planning in Chicago, but urban planning combined with the development of urban transportation. Pedway was constructed in 1950s. The aim is to connect the peripheral functions of the city with the subway station. The sidewalk connects about 40 blocks of the CBD and covers about five miles (Zhang, 2012) (Fig. 18.2). Then it opened the prelude of using underground space to build a pedestrian system in Chicago. In the 1960s, people began to focus on the environment of the central area, and compiled the comprehensive plan of Chicago (1966) (Wetmore, 1967) and the transportation planning study of central Chicago (1968). In 1972, according to the planning guidelines provided by the Urban Planning and Development Bureau, the underground pedestrian system of Illinois center was completed; in 1988, the Randolph pedestrian corridor was completed, and Pedway was further expanded (Gu &Yu, 2014).

- Regulations and Policies

The legal system of the United States, Australia and Canada originates from a mutual British legal source (Christie, 2007). However, it also inherited the concept of absolute land ownership in Roman law, holding that 'the owner of the surface also owns the sky and the deep'. After the invention and use of hot-air balloons in the early twentieth century, the legal circles in the United States began to recognize that human beings should have the right to fly in the air beyond the ownership of land. It is necessary to establish the principle of limited land ownership, which lays a theoretical foundation for the three-dimensional development and utilization of urban land in the United States. In the United States, the development of technology has led to the increasing exploitation of underground mineral resources, natural gas and other resources, and also triggered a debate about the restrictions on the rights of surface landowners to underground resources (Chew, 2017).

- Management System

In 1966, in order to adapt to the large-scale and fast development of the transportation industry, President Johnson proposed the establishment of the Department of transportation, which was formally established in April of the next year with the approval of Congress. Thus, the traffic management functions and related affairs originally scattered in eight ministries and departments, such as the Department of Commerce and the Department of finance, were centralized, and a unified management mode was established (Zhang et al., 2011).

Zoning is a significant basis for urban development control in the United States, which regulate the function, building types and development intensity of land. For the construction of the Pedway, zoning has clear requirements. The zoning of the United States stipulates the project approval process and floor area ratio reward procedure for

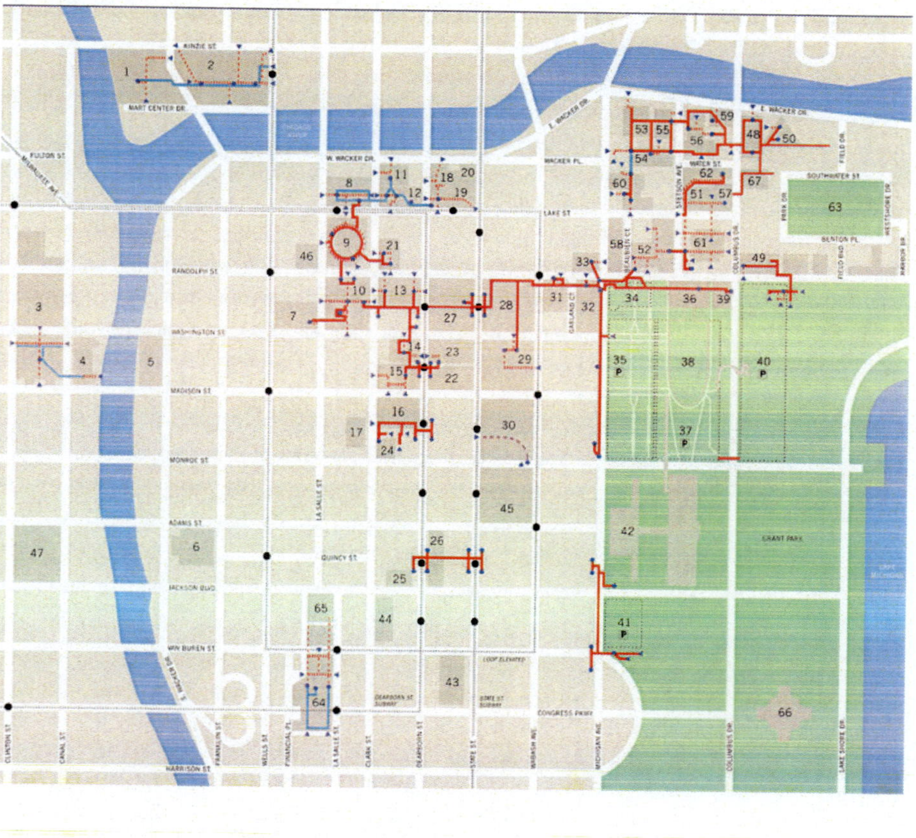

Fig. 18.2 Plan of Pedway in Chicago. *Source* https://www.chicago.gov/city/en.html

submitting the scheme first. For example, according to the Chicago zoning, the plot ratio for the construction of Pedway can be increased by 2.0% on the basis of the basic plot ratio. Developers can enjoy this preferential policy if they meet the relevant conditions.

4.1 Finland

Northern Europe has good geological conditions and is an advantageous area for the exploitation of underground space, especially in municipal facilities and public buildings. Since 1960, Helsinki has been adept at using the underground on a large scale. Its underground space has reached 10 million cubic meters. Its functions include the parking lot, sports facilities, oil and coal storage, subway, etc. (Li et al., 2016b).

- Planning

There are similarities between Finland and China in re-planning. Finland carried out research on "underground space planning and land use" in the 1990s. This project comprehensively reviews the current circumstance of urban underground space planning in Finland, and discusses the different levels of underground space planning and underground space construction permission procedures.

In Finland, the following several different scenarios are used to estimate when underground space needs to be developed with detailed development planning (Working Group No. 4, 2000) (Table 18.1).

In Helsinki, the master plan of underground space is compiled by the municipal planning department, which is an important part of the overall land-use planning. Helsinki's underground space master plan not only provides a framework for underground space construction, but also reserves space for long-term public utilities.

In 2012, the city planning department compiled and completed the "The underground master plan of Helsinki", including underground rock resource sorting and special facility planning (Vähäaho, 2011).

- Regulations and Policies

In Finnish legislation, there is not an exact description about the scope of land ownership-either upward or downward. The property use rights and land ownership differ from each other. The lower limit of the right to use property has been confined to the depth that can be used technically (Vähäaho, 2016). In fact, from a depth of 6 m below the building site, this acknowledges the development of public underground facilities below this depth without paying for space use fees (Sterling et al., 2012).

Finland's Ministry of the environment is an important management and guidance department for the planning and construction of underground space. "Environmental impact assessment" regulations focus on the impact of underground engineering on the environment in the process of construction and use, guiding and restricting developers to carry out underground engineering construction and utilization. Its large-scale underground caverns are mainly used to store important energy and build large-scale urban infrastructure.

If the owners of the land want to construct multiple underground floors on their building, they must have a construction license; on the other hand, the right to build a cellar must conform to the zoning. Therefore, the problem in the actual use of underground space is not the land ownership, but the right of the land construction usage (Vähäaho, 2016). With the guarantee of regulations and policies, the master plan for the space under the Helsinki base can be better pushed forward and implemented.

- Management System

In Finland, the Ministry of Environment designated a committee to examine the extant underground building planning system in 1988 (Working Group No. 4, 2000). The committee is responsible for investigating how underground space planning connects with urban planning and determining the planning system specifically needed for underground space development. In 1990, the committee made a report: it is better to adopt the existing laws for the

Table 18.1 When should underground space have a detailed development plan

• The project is significant in terms of scope and extent;
• People will be working full-time in the space;
• Underground construction will affect local housing, workplace and transport arrangements as well as the rights of the people in the area;
• Surface connections to the underground space affect traffic or parking arrangements;
• Underground construction has significant consequences with regard to arrangements above ground level;
• Underground construction is directly connected with the surface structures;
• Underground construction causes significant environmental impacts during the construction stage and during use (e.g. traffic emissions and exhaust gases from sewerage treatment plants).

Source Working Group No. 4 (2000)

construction of underground space facilities, expand and clarify the existing laws, establish a planning system for the development and utilization of underground space, obtain the right of underground space with the consent of the government, and establish a compensation system for the possession of underground space (Zhang et al., 2011).

5 Results

From the above case analysis, it can be seen that in these four areas, whether the planning of underground space has a strong effect depends on whether it has legal benefits. Different countries have adopted different management systems and management strategies for underground space because of different legal backgrounds. Therefore, in underground space planning, legal policies, and management systems, legal policies are the core. In terms of legal policies, the central issue is the study of underground ownership, including the ownership, use rights, and management rights of above and underground land. The case of Finland shows that not only ownership should be paid attention to, but in the use of underground space, the right to use is often more important than the ownership. In countries with sound legal systems, more attention is paid to rights. Therefore, the rights and interests of stakeholders can be effectively protected and underground space can be well developed.

Compared with Japan and other countries with more sound laws and regulations, China's systematic laws and regulations on underground space are still in the process of exploration. However, different from capitalist countries with private ownership as the core, in China, land ownership belongs to the nation, and underground space planning is a means for the Chinese government to manage and regulate the exploitation of underground space resources at a macro level. Therefore, the different ownership of underground space also brings the difference of management system.

In China, the administrative department of construction under The State Council is responsible for the exploitation and usage of urban underground space throughout the country. The construction administrative departments of the people's governments of provinces and autonomous regions shall be responsible for the administration of the exploitation and usage of urban underground space within their respective administrative regions. The administrative departments of construction and the administrative departments of city planning under the people's governments of municipalities directly under the Central Government, cities and counties shall, according to their respective functions and duties, be responsible for the administration of the exploitation and usage of urban underground space within their respective administrative areas.

Different cities in China have different forms of underground space exploitation and utilization organization and management. Taking Nanjing city as an example, the underground space development and utilization is mainly coordinated and organized by the Civil Air Defense Office (Fig. 18.3).

6 Suggestions

Through the above discussion, the paper can draw the following enlightenment for the underground space planning and governance in China.

1. Adhere to the Concept of Planning First, and Integrate Underground Space into the Master Plan.
   Currently in China, local governments usually develop underground spaces based on a master plan of physical space. Master planning is a major feature of China's planning system. In fact, only by incorporating underground space into the overall urban planning instead of independent underground space planning, can underground space be better integrated and developed with ground space. At the same time, in the planning process, we must adhere to the public interest and strengthen public participation in the formulation.
2. Formulate Special Legislation and Auxiliary Legislation to Clarify the Ownership of Underground Space.
   China's existing laws do not have many regulations on underground space. Although some cities are trying to legislate, it is more difficult to compound legislation. It is possible to clarify the related issues of underground ownership by choosing to formulate special legislation and auxiliary legislation.

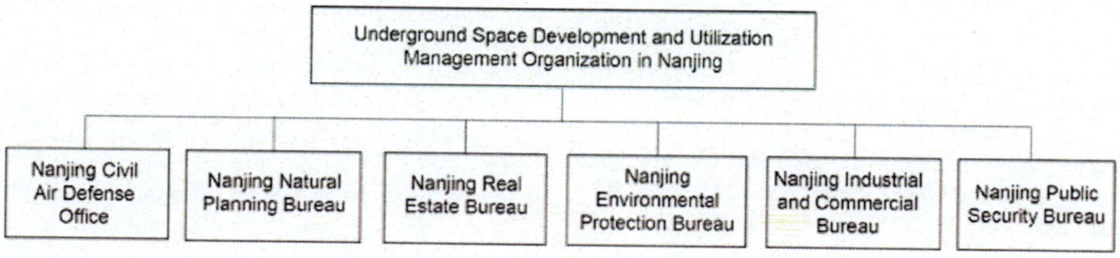

Fig. 18.3 Underground Space Development and Utilization Management Organization in Nanjing

3. Establish an Authoritative and Efficient Management System and Management Organization.
The development and utilization of underground space require efficient management means, and the core of management is coordination. It can be seen from the Canadian case that the nature of coordinating all parties is actually a kind of balance. The government needs to balance the interests of all parties on the premise of ensuring the public interest. This means that mechanisms such as the sharing mechanism of underground space management information, the monitoring mechanism for the safe use of underground engineering, and the investment and financing mechanism of underground space need to be established to stimulate and guarantee the development of underground space. A special underground space management committee shall be set up to coordinate the interests of all parties.

References

Admiraal, J. B. M. (2006). A bottom-up approach to the planning of underground space. *Tunnelling and Underground Space Technology, 21*(3), 464–465. https://doi.org/10.1016/j.tust.2005.12.102

Bazinet, G. (2004). The role of Montreal municipal government in development and underground pedestrian system. *Underground Space, 4*, 126–128.

Bélanger, P. (2007). Underground landscape: The urbanism and infrastructure of Toronto's downtown pedestrian network. *Tunnelling and Underground Space Technology, 22*(3), 272–292. https://doi.org/10.1016/j.tust.2006.07.005

Beroggi, G. E. (2000). Integrated safety planning for underground systems. *Journal of Hazardous Materials, 71*(1–3), 17–34. https://doi.org/10.1016/s0304-3894(99)00070-9

Boivin, D. J. (1991). Montreal's underground network: A study of the downtown pedestrian system. *Tunnelling and Underground Space Technology, 6*(1), 83–91.

Chew, E. (2017). Digging deep into the ownership of underground space recent changes in respect of subterranean land use. *Singapore Journal of Legal Studies, 1*, 1–17.

Christie, D. R. (2007). A tale of three takings: Taking analysis in land use regulation in the United States, Australia, and Canada. *Brooklyn Journal of International Law, 32*(2), 343.

Cui, J., Allan, A., Taylor, M. A. P., & Lin, D. (2013a). Underground pedestrian systems development in cities: Influencing factors and implications. *Tunnelling and Underground Space Technology, 35*, 152–160. https://doi.org/10.1016/j.tust.2012.12.009

Cui, J., Allan, A., & Lin, D. (2013b). The development of grade separation pedestrian system: A review. *Tunnelling and Underground Space Technology, 38*, 151–160. https://doi.org/10.1016/j.tust.2013.06.004

Delmastro, C., Lavagno, E., & Schranz, L. (2016a). Energy and underground. *Tunnelling and Underground Space Technology, 55*, 96–102. https://doi.org/10.1016/j.tust.2015.10.021

Delmastro, C., Lavagno, E., & Schranz, L. (2016b). Underground urbanism: Master Plans and Sectorial Plans. *TUNNELLING AND UNDERGROUND SPACE TECHNOLOGY, 55*, 103–111. https://doi.org/10.1016/j.tust.2016.01.001

Godard, J. P., & Sterling, R. L. (1995). General considerations in assessing the advantages of using underground space. *Tunnelling and Underground Space Technology, 10*(3), 287–297. https://doi.org/10.1016/0886-7798(95)00018-T

Gu, X., & Yu, W. (2014). *Planning and management of urban underground space utilization*. Southeast University Press. (in Chinese).

Hunt, D. V. L., Makana, L. O., Jefferson, I., & Rogers, C. D. F. (2016). Liveable cities and urban underground space. *Tunnelling and Underground Space Technology, 55*, 8–20. https://doi.org/10.1016/j.tust.2015.11.015

Jacques, B. (2007). A master plan or a regulatory approach for the urban underground space development: The Montreal case (B. Zhang, Trans.) *Urban Planning International, 6*, 16–20 (in Chinese).

Kaliampakos, D., Benardos, A., & Mavrikos, A. (2016). A review on the economics of underground space utilization. *Tunnelling and Underground Space Technology, 55*, 236–244. https://doi.org/10.1016/j.tust.2015.10.022

Kishii, T. (2016). Utilization of underground space in Japan. *Tunnelling and Underground Space Technology, 55*, 320–323. https://doi.org/10.1016/j.tust.2015.12.007

Labbé, M. (2016). Architecture of underground spaces: From isolated innovations to connected urbanism. *Tunnelling and Underground Space Technology, 55*, 153–175. https://doi.org/10.1016/j.tust.2016.01.004

Li, H. Q., Li, X. Z., & Soh, C. K. (2016a). An integrated strategy for sustainable development of the urban underground: From strategic, economic and societal aspects. *Tunnelling and Underground Space Technology, 55*, 67–82. https://doi.org/10.1016/j.tust.2015.12.011

Li, W., Chen, Z. L., & Guo, D. J. (2016b). Foreign urban underground space planning: A case study of Helsinki. *International Urban Planning, 31*, 119–124. (in Chinese).

Nishida, Y., & Uchiyama, N. (1993). Japan's use of underground space in urban development and redevelopment. *Tunnelling and Underground Space Technology, 8*(1), 41–45.

Sterling, R., Admiraal, H., Bobylev, N., Parker, H., Godard, J., Vähäaho, I., Rogers, C. D. F., Shi, X., & Hanamura, T. (2012). Sustainability issues for underground space in urban areas. *Proceedings of the Institution of Civil Engineers - Urban Design and Planning, 165*(4), 241–254. https://doi.org/10.1680/udap.10.00020

Takasaki, H., Chikahisa, H., & Yuasa, Y. (2000). Planning and mapping of subsurface space in Japan. *Tunnelling and Underground Space Technology, 15*(3), 287–301. https://doi.org/10.1016/S0886-7798(00)00057-2

Vähäaho, I. (2011). *Helsinki experience with master planning for use of underground space*. Proceedings of the Joint HKIE-HKIP Conference on Planning and Development of Underground Space. The Hong Kong Institution of Engineers & The Hong Kong Institute of Planners, pp. 23–24, 1–9.

Vähäaho, I. (2016). An introduction to the development for urban underground space in Helsinki. *Tunnelling and Underground Space Technology, 55*, 324–328. https://doi.org/10.1016/j.tust.2015.10.001

Wen, Y. M., et al. (2020). Environmental and health effects of ventilation in subway stations: A literature review. *International Journal of Environmental Research and Public Health, 17*(3), 1084. https://doi.org/10.3390/ijerph17031084

Wetmore, L. B. (1967). The comprehensive plan of Chicago prepared by the city of Chicago department of development and planning, December 1966. *Journal of the American Institute of Planners, 33* (5), 353–359. https://doi.org/10.1080/01944366708977942

Working Group No.4. (2000). Planning and mapping of underground space—an overview. *Tunnelling and Underground Space Technology, 15*(3), 271–286. https://doi.org/10.1016/s0886-7798(00)00056-0

Xiao, J. (2008) *Study on the legal system of urban underground space utilization*. Intellectual Property Press (in Chinese)

Xie, H. P., et al. (2020). A case study of development and utilization of urban underground space in Shenzhen and the Guangdong-Hong Kong-Macao Greater Bay Area. *Tunnelling and Underground Space Technology, 107*, 103651. https://doi.org/10.1016/j.tust.2020.103651

Yuan, H., He, Y., & Wu, Y. (2019). A comparative study on urban underground space planning system between China and Japan. *Sustainable Cities and Society, 48*, 101541.https://doi.org/10.1016/j.scs.2019.101541

Yuan, H., Zhao, W. M., & Zhao, S. C. (2014). Analysis of underground space utilization planning system in Japan. *Urban Development Studies, 21*, 112–118. (in Chinese).

Zhang, H. J. (2012) *Developing underground space to decrease urban sprawl: a case study of chicago* (master's thesis). Ball State University.

Zhang, J. C. et al. (2011) *Research on management mechanism and operation guarantee system of urban underground space development and construction*. Beijing: Science Press (in Chinese).

https://www.chicago.gov/city/en.html

A Chronological Exploration of Initiatives in an Automobile Cluster: A Case of Pithampur, Madhya Pradesh, India

Pratyoosh Madhavi, Binayak Choudhury, and Ankit Kumar

Abstract

Industrial clusters are often used as a tool for economic development. The world follows two models under strategies to support industrial clusters. One is Cluster-Oriented Economic Development Plans, much followed in western countries with an intent to integrate cluster strategies into broader Economic Development Plans. The other is Cluster initiatives which are more focused and tailored for specific clusters and are majorly followed in Asian countries. In India, there are many Industrial clusters. Some have historic origins and some are developed deliberately by the government to boost economy of an undeveloped region. Depending on Cluster's needs, the objectives of the Cluster initiatives differ for every Industrial Cluster. Pithampur, in Madhya Pradesh, the case discussed in this research paper, has been developed as an automobile hub in Central India. The development started in the year 1983 in a backward area consisting of villages with an agrarian economy. Apart from proximity to Indore which is a commercial capital of Madhya Pradesh and being logistically ahead, as is in Central India, Pithampur had nothing else as an endowed advantage. The automobile industry was chosen for development considering it to be one of the potential industries to leverage the regional economy. As Pithampur lacked basic infrastructure and had poor economies of scale to attract industries, a lot of effort has been consistently made by the state government to develop and grow it as an industrial town. Though initially, the cluster faced an unfavourable industrial scenario, this cluster survived three and a half decades and has attained a good status as an automobile manufacturing hub in India. Pithampur has seen a complete transformation from agriculture to an industrial economy. The objective of this research is to study the chronological development of Pithampur Automobile Cluster since its initiation, taking into account the different initiatives taken at national and state levels which have contributed directly and indirectly in development of the cluster. The methodology relies on extracting evidences from archival records and expert opinion surveys. This study documents evolution of an automobile cluster in a location that was initially unsuitable for industrial development.

Keywords

Pithampur automobile cluster • Cluster Initiatives • Chronological development • Industrial growth

1 Introduction

There are many forms of industrial clusters and they are present throughout the length and breadth of the habitable world. The performance of these clusters depends on the various interventions made at different levels. These interventions range from industry-led initiatives like collaborations with other industries or Institutes or government-level cluster interventions through policies and schemes made for growth and development of the cluster. Broadly for industrial cluster development the western part of the world follows market-driven policies in contrast to the Asian countries following model led by their government which usually focuses on specific industrial growth (Ketels, 2015a). The approach towards cluster development can be either top-down or bottom-up. The former implies major efforts by government authorities in the form of explicit cluster policy followed and implemented by local and regional authorities. The latter happens when major concerned industries themselves without the support of government lead their way through implicit initiatives (Roelandt et al., 1999). United

P. Madhavi (✉) · B. Choudhury
School of Planning and Architecture, Bhopal, India
e-mail: pratyoosh.madhavi@gmail.com

A. Kumar
School of Planning and Architecture, Delhi, India

H. Altan et al. (eds.), *Advances in Architecture, Engineering and Technology*,
Advances in Science, Technology & Innovation, https://doi.org/10.1007/978-3-031-11232-4_19

States has explored and exploited its industrial clusters to their optimum. It houses many successful clusters and has made tremendous efforts in catalyzing its development. In US there are three layers of government that governs these clusters, i.e. Federal, State or Regional and Local. Federal government acts as facilitator to State when it comes to economic development. It permits states to seek after their own economic development (ED) approaches as per current demands and accessible resources funds (Crec, 2015). Local governments also play a major role as they involve mayors and their administrations in different aspects of cluster initiatives (Baily & Montalbano, 2017). The success of industrial clusters in United States is an outcome of immense support of Federal, State and local governments. Since 2010, range of programmes has been introduced by the Federal government to support clusters. Some Federal agencies like Economic Development Administration (EDA) and the Small Business Administration (SBA) have collaborated for the same (Lindqvist et al., 2013).

Though European clusters are less prominent than the American ones, they too have many successful stories (Sopoligová & Pavelková, 2017). Europe has completely changed their strategy for dealing with industrial clusters. Before 2000 European countries took charge of cluster development through cluster organizations and national and regional cluster programmes, but now European Union (EU) has taken a lead and now the cluster facilitators have become managers contributing in the whole process of cluster development (Lindqvist et al., 2013). The EU offers help to clusters in various ways and incorporates them in some of their key policies. Most EU member countries and numerous EU sub-national regions have committed to cluster programmes. There are more than 2000 cluster organizations as reported by Cluster Observatory (Ketels, 2015b). This has resulted in the establishment of a number of programmes and initiatives like the PRO INNO Europe, the Europe INNOVA initiative, the Regions of Knowledge, etc. At national level, European countries manifest a good confluence of cluster approach with their economic policy. For example, Netherlands and Denmark have always employed clusters in their policy. Similarly, Ireland, United Kingdom, Portugal, Sweden, Finland, Germany and many other countries have made active cluster development efforts (Ketels, 2004).

'Asian cluster policies are as diversified as the Asian economies' (Mazurek, 2014). China, which is the world's biggest manufacturer follows a top-down approach for Industrial cluster development. Industrial clusters are integral part of industrial policy in China (Sopoligová & Pavelková, 2017). There are specialized towns that work as cluster for a particular product like socks, sweaters, kid's clothing, Footwear and so on. Japan too follows top-down approach for implementation of cluster policy. Cluster policy is well incorporated in the economic and industrial policy of Japan. Japan has nurtured many industrial clusters which came naturally because of some of the other factors. For example, Fukui and Ishikawa prefectures form a large cluster of textile manufacturers. Gifu City has worldwide recognition for apparels. Seto City excels in ceramics goods. Morodomi in Saga prefecture is currently a part of the furniture cluster in the region and similarly, Ota City is known for its automobile parts cluster since 1918 (Yamawaki, 2002). In India too, there are many clusters that fall under different cluster typologies. Some are locally concentrated whereas some are widely spread in a geographical area. Some clusters are dense in the sense that they handle large volumes or business through many small no. of firms running within it. In the contrary, some clusters portray sparse character and have few firms. According to Cluster Observatory, India website India has around 2488 Handicraft clusters, 1300 SME clusters and 566 Handloom clusters. Indian post-colonial industrialization history reveals that there were sporadic State level and Institute level initiatives to support industrial clusters till 2007 but substantial efforts were made after the coming of eleventh Five Year Plan (2007–2011). It gave much thrust to the development of industrial clusters. Since then a lot of schemes and programmes at State and Central level have been launched for industrial cluster development. They are in forms of various initiatives concerned with technical aid, technology advancement and marketing assistance to provide a competitive edge to Micro, Small & Medium Enterprises units in the global ecosystem. Industrial Infrastructure Upgradation Scheme (2003), Mega Food park Scheme (2008), Scheme for Integrated Textile Parks (2005), Scheme for Development of AYUSH Clusters (2007), Scheme of Fund for Regeneration of Traditional Industries (2005), Micro & Small Enterprise Cluster development programme (1998 renamed in 2007) are few of the major initiatives taken for Industrial cluster development in India. United Nations Industrial Development Organization's (UNIDO) has also played a major role in development of some selected industrial clusters in India. Figure 1 shows timeline of major national-level cluster initiatives taken in India for industrial cluster development. India also has many clusters which have been planned at different locations under decentralization or backward area development drive considering their potential to leverage local and regional economy. To a developing country like India which faces issues of unemployment, high rural–urban migration, shortage of housing, burdening of urban infrastructure, Planned Industrial Clusters at remote rural locations seem to solve such issues at the source. Pithampur automobile cluster (PAC), chosen for this study is one such case of planned industrial cluster in Madhya Pradesh developed in 1983 under Growth Centre Approach. The state of Madhya Pradesh displays agrarian status and is also characterized being industrially

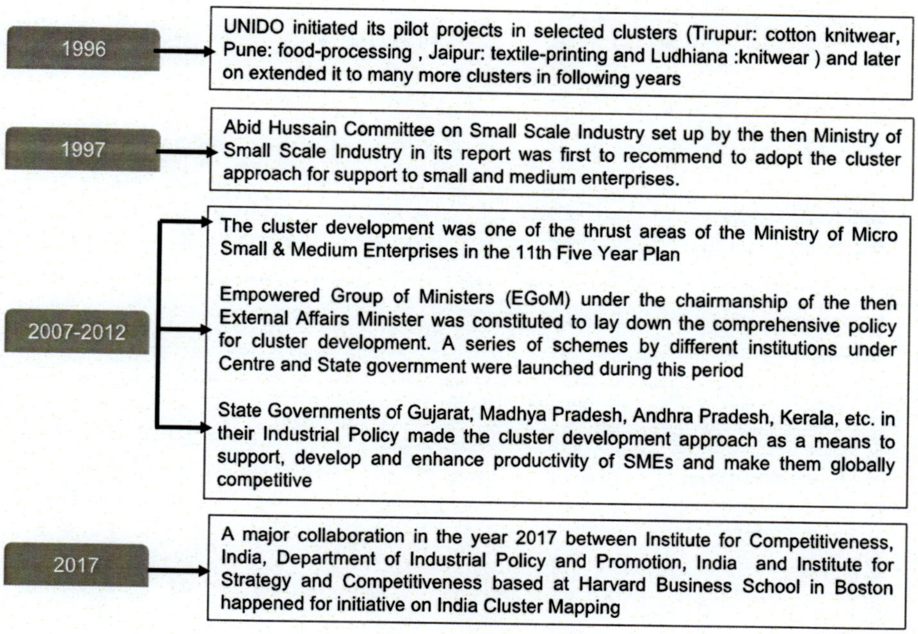

Fig. 1 Timeline of major National-level cluster initiatives, India. *Source* Generated by the authors

backward (Jalaja, 2005). Agriculture sector in Madhya Pradesh forms the backbone of its economy. For promotion of industries, Madhya Pradesh Industrial Development Corporation (MPIDC) has adopted the growth centre approach to promote and foster balanced development of industries. There are 26 industrial growth centres under different levels of development. Madhya Pradesh has developed parastatal agencies which facilitate industrial development in these growth centres falling in their jurisdiction. Development of Pithampur as an industrial town within an agrarian state has been a triumph for the state government. State industrial policies, Central government schemes and development of major infrastructure by Central government institutions and role of parastatal agency as an active stakeholder in Pithampur's development are major reasons behind Pithampur's transformation. This paper attempts to document the chronological evolution of Pithampur automobile cluster stressing on major landmark initiatives and developments which have happened in the past three and a half decades.

2 Pithampur: An Emerging Industrial Town

Through the growth centre approach, Pithampur has sustained three and a half decades of industrial development. A lot of industries have developed in Pithampur since its initiation. Most of them have developed in the form of clusters. There is total of 1402 industries in Pithampur with 42 large scale, 71 medium and 1289 small-scale industries. Automobile and Pharmaceuticals are the two major clusters that have developed in Pithampur. Figure 2, shows the timeline of industrial growth in Pithampur. It also indicates the dominance of automobile cluster in the region (MPAKVN, 2020a, b).

Automobile cluster is the largest cluster with many large, medium and small-scale industries. The cluster is specialized in commercial vehicles like Light/Heavy/ Ultra Heavy Commercial Vehicles, Two Wheelers, Earthmovers, etc. As the cluster depicts a Hub-and-Spoke model (Markusen, 1996a, b) there are many ancillaries that have developed around the major anchor industries. There are 15 Large scale, 22 Medium and 360 small-scale industries in the automobile cluster (MPAKVN, 2020a, b). The Giants which have been established in Pithampur are V. E. Commercial Vehicles Pvt Ltd., Force Motors Limited, Larsen & Toubro Limited, Liugong India Pvt. Ltd. and Bridge Stone, etc. With time these industries have expanded in production, area and are the highest employment generators in the region.

Figure 3, represents the growth of large and small industries within Pithampur automobile cluster. The graph shows that the mother units (Large scale units) in the automobile units have attracted many ancillaries (medium and small-scale units) leading to development of linkages in the cluster. The automobile cluster has witnessed many kinds of changes that have happened within the industry. Expansion of mother units, coming up of ancillaries and change of signboards due to change in collaborations are few to mention.

Along with industrial development, Pithampur has also witnessed a sweeping change in demographic figures as seen in Table 1. The Census data reveals a high annual population growth rate of 46.8% from 1991 to 2001 whereas for the district and the state it remained 2.7% and 2.4%,

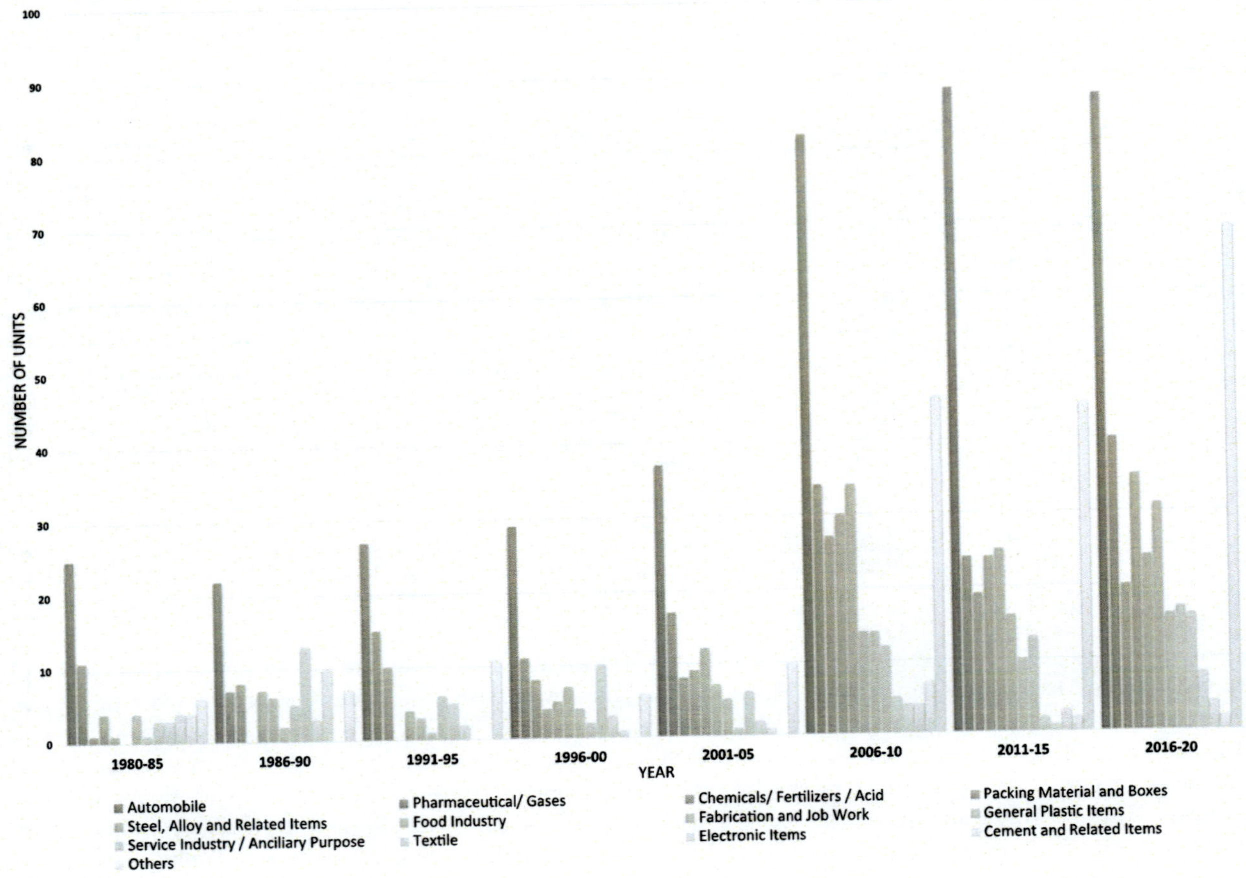

Fig. 2 Timeline of industrial growth in Pithampur. *Source* Generated by the authors

respectively. Industries also brought change in occupational structure of Pithampur. In 2011 around 84% working population fell under the category of other workers (those who had been engaged in some economic activity during the year preceding numeration and who were not cultivators or agricultural labourers or household industry workers) whereas in 1991 and 1981 it was 42.5% and 8.2%, respectively. The immigration of male migrant workers has also brought the sex ratio down to 796 in 2011 (Census of India, 1991, 2001 & 2011). Much transformation is anticipated in this region after development of Special Investment Region (SIR) under Delhi Mumbai Industrial Corridor (DMIC) whose early bird projects are in progress.

3 Chronological Development of Pithampur Automobile Cluster

3.1 Initial Years: 1984–1989

Strong 'Political will' played a major role in the development of Pithampur automobile cluster. It was Chief Minister of Madhya Pradesh, Arjun Singh, who pushed for auto complex in Pithampur in the early 1980s. At the time of its initiation, the district Dhar, where the industrial area lies owned an identity bearing high tribal population which was devoid of any industrial development. The State government through policies offered attractive incentives and subsidies to industries to invite them to Pithampur. Tax Holiday of 9 years was also offered to industries by the government. Small units had Central investment subsidy of 25% upto 25 lakhs. Electricity subsidy was provided by the government in the initial years to the industries. Land on lease was also given to industries for a period of 90 years.

To start the development, the government had certain goals to accomplish. Through the parastatal agency, i.e. AKVN Indore, land was made available to industries through land acquisition. The land was taken over by the government from the farmers who were given monetary compensation along with one job per family. Those who came got job but most of them used compensation money for starting some or the other kind of businesses. Some brought dumpers, trucks, tractors and started venturing into different businesses. (Kothari, 2019). The second goal was to

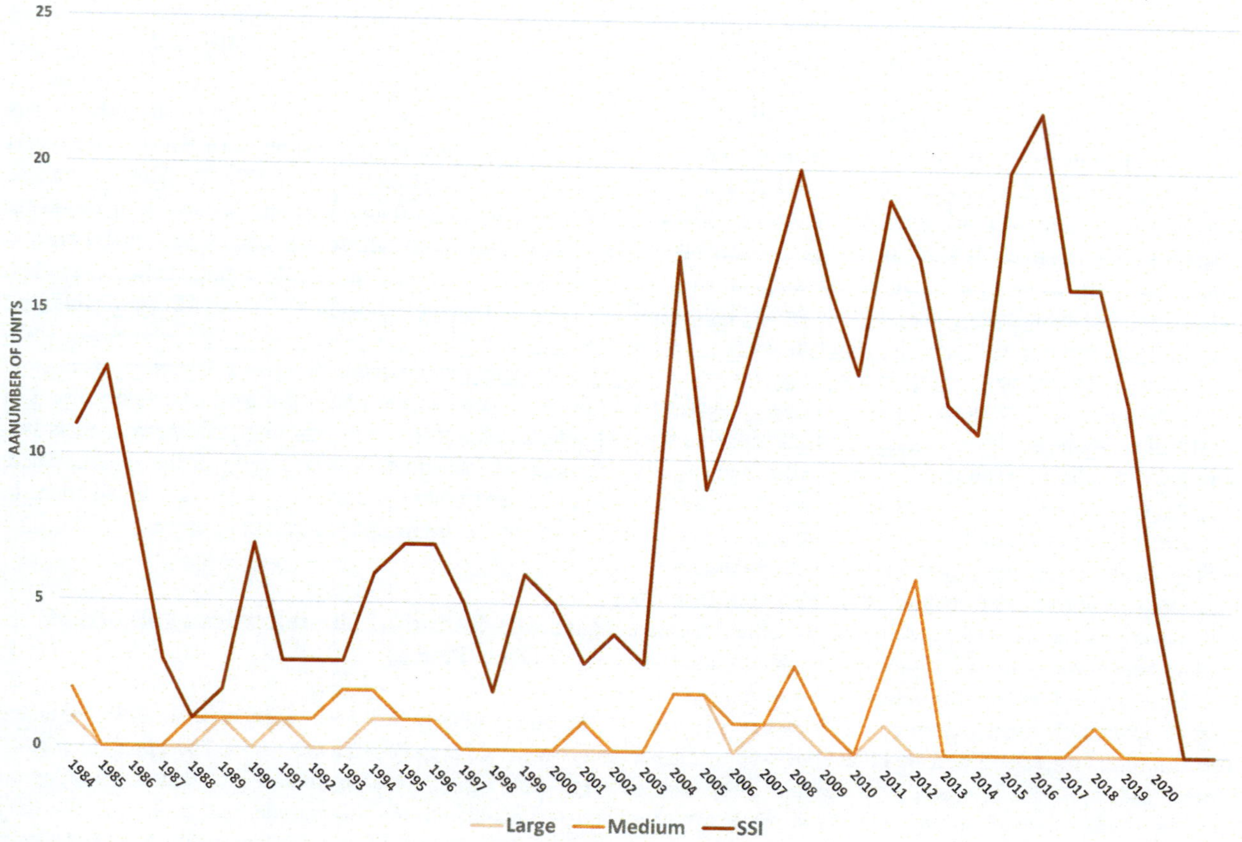

Fig. 3 Growth of Large, Medium and Small-scale industries in Pithampur automobile cluster. *Source* Generated by the authors

Table 1 Demographic change in Madhya Pradesh, Dhar district and Pithampur

	Year	1991	2001	2011
Madhya Pradesh	Population (Nos)	48,566,242	60,348,023	72,626,809
	Annual growth rate (%)	2.72	2.43	2.0
	Sex ratio	912	919	931
	Workforce participation ratio	42.8	42.7	46.6
	Percentage other workers	22.5	24.5	27.17%
Dhar district	Population (Nos)	1,367,412	1,740,577	2,185,793
	Annual growth rate (%)	2.93	2.73	2.56
	Sex ratio	951	955	964
	Workforce participation ratio	46.7	36.2	47%
	Percentage other workers	15.9	17.0	18.40%
Pithampur	Population (Nos)	11,980	68,080	126,200
	Annual growth rate (%)	2.2	46.8	8.5
	Sex ratio	789	714	796
	Workforce participation ratio	43	41	39.4
	Percentage other workers	42.5%	87%	84%

Source Census of India (1991, 2001, 2011)

get sufficient power. For this MPEB (Madhya Pradesh Electricity Board) was given free land to set up substations. The third goal was to get water in the region. Looking at the water scarcity in the Pithampur town, the government worked on two projects Sajay jalashay and Karam jalashay and reached a water capacity of 15 MLD in the region.

Simultaneously the road development works was initiated by the government to make the place logistically more sound. In the year 1986, Warehouse of capacity 5000MT was also developed by Central Warehousing Corporation, India. These major measures by government made a difference in the way Pithampur land was perceived before industrial growth was envisaged in the region. Pithampur Adyogik Sangathan, was also established in the year 1988, representing all large, medium and small-scale units situated in Pithampur. This association voices the concerns of industries to the government and also publishes English Weekly which includes news related to Pithampur industries.

In this period, a Phased Manufacturing Programme (PMP) to promote local manufacturing of components was introduced by the Government of India. According to this programme original equipment manufacturers (OEMs) had to increase domestic input in their output over a specific period. This encouraged development of ancillaries in Pithampur. A lot of joint ventures were also seen in India with several Japanese and foreign OEMs which enabled them to benefit from equity inflows and technology transfers (Miglani, 2019). Pithampur nurtured many such joint ventures in automobile manufacturing since its beginning years. The first big firm established in Pithampur was of Eicher Motors Ltd. in 1986. It was a joint venture of an Indian company Eicher Goodearth and a Japanese company Mitsubishi Motor Corporation. Eicher had already established itself in other parts of India like Faridabad (Haryana), Alwar (Rajasthan), Parwanoo (Himachal Pradesh) and so the development of Eicher in Pithampur was mainly dependent on Faridabad as most of the skilled and unskilled labours were procured from there to start the production process (Kothari, 2019) After successful transfer of technology and achieving total indigenization, the agreement between Eicher's and Mitsubishi ended in March 1994 (Buissness Standard, 2020). Eicher survived the competition of indigenization and it established its link with other Indian companies. It succeeded in localizing its product and even reduced its cost of production. Similarly, another collaboration that developed in Pithampur was Kinetic Honda Motor Ltd. It was also a collaboration between Indian Kinetic Engineering Company and Japanese Honda Motor Company. It manufactured two and three-wheelers. The third big entry in Pithampur was Hindustan Motors. It commenced production of petrol engines and transmissions at Pithampur, in collaboration with Isuzu Motor Company of Japan. Initially, its Pithampur branch heavily depended on its main branch in Kolkata for capital and other services. Bajaj Tempo which was well known for its two and three-wheelers in India established in Pithampur in 1987 with the intent to produce light commercial vehicles. It was a joint venture between an Indian automotive manufacturer Bachraj Trading Ltd. and Germany's Tempo. Due to some hurdles, its production started in 1989. Similarly, Larsen and Tubro also established their plant in Pithampur. Along with these main industries, there were nearly 51 ancillaries including small-scale and medium-scale automobile units developed to support the mother industries (Large scale automobile units) (MPAKVN, 2020a, b). During Pithampur's initial years of development as an industrial town, industries faced a lot of problems in increasing the pace of production. Due to lack of ancillary units, mother units had to depend on units outside the state for auto parts and components. Also, mother units brought labourers with them as the region was devoid of skilled labourers. The other challenge was inadequate infrastructure. Industries that planned to run three shifts had to curtail themselves at two because of inadequate power supply in the region. There were many other issues that industries faced due to bureaucratic systems and infrastructure issues.

3.2 The Period of Liberalization: 1990–1999 (First Phase)

The automobile policy announced in June 1993 contained measures, such as de-licensing, automatic approval for foreign holding of 51% in Indian companies, abolition of phased manufacturing programme, reduction of excise duty to 40% and import duties of 'Completely Knocked Down units (CKD)' to 50% and of 'Completely Built Up units CBU' to 110% and commitment to indigenization schedules (Muthumoni & Priya, 2015).

A major development that happened in Pithampur during this decade was development of Inland Container Depot (ICD) by Container Corporation of India in 1994. Being landlocked, Pithampur inevitably needed this facility. ICD provides logistics solutions to all importers and exporters. The catchment area for this Dry port covers many close-by industrial areas. It is spread in an area of 17acres and provides all the facilities provided by gateway ports for containerized cargo. Table 2 shows ICD's distance from various gateway ports of India. In this decade industries witnessed a challenge related to taxes. Industries in Pithampur enjoyed tax holidays till 1995–1996 but when the tax holiday was over, the industries were made to pay more taxes in comparison to taxes in other states of India. At that time nearly 70% of the materials (auto parts/components) were transported from other states due to lack of ancillary units within the state.

The mother unit heavily depended on purchases from outside the state. The central sales tax levied was 4% for this purchase. As 80 to 85% of finished products were sold outside the state, again industries had to pay 13.8% tax to cross the state boundary of Madhya Pradesh. In this whole process, the local tax which was charged was not

Table 2 Distance of major Indian gateway ports from Inland Container Depot, Indore

Ports	Distance (kms): Road + Rail
JNPT/NSICT/GTIL	874
Mundra	950/
Pipavao	977

Source EXIM—India (2018)

reimbursed. In comparison to other states, the industries in Pithampur had to pay 8% more tax. This brought a need to develop local ancillaries in the region.

During the decade expansion of Bajaj Tempo and Larsen & Toubro Ltd. happened in Pithampur. Bridgestone (India) Pvt Ltd., also established its branch in Pithampur. In 1998 Bridgestone, in its plant at Kheda sector in Pithampur started commercial production of tyres. In this decade nearly 51 ancillary units became part of the automobile cluster (MPAKVN, 2020a, b).

3.3 Post Liberalization: 2000–2009 (Second Phase)

The decade started with coming of government's strict regulation of emission standards on vehicles. This was Bharat stage emission standard II (BS II). The vehicles which were manufactured after the implementation of the norms had to comply with new rules and regulations. The upgradation of this standard till 2010 reached stage IV. Eicher had always been in social news for changing technology as per requirement and abiding by new emission standards. The other industries producing vehicles in Pithampur have also adapted to new technologies for better emissions as per government's regulations.

In this decade a lot of schemes and programmes were launched in India for industrial cluster development. One such scheme was Infrastructure Upgradation Scheme (IIUS) launched in 2003. A lot of development in Pithampur automobile cluster has happened under IIUS. It is a Central government initiative aimed to enhance industrial competitiveness by providing grant funding for quality infrastructure through public–private partnerships. Pithampur automobile cluster in 2004 got approval for this scheme. Under Companies Act 1956 an S.P.V. named Pithampur Auto Cluster Ltd. (PAC) was formed. The SPV has exclusively contributed in the automobile cluster development.The project cost was around INR 65.20 crore and the work under the scheme was completed by March 10, 2011 (Cluster, 2020). The P.A.C. has four major components, which include:

- Advanced technical and Industrial training Centre (ATITC)
- Hazardous Waste Disposal Facility,
- Tool Room & Technical Services Centre and,
- Road Infrastructure & Administrative Complex.

ATITC was established in the year 2007 and it runs courses recognized by National Council of Vocational Training. These are two-year and one-year courses specially programmed for the skill set required in the Pithampur Industrial area. These courses include Machinist, Electrician, Fitter, Welder and Computer Programmer and Programming Assistant courses with income ranges from INR 12,000–28,000. ATITC also runs free certificate courses for students under Scheduled tribes and Scheduled class categories. A Hazardous Waste Disposal Facility has also been developed in collaboration with M/s Ramky Enviro Engineers Ltd., Hyderabad. The waste is segregated and depending on the type of waste; it is either landfilled, stabilized followed by landfills or incinerated.

Around INR 30 Crores is the yearly tooling requirement of the Industrial Units at Pithampur automobile cluster. The small and medium enterprises in Pithampur generally outsource their tooling needs to Indore and also to Pune, whereas the large units in Pithampur have their own tool rooms. Therefore, to meet the requirements of tools of various industries P.A.C has set up modern tool rooms (P.A.C, 2020). Similarly, a wide range of technical services related to design and development, materials testing, calibration and technical & quality advice are also outsourced from service providers situated in locations such as Delhi, Pune and Kolkata. To eliminate this dependency P.A.C has established a Technical Service centre at Pithampur. The S.P.V. also undertook the responsibility of developing the road infrastructure. In this regard, road drainage and other site development work worth INR 2.50 crore have been completed. An administrative complex has also been developed under this programme (Cluster, 2020).

In another development by National Automotive Testing Research & Development Infrastructure Project (NATRiP) initiated by Government of India in the year 2005, Pithampur was chosen for development of Auto Testing Tracks and Lab (NATRiP, 2020). In 2007, 1080 Hectare of irrigated cultivable land was acquired by the state government from farmers of 10 villages. This project is not only contributing in giving an exclusive facility to Pithampur but also helping in brand building of Pithampur automobile cluster.

One of the world's largest infrastructure projects, the Delhi Mumbai Industrial Corridor project was launched in 2006. For this, an MOU was signed between the Government of India and the Government of Japan. The eight Investment Regions proposed in the corridor to be developed in the Phase I included Pithampur–Dhar–Mhow (in Madhya Pradesh). As Pithampur town had already developed an industrial base, the proposed SIR envisaged it as an epi-centre for the development. During this period 112 ancillary units came into existence to support mother units. V. E. Commercial Vehicles Pvt Ltd., Man Truck India Pvt Ltd., Force Motors Limited, Mahindra Two Wheelers Ltd., Avtec Ltd., Bridge Stone (India) Pvt Ltd., Liugong India Pvt. Ltd. were some major players which either expanded or established their firms in Pithampur (MPAKVN, 2020a, b).

3.4 2010 Onwards

The State Industrial Policy 2014 of Madhya Pradesh offered attractive investment promotion assistance, impetus on employment generation, reimbursement of expenditure on infrastructure development, etc. (M.P, 2014). These initiatives were to bring more ease and comfort to industries that were being developed in different industrial growth centres of Madhya Pradesh. Till 2010, Pithampur had attained a good name amongst industrial areas of India and also had become a growing automobile hub. Pithampur-Dhar-Mhow SIR had been identified as one of the projects under Phase I development of DMIC project. The master plan for the same was developed by Lea Associates South Asia Pvt Ltd. in association with Development and Research Service Pvt Ltd. Till 1984, Pithampur Industrial Area was of 2116 hectares which were proposed to grow to 37,200 Hectares under SIR. Now it was not just an industrial area, rather has become a region for investment and economic growth significant to the State and Nation. There were four early bird projects to start the development of SIR. These projects were development of Economic Corridor of 20.3 km in length along link road connecting Indore Airport to Pithampur (LEA Associate, 2010).Another project was continuous water supply for proposed SIR. The main components of this project include 90 MLD Raw Water Pump House, 60 MLD Water Treatment Plant, 114 km of MS & DI Pipeline, two Over Head Tanks and two Ground Level Sumps (PJPCL, 2016). Other two projects are the Development of Integrated Multimodal Logistic Hub and development of Knowledge City in Ujjain District. Another initiative by Container Corporation of India Ltd. facilitating Pithampur automobile cluster is the ongoing development of Multi-model logistic park to provide cost-effective and efficient logistic solutions to the export–import and the domestic cargo (EXIM—India, 2018) This development is taking advantage of the Indore-Tihi section of the rail line between Indore and Dahod.

A major challenge of this decade was holding land under National Automotive Test Tracks (NATRAX) in Pithampur without development. In 2016 The union cabinet had decided to return back the 478 hectares of land out of 1080 hectares of land to MP government, which they took in 2007 for developing auto testing track of NATRIP Project in Pithampur, Indore, MP. This happened due to many reasons.

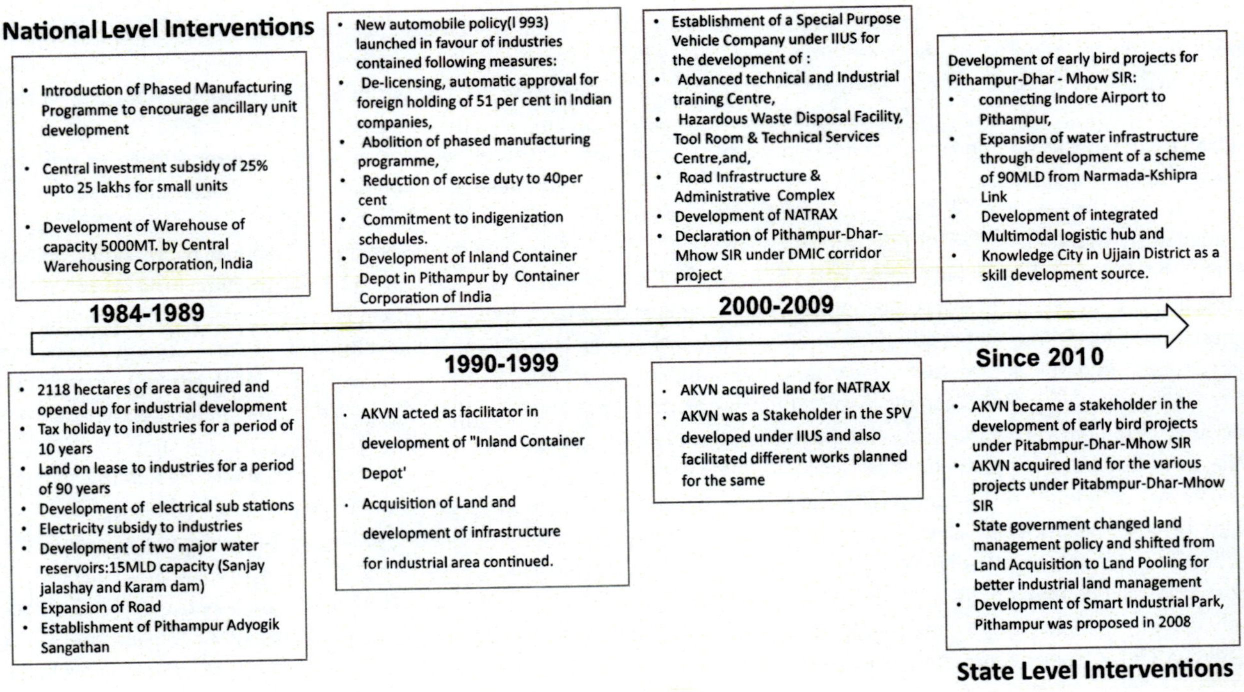

Fig. 4 Timeline Showing National and State Level Interventions Contributing directly and Indirectly in the development of Pithampur Automobile Cluster. *Source* Generated by the authors

The state government gave land to the project anticipating economic growth, better revenue and employment. The project didn't take off the way it was projected. The delay in completion of the project and more than 50% of land lying unused caused heavy loss to state government. This led to the decision of selling back of some parts of land to state government. This Denotified area of NATRAX was later targeted by the state government for attracting Japanese investment. Presently, Smart industrial park of an area 468 hectares is being developed where there is integrated development of industry along with housing.

This decade witnessed a major shift in land management tool used by the state government. Till 2013 most of the land was acquired in Pithampur through Land Acquisition Act 1984. The government dealt with lots of disputes in land acquisition which caused delay in many projects. The Land Acquisition bill 2013 asks for a lot of compensation to the seller and so the government found it difficult to acquire the land. The government has now switched to Industrial Land pooling where the participating landowners do not require to relinquish rights but instead become partners in the development process (Yadav, 2019). V. E. Commercial Vehicles Pvt Ltd., Bridge Stone (India) Pvt Ltd., Avtec Ltd. and 174 ancillary units also came in this decade (MPAKVN, 2020a, b). In this period for brand building of the State, there were initiatives like Make in India Campaign launched in 2014 which portrays Madhya Pradesh as a good investment destination for multinational companies to manufacture their products. Events like Global Investor Summit organized in October 2016 at Indore intended to attract investment in sectors like Agribusiness and food processing, automobile and Engineering, IT/ITES, Pharmaceuticals, Textile and Tourism. Figure 4 is a compilation of the initiatives taken at State and National levels since initiation of industrial development in Pithampur which helped in the overall development of Pithampur automobile cluster.

4 Conclusion

The study shows that Pithampur automobile cluster is an outcome of series of initiatives taken by the Central and State government. Through these initiatives, Pithampur has been transformed from bunch of villages to an industrial town. As Pithampur automobile cluster is a planned industrial cluster amid villages, the State government has made a lot of capital investment to push industrial growth. The State government through a dedicated parastatal agency, i.e. Adyogik Kendra Vikas Nigam (AKVN) has managed land and developed industry-specific infrastructure. It has been an active stakeholder in the development of Pithampur Industrial area. The efforts by AKVN are tailored as per the specific needs of Pithampur industries. To create an amicable industrial environment in an industrially backward region even the Central government had to come forward and develop major infrastructures like Inland Container Depot, Central Warehouse and other developments under Integrated Infrastructure Upgradation Scheme. These initiatives have helped Pithampur to grow and thrive as an industrial town. Attractive incentives and subsidies have played a major role in inviting industries to establish in Pithampur. Through these initiatives, Pithampur automobile cluster has witnessed immense growth as industries have grown manifolds over the decades. This scenario where government wins in calling industries on their land at the cost of paying heavily poses a very important question regarding choice of location of industries. The quest to search for is whether is it good to locate an industry like that of automobile in a rural location where everything has to be developed from scratch or should it be located in or close to an urban area where economics of urbanization plays a role in reducing cost of production as well as cost of investment on infrastructure development. This research also reveals the fact that there are only two initiatives specific to automobile cluster development taken by the Central government, in the three and a half decades of Pithampur automobile cluster's growth. These initiatives are development under IIUS and, the ongoing development of NATRAX. Unlike Europe or USA, there are no umbrella organizations at Central or even at State level to monitor industrial cluster development in India. This calls for a new hierarchal system in which there are different bodies for cluster development at apex and lower levels to take care of every cluster present in the country.

References

Baily, M. N., & Montalbano, N. (2017). *Clusters and innovation districts: Lessons from the United States experience*. The Brookings Institution.

Buissness Standard. (2020, September 25). Eicher Motors Ltd. (Eichermot)—Company history. Retrieved from Buissness Standard. https://www.business-standard.com/company/eicher-motors-151/information/company-history

Census of India (1991, 2001 & 2011). *District census handbook Dhar*. Directorate of Census Operations Madhya Pradesh.

Central Bank of India. (2020, September 15). *Central Bank of India*. Retrieved from State profile of Madhya Pradesh. http://slbcmadhyapradesh.in/agriculture.aspx

Cluster, P. A. (2020, August 5). *Auto cluster project-at a glance*. Retrieved from pithampurautocluster.com. https://www.pithampurautocluster.com/auto-cluster-project-at-a-glance/

Crec. (2015, December 9). *Center for regional economic competitiveness*. Retrieved September 28, 2018, from http://creconline.org; http://creconline.org/uncategorized/regional-innovation-clusters-federal-state-economic-development-collaborations/

EXIM—India. (2018). *EXIM—India year book 2017–18*. Team EXIM-India.

Goetz, S. J., Deller, S., & Harris, T. (2009). *Targeting regional economic development*. Routledge.

Government of Madhya Pradesh. (2016). *Invest Madhya Pradesh, global investors summit*. Government of Madhya Pradesh.
IBEF. (2013). *MSMEs and the growing role of industrial clusters*.
Jalaja, N. R. (2005). *Industrial growth in Madhya Pradesh structure and economic backwardness*. Department of Applied Economics, Cochin University of Science and Technology.
Jankowiak, A. H. (2012). Chinese industrial clusters. *Innovation Sources of Economies in Eastern Asia*, 164–173.
Ketels, C. (2004). European clusters. In *Structural change in Europe 3—Innovative city and business regions* (pp. 1–5). Hagbarth Publications.
Ketels, C. (2015a). *Clusters for competitiveness: Mobilising the power of modern cluster policy*. 18th TCI Global Conference. Daegu, South Korea.
Ketels, C. (2015b, February). *Competitiveness and clusters: Implication for a new European growth strategy*.
Kothari, G. (2019, November 17). *Land management in Pithampur*. (P. Madhavi, Interviewer)
LEA Associate. (2010). *Comprehensive development plan of economic corridor*. Indore.
Lindqvist, G., Sölvell, Ö., & Ketels, C. (2013). *The cluster initiative greenbook 2.0*. University of Southern Denmark, Kolding.
M.P, G. (2014). *Industrial promotion policy 2014*. M.P, Government of India.
Markusen, A. (1996a). Interaction between Regional and Industrial Policies : Evidences from four countries. *International Regional Science Review, 19*(1&2), 19–77.
Markusen, A. (1996b). Sticky places in slippery space: A typology of industrial districts. *Economic Geography, 72*, 293–313.
Mazurek, S. (2014). Diversity of cluster policies. *Redefinition of the Role of Asia-Pacific Region in the Global Economy, 61*.
Miglani, S. (2019). The growth of the Indian automobile industry: Analysis of the roles of government policy and other enabling factors. In R. U. Liu (Ed.), *Innovation, economic development, and intellectual property in India and China*. Springer.
MPAKVN. (2020a, September 15). *Functions*. Retrieved from Madhya Pradesh Audyogik Kendra Vikas Nigam (Indore) Limited. http://www.mpakvnindore.com/
MPAKVN. (2020b, September 11). *List of units*. Retrieved from Madhya Pradesh Audyogik Kendra Vikas Nigam (Indore) Limited. http://www.mpakvnindore.com/index.php?page=viewlist-unit
Muthumoni, A., & Priya, S. (2015). Automobile industry: Need of the hour. *International Journal of Management Research and Buisness strategy*, 177–179.
NATRiP. (2020, September 15). *National automotive testing and R&D infrastructure project (NATRiP)*. Retrieved from https://www.natrip.in/
P.A.C. (2020). *About Pithampur*. Retrieved November 21, 2020, from https://www.pithampurautocluster.com/
PJPCL. (2016). *Improvement of water supply system for Pithampur industrial area and phase—1 of Pithampur—Dhar—Mhow Investment Region, Madhya Pradesh*. Pithampur Jal Prabandhan Company Limited (PJPCL).
Porter, M. E. (2003). The economic performance of regions. *Regional Studies*, 549–578.
Ranawat, M., & Tiwari, R. (2009). Influence of government policies on industry development: The case of India's automotive industry. In *Technology and Innovation Management*. (University of Hamburg).
Roelandt, T., Hertog, P. D., Sinderen, J., & Hove, N. (1999). Cluster analysis and cluster policy in the Netherlands. In *Boosting innovation: The cluster approach*. OECD.
Sopoligová, M., & Pavelková, D. (2017). Cluster policy in Europe and Asia: A comparison using selected cluster policy characteristics. *Journal of International Studies*, 35–50.
Yadav, S. (2019, October 16). *The Hindu*. Retrieved September 2020, from Landowners to be part of development process: Kamal Nath. https://www.thehindu.com/news/national/landowners-to-be-part-of-development-process-kamal-nath/article29710610.ece
Yamawaki, H. (2002). The evolution and structure of industrial clusters in Japan. *Small Business Economics*, 121–140.

Implementation and Evaluation of a MLaaS for Document Classification with Continuous Deep Learning Models

Franz Frederik Walter Viktor Walter-Tscharf

Abstract

This paper indicates an approach of a continuous training pipeline to enhance deep learning models and assessing their feasibility based on an evaluation. The purpose of this research is to analyze the quality effect of a continuously learning neural network algorithm for document classification by taking user feedback into account. The hypothesis implies that user feedback through active learning increases the precision and thus makes the process of document classification more efficient. For this purpose, based on a utility analysis, the available technologies are identified, and necessary ones are selected for designing a software concept. TensorFlow as a deep learning framework, Tesseract as an OCR engine, and Apache Airflow for the life cycle management and for orchestrating the elements for the continuous training pipeline are used. This implementation of a machine learning as a service prototype allows for exploration into the synergistic effect between the use of active learning, in the form of user feedback, and the quality of document classification achieved by deep learning. In an experiment, the implemented service is used to analyze the models behavior based on three different states. This includes synthetic data and active learning in the form of user feedback through data from data augmentation and simulated realistic data. The result shows that active learning enhanced models indicate a higher accuracy than artificially generated models. The evaluation experiment confirms the hypothesis that user feedback with continuously learning models perform better in terms of generalizing within the document classification. In conclusion, the paper demonstrates the technical requirements for implementing a machine learning as a service and affirms that the use of active learning can be integrated into existing industrial systems.

Keywords

Active learning • Document classification • TensorFlow serving • Lifecycle management • Continuous integration • Deep learning • Continuous training pipeline • Machine learning as a service • MLOps • TFX

1 Introduction

Each year, approximately 500 billion electronic invoices and documents are sent to customers or consumers worldwide (Koch, 2019, p. 12). On average, the cost of processing each business transaction is estimated at 17 US dollars (Pezza & Jan, 2012, p. 6), and is partly due to the process of capturing and digitizing or processing documents, which is still manual today. Former Chief Scientist Andrew Ng of search engine company Baidu described this process as follows: "The industrial revolution freed humanity from much repetitive physical drudgery; I now want AI to free humanity from repetitive mental drudgery, such as driving in traffic" (Ng, 2017). With this, he expresses that artificial intelligence is intended to help relieve humans of repetitive tedious work. The authors Luger and Stubblefield also define artificial intelligence as "The branch of computer science that is concerned with the automation of intelligent behavior" (Luger & Stubblefield, 1998). They reference methods and technologies within information technology, which aim to perform tasks that require intelligence in their execution (Theobald, 2019, p. 115). According to a study by McKinsey, the industries with the highest potential for automation are tourism and hospitality with 73% and manufacturing with 60% (Manyika et al., 2017, p. 7). Companies are now investing in the automation of tasks that were originally assumed to be performed only by clerks. A key issue is

F. F. W. V. Walter-Tscharf (✉)
Department of Information Engineering and Computer Science, University of Trento, Trento, Italy
e-mail: viktor.waltertscharf@unitn.it

H. Altan et al. (eds.), *Advances in Architecture, Engineering and Technology*,
Advances in Science, Technology & Innovation, https://doi.org/10.1007/978-3-031-11232-4_20

information extraction or document recognition. Many of the use cases focus on data collection, processing, and consolidation from formats such as email, PDF, or fax (Denecken, 2018). Instead of the time-consuming manual extraction of information from documents before the transferal to an Enterprise Resource Planning–ERP–system, a clerical salesperson's work focus can be concentrated on more demanding activities through automation.

A deep learning model must be developed for recognizing key information within a document to enable generic document processing, therefore, requiring a large and annotated dataset. The downsides are addressed throughout this paper, for example in preparation, the data must be labeled, which requires time and work. This research deals with the incorporation of user feedback into deep learning models for document classification. As the concept of active learning is a new and promising topic, the goal is to continuously extend the models with user feedback through this method.

2 Literature Review

For the thematic classification of the present work, the state of the art research, related methods, and concepts of relevant scientific papers are briefly introduced and differentiated. This involves work on document recognition, extraction, and annotation. For several years, different methods have been developed focusing on research approaches for text classification, which concentrate on the interpretation of the content of a document, as well as object recognition algorithms, which identify the layout or important areas.

2.1 Theoretical Background

Chen et al. in their work on "Convolutional Neural Networks for Page Segmentation of Historical Document Images" (Chen et al., 2017) develop a page segmentation method based on a convolutional neural network—CNN. They focus the page segmentation problem on pixel labeling and propose to train features of raw image pixels through a CNN. Although the focus of their work is on recognizing handwritten documents, their approach implies that the basic idea of using a CNN to recognize important regions in PDF documents has potential. Research papers such as "Page Object Detection from PDF Document Images by Deep Structured Prediction and Supervised Clustering "(Li et al., 2018) or" ICDAR2017 Competition on Page Object Detection" (Gao et al., 2017) deal with structure and range detection of documents and also use a CNN in their approach. Their model focuses on the use of object recognition by deep learning algorithms in images of documents where—among others—tables, mathematical formulas, graphs, or figures are identified. Another approach to document classification is presented in the conference paper "Fast CNN-Based Document Layout Analysis" (Oliveira & Viana, 2017). The authors focus on document layout analysis to extract information from document images. Their model presents a one-dimensional approach for document layout analysis considering text, figures, and tables based on a CNN. They take the approach of reducing the data representativeness, of text and table blocks, to a one-dimensional CNN, which compared to the classical two-dimensional CNN approaches, should significantly improve the overall performance without affecting accuracy. This paper especially (Oliveira & Viana, 2017, Fig. 1) indicates an adequate abstract method to differentiate the structure of documents. The research paper "DeepDeSRT: Deep Learning for Detection and Structure Recognition of Tables in Document Images" (Schreiber et al., 2017) outlines a novel end-to-end system for identifying tables in images of documents through deep learning; rows, columns, and cell positions are identified. The DeepDeSRT model fundamentally consists of a CNN in conjunction with object recognition to identify the position of tables. For this, transfer learning with a pre-trained Faster R-CNN model is used.

For structure detection, DeepDeSRT uses a fully convolutional network (FCN)-based segmentation model, providing a reliable concept for document content classification. In the evaluation of the evaluation results, values between 91.44% and 96.77% are achieved for table and structure recognition.

2.2 Related Work

The topic of continuously augmenting neural networks with user feedback is sparsely covered in current literature. It was introduced and presented by Google in 2017 with the title "TFX: A TensorFlow-Based Production-Scale Machine Learning Platform" (Baylor, et al., 2017). TensorFlow Extended (TFX) is a platform developed by Google for data preparation, training, validation, and deployment of machine learning models in production environments. The platform is based on TensorFlow Core and takes as its initial idea the problem of maintaining a machine learning service. A pipeline is described consisting of different components including an analysis, transformation and validation of data, as well as a component for generating the machine learning model with the processed data. Likewise, a phase for the evaluation of the previously developed model exists, as well as a unit for the deployment of the model into the productive system. The integration of the platform and the accompanying orchestration of the components into a pipeline is

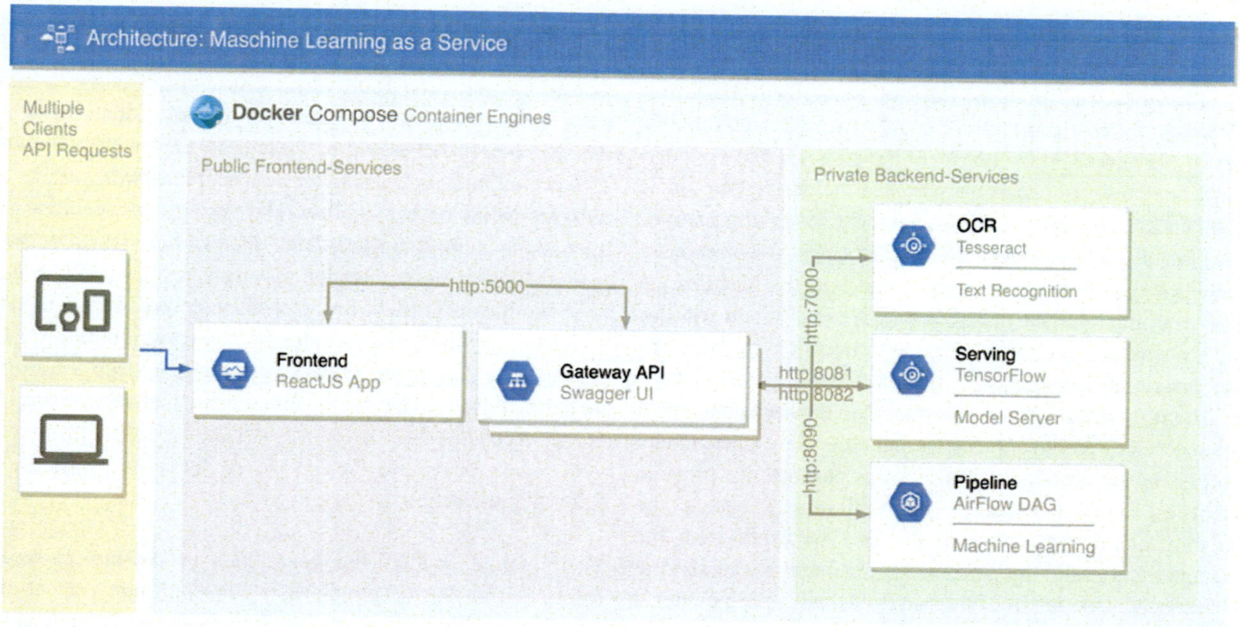

Fig. 1 Architecture of the MLaaS

intended to reduce development time and increase performance (Baylor et al., 2017, p. 1394). TFX is intended to provide an interactive platform for the user to afford deeper insight into the operation and decision path of a specific machine learning model. By building the pipeline (Baylor et al., 2017, p. 1389), the concept for a "continuously training pipeline" (Baylor et al., 2017, p. 1393) is defined.

Another scientific contribution is provided by the article "TensorFlow-Serving: Flexible and High-Performance ML Serving" (Olston et al., 2017). The authors describe an approach to use machine models productively and provide ways to feed TensorFlow with new data for training a new machine learning model. The main task of TensorFlow Serving (TFS) is to be able to productively deploy different machine learning models. RPC and HTTP interfaces are provided to interact with the model. Through TFS, execution paths for discovering new machine learning models and performing predictions of result values have been optimized to avoid performance issues of the native implementation through TensorFlow Core (Olston et al., 2017, p. 1). Routine tasks such as adding, removing, and updating a machine learning model are also handled. Additionally, rollback and canary requests, which are central to resilience, are supported. TFS is part of the end-to-end machine learning pipelines and is actively used in TFX (Olston et al., 2017, p. 6). Concepts for incorporating user feedback into the training process is provided in "Active Learning Literature Survey" (Settles, 2010). The topic is classified under the term active learning and the paper outlines its application scenarios. The implementation of these individual methods is based on the mathematical logic of (Settles, 2010, Fig. 2—1, 2, and 3). Of central importance is the "membership query" type. In this method, the artificial intelligence algorithm proposes a classification and the oracle or the user can agree or correct it (Fischer, 2000, p. 6). The scientific article "From Theories to Queries: Active Learning in Practice" (Settles, 2011) shows the practical implementation of these individual methods.

A large number of scientific papers address the topic of artificial intelligence in the area of machine learning and deep learning. The current scientific consensus is that the performance of deep learning models improves by increasing the amount of data compared to traditional machine learning algorithms. Object recognition methods based on artificial neural networks are successfully used for document classification. However, a new approach is the combination of transfer learning and active learning. By using both concepts, an autonomous self-learning system for document classification is developed utilizing user feedback.

3 Concept and Design

A clear concept and design are essential when creating a deep learning model and continuously integrating user feedback. An OCR technology which converts PDF documents into machine-readable code is necessary. Far more important, however, is a machine learning framework that can use a model to detect where critical information is located in the document. The model should be able to be improved by feedback from the user and continuously deploy new advanced deep learning models. This requires a pipeline that can process, validate, share, and transform feedback data. It must also be able to train and monitor neural networks. From these requirements for the system's functionalities, the necessary parts can be derived; an OCR engine that converts the PDF document, a serving component that provides the deep learning model and continuously waits for the latest version, and a pipeline that handles the deep learning lifecycle management. These different services are difficult to orchestrate. Figure 1 shows an overview of the described components.

The task of the serving component in the container-based microservices architecture is to make the developed model accessible via Web interfaces. For this purpose, the pipeline provides the functionalities to continuously extend the existing neural network with the user feedback from the frontend and make the new model available to the serving component. The output of the current deep learning model is processed by the OCR component after transformations have been performed in the API gateway and thus the required information is determined.

3.1 Serving

The serving component with TFS is used to address the deep learning model via Web http requests. In addition, a service is needed to continuously wait for a newly trained deep learning model in order to incorporate it at runtime when it arrives and to use it via the interfaces. Currently, only one deep learning model is needed to perform the structure determination of the document, however, TFS can manage multiple versions of the model. Through the API interface, the neural network can be integrated into the target environment of the productive system quickly, flexibly, and with high performance. For this purpose, TFS provides a REST Web API and a gRPC interface for integration. Both can be accessed independently of the programming language. For setting up TFS and the developed deep learning model, Google provides an official Docker image that includes all required resources. The drawbacks are minor; the additional software requires packages, dependencies, or libraries, which increase complexity and maintenance efforts. Moreover, an additional service can decrease the comprehensibility of the developed code, however, the effort to send an image file to a Web API endpoint would be significantly greater when implementing the interfaces natively or without TFS. In the architecture, TFS is used to address a deep learning model and identify ROIs—Region of Interests—based on images. This refers to the structure determination of the uploaded image document using object recognition in images. The return values of the ROIs are the coordinates and the designation of areas which contain important information from the document. Also returned is the precision of the particular ROI, which distinguishes between the areas Header, Content, Customer, Type, and Identification. Another aspect that would be difficult to implement without TFS is continuously waiting for new neural networks trained by the pipeline or TensorFlow. The versioning capability is intended to support jumping back to an earlier version of the model. Versioning is intended to ensure resilience and prevent significant impacts on mean average precision—mAP — which may have occurred due to the evaluation of the test datasets, the use of the MLaaS with the new deep learning model or limits to the functional capability of the service.

3.2 Pipeline

The pipeline enables the continuous processing of user feedback and thus accomplishes a constant improvement of the system. The amount of data that can be used for a new deep learning model increases with the user feedback which, along with the object classes and their coordinates, are stored in a CSV file and are then used to continuously train new versions of the model. The machine learning pipeline has eight phases, which allow for the required workflow for the training process and continuous integration (Google, 2019c). The data collection phase is responsible for preparing the collected data of the user feedback for the upcoming training process. The data preprocessing phase is used to measure and analyze the collected data. Based on the amount of data, a random determination of a dataset for evaluation and actual training is performed. In the third phase, the TensorFlow record files are created from these datasets. Phase four is used to monitor the evolution of the neural network and assess the model at runtime of the training process, based on the test dataset, using metrics for accuracy. The central utility for this purpose is TensorBoard, which is integrated in TensorFlow. The tool can be used for validating and analyzing deep learning specific parameters. Important parameters are mAP, Total Loss, Localization, Precision, and Recall. Phase five executes the training process, which includes developing a new model graph that is responsible for the prediction. The TensorFlow API provides the necessary Estimator, which represents the code for our model and manages the training process. The Estimator can be seen as a pre-built framework for training the neural network. Among other things, it also supports routine tasks like saving the learning progress of the training into so-called "check-points" or exporting the trained graph as a SavedModel, which can then be used in production environments. These steps are part of the "Save Training" and the "Export Interference Graph" phase. The final phase involves making the new deep learning model available through interaction

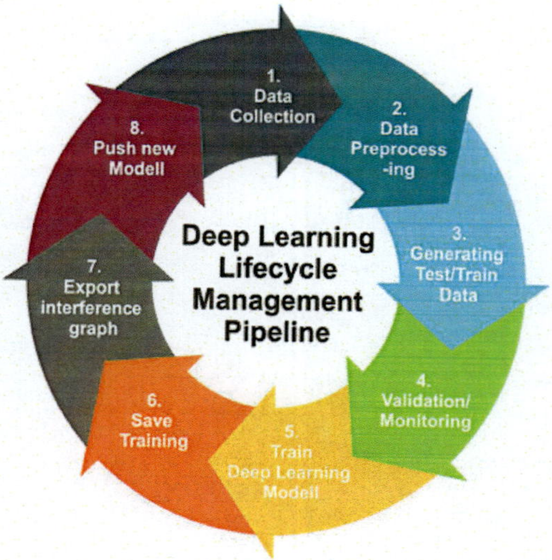

Fig. 2 Deep learning lifecycle management pipeline

with the serving component. The model is initiated dynamically without interrupting the service at runtime. After all phases of the pipeline have been executed, the result is a new deep learning model available at the serving component which has been augmented with user feedback from the previous queries. An overview and abstract framework of the flow is described in Fig. 2.

The phases must be combined to form an overall structure, the result of which is the Continuous Training Pipeline. This monitors the correct progress of the individual phases in a targeted manner takes over the orchestration or arrangement of the phases and merges them into an overall construct. Google recommends Apache AirFlow (Google, 2019a) for these requirements. In Airflow, the combined flow or pipeline is called a DAG— directed acyclic graph. Here, the DAG consists of a collection of all the tasks required to fulfill the workflow (Apache Airflow, 2020). In practice, DAGs are Python files, which in turn contain multiple operators describing a single task in a workflow. In the described pipeline, the operators represent the different phases and, using AirFlow, the logs and code of these can be viewed individually. In addition, Airflow provides interfaces to trigger the modeled pipeline or the DAG via Web requests. The REST API can also be used to activate/deactivate the DAG, query the status, and retrieve execution time or task information. These functionalities are only indirectly accessible from the client. The gateway component manages this REST API and corresponding requests are only forwarded if required. This approach is chosen to comply with the architecture guidelines and the SoC.

The Web application is developed with the frameworks ReactJS and UIKit as demonstrated in Fig. 3. The user has the possibility to define the content of specific tags by moving the rectangles and thus providing corrections which form the necessary feedback. For the implementation of the animation and graphical functionality, the D3.JS framework

Fig. 3 Adding feedback to the suggested tags of a document in form of content areas, movable as rectangles

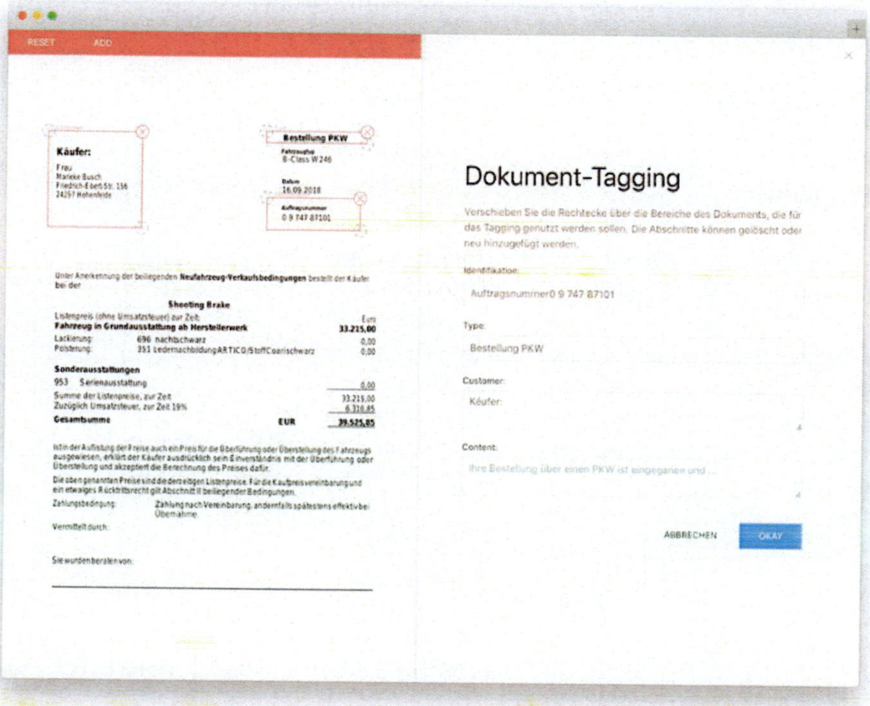

is used and communication with the backend utilizes the JS Fetch API. The gateway is implemented using FlaskREST-Plus. The serving component in the container-based microservices architecture allows the developed model to be accessible via Web interfaces and is implemented using TensorFlow Serving. The pipeline continuously augments the existing neural network with user feedback and provides the new model to the Serving component. The deep learning model will be developed using TensorFlow and for the orchestration, AirFlow is used. A translation of the images into text is achieved with the OCR component by Tesseract and Flask. In combining these subcomponents, a document classification service through a continuous deep learning training pipeline is outlined and available (Walter-Tscharf, 2021).

4 Evaluation

An assessment and evaluation of the benefits of user feedback as well as the continuous expansion of the deep learning models are carried out via an experiment with the developed prototype. The basis of the assessment is a model that is continuously extended through active learning and is available in a total of three different states, which compared using an evaluation dataset.

- Model 1: Faster R-CNN and synthetic data
- Model 2: Model 1 and active learning in the form of data augmentation
- Model 3: Model 2 and active learning in the form of data simulation

The Faster R-CNN model acts as the foundation.

4.1 Metric

The methodology behind the assessment follows the approach of the Pascal Challenge (Everingham et al., 2010, pp. 313–314). The research paper "A Comparative Analysis of Object Detection Metrics with a Companion OpenSource Toolkit" (Padilla, 2019) shows an applicable approach for evaluating the deep learning models. The focuses are the metrics Accuracy, Precision, Recall, Loss, and the $F1$-score demonstrated in Eqs. 1–6.

$$\mathrm{IOU} = \frac{\text{area of overlap}}{\text{area of union}} = \frac{\mathrm{area}(B_p \cap B_{gl})}{\mathrm{area}(B_p \cup B_{gt})} = \frac{|B_p \cap B_{gr}|}{|B_p \cup B_{gt}|} \begin{cases} \mathrm{TP}, \mathrm{IOU} \geq \text{ threshold} \\ \mathrm{FP}, \mathrm{IOU} < \text{ threshold} \\ \mathrm{FN}, B_{gl} = \emptyset \end{cases} \quad (1)$$

$$\text{accuracy} = \frac{\text{true positives } + \text{ true negatives}}{\text{all samples}} = \frac{\mathrm{TP}+\mathrm{TN}}{\mathrm{TP}+\mathrm{TN}+\mathrm{FP}+\mathrm{FN}} \quad (2)$$

$$\text{precision} = \frac{\text{true positives}}{\text{all detections}} = \frac{\mathrm{TP}}{\mathrm{TP} + \mathrm{FP}} \quad (3)$$

$$\text{recall} = \frac{\text{true positives}}{\text{all ground truths}} = \frac{\mathrm{TP}}{\mathrm{TP}+\mathrm{FN}} \quad (4)$$

$$\mathrm{mAP} = \frac{1}{N}\sum_{j=1}^{n} AP(j) \qquad \mathrm{AP} = \frac{1}{M}\sum_{j=1}^{N} \mathrm{precision}(k) \quad (5)$$

$$F_1 \text{ - score } = 2 \cdot \frac{\text{precision } \cdot \text{ recall}}{\text{precsion } + \text{ recall}} \quad (6)$$

In practice, these values were determined by combining the Toolkit and the TensorBoard.

4.2 Experiment

Model 1 is developed using only synthetic data and transfer learning. Models 2 and 3 represent a use of the developed prototype to directly extract user feedback from the document classification system to extend the neural network. Model 2 is intended to assess what impact user feedback with data augmentation has on document classification. Model 3 will be used to investigate the impact of using data with simulated real-world scenarios.

The different training patterns are shown in Figs. 4, 5, 6, 7, 8, and 9.

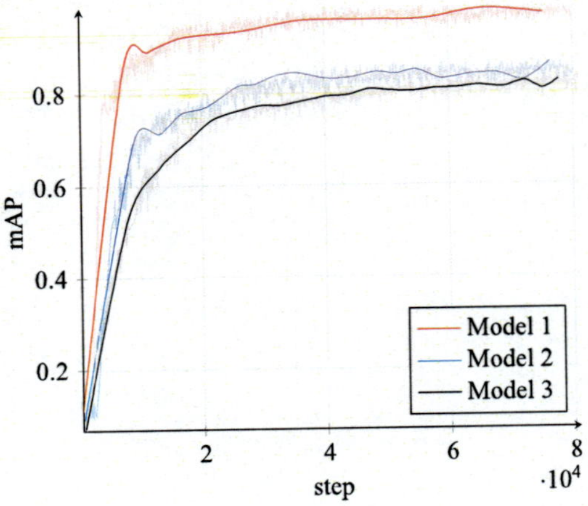

Fig. 4 Detection boxes precision/mAP of experiments

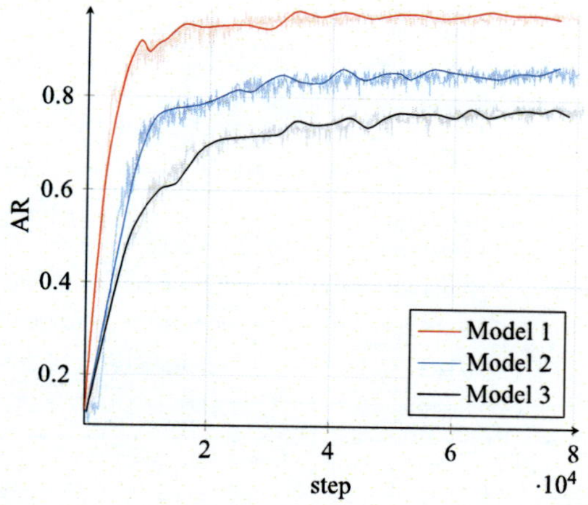

Fig. 5 Detection boxes recall/AR of experiments

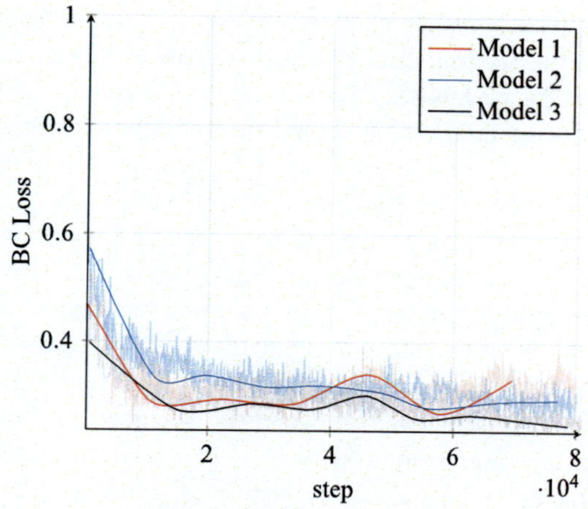

Fig. 6 BC Loss/localization loss of experiments

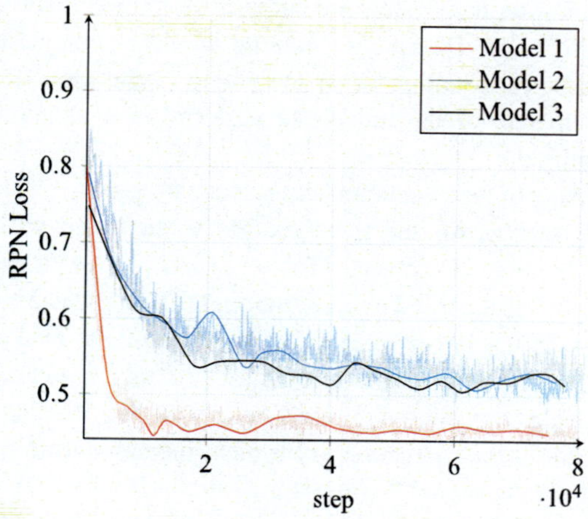

Fig. 7 RPN Loss/localization loss of experiments

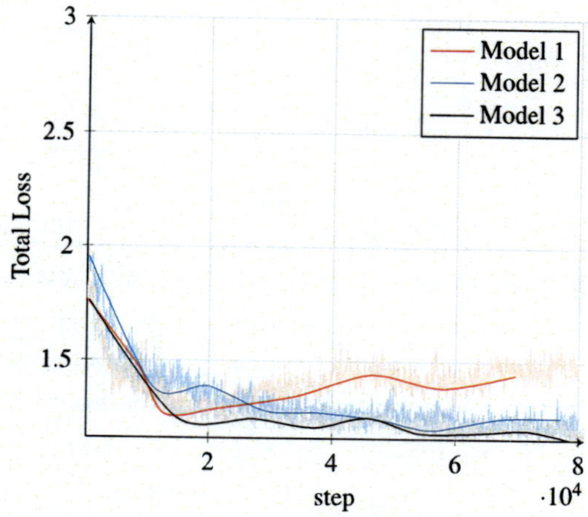

Fig. 8 Total loss of experiments

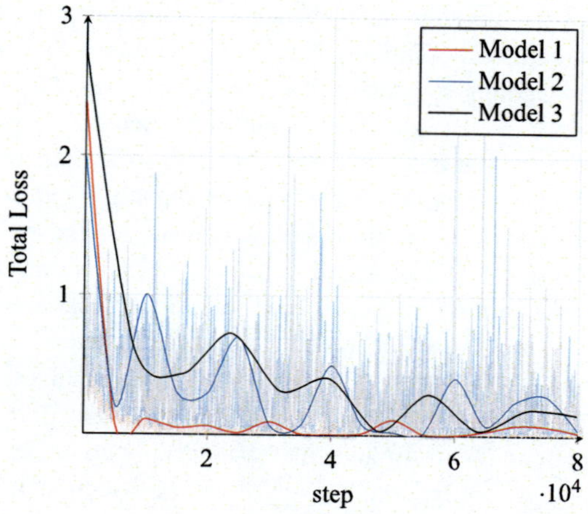

Fig. 9 Train total loss of experiments

After the performed training of the models, the three states of the model exhibit the characteristic values given in Table 1.

It is remarkable that for Models 2 and 3, which both use user feedback, the mAP decreases. Further analysis is required into the behavior of the models when an evaluation dataset from a real-world scenario is present, which the models are not trained for. That is to say, if the generalization for Models 2 and 3 increased. In addition, the new evaluation dataset also includes new types of documents that were not previously considered. Relevant to this is the use of documents without content or arbitrary documents and without object classes or labels. Each of the mentioned models is run with the evaluation dataset to test the hypothesis of the impact of user feedback. Central to this is the analysis of the different states of the models using the metrics, see Table 2.

Table 1 Overview of the results from the three models in the different stages

Model	Iteration	Recall (AR)	Precision (mAP)	*F*1-score	Total loss
Model 1	82,532	0.9777	0.9693	0.9735	1.4017
Model 2	80,150	0.8617	0.8451	0.8533	1.2648
Model 3	80,925	0.7852	0.8221	0.8032	1.1916

Table 2 Evaluation of the different models

Model	Iteration	Recall (AR)	Precision (mAP)	*F*1-score	Total loss
Model 1	82,532	0.9777	0.9693	0.9735	1.4017
Model 2	80,150	0.8617	0.8451	0.8533	1.2648
Model 3	80,925	0.7852	0.8221	0.8032	1.1916

5 Results

Discussing the results, a distinction must be made between training the models and running the experiment with an evaluation dataset. Training the models assesses whether the transfer learning is performed correctly, whereas the evaluation assesses the quality in terms of the respective generalizability of the three models as a result of an experiment with the same dataset— the evaluation dataset. Figures 4, 5, 6, 7, 8, and 9. show the metrics relevant for the training process; mAP, Recall/AR, BC Loss, RPN Loss and Total Loss refer to the test dataset, and train Total Loss refers to the training dataset. For model 1, the mAP increases steeply up to the 10,000– 15,000 iteration line and then converges to a value between 0.95 and 0.98. Models 2 and 3 have a similar trend; the mAP level increases steeply up to the value 15,000 and converges to a range of values from 0.81 to 0.87 and 0.78 to 0.82, respectively. This indicates a lack of diversity or the greater homogeneity of the training dataset in Model 1 compared to Models 2 and 3. AR and mAP are similar in the course and with respect to their values. It can be inferred that FN and FP have similar values. The values of mAP and AR of 0.96–0.80 from Table 1 conclude that only low values for FN and FP occur. All values of the Failure or Loss functions are absolute values and do not represent percentages. The value of the Box Classifier Localization Loss of all models varies between 0.15 and 0.35, implying that all models are at a similar level with respect to the different test datasets and the region of interest detected. Here, the BC Loss of the localization refers to the last layer of the Faster R-CNN model, which is responsible for determining the position of a bounding box (Lee et al., 2019, p. 1). The RPN Loss of the models is in a comparable range; the converging values are at 0.45 for model 1 and 0.52 for Models 2 and 3. The low first value could be due to the small horizontal and vertical displacement of the data. The Total Loss of the three models is within the range of 1.2 and 1.5 after the 20,000 iteration. The low variation could be on account of the minimum of a loss function not being found among the summed Total Loss. To evaluate the training process, the Total Loss of the training dataset is described in addition. A low value here corresponds to a model with a higher quality— unless the model tends to overfit the training data (Arsalan Soltani & Chen, 2015). Each of the models has a value below 0.2, therefore, the loss functions of the training should be sufficiently minimized. Crucial for the final assessment of the training quality of the models is the *F*1-score, presented in Table 1; the values range from 0.97 for model 1, 0.85 for model 2, and 0.80 for model 3 for a comparable number of iterations. The predictive ability of model 1 result values is the highest, however, this is accompanied by an increasing Total Loss— 1.40 for model 1 compared to 1.26 for model 2 and 1.19 for model 3. This means that while the accuracy of the first model increases compared to Models 2 and 3, the misses also increase. This indicates increasing generalizability of Model 2 and 3 compared to Model 1. The values of the three models show little change with further iterations. The improvement of the mAP and AR values are minimal. The values of the loss functions are minimized or have a small increasing tendency.

Therefore, the models are sufficiently trained (Srivastava et al., 2014, p. 1). Another training would not increase the performance, resulting in an overfitting of the models. The requirements for the models to perform an evaluation are thus fulfilled.

Results of the evaluation experiment are given in Table 2. The *F*1-scores for the three models when tested with the same evaluation dataset (Table 1) are lower than the *F*1-scores of the test dataset in training (Table 2). The reason for this is the larger variance of the data of the evaluation dataset compared to the test dataset of the training. It is noticeable that contrary to the results of the training, the *F*1-score in the evaluation experiment increases from model 1 with 0.6, to model 2 with 0.65, to model 3 with 0.72. This is evidence that the predictive ability of the models trained using active learning is greater than the first model trained using only synthetic data Model 3, which was trained based on data augmentation and data simulations using active learning, achieves the highest results. The lower total loss of Model 1 (2.23), Model 2 (1.55), and Model 3 (1.19) also illustrates increasing quality by incorporating user feedback.

6 Discussion and Limitations

The experiments and results confirm that active learning improves the quality of models in terms of diversification and applied generalization. It also reduces the effort of the manual labeling of data. The results of increasing accuracy of document recognition by the developed MLaaS show that systems which learn and continuously improve with active learning through user feedback have a higher probability of achieving a desired prediction of results than alternative systems. The result of the evaluation is the behavior of a deep learning model studied in three different states. These are created by continuous training with different data; the basis is a model based on synthetic data, and the other two models—for investigating the continuous augmentation of deep learning models—are implemented by incorporating user feedback. The data used is generated using data augmentation and a simulation of real data. The result of the training is that the $F1$-score for the models with active learning decreases due to the homogeneity between training and test datasets and the different structure of the datasets. The evaluation experiment confirms the hypothesis that the continuously extended models have a better generalizability with regard to document recognition.

The limitations and opportunities of the present prototype naturally span across different domains. Enhancements in the area of text classification, a model metadata database, or the development of an outsourced service for autonomous model validation are considerable. The scope of the current implementation does not include functionalities for user authentication. The service OCR, serving, and pipeline are not able to simply adapt authentication methods— such as a "bearer token". A limitation occurs due to the storage of user feedback in a CSV list, which compromises the ACID principle (atomicity, consistency, isolation, and permanence or persistence), which is supposed to ensure transaction security. Therefore, a database like SQLite or PostgreSQL would be useful for the metadata of the model. Google has introduced the MLMD—ML Metadata— concept for this purpose, which makes all relevant information for the model accessible (Google, 2010b). Another significant limitation is associated to the belief that the user will provide qualified valuable feedback. The pipeline would continue to produce new models, however, the models would be redundant, missing an improved increase in the precision An interesting approach to solve this would be by taking a psychological factor into account, for instance a reward system for the user. Even though these limitations exist, the developed prototype can simply be transferred to other related use cases without significant challenges. Apart from the defined object classes or label names, all approaches presented in the concept are free of context-specific algorithms. Essentially, they are defined by the data used for training. If a comparable dataset was available for a similar use case, the overall system could be transferred with little effort.

7 Conclusion

The objective of this paper is to develop a generalized scientific concept as a basis for the further development of deep learning models using user feedback. The working hypothesis is the assumption that a software learns via user feedback and thus improves the system's efficiency. The topic of artificial intelligence in relation to machine learning and deep learning is covered increasingly extensively in the current research literature. Deep learning is well suited for object recognition tasks, as the performance of the models and the algorithms improve as the size of the data increases. Relevant error tolerances for the training process of deep learning models are overfitting and underfitting. The paper shows that object recognition methods based on CNNs can be successfully used for document classification. In order to create the possibility to continuously extend a deep learning model, a pipeline and a lifecycle management are necessary. The combination of transfer learning and active learning is novel in this context. It is used to develop an autonomous and self-learning system for document classification based on user feedback data. The practical implementation of single partial solutions is already presented in the literature: however, the combination as a complete system exists only as a theoretical approach.

The software solution developed in this paper is able to capture documents and independently determine the content relevant to the business process. The user can supplement the data determined and suggested by the system as needed and furthermore has the possibility to adjust the data. The development concept of a prototype requires two basic technologies; an OCR engine and a machine learning framework to identify content areas using the deep learning model. TensorFlow as a deep learning framework and Tesseract for text recognition are selected. The container-based microservice architecture consists of the components frontend, middleware (API gateway), and backend services (Serving, Pipeline and OCR). For the frontend, React is provided as the Web framework and UIkit and D3 for the design. The API gateway in the software solution is based on Python in combination with Flask for the required interfaces. The API is designed and documented in the software solution using a Swagger UI. Serving, as a component of the backend service, allows the deep learning algorithm to be accessible via the Web. A pipeline allows for the existing neural network to be continuously extended with user feedback from the frontend and provides the serving component with the new model. Apache Airflow is used for lifecycle management and the accompanying orchestration of the elements of the pipeline. The OCR microservice uses Tesseract to translate images into text. As a result, the implementation of the software concept has created a prototype that is used to classify documents, and the system is continuously being extended as part of active learning with the Faster R-CNN deep learning model. The quality of the deep learning models can be technically monitored and compared using the TensorBoard. Metrics for validation include Mean Average Precision, Average Recall, Total Loss, and $F1$-score.

Via the frontend, a user can easily move the recommendation of the recognized content area—visualized by a rectangle in the image of the document—and thus correct the classification of the uploaded document. Challenges in the implementation are the conversion of relations with regard to the selection mask, the displayed images of the document, and the original sizes of the document.

The subject of the evaluation is an experiment with a deep learning model using the newly developed machine learning as a service. This uses active learning with user feedback from the document classification system directly, thereby continuously expanding the neural network. The behavior of the model is analyzed based on three states which are created by continuous training with different data. The first state is a model based on synthetic data and the other two models are implemented by using the service and considering user feedback—for the second state with data from data augmentation and for the third state additionally with real data generated in a simulation. The experiment consists of both training and experimental components, which gather opposite results. The training shows that the $F1$-score for the models with active learning decreases. This is caused by the homogeneity between training and test datasets and the different structure of the datasets for each model. However, in the experiment with a test dataset of potentially simulated realistic images, the recognition quality of the models increases when using active learning. Hence, a deep learning model trained with only synthetic data performs worse in evaluation and practice than enhanced models trained with user feedback through active learning. The evaluation experiment, therefore, confirms the hypothesis that the continuously augmented models have improved generalizability and a continuous training pipeline increases accuracy with respect to document recognition.

Acknowledgements The research for this paper was financially supported by Exprivia S.p.A, University of Trento (UniTrento), Fondazione Bruno Kessler (FBK) and the European Institute of Innovation & Technology (EIT).

References

Apache Airflow. (2020). *Airflow documentation concepts*. Retrieved from Airflow Documentation Concepts: https://airflow.apache.org/concepts.html

Arsalan Soltani, A., & Chen, K. (2015). *Neural network—How to interpret loss and accuracy for a machine learning model*. Retrieved from Neural network – How to interpret loss and accuracy for a machine learning model: https://stackoverflow.com/questions/34518656/how-to-interpret-loss-and-accuracy-for-a-machine-learning-model

Baylor, D., Koc, L., Koo, C. Y., Lew, L., Mewald, C., Modi, A. N., Jain, V. (2017). *TFX: A tensorFlow-based production-scale machine learning platform*. Proceedings of the 23rd ACM SIGKDD International Conference on Knowledge Discovery and Data Mining—KDD '17, pp 1387–1395. ACM Press. https://doi.org/10.1145/3097983.3098021

Chen, K., Seuret, M., Hennebert, J., & Ingold, R. (2017). *Convolutional neural networks for page segmentation of historical document images*. 2017 14th IAPR International Conference on Document Analysis and Recognition (ICDAR), pp. 965–970. IEEE. doi: https://doi.org/10.1109/ICDAR.2017.161

Denecken, S. (2018). *SAP intelligent robotic process automation—Adding the "Brain" and "Skillsets" to your digital workers*. Retrieved from SAP Intelligent Robotic Process Automation—adding the "Brain" and "Skillsets" to your digital workers: https://blogs.sap.com/2018/09/17/sap-intelligent-robotic-process-automation-adding-the-brain-and-skillsets-to-your-digital-workers/

Everingham, M., Van Gool, L., Williams, C. K., Winn, J., & Zisserman, A. (2010). *The Pascal Visual Object Classes (VOC) challenge*, pp. 303–338. Retrieved from https://doi.org/10.1007/s11263-009-0275-4

Fischer, P. (2000). *Algorithmisches Lernen*. Teubner.

Gao, L., Yi, X., Jiang, Z., Hao, L., & Tang, Z. (2017). *ICDAR2017 competition on page object detection*. 2017 14th IAPR International Conference on Document Analysis and Recognition (ICDAR), pp. 1417–1422. IEEE. https://doi.org/10.1109/ICDAR.2017.231

Google. (2019a). *Orchestrating TFX pipelines*. Retrieved from Orchestrating TFX Pipelines: https://www.tensorflow.org/tfx/guide/airflow

Google. (2019b). *ML metadata TFX*. Retrieved from ML Metadata TFX: https://www.tensorflow.org/tfx/guide/mlmd

Google. (2019c). *The TFX user guide*. Retrieved from The TFX User Guide: https://www.tensorflow.org/tfx/guide

Koch, B. (2019). *International market overview & forecast*, p. 23.

Lee, S., Kwak, S., & Cho, M. (2019). *Universal bounding box regression and its applications*. Retrieved from http://arxiv.org/abs/1904.06805

Li, X.-H., Yin, F., & Liu, C.-L. (2018, August). *Page object detection from PDF document images by deep structured prediction and supervised clustering*. 2018 24th International Conference on Pattern Recognition (ICPR), pp. 3627–3632. IEEE. doi:https://doi.org/10.1109/ICPR.2018.8546073

Luger, G. F., & Stubblefield, W. A. (1998). *Artificial intelligence: structures and strategies for complex problem solving* (3rd ed ed.). Addison-Wesley.

Manyika, J., Chui, M., Miremadi, M., Bughin, J., George, K., Willmott, P. J., & Dewhurst, M. (2017). *A future that works: automation, employment and productivity*.

Ng, A. (2017). *Opening a new chapter of my work in AI.* Retrieved from Opening a new chapter of my work in AI: https://medium.com/andrewng/opening-a-new-chapter-of-my-work-in-ai-c6a4d1595d7b

Oliveira, D. A., & Viana, M. P. (2017). *Fast CNN-based document layout analysis*. 2017 IEEE International Conference on Computer Vision Workshops (ICCVW), pp. 1173–1180. IEEE. doi:https://doi.org/10.1109/ICCVW.2017.142

Olston, C., Fiedel, N., & Gorovoy, K. (2017). *TensorFlow-serving: Flexible, high-performance ML serving*, 2, 8. Retrieved from http://learningsys.org/nips17/assets/papers/paper_1.pdf

Padilla, R. (2019). *Object-detection-metrics*. Retrieved from Object-Detection-Metrics: https://github.com/rafaelpadilla/Object-Detection-Metrics

Pezza, S., & Jan, W. (2012). AP invoice management in a networked economy. *AP Invoice Management in a Networked Economy*, 25.

Schreiber, S., Agne, S., Wolf, I., Dengel, A., & Ahmed, S. (2017). DeepDeSRT: Deep learning for detection and structure recognition of tables in document images. *2017 14th IAPR International Conference on Document Analysis and Recognition (ICDAR)*, pp. 1162–1167. IEEE. https://doi.org/10.1109/ICDAR.2017.192

Settles, B. (2010). *Active learning literature survey*, p. 67.

Settles, B. (2011). *From theories to queries: Active learning in practice*, p. 18.

Srivastava, N., Hinton, G., Krizhevsky, A., Sutskever, I., & Salakhutdinov, R. (2014). *Dropout: A simple way to prevent neural networks from overfitting*, p. 30.

Theobald, E. (2019). *Marketing Intelligence: ein Lehrbuch für die Praxis* (1. Auflage ed.). Verlag W. Kohlhammer.

Walter-Tscharf, V. (2021). *Source code implementation and evaluation of a MLaaS for document classification with continuous deep learning models*. Retrieved from Source code implementation and evaluation of a MLaaS for document classification with continuous deep learning models: https://github.com/FranzTscharf/ml-pipeline

Reflections on New Urbanism at the COVID-19 Pandemic Background—Urban Health Research

Junyu Hu and Xinyi Hu

Abstract

The theory of New Urbanism, which promotes high-density, mixed zoning, and high walkable communities, has had a profound influence on the design of many cities globally and played a certain positive role in urban health. However, under the current COVID-19 pandemic, this theory faces new challenges. This paper aims to analyze the influence of New Urbanism on urban health from different spacial scales and reflect on whether these design methods still apply to the post-pandemic era. A series of recommendations for urban systems at different levels are proposed for further improving health of cities and residents. A real-world scenario-based design case is used to demonstrate ideas for optimizing public spaces in the districts. This paper is expected to provide ideas for future urban development and renewal design leading to a healthier and more sustainable city.

Keywords

New urbanism • COVID-19 • Urban health • Public open space

1 Introduction

The New Urbanism is the urban renewal theory that emerged in the late 1980s, reflecting deeply on the rapid expansion mode of Suburbia. To some extent, it represents the development direction of modern urban design theory. After World War II, under the influence of industrial cities, architects and planners, led by Le Corbusier, proposed the idea of fleeing the city and expanding into the rural areas to realize the "American Dream", setting off a suburbanization movement (Clements, 2017). This trend temporarily allowed city dwellers to escape the congestion of the metropolis. However, with the disorderly expansion into rural areas, large amounts of land were changed into low-density blocks termed "leapfrog pattern" (Frumkin, 2002). Different land functions are far from each other, forming an automobile-dominated commuting system, which led to city center deterioration, environmental pollution, frequent traffic accidents, and health problems, making people reflect on this development pattern. In 1994, the Charter of the New Urbanism was agreed and signed on the fourth Congress for the New Urbanism. It advocates creating a high-density, walkable, and vibrant community and has profoundly influenced the development of many cities, towns, and suburbs (Congress for the New Urbanism, 1996; Trudeau, 2013). The previous literature mainly focused on design methods and public policies about the community and local urban space under the guidance of the New Urbanism principle, including multi-functional mixed zoning (Lucchesi et al., 2020) and multi-level street system (Iravani & Rao, 2020), in order to improve the walkability and vitality of the community and increase the physical activity of the residents. There is not much research on the impact of New Urbanism on urban health from different spatial scales.

The COVID-19 pandemic has threatened public health and swept the whole world since last winter, directly or indirectly affecting everybody, which has aroused people's great concern about urban health. During this period, many countries adopted different degrees of restrictions to control the pandemic. Residents' outings were restricted to some extent (Zaremba et al., 2020). Because of the lack of outdoor sports, other physical health problems such as obesity have also become a "pandemic" (Chua, 2020). Simultaneously, insomnia, anxiety disorders, depression, and other mental

J. Hu (✉)
Department of Architecture, Civil Engineering and Environmental Sciences, Technical University of Braunschweig, 38106 Braunschweig, Germany
e-mail: j.hu@tu-braunschweig.de

X. Hu
School of Architecture, Southeast University, Nanjing, 210096, China

H. Altan et al. (eds.), *Advances in Architecture, Engineering and Technology*,
Advances in Science, Technology & Innovation, https://doi.org/10.1007/978-3-031-11232-4_21

health problems increased with home isolation (Wang et al., 2020). However, the spread of COVID-19 will not disappear in the near future but is a long-term challenge (Pak et al., 2020). Whether the principles of New Urbanism can still have a positive impact on urban health requires further research.

In the following lines, the influence of New Urbanism on urban health will be studied from different spatial scales. Then the reflection on the New Urbanism will be conducted about whether it is still applicable in the background of the post-pandemic era and whether any improvements are required to reduce the risk of epidemic re-emergence and increase the speed of emergency response in the future. Based on the reflection results at the district level, a design will be illustrated to state a specific design idea, which embodies the understanding of New Urbanism and urban health.

2 The Influence of New Urbanism on Urban Health

This section analyzes the influence of New Urbanism on urban health from the three different spatial scales: the overall urban planning, the districts, the public open space, and buildings inside the districts.

2.1 Overall Urban Planning

One of the main concerns of New Urbanism is the urban transportation system. With the allocation of land resources, the compact city pattern is promoted, where the transportation mode is changing from private automobiles to non-motor (walking or cycling) and public transportation systems. This trend also has a positive impact on urban health.

The former suburbia advocated an idealized artificial system in which widening car roads and compressed sidewalks were becoming the external manifestations of the modern traffic system. "Becoming a Showcase: Virginia Beach Boulevard-Phase I celebrated its completion…" (Duany et al., 2001) People in the city center were proud to achieve a concrete life vision: an 11-lane highway and plenty of parking area. This city was the practice result of modern engineering and regulation and the microcosm of a modern city. However, the increasing number of private cars led to a higher incidence of road congestion and traffic accidents. Therefore, old roads were widened, and more new ones were built, further promoting private automobiles and making the city's non-proliferation situation more severe and finally falling into a vicious circle. Although this kind of transportation system was first built on human demand, ultimately, it became the system for transporting cars instead of people. In this automobile-dominated city, people worried about parking spaces every day and had little time to exercise. People's health needs were neglected. The city eventually became one space designed for cars.

The use frequency of private automobiles is closely related to the users' health. A study of Atlanta-area showed that residents were 3% more likely to be obese with every five-minute increase in daily driving time (Speak, 2013). The impact of daily commuting on residents' mental health is also significant. It is reported that increasing commuting time spent on congested roads eventually raises car drivers' mental health problems, such as road rage (Frumkin, 2002). On the contrary, the non-motorized commuting model and the "public transport first" principle encouraged by New Urbanism can help weaken the adverse effects of private automobile commuting on residents' physical and mental health. Walking and cycling increase the amount and time of people's exercise to some extent. Public transport vehicles consume energy more centrally and efficiently than private cars, reducing carbon emissions and external transportation costs (Wu, 2006).

2.2 Districts

Many factors affect residents' healthy lives, including social factors, external and internal motivations, indoor and outdoor environments. It is reported that except for keeping pets, good urban design is also one of the main motivations for a healthy life (Katapally et al., 2018). The New Urbanism aims to explore how to build vibrant and walkable communities to guide people to travel more frequently on foot or by bike (Clements, 2017).

In terms of land function, New Urbanism advocates the land zoned with multiple uses to increase diversity. In urban design, living and public space diversity play a crucial role in residents' healthy living. As illustrated in Fig. 2, the left plan follows the typical Euclidean Zoning pattern used during suburban sprawl with wide roadways and large single-use zones. The right one is the layout with traditional small-scale blocks, where the building mass is smaller than the left one, and the multiple functions are mixed in each zone. The dotted circles in two images cover areas of the same size. In traditional layouts, the mixed zoning development enhances the close connection between functions with reduced travel distances. The demand for private cars is minimized, and people prefer walking or cycling. It is easy to understand that in mixed-use communities, daily-life facilities are scattered around the residential areas. When people travel with utilitarian purposes, if the distance between the starting point and the destination is relatively close, it will give people the reason to increase the possibility of walking or cycling (Rodríguez et al., 2006). The research conducted in 13 cities of the Atlanta region metropolitan area also confirms this argument. It is investigated that there is a direct linkage between the diversity of land function and the obesity rate of residents. The mixer the land use is, the more social activities can be completed by walking or public transport over short distances, and the lower obesity rate will be observed (Frank et al., 2004).

The mixed-use zoning also reduces commuting distances, and the time saved can be supplemented by sleep and physical activity, thereby reducing the incidence of chronic diseases and further enhancing public health. As the commuting distance reducing, people prefer taking public vehicles or riding bicycles, or even walking, and air quality can be improved with a gradual process that will eventually create a virtuous circle.

In terms of block connections, New Urbanism advocates roadway networks with improved interconnectivity and walkability. In the left image of Fig. 1, the block is bigger, and roads are wider than the right one, leading to the poor connecting road network. While the road layout in the right image is the one people prefer to walk. Because the smaller block scale and higher dense road network represent more intersections and more robust connectivity, which provide more route options, people can choose to take different routes between the starting point and the endpoint, improving the interest in walking. This road network will enhance people's willingness to walk and facilitate the number of their physical activities (Iravani & Rao, 2020). Due to the smaller blocks in the right image and compact layout, the streets are narrower, reducing the speed of automobiles. It means that the likelihood of a car accident can be decreased to ensure the safety of pedestrians.

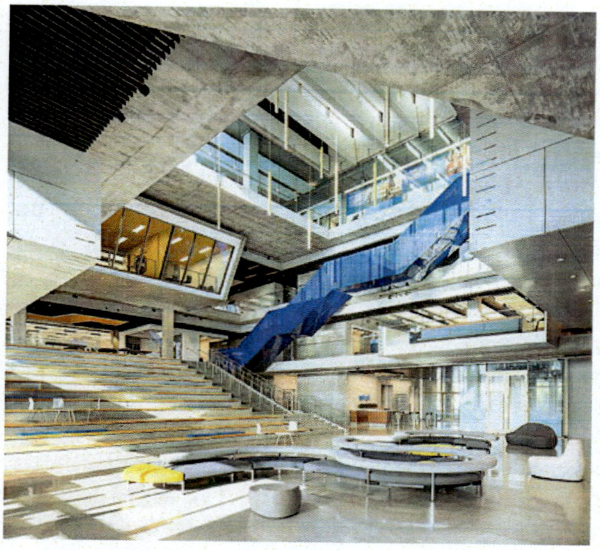

Fig. 2 Open stairs in an office building (Cohen, 2018)

2.3 The Public Open Space and Buildings in Districts

The New Urbanism also pays attention to public open space design such as squares and green spaces in blocks. Their relationship with surrounding areas should be organized well to ensure their accessibility and openness. In the previous automobile-dominated traffic mode, it was easy to form apathetic, low-dense urban space that transcended the scale of human nature. The lack of opportunities to communicate in public open spaces resulted in weak links between residents. While in a compact city, favorable design of the open space can attract residents, promote mutual acquaintance,

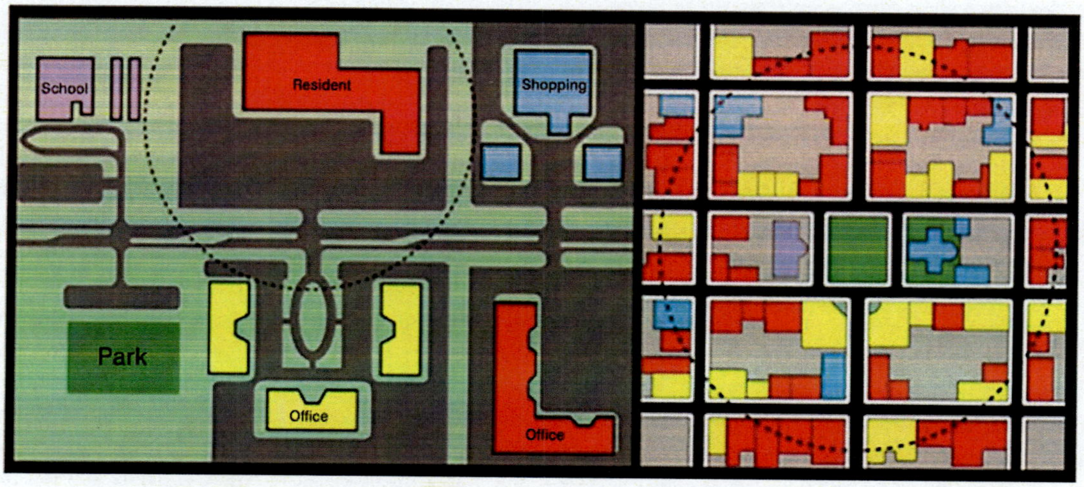

Fig. 1 Comparison of typical euclidean zoning and traditional small-scale zoning (Clements, 2017)

and communication between the residents, which can enhance the harmony and cohesion of the community and contribute to the mental health of the residents. In addition to being a social setting, public open space can encourage public health in two other ways. The green space environment is suitable for residents to conduct physical activities. In addition, it can be crossed by people to reach other destinations or be part of the walking or running route (Koohsari et al., 2015). Similarly, attention should also be paid to the design of the traffic space inside the buildings, such as good flow organization and staircase design, to increase people's physical activities in the building and reduce the dependence on elevators. For example, the open stairs in Fig. 2 are set in a public space without sight obstacles, which is more attractive than a traditional closed staircase.

The New Urbanism proposes design suggestions from macro to micro aspects of the city, which not only save the whole city from automobiles and free the residents, but also has a positive impact on the urban health.

3 Reflection on the New Urbanism Under the COVID-19 Pandemic

Since the COVID-19 outbreak, it has rapidly risen to a global pandemic prevention incident. People have experienced many changes, from the initial panic consumption causing insufficient supply and the spread of false news to the substantial blow to the global financial market and various industries (Ibn-Mohammed et al., 2021). The pandemic has destroyed the harmonious and modern society, and the flourishing market economy that we worked hard to create. In urban planning, the design trends about interactions that promote communication have become less critical or even excluded. However, the COVID-19 pandemic is impossible to end suddenly in a short time. Instead, this is likely to be a long-term challenge. Then, we cannot help thinking whether the New Urbanism principles are still appropriate in this context? In terms of contributing to urban health, what adjustments or improvements should New Urbanism make?

3.1 Urban Transportation System During Pandemic

The urban transportation system plays a crucial role in the virus spread process. Roads are like the bloodlines, connecting the economic and cultural lifeblood of the entire city. The primary and secondary roads guide the tertiary streets and then toward different blocks. This relationship can be analogized as the aorta, the secondary artery, and the capillaries in a human body. Oxygen and nutrients in the blood are transported through blood vessels throughout the body, penetrating capillaries into every tiny tissue, eventually feeding each cell. The vascular system that the human body has evolved over hundreds of millions of years can be used as a philosophical explanation for the segmentation of the transportation system and the city walkability in New Urbanism.

The way people travel has changed accordingly. The accessibility to medical facilities has become even more important for people in need of medical assistance. However, to avoid face-to-face interaction for the healthy ones, they would abandon taking public vehicles and prefer to travel by walking or cycling (Seidlein et al., 2020). Under the instructions of multi-functional mixed zoning by New Urbanism, people's daily needs can be met in the vicinity of residential areas, which could greatly reduce the travel distance and increase the frequency of people choosing to walk or ride, thus reducing the possibility of widespread of the virus across regions and facilitating the pandemic control.

3.2 Public Open Space During Pandemic

In addition to directly serving medical facilities, public open space can also benefit residents' mental health. During the pandemic, face-to-face interaction has been greatly reduced due to the block-down policy and public fear of the virus spread. However, the need for the accessibility of public open space is always there, especially for green space (Zhu & Xu, 2020). Prolonged home isolation not only affects residents' mental health but also may result in serious mental problems such as anxiety disorders. In the vicinity of urban residential areas, ensuring enough walkable green spaces for outdoor activities can reduce the risk of anxiety disorders. It is reported that as the distance between available green space and small settlements decreases, or the affection of green space in large residential areas increases, the number of treatments for anxiety disorders required per capita decreases (Nutsford et al., 2013). Therefore, the requirement to going outside should not be prohibited or ignored. It is expected that the reasonable designs of public open space could help people not only exercise the right to enjoy the fresh air and sunshine, the beautiful scenery, and communicate with others, but also protect them from the virus invasion.

During the block-down, Italian residents held concerts on their balconies, playing instruments on their own and cooperating with neighbors on different balconies, while others listened quietly. Then people would applaud and cheer for the performers across the balconies or the streets (Grigoriadou, 2020). The emergence of this phenomenon is the embodiment of people's social needs and the functional transformation case of the building as well. As a part of many families, the balcony is usually used for sunbathing, potted plants, and other leisure activities. The balcony is an open space that people can enjoy in their own homes. It is not comparable to the squares or parks with larger areas and richer plants. However, due to people's desire for outdoor activities and the boredom of home isolation, a functional transformation of the balcony is naturally achieved, making it change from a family leisure space to a social platform and satisfy the requirement of a safe distance. Such a transformation pattern can be considered to be applied under the circumstance of public open space, allowing residents to have more options in the post-pandemic era rather than just staying at home.

3.3 Architectural Design During Pandemic

Looking back at the story of the balcony concert in Italy during the outbreak, it is not difficult to recall the microscopic part of the New Urbanism, namely what can we do about the architectural design during the pandemic?

In buildings, especially residential buildings, the central spot of the interaction is the traffic space. Stairs and elevators are one of the significant parts of the virus's spreading path. When people operate the elevator, they make direct contact with the buttons, and then when the next person operates the elevator, an indirect contact occurs (Hasanreisoglu & Hasanreisoglu, 2020). The air quality in the confined elevator spaces is hard to ensure due to lack of ventilation, providing opportunities for the virus to spread. We can only find another way to replace it for this inevitable contact in a confined space.

Staircase, during the pandemic, is a better travel space because of its good lighting and ventilation. However, due to the height and physical power constraints, people prefer elevators. In order to reduce the frequency of elevator use and reduce the probability of infection in the elevator, it can be stipulated that the elderly and other people under particular circumstances can take priority elevator; other residents are not recommended to use elevators. The protection measures should be increased, and the number of people should be limited in the elevator. For new buildings in the future, it is recommended to create a more comfortable staircase environment and provide the staircase with richer functions. For example, a rest platform or common area can be set up in the staircase to carry certain indoor functions like reading, discussion, and planting to attract more people to take the stairs and increase their physical activities.

4 Design Case

Based on the reflection mentioned above on New Urbanism during the pandemic, a landscape architectural design is proposed as an example. The protection requirements are taken as the design precondition. This design is determined to provide residents with appropriate public areas for activities and communication, both physically and psychologically, to promote healthy lives. The main issues that people may encounter when isolated at home during the pandemic are analyzed in Fig. 3, and they are also considered the primary motivation of this design. As shown in Fig. 4, they are further subdivided into six specific functions.

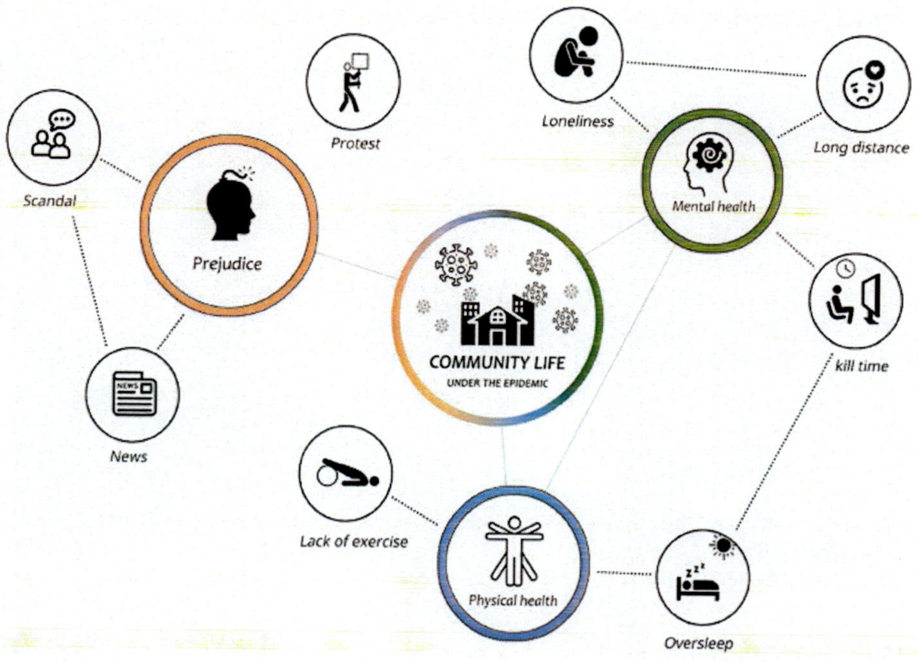

Fig. 3 Main problems that may arise from home quarantine during pandemic

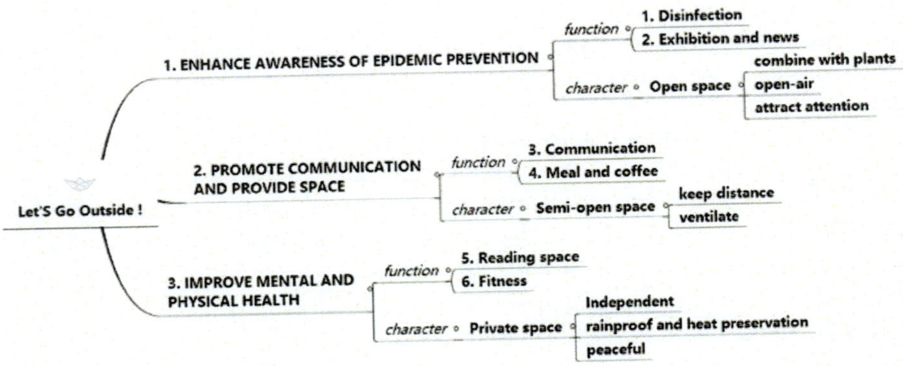

Fig. 4 Primary design motivations and main functions of the pavilions

4.1 The Pavilion as the Information Platform

As shown in Fig. 5, the function of this pavilion mainly contains information display and art expression. The first is to update the pandemic information, pass it on to the residents, and provide related protection recommendations. The information is spread through the network media and monitors. The key information, such as the number of daily new patients, can be illustrated with a separate digital board for warning function. This pavilion is recommended where people pass by every day, such as the street corner or the parking lot. The other function of this pavilion is to convey people's emotions about the pandemic through art methods. The mental problems that arise during the pandemic, such as fear of the virus and distrust of human interaction can be expressed more artistically. It is essential to provide a platform for the display of our ideas and emotions.

This pavilion does not take up much public space and is set like a small module in the city to complement the missing features. After the pandemic, it can be transformed into other functions, such as a small shop, advertising location, or art exhibition space.

4.2 The Pavilion as Public Communication Space Meeting the Protection Requirements

This pavilion presented in Fig. 6 is the same size as the previous one, but their difference is in the internal function and spatial form. In this case, a more private space is created for two people (friends or strangers) to communicate. Due to the fixed seats and narrow tables in this pavilion, two people can only sit opposite each other and are separated by 1.6 m (1.5 m is the minimum safe distance) with a small window in front of them. If they want to initiate a conversation, they can open the door panel and chat face-to-face. They can also remain relatively independent if they want to do something for themselves. Each side is equipped with a disinfectant

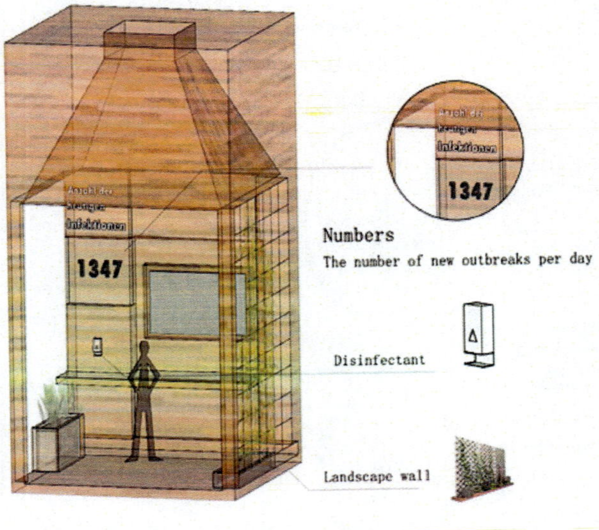

Fig. 5 Pavilion as the information platform

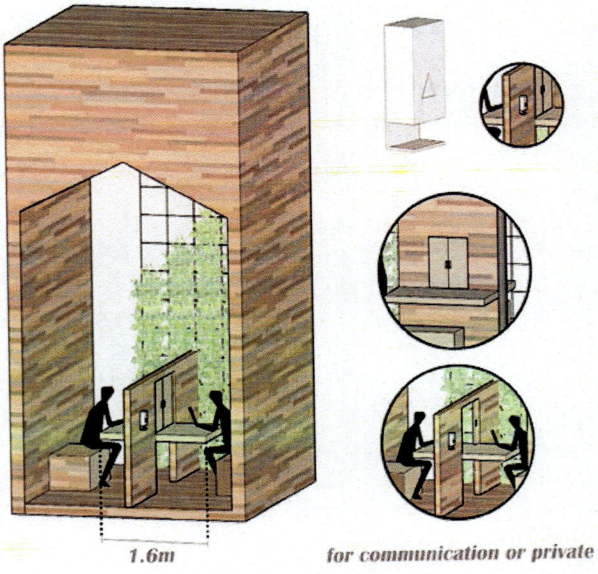

Fig. 6 Pavilion as public communication space

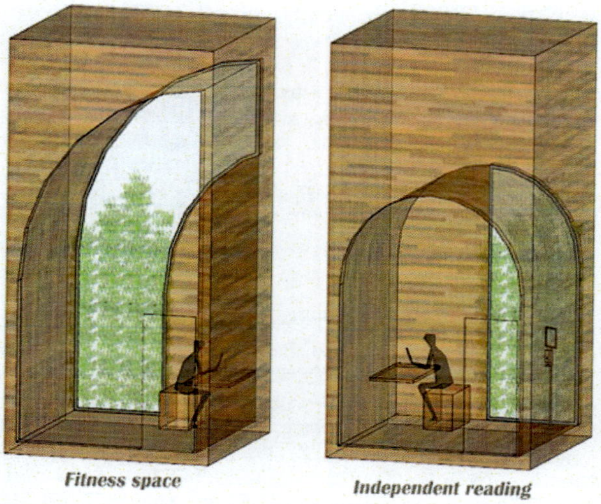

Fig. 7 Pavilion used for independent reading and exercising

hand sanitizer to keep the furniture clean during use and prevent the virus spread caused by indirect contact.

4.3 The Pavilion Used for Independent Reading and Exercising

As illustrated in Fig. 7, the arched element is introduced into this pavilion's interior space to create a bright and welcoming indoor environment (niche). Arches have beautiful continuous curves, which often play a role in regulating emotions in architectural elements. This relatively independent space can be used for residents to read or exercise. A confined space, in principle, can only serve one person at present. However, when combined with the two types of pavilions mentioned above, small groups can be created in cities to achieve various public needs.

The above three kinds of pavilions can be combined to meet residents' psychological and physiological spatial needs to some extent, strengthening people's understanding of the pandemic simultaneously. Following the principles of New Urbanism, appropriate public space within walking distance should be selected to add these pavilions. It was mentioned earlier that the conversion capacity of public open space between streets and groups should be enhanced, leaving room for the additions during the pandemic and space for structures similar to these pavilions.

5 Conclusion

This study set out to study the influence of New Urbanism on urban health from different spatial scales (from the urban transportation system to district level and architectural design). In light of the current pandemic, it is considered systematically whether the New Urbanism principles are still a boost to urban health or areas that need improvements. Finally, through an experimental architectural design case, one of the ideas to improve the residents' physical and mental health is proposed at the district level.

After studying the theory of New Urbanism at different spatial levels, it is concluded that these principles still have a promoting effect on urban health. They are also conducive to the overall framework for pandemic control by increasing community density and functional diversity, encouraging residents to travel on foot or by bike, and encouraging people to use open stairs rather than closed elevators. However, there are still areas that need to be reconsidered and redesigned. The public vehicles and buildings need to increase facilities to meet the pandemic's prevention and control requirements. Public open space should be thoughtfully redesigned to meet the residents' spiritual needs of green space and social interactions, as well as the physiological needs of outdoor activities, to ensure people's healthy life.

In the urban development during the post-pandemic era, some of the New Urbanism principles are worthy of references, such as public space walkability and multi-functional mixed zoning of districts. The pandemic will be a long-term battle, not only now, but something similar will happen again in the future. The learning, reference, and reflection of the previous design thoughts will guide us to a more sustainable and healthy city.

References

Chua, W. J. M. (2020). Managing patients with obesity in the post COVID-19 world: time to sharpen the saw. *Obesity Research & Clinical Practice*. https://doi.org/10.1016/j.orcp.2020.11.008.

Clements, J. (2017). *New urbanism: How urban design impacts health, wealth, and sustainability of our cities*. https://epublications.regis.edu/theses/806

Cohen, E. (2018, January 4). *Intuit by clive Wilkinson architects and WRNS studio: 2017 Best of year winner for large creative/tech office*. https://www.interiordesign.net/projects/14280-intuit-by-clive-wilkinson-architects-and-wrns-studio-2017-best-of-year-winner-for-large-creative-tech-office/

Congress for the New Urbanism. (1996). *The charter of the new urbanism*. https://www.cnu.org/who-we-are/charter-new-urbanism

Duany, A., Plater-Zyberk, E., & Speak, J. (2001). *Suburban nation: The rise of Sprawl and the decline of the American dream*. Farrar, Straus and Giroux.

Frank, L. D., Andresen, M. A., & Schmid, T. L. (2004). Obesity relationships with community design, physical activity, and time spent in cars. *American Journal of Preventive Medicine, 27*(2), 87–96. https://doi.org/10.1016/j.amepre.2004.04.011

Frumkin, H. (2002). Urban sprawl and public health. *Public Health Reports, 117*(3), 201–217. https://doi.org/10.1093/phr/117.3.201

Grigoriadou, E. T. (2020). The urban balcony as the new public space for well-being in times of social distancing. *Cities & Health*, 1–4. https://doi.org/10.1080/23748834.2020.1795405

Hasanreisoglu, M., & Hasanreisoglu, B. (2020). COVID-19 Pandemia-uveitis specialist perspective. *Retina-Vitreus, 29*(2), 166–169. https://doi.org/10.37845/RET.VIT.2020.29.30

Ibn-Mohammed, T., Mustapha, K. B., Godsell, J. M., Adamu, Z., Babatunde, K. A., Akintade, D. D., Acquaye, A., Fujii, H., Ndiaye, M. M., & Yamoah, F. A. (2021). A critical analysis of the impacts of COVID-19 on the global economy and ecosystems and opportunities for circular economy strategies. *Resources, Conservation and Recycling, 164*, 105169. https://doi.org/10.1016/j.resconrec.2020.105169.

Iravani, H., & Rao, V. (2020). The effects of new urbanism on public health. *Journal of Urban Design, 25*(2), 218–235. https://doi.org/10.1080/13574809.2018.1554997

Katapally, T. R., Bhawra, J., Leatherdale, S. T., Ferguson, L., Longo, J., Rainham, D., Larouche, R., & Osgood, N. (2018). The SMART study, a mobile health and citizen science methodological platform for active living surveillance, integrated knowledge translation, and policy interventions: longitudinal study. *JMIR Public Health and Surveillance, 4*(1). https://doi.org/10.2196/publichealth.8953

Koohsari, M. J., Mavoa, S., Villanueva, K., Sugiyama, T., Badland, H., Kaczynski, A. T., Owen, N., & Giles-Corti, B. (2015). Public open space, physical activity, urban design and public health: Concepts, methods and research agenda. *Health & Place, 33*, 75–82. https://doi.org/10.1016/j.healthplace.2015.02.009

Lucchesi, S. T., Larranaga, A. M., Cybis, H. B. B., e Silva, J. A. de A., & Arellana, J. A. (2020). Are people willing to pay more to live in a walking environment? A multigroup analysis of the impact of walkability on real estate values and their moderation effects in two Global South cities. *Research in Transportation Economics*, 100976. https://doi.org/10.1016/j.retrec.2020.100976

Nutsford, D., Pearson, A. L., & Kingham, S. (2013). An ecological study investigating the association between access to urban green space and mental health. *Public Health, 127*(11), 1005–1011. https://doi.org/10.1016/j.puhe.2013.08.016

Pak, A., Adegboye, O. A., Adekunle, A. I., Rahman, K. M., McBryde, E. S., & Eisen, D. P. (2020). Economic consequences of the COVID-19 outbreak: The need for epidemic preparedness. *Frontiers in Public Health, 8*.

Rodríguez, D. A., Khattak, A. J., & Evenson, K. R. (2006). Can new urbanism encourage physical activity? Comparing a new urbanist neighborhood with conventional suburbs. *Journal of the American Planning Association, 72*(1), 43–54. https://doi.org/10.1080/01944360608976723

Speak, J. (2013). *Walkable city: How downtown can save America*. North Point Press.

Trudeau, D. (2013). A typology of New Urbanism neighborhoods. *Journal of Urbanism, 6*(2), 113–138. https://doi.org/10.1080/17549175.2013.771695

von Seidlein, L., Alabaster, G., Deen, J., & Knudsen, J. (2020). Crowding has consequences: Prevention and management of COVID-19 in informal urban settlements. *Building and Environment*, 107472. https://doi.org/10.1016/j.buildenv.2020.107472

Wang, M., Zhao, Q., Hu, C., Wang, Y., Cao, J., Huang, S., Li, J., Huang, Y., Liang, Q., & Guo, Z. (2020). Prevalence of psychological disorders in the COVID-19 epidemic in China: A real world cross-sectional study. *Journal of Affective Disorders*. https://doi.org/10.1016/j.jad.2020.11.118

Wu, R. (2006). *Compact urban space development pattern research—Construct the sustainable development city of China in the new time*. https://kns.cnki.net/KCMS/detail/detail.aspx?dbname=CMFD2008&filename=2008024067.nh

Zaremba, A., Aharon, D. Y., Demir, E., Kizys, R., & Zawadka, D. (2020). COVID-19, government policy responses, and stock market liquidity around the world: A note. *Research in International Business and Finance*, 101359. https://doi.org/10.1016/j.ribaf.2020.101359

Zhu, J., & Xu, C. (2020). Sina microblog sentiment in Beijing city parks as measure of demand for urban green space during the COVID-19. *Urban Forestry & Urban Greening*. https://doi.org/10.1016/j.ufug.2020.126913

Additive Formwork: Examining Design, Fabrication Space and Resolution for Bespoke Concrete Elements

Roberto Naboni and Luca Breseghello

Abstract

In the era when the construction industry urges a paradigm shift towards more sustainable and efficient building solutions, the paper describes the design scenarios opened by 3D printed reusable formworks in the production of bespoke building scale concrete elements as a solution to the formal and technical limitations of the conventional manufacturing processes. Combining the latest advances in computational design and Fused Deposition Modelling (FDM) 3D printing technology, the paper presents and discusses an exploration and assessment of the higher design freedom given by the use of Additive Formwork for architecture through an experimental setup, where the design and fabrication of a series of morphologically diverse concrete panels were carried out. As a method of assessing the geometric freedom enabled by the employed fabrication technology, it is presented as an explorative design strategy that translates into multiple outputs with distinct and complementary features within the manufacturing process's characteristics high-precision control over section, curvature, inclination and detailing. Through an analysis of the digital models' geometric features and an assessment of the precision of the manufactured artefacts, the results prove that combining FDM 3D Printing and concrete is a viable fabrication technique for bespoke elements, opening to unexplored aesthetics and design solutions that can potentially improve structural and material efficiency of concrete construction.

Keywords

Additive Formwork • Ultra-High-Performance Fibre Reinforced Concrete (UHPFRC) • Design for manufacturing • Geometric analysis • High-resolution

1 Introduction

The need to reduce carbon emissions and material consumption has motivated the recent focus on evolving concrete construction technology towards more flexible, material-efficient and high-performance methods of production, with environmentally and economically sustainable design solutions. As a semi-fluid material, concrete has indeed the capacity to be used for freeform, geometrically complex and high-resolution construction that can respond to the aforementioned design shift. However, technical, material and economic constraints in manufacturing formworks limit the adoption of high-resolution and non-standard architectural elements. On the one hand, in simple concrete construction, formwork materials and related labour account for over 50% of the total cost of concrete construction (Lab, 2007), which increases to over 83% for doubly-curved geometries (De Soto et al., 2018). Nonetheless, it is well-known that a gap exists between design and making as the latest advances in digital design and modelling tools do not correspond to similar growth in building and fabrication techniques (Lloret et al., 2015). The flexibility and sub-millimetre details that digital design tools enable are strongly bound by the production constraints, and a great effort in the building design processes is often put in the rationalisation of complex geometries to respond to these limitations (Naboni & Paoletti, 2015; Pottmann, 2013).

R. Naboni (✉) · L. Breseghello
CREATE—University of Southern Denmark (SDU), Section for Civil and Architectural Engineering, Odense, Denmark
e-mail: ron@iti.sdu.dk

H. Altan et al. (eds.), *Advances in Architecture, Engineering and Technology*,
Advances in Science, Technology & Innovation, https://doi.org/10.1007/978-3-031-11232-4_22

1.1 Research Aim

While innovative processes such as Computer Numerical Control (CNC) are already employed to manufacture custom formworks, this paper focuses on using Fused Deposition Modelling (FDM) 3D printing to produce formworks for high-resolution complex concrete elements. From a theoretical point of view, Additive Manufacturing (AM) offers the potential to match contemporary modelling tools' design freedom while minimising the need for discretisation processes. Moreover, AM has the flexibility to allow for geometric features hardly achievable with alternative fabrication methods. A limit to using this technology for the production of concrete formworks at an industrial scale is the development of a consistent method for designing, analysing and manufacturing formworks with AM. The overarching goal of this research is reducing the existing geometric gap between digital modelling and manufacturing of the formworks and enabling the full morphological potential of concrete. In this context, we conduct a design-to-manufacturing experiment to realise high-resolution concrete panels with additively manufactured reusable formworks (Fig. 1). This paper: (i) tests an integrated workflow for the design and fabrication through a design experiment of eight highly unique tiles; (ii) describes and analyses the geometric freedom provided by the proposed manufacturing technique, defining a set of objective parameters and analysis tools that evaluate the complexity of the elements.

2 Background

2.1 Geometrical and Technical Limitations in Concrete Construction

To understand the potential impact of AM on the architectural design and construction industry, it is helpful to refer to the evolution of constructions in history. Architectural geometry has been closely linked to innovation in manufacturing and construction technology in a mutual problem-setting and problem-solving relation and played a fundamental role in developing the field (Fischer, 2012). In Roman and Gothic architecture, as much as in the work of Nervi, Isler and Otto, geometric constructs were exploited to conceive efficient structures. Works such as the Philips Pavilion by Xenakis and Le Corbusier and Utzon's Sydney Opera marked the role of geometry in rationalising an original design (Austern et al., 2018). With the development in digital modelling tools of the late twentieth century and the consequent emphasis on complex shapes in architectural design, rationalisation became more and more relevant over the years (Flöry & Pottmann, 2010). More than sixty years

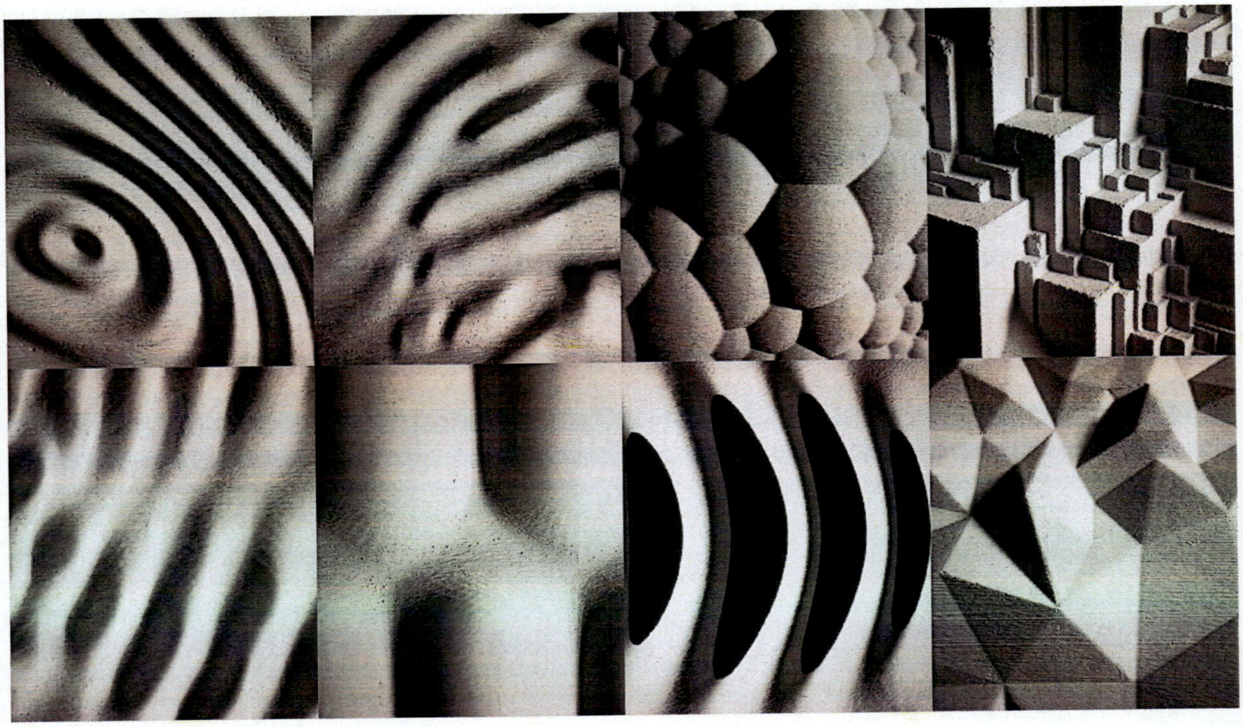

Fig. 1 Close-up views of eight experimental concrete panels designed and fabricated as a demonstration of the proposed design and manufacturing method. Image by CREATE/University of Southern Denmark

ago, Nervi observed that concrete construction was not exploited to its full potential due to technical limitations in preparing adequate formworks to contain the fluid material (Nervi, 1956). While the use of steel reinforcement is considered a milestone in the history of construction, Collins argued that the central characteristic of architectural concrete is not its plasticity but rather the way it is crafted by its mould (Collins, 1959). Still today, challenges and limitations posed by manufacturing processes are particularly relevant for constructions in concrete, still restricted by technical limitations. This paper investigates the potential of AM in expanding the design space of concrete construction, considering its level of detail and geometric flexibility.

2.2 State-of-the-Art in Digital Formwork

Due to its impact on construction management and the economy, the making of moulds for non-standard concrete elements is being increasingly explored in practice and academia (Clifford et al., 2014). CNC milling of plywood is a well-established process in the industry for bespoke formwork manufacturing, as it relies on a consolidated material and a technology that offers high design freedom and precision and consistency (Liew et al., 2017). However, inherent in the subtractive milling process are the long machining time and the waste of material, which affects both the environmental and economic sustainability of the operation. Alternatively, milled Expanded Polystyrene (EPS) is employed. Large-scale-robotic hot-wire cutting of EPS has recently gained acceptance, as it is up to 126 times faster than comparable CNC and it provides a smooth surface finishing (Brander et al., 2016). Due to the nature of the process, the design freedom is limited to ruled surfaces, demanding heavy geometric control and rationalisation. Several alternative methods are being investigated in professional practice and the academic context to reduce material waste and production time. With roots that can be traced back to the Romans, fabric formworks have a vast taxonomy of examples through history, but a small amount of scientific literature can be found (Veenendaal et al., 2011). The lightweight nature and high-quality surface finishing are the benefits of such technology; however, its implementation requires complex calculations and the formal possibilities are still limited today. Dynamic formworks such as the slipforming technology (Lloret et al., 2014) and pneumatic systems (Kromoser & Huber, 2016; Adapa, 2020), as well as reinforcement cage formworks (Hack et al., 2017) are also being investigated. In the last decade, a particular interest for AM in concrete construction arose, with most studies (Labannote et al., 2016) and industrial projects focusing on direct 3D Concrete Printing (3DCP) where the need for any mould is eliminated. To date, despite its disruptive potential, 3DCP still carries several geometric limitations and unknown performances. Alternatively, there is a growing body of research on the use of AM technologies to produce formworks to take advantage of the flexibility of the production process and rely on well-consolidated knowledge of conventional concrete (Naboni & Breseghello, 2018).

2.3 Features of Additive Formwork Manufacturing

The use of FDM technology for the fabrication of formworks presents several advantages compared to subtractive methods. Firstly, thanks to the inherent additive nature of the process, which allows depositing material only where required and to the mechanical properties of thermoplastics, the moulds can be produced as thin shell elements, saving in volume and material waste. Secondly, the programmed deposition of material guarantees high geometric freedom, with the only limitation coming from the printing overhangs and a high-resolution, dependent on the extruder dimension (Naboni & Breseghello, 2019). Furthermore, the strength of the plastic material makes the moulds suitable for reuse in multiple casts without losing their functional and mechanical attributes (Naboni & Paparella, 2020). Finally, the high degree of process automation of FDM 3D printing makes it an accessible, safe and sustainable formwork technology.

3 Methods

This research is designed to assess and quantify the geometric freedom provided by the Additive Formwork fabrication technique for concrete panels. This application is evaluated through the development and experimental testing of a workflow for the design, digital tooling and fabrication analysis of eight individual and geometrically differentiated panels. In the first phase, based on previous research by the authors (Naboni & Breseghello, 2020), this study implements an integrative design-to-manufacturing workflow (Fig. 2) for complex concrete elements realised with reusable additive formworks. The workflow encompasses four phases: (i) *Computational Modelling* of the panels, which includes geometric analysis and evaluation of their formal complexity and manufacturability; (ii) *Digital Tooling for Geometry and Process Optimization,* where digital tools are developed and employed to inform the design process with critical data regarding geometry and fabrication; (iii) *Formwork 3D Printing*, including the design and detailing of the moulds, optimisation of the machining code and production by means of FDM; (iv) *Concrete Casting,* which involves the preparation of the moulds and of an appropriate concrete

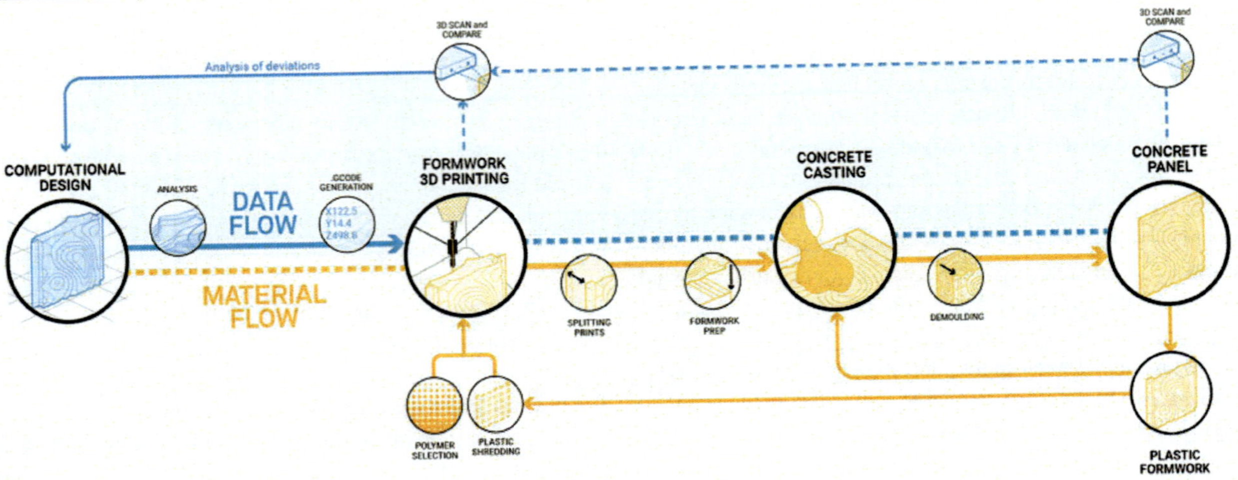

Fig. 2 Design-to-manufacturing workflow to produce concrete panels through additive formwork. Image by CREATE/University of Southern Denmark

mixture, the casting process and demoulding process, where the final concrete panels are obtained and the plastic formwork is removed in a non-destructive way. In the second phase, the digital models are analysed in their geometric characteristics, i.e. mean curvature, thickness, printing angle, demoulding angle. Their geometric accuracy is evaluated with high-precision 3D scanning to evaluate the production accuracy on both the formwork and the final concrete panel.

The workflow integrates two main streams of information: on the one hand, a data flow refers to the digital information which is transferred from the digital modelling environment to the 3D printing process, consequently informing the final concrete panel and getting the feedback information from the 3D scanning process; on the other hand, a material flow informs the computational design and optimisation, where material characteristics and fabrication constraints are embedded, becoming explicit in the Formwork 3D Printing, Concrete Casting and Demoulding, where the digital geometric code (*gcode*) is translated into a machine toolpath, the concrete is poured and the panels are extracted and directly reused or shredded and recycled for the 3D printing of new formworks.

3.1 Computational Modelling

Modelled to be produced in a squared dimension of 500 × 500 mm, eight panels with highly unique designs are conceived to maximise the variation of their geometric characteristics to emphasise and test the flexibility provided

Fig. 3 The design of eight prototypical concrete panels. Image by CREATE/University of Southern Denmark

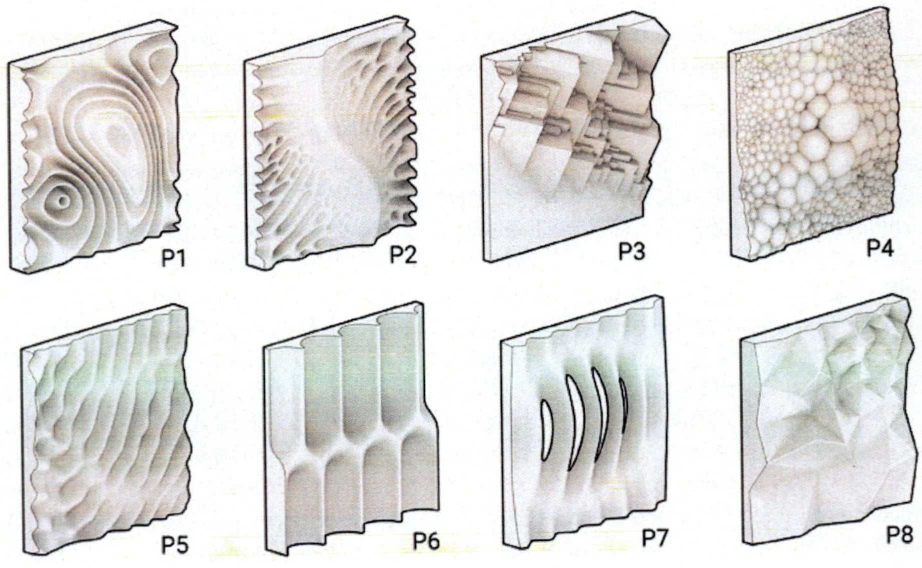

by AM (Fig. 3). All the designs were developed in the Rhinoceros environment, utilising various algorithmic processes based on Grasshopper and Python. The eight panels are modelled and described through different geometric approaches, i.e. mesh, NURBS and SubD. The first panel, *P1*, exploits a distance-based sine function to produce a corrugated mesh; *P2* is modelled as a ribbed mesh with a distance-based curve attractor onto which is applied a Reaction–Diffusion algorithm based on the Gray-Scott model (Gray & Scott, 1984); *P3* is modelled through a series of semi-randomly positioned and scaled parallelepipeds, each placed with a 45° rotation on the vertical axis and 25° rotation on the axis perpendicular to the panel; the geometry of *P4* emerges from a steered circle-packing operation, where the scale of the circles follows three curve axis; *P5* is built through a wave simulation developed with the live physics engine of Kangaroo for Grasshopper (Piker, 2013); *P6* is designed using Subdivision Surface Modelling (SubD) (Peters & Reif, 2008) to achieve a periodic smooth surface; *P7* is modelled to test topological discontinuity and the application of internal openings in the panel; *P8* presents a faceted surface through adaptive re-meshing and a controlled corrugation.

3.2 Digital Tooling for Geometry and Process Optimisation

A set of analytical tools is developed within the Grasshopper environment to test and inform the design process through visual and numerical feedback. On the one hand, the geometric features of the panels are evaluated through an analysis of intrinsic characteristics of the shapes: curvature, design resolution, local panel thickness, surface area; on the other hand, a set of computational routines is used to assess and integrate on the geometric aspects related to the fabrication processes: printing angle, toolpath optimisation, demoulding angle.

Geometry. A mapping of the mean curvature of the eight designs is performed to analyse the geometric variation and its extent along each surface and assess the geometric differences between the panels. For consistency, all the digital models are converted into triangular meshes. In the modelling phase, a live visual and numerical feedback of the thickness of the designed panels is provided. A minimum threshold is set at 25 mm as it is considered the minimum thickness that the concrete used in the experiment can be used at and a maximum thickness of 140 mm is considered.

Printing. The 3D modelled designs are then translated into *gcode* using a custom slicing tool developed through Python in Grasshopper. While different programmes to translate models into machining code are commercially available, using a tool integrated into the modelling environment brings several advantages. Firstly, it guarantees complete control over the machining toolpath, allowing for time reduction and optimisation of the travel movements and interruptions of the printing flow. This is particularly relevant in large-scale extrusion processes to prevent imprecisions due to material leakage when the print is interrupted and restarted. Secondly, within the custom toolpath, a series of geometric manipulations can be controlled parametrically: for printing purposes, the panels are joined in couples using curves boolean operations on the sliced layers; two lateral surfaces useful in the casting procedure were added with the same method; moreover, an adhesion surface, i.e. brim, of width *d2* and height *h2* of 0.3 mm is generated as a result of a series of offsets of the first printing layer. Thirdly, the custom slicing tool gives live feedback during the design phase on the printability of the digital model. The stacked layers need a good contact surface with the layer below to produce a qualitative and watertight print. For this purpose, the slicing tool calculates the horizontal deviation between corresponding points in consecutive layers to provide feedback on the printing build-up in the form of a coloured mesh with a resolution of 0.1 mm (Fig. 4). This analysis generates a seamless interaction between design and analysis of the fabrication feasibility, anticipating possible problems and optimising the design process.

Casting. To preserve the rigid formworks intact to be reused, a geometric analysis of the demoulding process was performed. The normal at every point of the panel is used to calculate the angle occurring with the vector normal to each panel's planar back face, considered as the one through which the mould is extracted from the concrete (Fig. 5). This provides feedback during the design process, where the most critical areas of the demoulding are highlighted for closer examination.

3.3 Formwork 3D Printing

The slicing procedure takes into account the different parameters of the 3D printing process. The layer height is set to 0.6 mm, balancing between the speed of the process, precision and design flexibility. The higher the layer, the coarser the resolution in the vertical dimension and the smaller the possible inclination of consecutive layers. The width of the layers is set to 4.5 mm following preliminary casting tests, as it demonstrated to be sufficient to withstand the pressure exerted by concrete

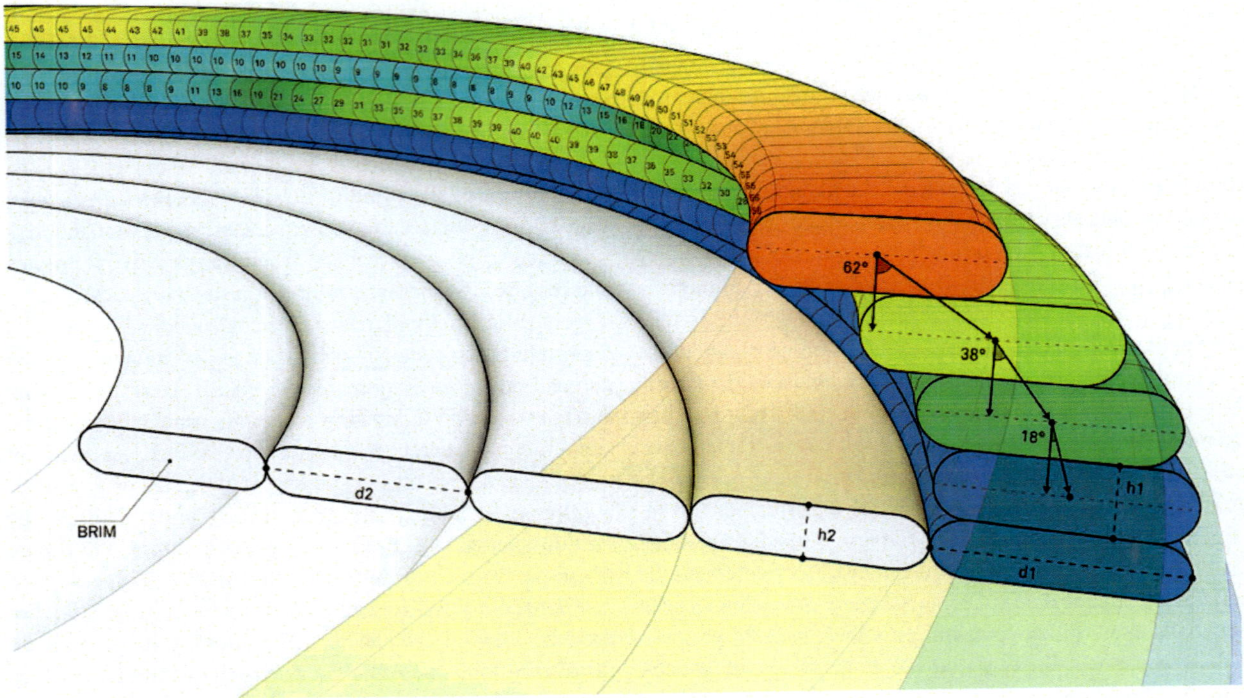

Fig. 4 Simulation and analysis of the printing process evaluate the material overhangs and possible printing angles. Image by CREATE/University of Southern Denmark

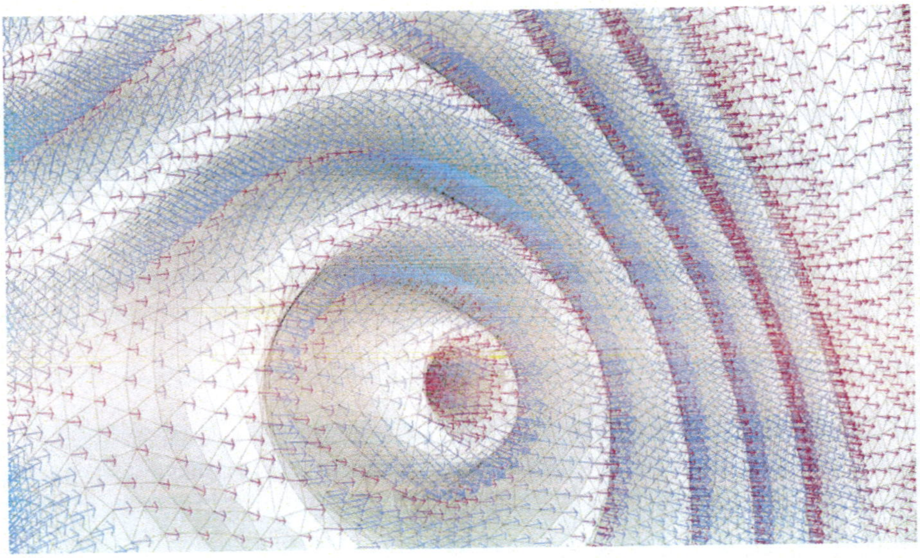

Fig. 5 Visualisation of the demoulding vectors across the formwork surface. Image by CREATE/University of Southern Denmark

with the employed formwork material. The moulds are produced using bio-polymer Polylactic Acid (PLA) pellets obtained from shredded waste from industrial production. The moulds are printed at an average speed of 80 mm/s in a delta WASP 3MT Industrial printer with a cylindrical printing volume of 1000 mm diameter and 1200 mm height (Fig. 6). The printing temperature is constantly kept at 195 C° and an overflow of the material of 30% is given to increase the extrusion pressure and the interlayer bonding strength. The 834 layers of the four prints were produced in an average of 4.5 h per panel, with an average toolpath length of 1312 m and weights ranging between 1.95 and 2.12 kg for each of the eight panels (Fig. 7).

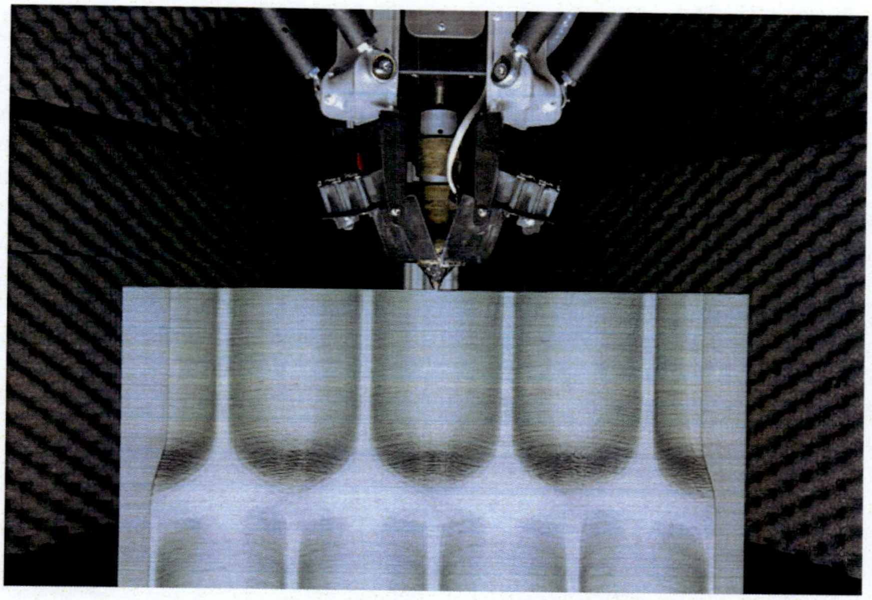

Fig. 6 FDM process of the formworks with a Delta 3 M Industrial printer. Image by CREATE/University of Southern Denmark

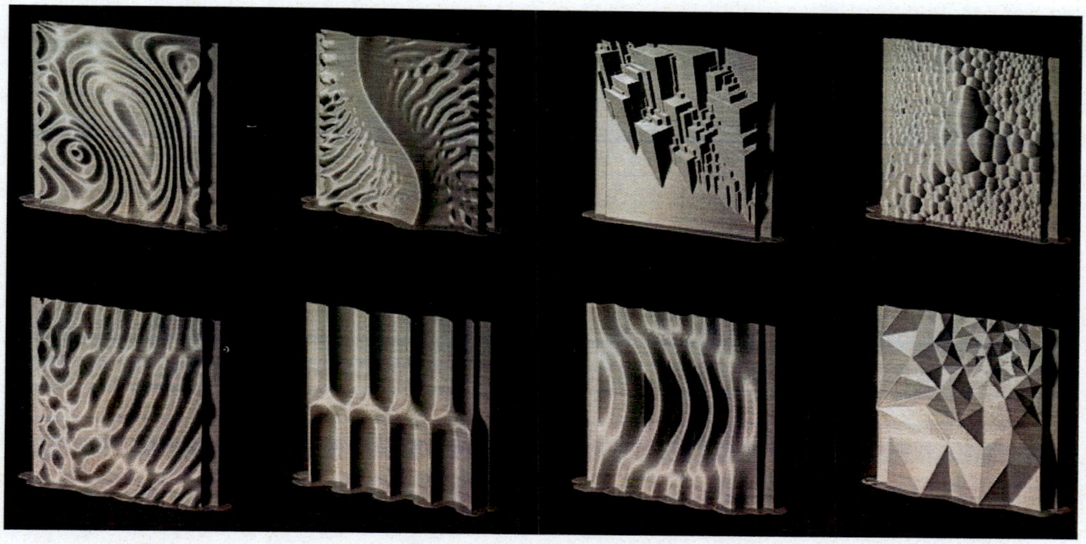

Fig. 7 Eight 3D printed moulds for concrete panels. Image by CREATE/University of Southern Denmark

3.4 Concrete Casting

The formworks are bolted to reusable modular wooden frames and oriented horizontally. This orientation of the moulds is preferred over a vertical positioning to minimise the hydrostatic pressure exerted by the fluid concrete over the mould, which is a function of the density of the fluid material and the height of the formwork. The material employed in this experiment is a Compact Reinforced Concrete (CRC) type of Ultra-High-Performance Fibre Reinforced Concrete (UHPFRC), with a self-compacting matrix suitable for the high resolution required in this application. The binder is mixed with 12% of water and 12% of short steel fibres. A release agent is applied to the moulds. The concrete is then cast and left to cure for 72 h before demoulding. The formworks are manually separated from the dry concrete. Once the mould is detached, it is washed for producing subsequent elements or eventually recycled for another cycle of 3D printing. All the concrete panels are scanned and compared to their respective digital model and the formwork point clouds, with the same procedure used for the formworks.

4 Results and Reflection

The physical output of this experiment consists of eight concrete panels with highly unique surface texture (Fig. 8). This is achieved with the use of thin reusable formworks fabricated with FDM additive manufacturing technology. The prototypes highlight the flexibility of the employed fabrication workflow.

4.1 Geometric Complexity Analysis

Mean Curvature and Thickness. The geometric analysis performed during the experiment highlighted the complexity of the design features achievable with the proposed fabrication approach. The realised panels present a mean curvature ranging from -0.59 in the designs with planar faces and sharp edges *(P3, P8)* to 0.65 in doubly-curved geometries with variable radii along their surface, with the minimum radius of curvature produced in *P1*. *P1, P2* and *P5* present significant variations in curvature, i.e. the rate of change of curvature per unit of area, compared to panels P4, P6 and P7, which have areas of small radius of curvature but smaller variations (Fig. 10a). All the manufactured panels have variable volume thickness, with sections ranging between 128 mm in the thickest area of *P1* and 25 mm in the shallow parts of *P2, P3, P6* and *P7* (Fig. 10b). Thanks to the adopted UHPFRC compound's tensional capacity, all the slender panels can be handled without risk of breaking (Fig. 9). The described geometric features and design resolution defined in the design are hardly compatible with any other fabrication method. In particular, the variation of the above-described parameters in the single panels and across the different panels is an unparalleled feature of AM without any additional fabrication step or cost.

Formwork 3D Printing Features. The 3D printing of the formworks proved to be consistent and met the expected design flexibility. The panels' design uses *printing angles* varying between 0° and 74°, beyond the typical limits imposed by FDM 3D printing technology. These were printed successfully as watertight geometries due to the optimisation of the height/width layer proportion, reduced from 1.0/3.0 to 0.6/4.5 and the use of cooling (Fig. 10c). The experiment shows the influence of the printing angle in the demoulding process, as more overhanging material corresponds to a rougher surface finishing, which creates additional friction between the plastic form and the concrete. From visual inspection, it is possible to assess that large overhangs towards the inner side of the mould cause irregularities and in turn, enhance the grip between the concrete and the mould. Printing angles are one of the main shortcomings of FDM 3D printing. However, the successful

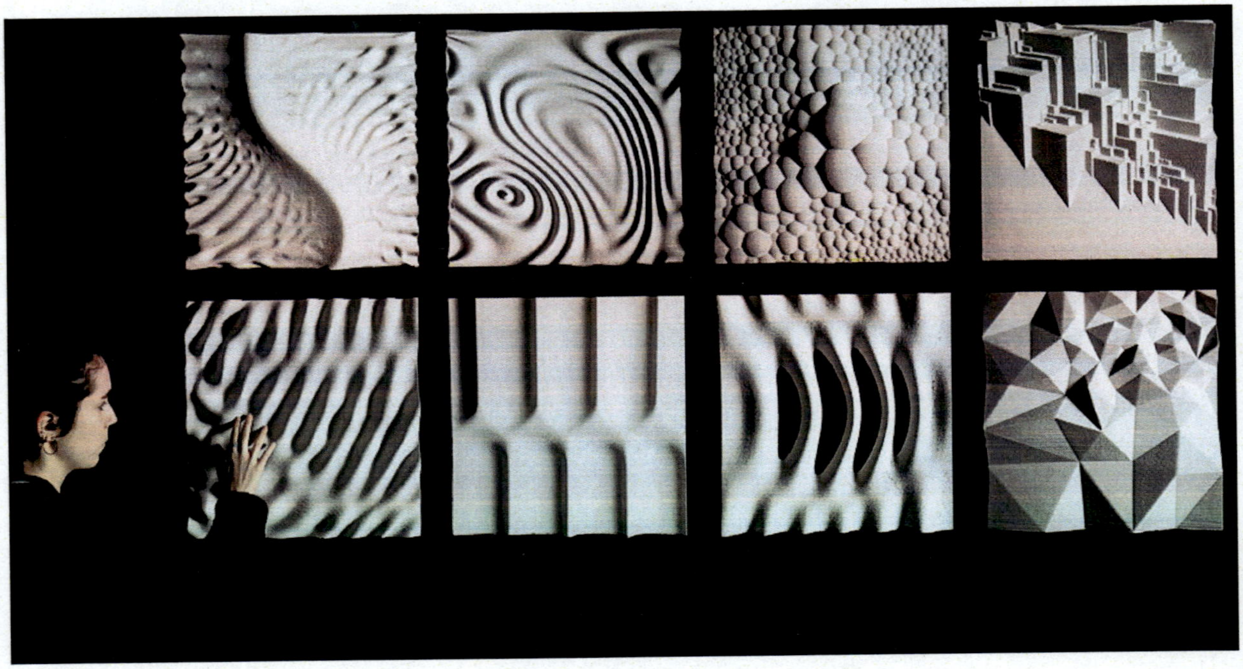

Fig. 8 The eight resulting concrete panels. Image by CREATE/University of Southern Denmark

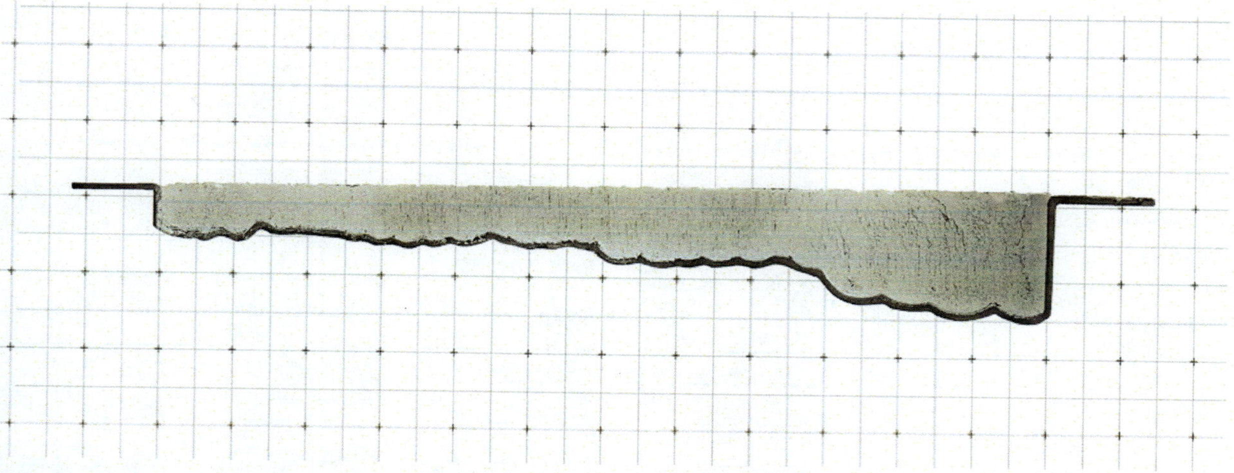

Fig. 9 Section of P4 shows the concrete element and the plastic formwork and highlights the thickness variation. Image by CREATE/University of Southern Denmark

production of these moulds and concrete elements, with an optimisation of the layer proportion, reduces the limitations.

Demoulding Features. Demoulding the panels to reuse the rigid formworks was carried out successfully for most of the elements. On a theoretical level, any angle lower than 90° between the direction of extraction of the mould and the surface should be possible. However, we observed an influence of the friction of the layered concrete on the extraction process proportional to the demoulding angle (Fig. 10d). The analysis shows the maximum and average inclination for each panel: the larger angles are registered in *P3*, which has a maximum inclination of 79° and an average of 50.9°, whereas the lowest is seen in *P8* presents a maximum angle of 61° and an average of 31.5°. While all the other panels were smoothly demoulded, *P3* and partially *P1* and *P2*, required using a heat source to deform the formwork for removal, making it not suitable for reuse. The demoulding of *P1* and *P2* was particularly difficult in the areas with higher curvature and printing angles, suggesting an interdependent relationship between these parameters and the operation of demoulding. The demoulding of the panels highlighted the importance of accounting for the demoulding angles in the design phase as it constraints the fabrication and the possibility of reusing the 3D printed formworks.

4.2 Fabrication Accuracy

All the formworks were scanned and digitised using an optical structured light GOM ATOS 3D scanner. Detailed adaptive point clouds were created, with a range of 1.5–2 million points and an accuracy of 0.06 mm. Imported in Grasshopper through Volvox (Zwierzycki et al., 2016), the deviation from the original 3D digital models was analysed for each of the forms. The use of 3D scanning to digitise the thermoplastic formworks and the concrete panels allowed to provide feedback on the relation between digital and physical prototype results and between the results of the 3D printing and casting operations. Scanning the plastic formworks before casting and testing them against the digital 3D models shows an average deviation on all the moulds of 1.07 mm. Caused by the thermoplastic shrinkage, a deformation towards their centre can be consistently observed in most of the printed moulds. From the 3D scanning of the concrete panels and analysis of the deviations from the digital 3D models (Fig. 11), a total average deviation of 1.52 mm is measured. While the printing angles have an influence at the microscale, from the scans no direct link with deviations can be outlined. This suggests that the mechanical capacity of the mould is homogeneous and there is no relation with the printing angles. The geometric differences that characterise the designs do not influence the production process and have only a minor influence on the quality of the final elements.

5 Conclusion

Overcoming the geometric limitations imposed by conventional formwork manufacturing techniques is a relevant challenge to fully unlock the potential of concrete and allow the design and production of customised, material-efficient, high-performance concrete architectural elements. Investigating and analysing the geometric freedom and fabrication precision of AM is a crucial step towards new applications in the construction industry. This study has proved the viability of the fabrication method and provided a numerical understanding of the geometric and fabrication domain it is bound to. Moreover, the design-to-manufacturing experiment has suggested the flexibility of the manufacturing method in

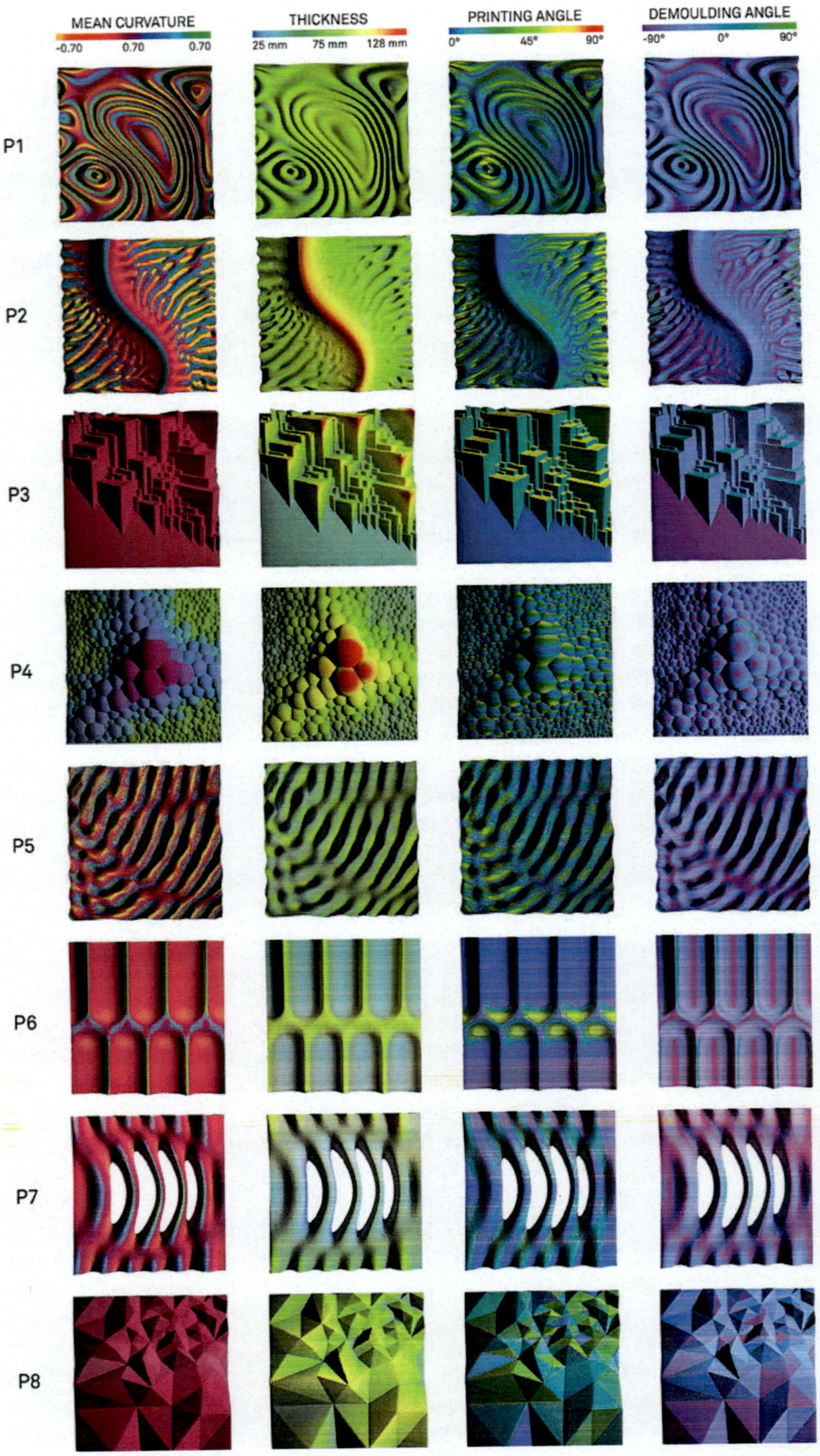

Fig. 10 Computational simulation and analysis of the eight panels with regard to Mean curvature, Thickness, Printing angle and Demoulding angle. Image by CREATE/University of Southern Denmark

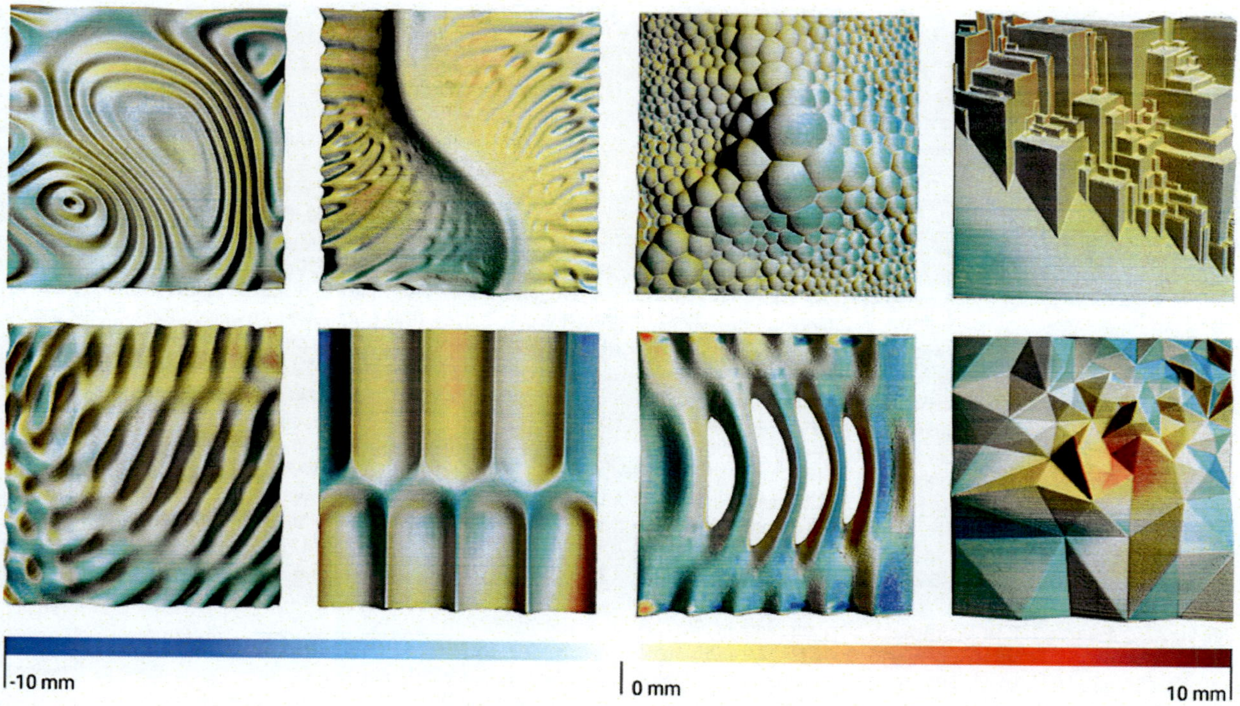

Fig. 11 Eight concrete panels with overlaid information from the 3D scanning process, showing the deviation of the resulting concrete panel compared to the digital model. Image by CREATE/University of Southern Denmark

addressing different design and geometric scales. While other techniques might achieve comparable results in some instances, our fabrication approach has revealed versatility in manufacturing a wide range of design outputs. Such a geometric resolution and production flexibility can open prospective applications for indoor surfaces and facade elements with bespoke textural and ornamental effects. Future works will engage with: the development of automated processes of optimisation of the design and formwork geometry for consistent, precise and accurate results; the application and testing of the proposed methods against larger-scale concrete elements; a systematic study on the deformations caused by the hydrostatic pressure exerted by the semi-fluid concrete.

Acknowledgements This work was carried out at the CREATE Lab at the University of Southern Denmark—Section for Civil and Architectural Engineering. The authors wish to thank David W. Jokszies for the assistance, Hi-Con for in-kind support of concrete material and Ass. Prof. Knud Bjørnholt for the 3D Scanning at the SDU Prototyping Lab.

References

Adapa. (2020, February 27). *The adaptive mould*. www.adapa.dk.

Austern, G., Elber, G., Capeluto, I.G., Grobman, Y.J. (2018). Adapting architectural form to digital fabrication constraints. In S. Adriaenssens, F. Gramazio, M. Kohler, A. Menges, & M. Pauly (Eds.), *Advances in architectural geometry* (pp. 10–33).

Brander, D., Bærentzen, J. A., Clausen, K., Fisker, A-S., Graversen, J., Lund, M. N., Nørbjerg, T. B., Steenstrup, K. H., & Søndergaard, A. (2016). Designing for hot-blade cutting: Geometric approaches for high-speed manufacturing of doubly-curved architectural surfaces. In S. Adriaenssens, F. Gramazio, M. Kohler, A. Menges, & M. Pauly (Eds.), *Advances in architectural geometry* (pp. 306–327). https://doi.org/10.3218/3778-4

Clifford B., Ekmekjian N., Little P., & Manto A. (2014). Variable carving volume casting. In P. L. de McGee (Eds.), *Robotic fabrication in architecture, art and design* (pp. 3–15). https://doi.org/10.1007/978-3-319-04663-1_1

Collins, P. (1959). *Concrete: The vision of a new architecture*. Horizon Press.

de Soto, B. G., Agustí-Juan, I., Hunhevicz, J., Joss, S., Graser, K., Habert, G., & Adey, B. T. (2018). Productivity of digital fabrication in construction: Cost and time analysis of a robotically built wall. *Automation in Construction. 92*, 297–311. https://doi.org/10.1016/j.autcon.2018.04.004

Fischer, T. (2012). Geometry rationalisation for non-standard architecture. *Architecture Science., 5*(9), 25–47.

Flöry, S., & Pottmann, H. (2010). *Ruled surfaces for rationalisation and design in architecture.* ACADIA 10: LIFE information, on Responsive Information and Variations in Architecture. Proceedings of the 30th Annual Conference of the Association for Computer Aided Design in Architecture, pp. 103–109.

Gray, P., & Scott, S. K. (1984). Autocatalytic reactions in the isothermal continuous stirred tank reactor: Oscillations and instabilities in the system A + 2B → 3B, B → C. *Chemical Engineering Science*, pp. 1087–1097.

Hack, N., Wangler, T., Mata-Falcón, J., Dörfler, K., Kumar, N., Walzer, A.N., Graser, K., Reiter, L., Richner, H., Buchli, J., Kaufmanc, W., Flatt, R.J., Gramazio, F., Kohler, M. (2017). *Mesh mould: An on site, robotically fabricated, functional formwork.*

Second Concrete Innovation Conference (2nd CIC), Tromsø, Norway.

Kromoser, B., & Huber, P. (2016). Pneumatic formwork systems in structural engineering. *Advances in Materials Science and Engineering, 2016*(6), 1–13. https://doi.org/10.1155/2016/4724036

Lab, R. (2007, April). Think formwork—Reduce costs. *Structure Magazine*, pp. 14–16.

Labonnote, N., Rønnquist, A., Manum, B., & Rüther, P. (2016). Additive construction: State-of-the-art, challenges and opportunities. *Automation in Construction, 72*, 347–366. https://doi.org/10.1016/j.autcon.2016.08.026

Liew, A, López López, D., Tom Van Mele, T., & Block, T. (2017). Design, fabrication and testing of a prototype, thin-vaulted, unreinforced concrete floor. *Engineering Structures, 137*, 323–335. https://doi.org/10.1016/j.engstruct.2017.01.075

Lloret, E., Mettler, L. K., Shahab, A. R., Gramazio, F., Kohler, M., Flatt, R. J. (2014). *Smart dynamic casting: a robotic fabrication system for complex structures.* Proceedings of 1st Concrete Innovation Conference, Oslo, Norway.

Lloret, E., Shahab, A. R., Linus, M., Flatt, R. J., Gramazio, F., Kohler, M., & Langenberg, S. (2015). Complex concrete structures: Merging existing casting techniques with digital fabrication. *Computer-Aided Design, 60*, 40–49. https://doi.org/10.1016/j.cad.2014.02.011

Naboni, R., & Paoletti, I. (2015). Advanced customization in architectural design and construction (9783319044224 ed.). Springer: SpringerBriefs in Applied Sciences and Technologies PoliMI SpringerBriefs https://doi.org/10.1007/978-3-319-04423-1

Naboni, R., & Breseghello, L. (2018). *Fused deposition modelling formworks for complex concrete constructions.* Proceedings of the XXII Congresso Internacional da Sociedade Ibero-americana de Gráfica Digital. Blucher Design Proceedings, 5, 700–707. https://doi.org/10.5151/sigradi2018-1648

Naboni, R., Breseghello, L. (2019). *Additive formwork for concrete shell constructions.* In C. Lázaro, K. U. Bletzinger, & E. Oñate (Eds.), Form and Force IASS Symposium 2019 Conference Proceedings, pp. 87–94.

Naboni, R., & Breseghello, L. (2020). High-resolution additive formwork for building-scale concrete panels. In F. P. Bos, S. S. Lucas, R. J. M. Wolfs, & T. A. M. Salet (Eds.), Second RILEM International Conference on Concrete and Digital Fabrication—Digital Concrete 2020. DC 2020. RILEM Bookseries, 28. Springer, Cham. https://doi.org/10.1007/978-3-030-49916-7_91

Naboni, R., & Paparella, G. (2020). Circular concrete construction through additive FDM formwork. In D. Holzer, W. Nakapan, A. Globa, & I. Koh (Eds.), *RE: Anthropocene, design in the age of humans—Proceedings of the 25th International Conference on Computer-Aided Architectural Design Research in Asia, CAADRIA 2020* (pp. 233–242). The Association for Computer-Aided Architectural Design Research in Asia (CAADRIA).

Nervi, P. L. (1956). *Structures*. F. W. Dodge.

Peters, J., & Reif, U. (2008). *Subdivision surfaces*. Springer series Geometry and Computing monograph, 3.

Piker, D. (2013). Kangaroo: Form finding with computational physics. *Architectural Design, 83*, 136–137.

Pottmann, H. (2013). Architectural geometry and fabrication-aware design. *Nexus Network Journal, 15*(2), 195–208. https://doi.org/10.1007/s00004-013-0149-5

Veenendaal, D., West, M., & Block, P. (2011). History and overview of fabric formwork: Using fabrics for concrete casting. *Structural Concrete, 12*(3), 164–177. https://doi.org/10.1002/suco.201100014

Zwierzycki, M., Evers, H. L., & Tamke, M. (2016). Parametric architectural design with point-clouds—Volvox. In A. Herneoja, T. Österlund, & P. Markkanen (Eds.), *Complexity & Simplicity—Proceedings of the 34th eCAADe Conference*, 2, pp. 673–682.

Ceiling Sound Reflectors to Optimize Acoustic Performance Using Parametric Tools

Doris A. Chi Pool, Jaqueline S. McNamara Trujillo, and Edwin González Meza

Abstract

This paper aims to integrate acoustic simulations and parametric geometry modelling into the design process of architectural acoustic design. An existing conference room of 226 m^3 was selected as the case study. Different geometries of sound reflectors were mounted under the ceiling, to be tested and compared in order to identify the best acoustical performance. With this purpose, a methodology consisting of three main steps was developed. First, the room was digitally modelled, and the geometry of the reflectors was parameterized through Snail and Silvereye—flat and curved surfaces were evaluated, along with different materials of the sound reflectors. The fitness function was to maximize the reflections falling into the seating area. Once the optimized geometries were identified, namely those that get the most number of reflections falling into the analysis surface, acoustic simulations were run with Pachyderm. Sound pressure level, reverberation time, and clarity sound were the indicators used for profound comparisons. Results showed a significant improvement in the acoustic performance of the auditorium from the current state to the optimized geometries. This methodology could be helpful to provide architects with acoustic feedback during the initial stages of building design and to improve the acoustic performance in their projects.

Keywords

Architectural acoustics • Parametric design • Geometry optimization • Acoustic simulation

D. A. C. Pool · J. S. M. Trujillo · E. G. Meza (✉)
Department of Architecture, Universidad de las Americas Puebla, ExHacienda Santa Catarina Martir S/N, San Andres, 72810 Cholula, Mexico
e-mail: edwin.gonzalez@udlap.mx

1 Introduction

Architectural acoustic design is a fundamental part of buildings performance and indoor environmental comfort. A proper design should provide an acoustic environment that maintains satisfaction and well-being among occupants. Research has shown that well-designed sound environments in offices or schools can help to improve concentration and enable better communication (Klatte et al., 2013; Banbury & Berry, 2005).

Towards decades, architectural acoustic design has evolved from drawing methods and calculation processes (based on mathematical and physical theory) to design experimentation progressively assisted by digital simulation tools (Milo, 2020). In comparison with room acoustic scale modelling, a traditional technique developed over the last 100 years, the costs and the time needed for room acoustic computer models are significantly cheaper and faster (Rindel, 2002). Digital models started around 1970 with the new advances in computer power and the development of specific software that allowed the prediction of the acoustic performance of spaces through digital technologies. Computer models allow gathering important information that cannot be obtained with the scale modelling technique (Rindel, 2011). Besides, they allow quick comparisons among results and practical organization of results in tables and graphs (Mason, 2020). As a result, the time to complete performance-based acoustic design is significantly more efficient than traditional methods. Moreover, the results can be more precise and improve the interior acoustic performance of architectural projects.

Computer modelling of acoustic enclosures has been developed in several forms. On the one hand, various methods for optimising acoustic environments through computer-based acoustic simulations exist as commercial packages that intend to clarify the sound performance of pre-conceived designs (Badino et al., 2020; Vorländer, 1995). These acoustic simulations require known factors and

H. Altan et al. (eds.), *Advances in Architecture, Engineering and Technology*,
Advances in Science, Technology & Innovation, https://doi.org/10.1007/978-3-031-11232-4_23

equations for acoustic evaluation. Moreover, they combine multiple simulation strategies, such as radiosity, raytracing, image source method, etc. (Krokstad et al., 1968; Vorländer, 1989).

On the other hand, parametric computer-aided modelling techniques are thought to let the machine create a computational search for an optimized acoustical performance, so the design can be altered until the desired performance level has been reached (Foged et al., 2012; Parigi et al., 2017; Vlaun, 2015). The optimization through parametric techniques involves several steps that include identifying variables and constraints, selecting tools, determining goals, choosing the optimization algorithm, running simulations, and presenting the final results (Østergård et al., 2016). Different optimization algorithms can be mentioned, such as simulated annealing, neural networks, and genetic and evolutionary algorithms (Brownlee, 2011). Much research has illustrated the diversity on the application of several search methods and the growing importance as a probabilistic solver for singular and multi-objective problems (Foged et al., 2012; Sastry & Goldberg, 2005; Sato et al., 2004; Wang et al., 2018).

A survey of existing design tools shows that no architectural software currently exists that combines both sound and geometry (DeBodt, 2006). Therefore, design methods and collaborative approaches that integrate both types of the mentioned digital tools are required for increasing efficiency (Scelo, 2015). For example, acoustic surfaces have been designed to improve the interior acoustics of building spaces in Politecnico Di Milano (Giglio et al., 2020). Through multi-objective optimization algorithms (Octopus add-on) and the ray tracing method (Snail add-on), the authors were able to produce a range of candidate solutions and one with most optimal acOustic values. Also, maximum intersection values from the sources to the receivers were obtained.

In another work, ray tracing analysis was implemented together with parametric tools to determine accuracy and reliability in an orchestra ensemble. Using Rhinoceros and Grasshopper, the study focuses on early energy distribution on stage. This work also reviews how parametric design enables a workflow between architectural and acoustic design (Soriano et al., 2019; Wright et al., 2016).

In another investigation, the authors gave an overview of processes used today by acousticians to refine their designs either through iterative design, auralization, optimization algorithms, and real-time computer modelling (Bassuet et al., 2014). The authors also discussed the relevance of computational and optimization tools in the design process of shaping rooms, walls, and reflectors in acoustical rooms to support creativity with complex architectural forms and accelerating solution-finding.

Different approaches to understand sound phenomena were investigated in another building with different acoustic ceilings (Rumpf et al., 2018). The authors used custom-made Grasshopper tools and the application of Pachyderm for simple three-dimensional visualizations of sound emittance. These visualizations were based on the ray characteristics of sound rather than on the wave-like properties. Thus, the results were driven by geometric principles where the angle of reflection equals the angle of incidence. Comparing different ceiling solutions with the same area of absorption material revealed that the adapted ceiling outperformed the flat ceiling, with more evenly distributed improvement.

In a different work (Elorza, 2005), the author discussed room acoustics modelling and the basics of different methods, comparing empirical and statistical methods, wave-based methods, and geometrical acoustics methods (ray tracing method, image method, particles, etc.). Then, the author concentrated on the mathematical implementation of the ray tracing method applied in room acoustics modelling. A geometrical and acoustical description of three enclosures was prepared to compare simulated and measured data. The findings showed that the raytracing algorithm can represent room acoustics with acceptable accuracy, in terms of the sound pressure level.

Integrating parametric models and acoustic simulation was the aim of an study developed to improve the acoustical performance of auditoriums (Lu, et al.). By using the component-based method that can generate more varied designs and by developing an interface connecting Rhinoceros with CATT (an acoustic simulation software), the authors manipulated the auditorium designs. At the end, the authors concluded that parametric models can save the time and efforts of architects as well as test technical performances.

In this work, digital acoustic simulations are integrated with parametric geometry modelling to develop a workflow that allows improving acoustical building performance. First, new panels were integrated into the case study (Fig. 1) and tested in terms of their shape, hanging distance, rotation angle, and materials. All the design variables were modelled by using Grasshopper, a well-known parametric software that enabled the integration of acoustical parametric tools such as Snail and Silvereye. Then, the models were tested through acoustic simulations with Pachyderm software in order to improve the sound performance: reverberation time (RT), sound pressure level (SPL), and sound clarity (C80) were the indicators selected to evaluate the acoustical quality. At the end, several combinations of the four design variables were identified as the optimal panel configuration since they significantly improved the three acoustical parameters (RT, SPL, and C80), in comparison with the auditorium's current state. The main advantage of the proposed workflow is that the integration of such computational tools can provide acoustical feedback during the initial stages of building projects. This means that many design

Fig. 1 Auditorium selected for acoustic evaluations. *Source* UDLAP

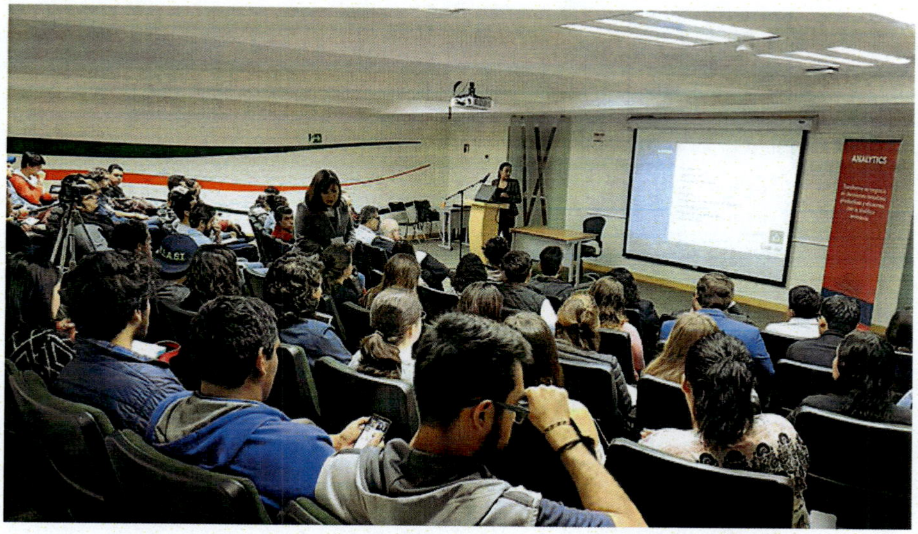

ideas can be quickly generated and tested with a computer, thus the time spent on modifying the design variables and accomplishing the acoustical targets are considerably reduced. Besides, architects can test more options following continuous design thinking.

2 Project Context

The case study chosen for acoustic evaluation is an existing auditorium (Fig. 1) located in the Universidad de las Americas Puebla, in Cholula, Mexico. The selected space is used as a master classroom and lecture hall. It has a ~226 m^3 volume, from a space measuring 11.54 m-wide, 8.46 m-deep, and 2.35 m-height. The seating area has a capacity of 70 seats distributed in five rows, each with 14 places. Table 1 summarizes the sound absorption coefficients for the existing materials of the auditorium. The walls are made of gypsum board and the floor is made of concrete covered with carpet. Ceiling tiles and gypsum boards are fixed on the ceiling as absorbent materials. Besides, two vertical glass panels are delimiting the speaker area in an attempt to redistribute sound to the seating area. Nevertheless, the audibility of this space has been perceived as poor and hindering communication.

3 Parametric and Acoustic Design

The design workflow (Fig. 2) primarily aims to improve the acoustic performance of the auditorium choosen as the case study. However, the method here proposed can be applied to other rooms and environments. To accomplish that purpose, the design process involved digital tools for: (a) parametric geometry modelling and (b) acoustic simulations.

3.1 Digital Tools

Rhinoceros and Grasshopper were the chosen platforms for modelling the geometry of the existing room and parameterising the new acoustical reflectors proposed for the design experiment. Rhinoceros is a software tool for three-dimensional modelling based on NURBS while Grasshopper

Table 1 Sound absorption coefficients by frequency band

Material	63 Hz	125 Hz	250 Hz	500 Hz	1 kHz	2 kHz	4 kHz	8 kHz
Ceiling tiles	70	70	85	75	85	90	90	90
Large ¼" plate glass	18	18	6	4	3	2	2	2
Carpet in concrete	8	8	24	56	69	71	73	73
Gypsum board	28	28	10	5	4	7	9	9
Plasterboard ceiling	18	18	6	4	3	2	2	2
Empty chairs, upholstered with cloth cover	44	44	60	77	89	82	70	70

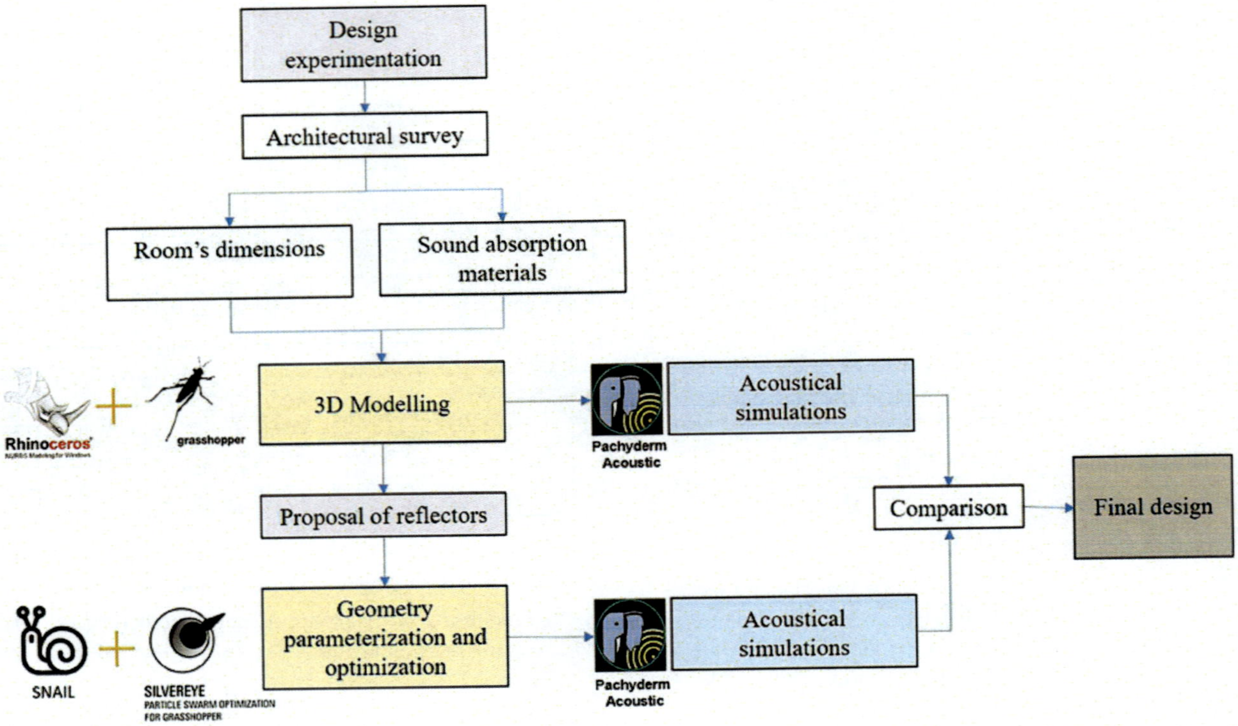

Fig. 2 Workflow for the acoustic design. *Source* Authors

is a graphical algorithm editor tightly integrated with Rhino's 3D modelling. These tools can be linked with different add-ons specialized in sound emittance visualization to deepen understand the sound propagation.

Within this framework, it should be noticed that the computational technique known as ray tracing is widely used to calculate the trajectories of the sound, viewing them as ray vectors travelling through the air with constant speed, from the sound source to the room boundaries. As the vectors collide with the interior surfaces, such as walls, furniture, ceiling, etc., they lose energy as a consequence of sound absorption. Along these travels, the rays intercept different receivers where the energy is stored until it falls under a certain limit (e.g., sound pressure level is desired, a decay of 20 or 30 dB is enough) (Krokstad et al., 1968). Ray tracing is considered one of the most elegant and accurate methods to represent sound reflections (Hongisto et al., 2004).

As observed, geometry laws are the basis to follow the paths of all the sound rays. In this work, Snail and Silvereye were the specialized plugins chosen to study sound propagation and to find the optimal design solution for the acoustical room. Snail is a geometric acoustic ray simulation tool linked to parametric shapes modelled with Rhino (Grasshopper, 2019). Snail allows visualizing sound propagation as vector rays within 'pre-defined geometry rooms'; however, it does not quantify sound metrics. Hence, Silvereye, an optimization and evolutionary algorithm based on the particle swarm optimization (PSO) was coupled with Snail to find the best performance of the sound propagation for the case study.

PSO simulates animal's social behaviour, including insects, herds, birds, and fishes. These swarms conform to a cooperative way to find food, and each member in the swarms keeps changing the search pattern according to the learning experiences of its own and other members. The main design idea of the PSO is to use a swarm mode which makes it simultaneously search large regions in the solution space of the optimized objective or fitness function (Wang et al., 2018). As a result, Silvereye is dedicated for single-objective searches and can be used for solving complex and real-world optimization problems in the design space (Cichocka et al., 2017).

Once the room shape and panels were geometrically optimized, the Pachyderm acoustic simulation was used to evaluate the acoustical performance of the case study, enabling to calculate the sound metrics, namely SPL, RT, and C80. Pachyderm is an open-source acoustics simulation plugin for Rhinoceros that uses IronPython (van der Harten, 2011), a new scripting language. Pachyderm has been used to customize the acoustics simulation algorithms in order to obtain specialized predictions by using conventional acoustics simulation algorithms, including geometrical, finite volume, transfer matrix, etc. (Van der Harten, 2019).

Pachyderm has allowed applying acoustic algorithms in ways that are less conventional by certain creative individuals (Van der Harten, 2019). Different authors have used this plugin to demonstrate how new parameter simulation types and customized scripted simulations can acoustically optimize the designs (Bork, 2000; van der Harten, 2011). Pachyderm has also been employed to test different acoustical rooms, such as meeting rooms, offices, and concert halls, showing that it is an intuitive tool that allows analysing different acoustic solutions while imagining new spatialities and interpreting the complexity in sound propagation (Milo, 2020; Peters, 2015; Rumpf et al., 2018).

3.2 Design Experiment

The relationship between the room geometry, its volume, and its surface acoustic properties is known as the acoustic environment. The treatment of floors, walls, and ceilings determine the way sound energy travels through the room: it can be absorbed, redirected by flat surfaces -acoustic reflectors- or scattered by diffusers (Cox & D'Antonio, 2004). Usually, the floor acts as an absorber by using carpets and having the audience, while walls finishes and ceiling panels play roles among absorbers, diffusers, and reflectors, according to the room specific needs.

In the specific case of a small auditorium or lecture room, the high use of absorbing elements could produce a decrease in the reverberation time and sound strength, leading to poorer speech intelligibility (Pilch & Kamisiński, 2011). Using diffusers can be detrimental in small rooms because they need to be narrow and deep, a condition that makes them expensive and more absorbing for this type of scenario (Cox & D'Antonio, 2006). Instead, reflectors only change the direction of sound waves without modifying them, so this would be the best option for small rooms where intelligibility is important (Cox & D'Antonio, 2006). Then, it is necessary to be careful about two aspects so reflectors can work properly: the sound reflected should never return to the sender and the audience needs to receive the most direct sound possible (Peters, 2009; Pilch & Kamisiński, 2011).

The position between the reflector, the sender, and the receiver is the key to acoustic performance. Plane surfaces work like a mirror, the best reflection will come to the receiver that has the same angle from the reflector to the sender, whereas having an oblique angle can be detrimental (Cox & D'Antonio, 2004). Besides, the size of the panel in an array and the density of the material determines the reflector's performance, being the material one parameter, this paper will discuss. Furthermore, the geometry of the reflectors is also an important parameter that should be considered. Usually, concave arcs are not an optimum shape, but it depends on the arc radius (Cox & D'Antonio, 2004), another parameter considered in this paper, including the options of a single-arched panel and a double-arched panel.

3.2.1 Geometry Parameterization

The case study was modelled and characterized as it was previously set in Sect. 2, in order to test its current acoustic performance. Then, new acoustic panels were proposed to better distribute the sound propagation within the room. They were mounted under the ceiling, as Fig. 3 illustrates.

Four main variables that lead the design of the panels were simultaneously parameterized, as Fig. 4 illustrates. The variables are explained below:

- Geometry: Three different shapes were tested: 1) flat surface, 2) single-curved surface, and 3) double-curved surface. In the curved panels, the sagitta length of the arc (s) was set variable from 0 to 30 cm.
- Hanging distance: A range from 0 to 30 cm was considered between the ceiling and the panels.
- Rotation angle: An inclination angle from -20° to 20° was parameterized.
- Materials: Conventional materials used for acoustic panels were also tested: (1) gypsum board, (2) plywood, (3) glass, and (4) metal. Table 2 summarizes the sound absorption coefficients considered for these materials. The absortion coefficients could be defined as the ratio of the acoustic energy that is absorbed to the total incident energy that impinges on the material. The coefficients are widely used to described the acoustic properties of materials that absorb sound (Deaconu et al., 2017; McGrory et al., 2012).

3.2.2 Fitness Function and Acoustic Metrics

The fitness function for the panels design optimization was to maximize the reflections falling into the seating area. The sound source was placed at the speaker position, for which a 60 dB–1 kHz and 500 emitting rays were considered. Four receivers were distributed into the seating area, as Fig. 3 illustrated. At each receiver, the following acoustical metrics were simulated:

- SPL: It can be defined as the experimental measurement of the amount of pressure in the air being sensed at a given location. SPL is expressed in decibels, dB.
- RT: It is the time expressed in seconds, that would be required for the SPL to decrease by 60 dB after the sound source has stopped. Since the approximate volume of the case of study is 226 m^3, it is estimated a recommended time of approximately 0.7 s for a speech presentation and 1.05 s for a music presentation as recommended RT values (Kinsler et al., 2000).

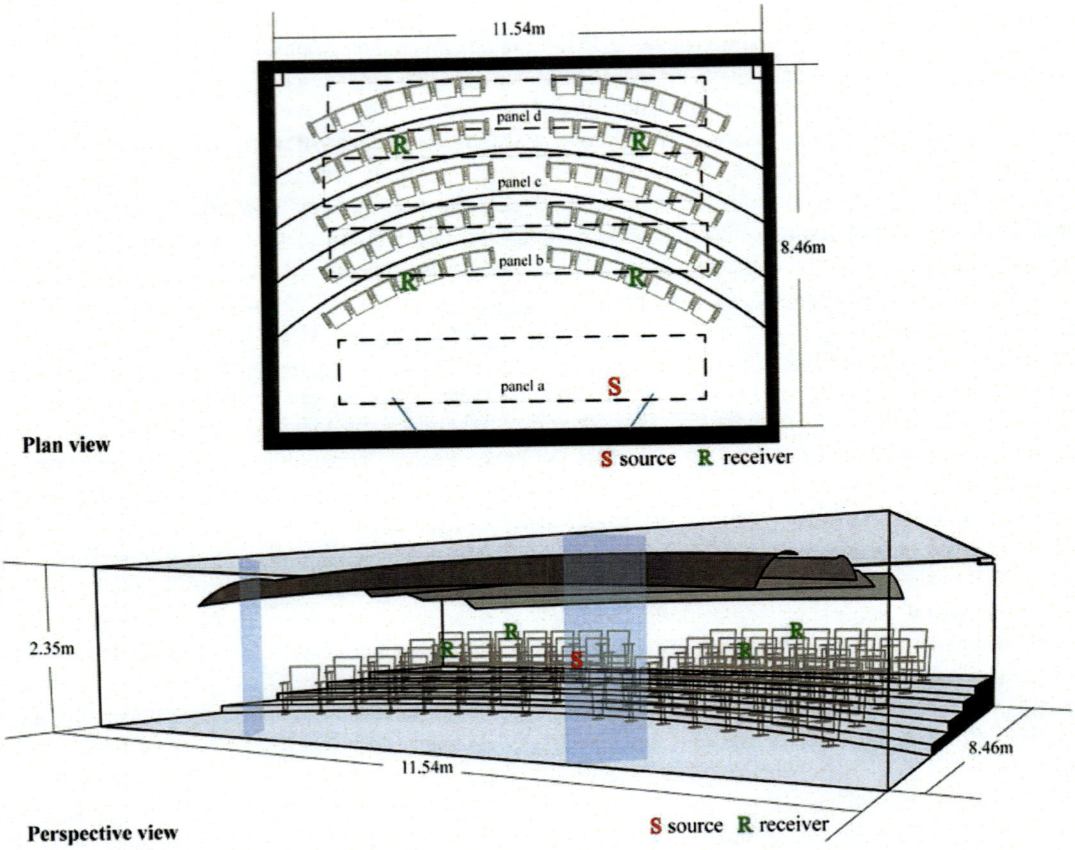

Fig. 3 Plan and perspective views of the auditorium modelled with Rhinoceros. Four panels were mounted under the ceiling. *Source* Authors

- C80: It is related to the precision with which the details of sound can be heard, namely the intelligibility of speech. In objective measurements, the clarity index, C80, has been defined as the logarithmic ratio of sound energy during the first 80 ms in the impulse response to the energy after 80 ms, expressed in dB; three categories of clarity can be distinguished: C80 < 1.6 dB or bad clarity, C80 1.6 or border clarity, and C80 > 1.6 or good clarity (Reichardt et al., 1975).

4 Results and Discussion

The results and the findings drawn from the comparisons are presented in the sections below.

4.1 The Current State of the Auditorium

Pachyderm was used to investigate the acoustical performance of the auditorium in its current state, with no reflectors in place. Figure 5 depicts the acoustic simulation results. As observed, the four sensors reached averages of 41.68 dB for SPL and 6.39 for C80. What is striking about the Figure is that the RT values were very high in the four sensors, achieving a mean of 2.29 s and causing a not adequate reverberation, the auditorium is used for conferences or classes for which an RT 0.7 s is recommended. Overall, the audibility of the space is perceived as poor, hindering communication.

4.2 Geometry Optimization of the Panels

Figure 6 presents an example of the optimization process performed with Snail and Silvereye. Basically, 500 rays were sent from the sound source to the room. As the vectors travelled, they collided with walls, floor, ceiling, the new panels here proposed, and so on. Moreover, they passed through an imaginary plane representing the seating area. In the meantime, the five panels were simultaneously rotating, displacing from the ceiling and changing its sagitta arc length. After 20 min of the PSO optimization process, specific sets of all these variables were established as the optimal design solutions since they maximized the fitness

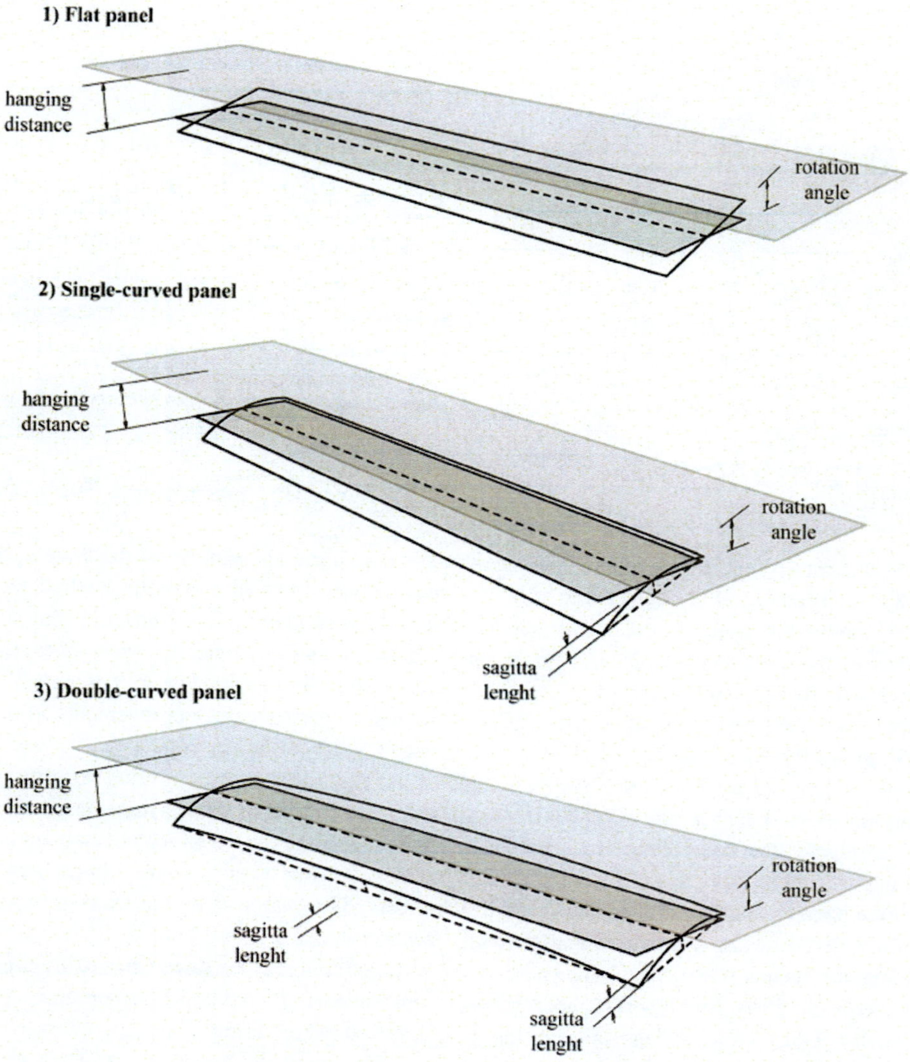

Fig. 4 Four panels parameterized by using Snail and Silvereye. *Source* Authors

Table 2 Sound absorption coefficients by frequency band, for the reflector materials

Material	63 Hz	125 Hz	250 Hz	500 Hz	1 kHz	2 kHz	4 kHz	8 kHz
Gypsum board	28	28	10	5	4	7	9	9
Plywood	32	28	22	17	9	10	11	13
Glass	18	18	6	4	3	2	2	2
Metal panel	59	59	80	82	65	27	23	23

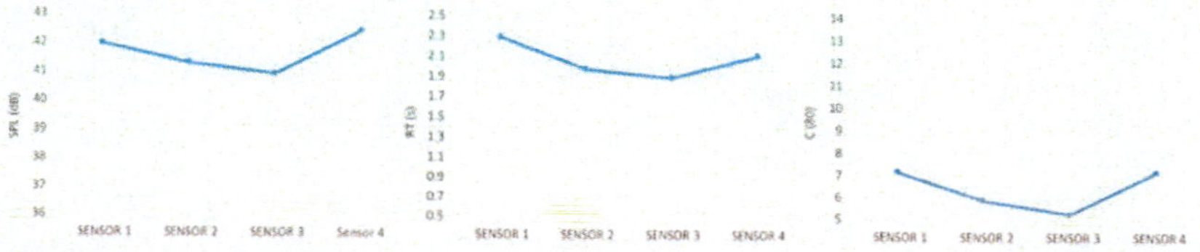

Fig. 5 Results from acoustic simulations performed with Pachyderm: Current state of the auditorium. *Source* Authors

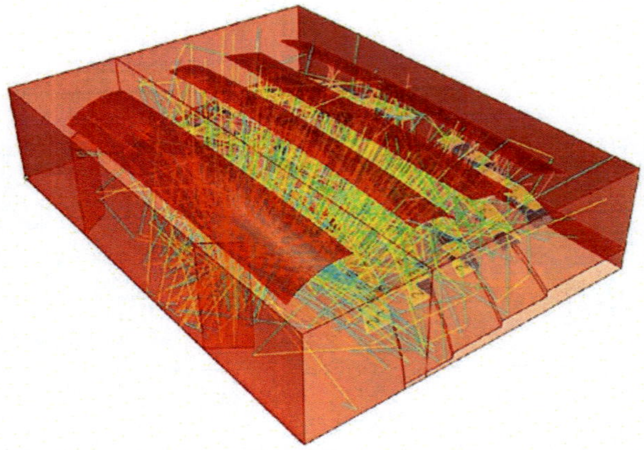

Fig. 6 Example of the optimisation process performed with Silvereye and Snail. *Source* Authors

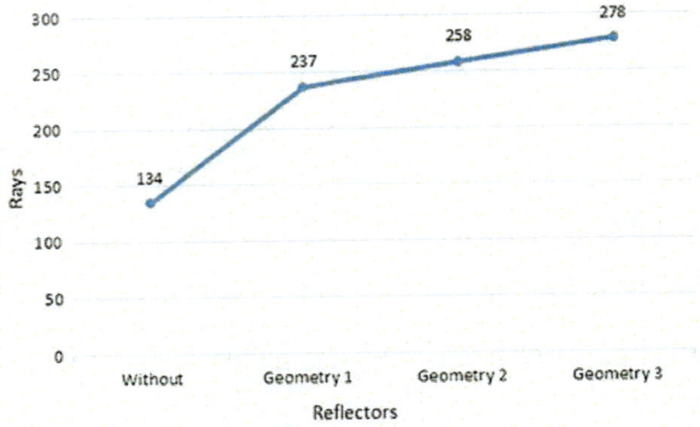

Fig. 7 Comparison of the geometry optimizations according to the fitness function. *Source* Authors

function, that is to say, these sets increased the number of rays falling into the seating area. It is important to mention here that after the simulation time, an statistical stability trend was identified in the results curve. So, no significant changes were observed after 20 min simulation.

Table 3 summarizes the three sets of design values derived from the geometry optimization with Snail and Silvereye. As observed, for every panel, specific hanging distances, rotation angles, and sagitta arc lengths were recognized. Figure 7 compares the three optimized geometries in terms of the fitness function. From this comparison, it can be seen that the three panels contributed to sending a higher number of reflections to the seating area, in comparisonwith the current state without panels. Geometry 3 increased the reflections by 107.46%, obtaining the highest number. It was followed by geometry 2 that increased the bounces up to 92.54% and geometry 1 that reached an improvement of 76.87%.

4.3 Acoustic Performance Evaluation

Then, acoustic simulations of the three optimized geometry panels were carried out with Pachyderm to assess their performance. Figure 8 summarizes the simulation results, where the tested materials were also considered. When comparing Fig. 8 with Fig. 5, it is possible to visualize the acoustical performance improvements from the optimization with respect to the current state.

As regards the SPL, results from the three geometries were very close to those from the existing room with no panels. Among the tested materials, glass and gypsum board achieved the best performances. In contrast, metal panels got lower SPL values than the current state, particularly at sensors 2 and 3.

Regarding RT, the three geometries got a reduction at the four sensors. Particularly, the optimal combinations were achieved when using glass for geometries 1 and 2, and gypsum board for geometry 3. Plywood got a standard performance and the metal panels achieved the worst results.

Table 3 Set of values established for the optimized panels

Design parameter	Panel	Geometry 1	Geometry 2	Geometry 3
Hanging distance, mm	a	159	209	243
	b	48	159	151
	c	186	0	287
	d	61	237	388
Rotation angle, °	a	16.37	15.84	16.17
	b	−7.19	−17.56	−12.57
	c	−6.82	−10.53	−10.39
	d	2.78	10.05	−20
Sagitta arc length, mm	a	–	170	160
	b	–	111	137
	c	–	61	18
	d	–	67	60

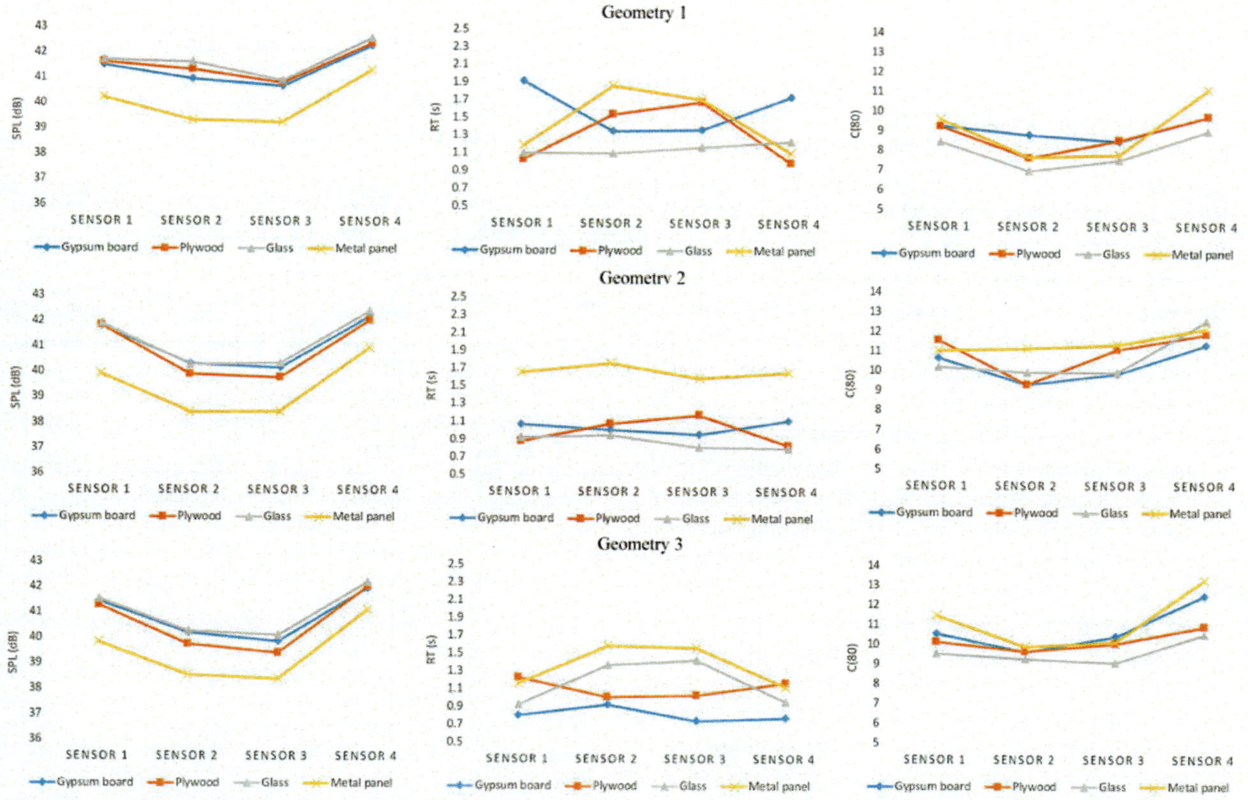

Fig. 8 Results from acoustic simulations performed with Pachyderm: Three optimized designs and four different materials were tested. *Source* Authors

Overall, geometry 3 was the best one, achieving an RT ~0.8 s when using gypsum board.

As regards C80, the three panels reached a better performance than the current state. Besides, C80 was considerably increased at the four sensors with all tested materials. Among them, glass was the best for geometry 1, metal and plywood for geometry 2, and metal and gypsum board for geometry 3.

From the above findings, it is clear that ray tracing simulations can successfully guide the design process of acoustical rooms from the starting stages of the architectural projects (Soriano et al., 2019; Scelo, 2015). In this work, ray tracing allowed optimizing the number of rays being directed towards the seating area, which in this work was set as the evaluation area. After the geometrical optimization, an acoustic specialized software was used to test the optimized forms, positions, and materials for the new panels. From all the cases, the optimized geometries significantly improved the acoustic performance of the auditorium's current state.

Therefore, this study supports evidence from previous studies about the reliability of applying ray tracing in acoustic design (Hongisto et al., 2004). Here, it is interesting to note that this research proposes ray tracing as an important step in the design process; however, acoustical simulations with specialized software are also recommended since they provide better feedback with quantifiable results. Thus, the quality of the acoustic rooms can be significantly enhanced by following the workflow proposed in this study.

Furthermore, the findings here presented also agreed with prior studies that noted the importance of employing computers to assist designers. In practice, the architects take a lot of time and effort to turn their ideas into designs, due to the complexity of acoustic rooms (Lu et al., 2016). Thus, the implementation of optimization algorithms can reduce time and costs in the architectural design process. With the progression of digital technologies more and more widely involved in the different stages of the projects, the exploration of new methods and algorithms needs to be further explored and compared. For example, Octopus and Silvereye algorithms can also be employed to optimize the geometry and rotation of the acoustic panels. In (Giglio et al., 2020), for example, the authors used the mentioned algorithms to improve the SPL and C80.

5 Conclusions

This paper applied parametric tools with acoustic simulations to improve the sound performance of an existing room. This study is not intended to be particular to the case study—the purpose of the space used is simply to offer an example to apply the method here proposed and the use of the digital tools.

The workflow demonstrated to be useful for architects when designing acoustical projects, particularly in terms of the time needed to perform acoustic simulations. Through this method, the simulation time was considerably reduced since only three panel geometries were finally tested. This was possible after performing the geometry optimization with parametric tools that allowed reducing the number of all possible and almost endless design options to only three. Therefore, the application of the workflow during the initial stages of the architectural projects will allow design proposals in a short time and with high acoustic performance.

When comparing the simulation results obtained from the current state vs those from the room with the optimized panels, a significant improvement was observed for the RT and C80 values while a similar result for SPL. Therefore, it is expected that the sound quality and the audibility of the auditorium will be better after a proper retrofit. Further research is, however, planned to assess these improvements in the real space through measurements on-site. The physical experimentation results can then be further explored and compared against the digital results. Important findings of the software reliability and possible errors from diverse plugins can be identified from this comparison.

Regarding the different geometries, results showed that curved panels were the best suited to obtain lower values for RT and higher values for C80. Depending on the material used, geometry 2 or 3 could work better. For example, optimal combinations were geometry 2 with glass and geometry 3 with gypsum board. Overall, glass was the material with the best overall performance because it reduced RT and increased C80. Instead, metal panels were the worst suited.

Future work can also test other acoustical parameters, such as early decay time, strength of the sound, binaural quality index, initial-time-decay, etc. Regarding the acoustic parameters tested in this work (SPL, RT, and C80), they allowed investigating the acoustic quality of a small room. However, it is expected that the parametric acoustic workflow here proposed could be applied in larger acoustic rooms, such as theatres or auditoriums. Here, not only single-objective optimization algorithms but also multi-objective optimization algorithms can be compared together with different acoustic simulation engines in order to identify which one works better according to the architectural space and the design objectives. Considerably more work will need to be done to explore this approach.

Acknowledgements The author acknowledges the support of Universidad de las Americas Puebla and CONACYT.

References

Badino, E., Shtrepi, L., & Astolfi, A. (2020). Acoustic performance-based design: A brief overview of the opportunities and limits in current practice. In *Acoustics* (Vol. 2, Issue 2). https://doi.org/10.3390/acoustics2020016

Banbury, S. P., & Berry, D. C. (2005). Office noise and employee concentration: Identifying causes of disruption and potential improvements. *Ergonomics, 48*(1), 25–37.

Bassuet, A., Rife, D., & Dellatorre, L. (2014). Computational and optimization design in geometric acoustics. *Building Acoustics, 21* (1), 75–85. https://doi.org/10.1260/1351-010X.21.1.75

Bork, I. (2000). A comparison of room simulation software—The 2nd round robin on room acoustical computer simulation. *Acústica—Acta Acústica, 86*(6), 943–956.

Brownlee, J. (2011). *Clever algorithms—Nature inspired programming recipes*. Lulu Press.

Cichocka, J. M., Migalska, A., Browne, W. N., & Rodriguez, E. (2017). *SILVEREYE–the implementation of Particle Swarm Optimization algorithm in a design optimization tool*. Proceedings of the International Conference on Computer-Aided Architectural Design Studies, 151–169.

Cox, T. J., & D'Antonio, P. (2004). Acoustic absorbers and diffusers. Theory, design and application. In *Acoustic absorbers and diffusers*. Spon Press—Taylor & Francis. https://doi.org/10.1201/9781482288254

Cox, T. J., & D'Antonio, P. (2006). A brief history of room acoustic diffusers. *Acoustics Today, 2*(3), 18–25.

Deaconu, M., Toma, A. C., Dragasanu, L. I., & Mihai, D. (2017). Comparative study of sound absorption coefficient determination using FEM method and experimental tests on Kundt's tube. *AIP Conference Proceedings, 1836*(1), 20061. https://doi.org/10.1063/1.4982001

DeBodt, K. (2006). Digital resonance: Sound parameters in architectural design game, set and match. In K. Oosterhuis & L. Feireiss (Eds.), *Computer games, advanced geometries, and digital technologies*.

Elorza, D. O. (2005). *Room acoustics modeling using the ray-tracing method: Implementation and evaluation*. University of Turku. https://citeseerx.ist.psu.edu/viewdoc/download?doi=10.1.1.112.43&rep=rep1&type=pdf

Foged, I. W., Pasold, A., Jensen, M. B., & Poulsen, E. S. (2012). *Acoustic environments: Applying evolutionary algorithms for sound based morphogenesis*. Proceedings of 30th International Conference on Education and Research in Computer Aided Architectural Design in Europe (ECAADe), pp. 347–353.

Giglio, A., Paoletti, I., & Zheliazkova, M. (2020). Performance-based design approach for tailored acoustic surfaces. In E. Bartezzagui, E. Bracchi, G. Del Bo, T. Sagarra, F. Stellacci, & E. Zio (Eds.), *Digital transformation of the design, construction and management processes of the built environment* (pp. 137–148). SpringerOpen.

Grasshopper. (2019). *Snail V.01* (No. 01). Food4Rhino.

Hongisto, V., Keränen, J., & Larm, P. (2004). Simple model for the acoustical design of open-plan offices. *Acustica - Acta Acústica, 90*, 481–495.

Kinsler, L., Frey, A. R., Coppens, A., & Sanders, J. (2000). *Fundamentals of acoustics* (4th ed.). Wiley.

Klatte, M., Bergstrom, K., & Lachmann, T. (2013). Does noise affect learning? A short review on noise effects on cognitive performance in children. *Frontiers in Psychology, 4*.

Krokstad, A., Strom, S., & Sørsdal, S. (1968). Calculating the acoustical room response by the use of a ray tracing technique. *Journal of Sound and Vibration*. https://doi.org/10.1016/0022-460X(68)90198-3

Lu, S., Yan, X., Xu, W., Chen, Y., & Liu, J. (2016). Improving auditorium designs with rapid feedback by integrating parametric models and acoustic simulation. *Building Simulation, 9*(3), 235–250. https://doi.org/10.1007/s12273-015-0268-x

Mason, S. (2020). *Simplified Architectural Models and Affected Acoustic Parameters: Creation of a method to simplify architectural models and the effects of the simplified models on various room acoustic parameters*. Chalmers University of Technology. https://hdl.handle.net/20.500.12380/301869

McGrory, M., Cirac, D. C., Gaussen, O., Cabrera, D., & Engineers, G. (2012). Sound absorption coefficient measurement: Re-examining the relationship between impedance tube and reverberant room methods. *Proceedings of Acoustics*.

Milo, A. (2020). The acoustic designer: Joining soundscape and architectural acoustics in architectural design education. *Building Acoustics, 27*(2), 83–112. https://doi.org/10.1177/1351010X19893593

Østergård, T., Jensen, R. L., & Maagaard, S. E. (2016). Building simulations supporting decision making in early design—A review. *Renewable and Sustainable Energy Reviews, 61*, 187–201. https://doi.org/10.1016/j.rser.2016.03.045

Parigi, D., Svidt, K., Molin, E., & Bard, D. (2017). *Parametric Room Acoustic workflows-Review and future perspectives*. Proceedings of the 35th ECAADe Conference, pp. 603–610.

Peters, B. (2015). Integrating acoustic simulation in architectural design workflows: The FabPod meeting room prototype. *SIMULATION, 91*(9), 787–808. https://doi.org/10.1177/0037549715603480

Peters, B. (2009). *Parametric acoustic surfaces*. ACADIA 09: ReForm: Proceedings of the 29th Annual Conference of the Association for Computer Aided Design in Architecture (ACADIA), pp. 174–181.

Pilch, A., & Kamisiński, T. (2011). The effect of geometrical and material modification of sound diffusers on their acoustic parameters. *Archives of Acoustics, 36*(4), 955–966. https://doi.org/10.2478/V10168-011-0065-1

Reichardt, W., Abdel, A. O., & Schmidt, W. (1975). Definition und Messgrundlage eines objektiven Masses zur Ermittlung der Grenze zwischen brauchbarer und unbrauchbarer Durchsichtigkeit beim Musikdarbietung. *Acustica, 32*, 126–137.

Rindel, J. H. (2002). Modelling in auditorium acoustics. From ripple tank and scale models to computer simulations. *Revista de Acústica, 32*, 31–35.

Rindel, J. H. (2011). Room acoustic modelling techniques: A comparison of a scale model and a computer model for a new opera theatre. *Building Acoustics, 18*(3–4), 259–280. https://doi.org/10.1260/1351-010X.18.3-4.259

Rumpf, M., Schein, M., & Kuhnen, J. (2018). Humanizing Digital Reality. *Humanizing Digital Reality*. https://doi.org/10.1007/978-981-10-6611-5

Sastry, K., & Goldberg, D. (2005). *Genetic algorithms*. Springer.

Sato, S. I., Hayashi, T., Takizawa, A., Tani, A., Kawamura, H., & Ando, Y. (2004). Acoustic design of theatres applying genetic algorithms. *Journal of Temporal Design in Architeture and the Environment, 4*(1), 41–51.

Scelo, T. (2015). Integration of acoustics in parametric architectural design. *Acoustics Australia, 43*(1), 59–67. https://doi.org/10.1007/s40857-015-0014-7

Soriano, J. S., Wright, O., van den Braak, E., & Day, C. (2019). *Exploration of stage acoustic considerations with parametric tools during early design stages*. International Symposium on Room Acoustics. https://www.isra2019.eu/sites/isra2019.eu/files/proceedings/articles/000033.pdf

Van der Harten, A. (2019). Novel applications of re-tooled open-source acoustic simulation algorithms. *The Journal of the Acoustical Society of America, 146*(4), 2801–2811. https://doi.org/10.1121/1.5136701

van der Harten, A. W. (2011). Customized room acoustics simulations using scripting interfaces. *The Journal of the Acoustical Society of America, 129*(4), 2366–2366. https://doi.org/10.1121/1.3587656

Vlaun, N. J. V. (2015). *Sound working environments: Optimizing the acoustic properties of open plan workspaces using parametric models*. TU Delft.

Vorländer, M. (1989). Simulation of the transient and steady-state sound propagation in rooms using a new combined ray-tracing/image-source algorithm. *Journal of Acoustics Society of America, 86*, 172.

Vorländer, M. (1995). *International round robin on room acoustical computer simulations*. Proceedings of the 15th ICA, pp. 689–692.

Wang, D., Tan, D., & Liu, L. (2018). Particle swarm optimization algorithm: An overview. *Soft Computing, 22*(2), 387–408. https://doi.org/10.1007/s00500-016-2474-6

Wright, O., Perkins, N., Donn, M., & Halstead, M. (2016). *Parametric implementation of café acoustics*. Proceedings of ACOUSTICS 2016, pp. 1–6. https://www.acoustics.asn.au/conference_proceedings/AASNZ2016/papers/p13.pdf